PRELIMINARY MATHEMATICS REVIEW GUIDE

By ISIDORE DRESSLER

Author of

Current Mathematics: A Work-Text, Book I
Geometry
Algebra I
Algebra Two
Algebra Two and Trigonometry
Trigonometry
Integrated Mathematics: Course I
Integrated Mathematics: Course II

Dedicated to serving

AMSCO

our nation's youth

When ordering this book, please specify:
either R 30 P
or PRELIMINARY MATHEMATICS REVIEW GUIDE

AMSCO SCHOOL PUBLICATIONS, Inc.

315 Hudson Street New York, N. Y. 10013

ISBN 0-87720-205-2

PRINTED IN THE UNITED STATES OF AMERICA

PREFACE

Preliminary Mathematics Review Guide treats comprehensively the traditional topics contained in a mathematics program for years 1 through 8. As a supplement to the class textbook or as a text for remedial work, it provides abundant material to aid students in their mastery of the subject matter of the program.

During the past few years, teachers of mathematics have concerned themselves with the problem of modernizing the mathematics to be taught in our schools. Different groups have been trying to develop new programs of instruction. At the present time, however, these groups have not yet agreed upon the changes that should be made in content, symbolism, and methods of instruction. During this transition period, teachers who wish to introduce some of the topics from modern mathematics in their traditional courses are sorely in need of appropriate instructional materials that their textbooks may lack. *Preliminary Mathematics Review Guide* aims to meet the needs of these teachers.

The supplementary materials dealing with the topics from modern mathematics are introduced in such a way that they will provide enrichment for those who wish to use them. However, the inclusion of these materials will not in any way interfere with the effective use of the book by those teaching a traditional mathematics course.

An unusual feature of the book is its method of organization. Each chapter contains a series of learning units which, with proper application, the student can master by himself. Units are introduced with problems that are meaningful to the student. The basic concepts of the unit are carefully developed with the use of simple language. Pictures are used to help represent mathematical concepts. New words are clearly defined. Explanations and teaching problems lead to the statement of important general principles and procedures, which are stated clearly and concisely. Model problems, whose solutions are accompanied by detailed explanations, teach the student how to apply the principles and follow the procedures. The student can complete his mastery of the unit by doing a series of well-graded exercises covering almost every type of difficulty and testing his competence both in manipulative skills and in the solution of real problems.

The author hopes that his full treatment of the topics of algebra and geometry will supply the teacher with materials and teaching aids that are helpful in developing concepts and skills in these areas of mathematics. Some optional topics are included for the benefit of the more rapid learners.

Since some pupils enter the seventh and eighth years without adequate competence in fundamental arithmetic skills and their application in social arithmetic and business arithmetic, remedial work must be provided for these pupils. The teacher will find in *Preliminary Mathematics Review Guide* many teaching materials that can help these pupils achieve the necessary mastery of the units in which they are deficient. With the expert guidance of the teacher, the book will make it possible for the pupils to do much of this corrective work by themselves.

The author is deeply grateful to Mr. Samuel Welkowitz, former Chairman of the Mathematics Department, James Madison High School, Brooklyn, New York, for having read the manuscript and having made helpful criticisms and valuable suggestions.

I. D.

CONTENTS

Chapter I

ARITHMETIC OF WHOLE NUMBERS

1. USING NUMBER SYMBOLS AND NUMBER NAMES

Thousands of years ago, man began to represent numbers by the use of number symbols. Primitive man kept a record of a number by making scratches on a stone or in the dirt, by cutting notches in a stick, or by tying knots in a rope. He also kept a count of his animals by placing in a pile one pebble for each animal.

In each of these methods of recording numbers, one mark stood for one object, and there were as many marks as there were objects. In those early days of recording numbers, there was no one mark to stand for several objects.

After many years, man learned to represent several objects by using a single symbol. This was a great step forward in the development of number symbols.

Different peoples in ancient civilizations created number symbols of various kinds. The Egyptians used the symbol | , resembling a stroke, to represent the number 1. They devised the symbol ∩ , which resembles a heel bone, to represent |||||||||| (10), and the symbol 𓍢 , which resembles a coiled rope, to represent the number 100. The Babylonians also invented symbols to represent numbers. Their symbol for 1 was ▼ and their symbol for 10 was ＜ . The Romans used letters of their alphabet as symbols to represent numbers: I to represent 1, V to represent 5, and X to represent 10. As we can see, the same number may be represented by many different symbols.

The number symbol or number name that is used to represent a number is called a ***numeral.***

We must be aware of the fact that a numeral and the number it represents are not the same thing. The numeral is only the symbol for the number, not the number itself. We know, for example, that the number 4 is larger than the number 3. At the right, however, the numeral 3 is printed larger than the numeral 4.

3 4

In our mathematics work, we shall add, subtract, multiply, and divide numbers, not numerals. At times, however, to avoid awkward and cumbersome language expressions, we will use the word *number* when we should really use the word *numeral.* For example, we may say "write the number" instead of saying "write the numeral that represents the number." This is commonly accepted practice.

2. UNDERSTANDING THE MEANING OF NUMBER

THE NUMBER OF A SET

At the right are pictured a collection, or a ***set***, of tennis rackets, set T, and a set of balls, set B. Each member of set B has been paired with a member of set T, and each member of set T has been paired with a member of set B. Both set B and set T have the property of containing the same number of members. This "sameness" is called ***number***. Whenever two sets of things can be matched in the manner used to match set T and set B, the two sets have the same number property. The number property of set T and set B is described by the number name "two" and by the number symbol "2."

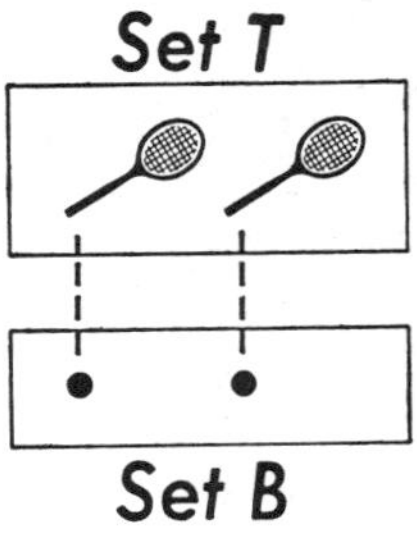

We have learned the number names for many different collections. These collections can be arranged in an order, each number being exactly one more than the one before it.

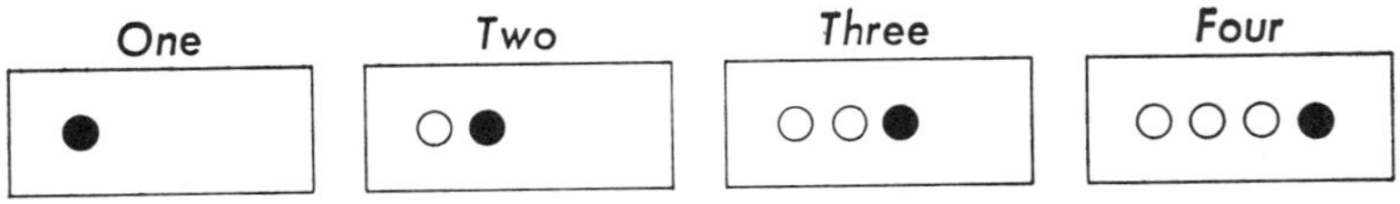

Beginning with one, we see that one and one more is two, two and one more is three, three and one more is four, etc. These arranged numbers are the ones we use in counting. They are called ***counting numbers,*** or ***natural numbers.*** The standard set of counting numbers is represented by the numerals 1, 2, 3, 4, 5, 6, 7, etc. When we refer to all the counting numbers *and zero,* we call this set of numbers the ***whole numbers,*** or ***integers.***

CARDINAL NUMBERS AND ORDINAL NUMBERS

If we cannot tell at a glance how many triangles are pictured below, we can count them from left to right. When we have finished counting, the last number word we said, "five," tells the number of triangles in the group. A number like "five," which is used to tell how many objects there are in a set of objects, is called a ***cardinal number.***

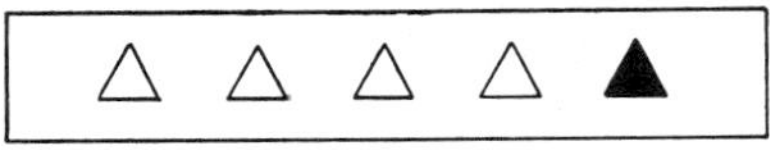

If we wish to describe the position of the darkened triangle as we count from left to right, we would use the number word "fifth." A number like "fifth," which is used to describe the position of an object when it is in a set of objects, is called an ***ordinal number.***

EXERCISES

You should be able to recognize the number of objects in each set without counting. Write the name of the number of objects in each set.

1. 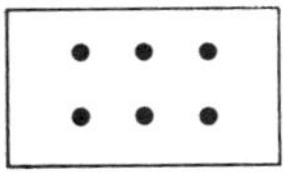 **2.** 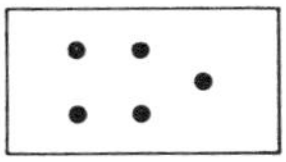 **3.** 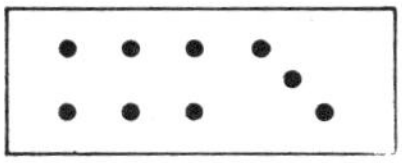

Tell whether each pair of sets has the same cardinal number. Do not count the number of objects in each set. Match the symbols in each pair of sets.

4. 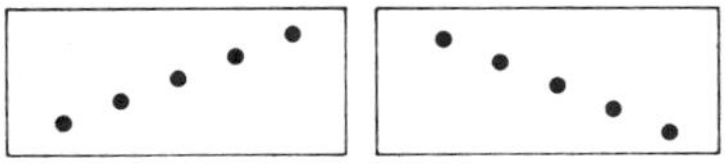 **5.** 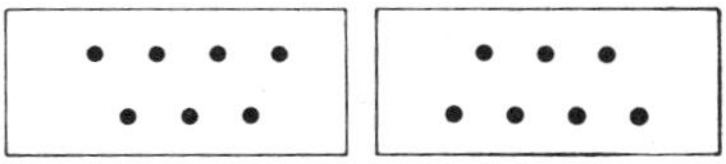

6. 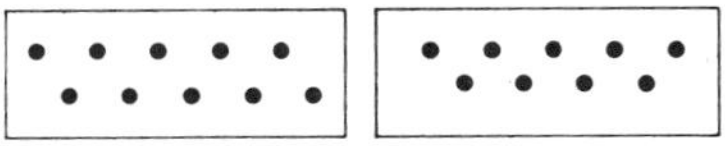 **7.** 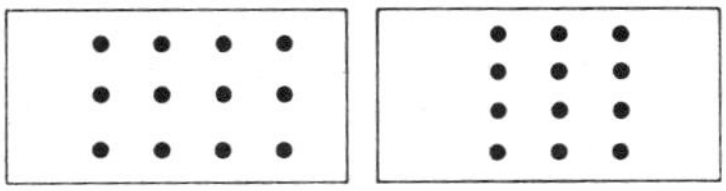

8. Name the set with the larger cardinal number. Do not count the number of objects in each set. Match the symbols in each pair of sets.

Set A

Set B

9. Tell how many more symbols there are in one set than in the other. Do not count the number of objects in each set.

10. *a.* Write the number name that tells how many diamonds appear in the figure.

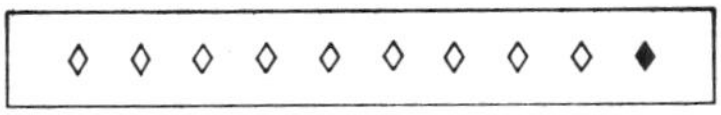

b. Write the number word that describes the position of the darkened diamond, counting from left to right.

Tell whether the number name in each sentence describes a cardinal number or an ordinal number:

11. Sam walked three miles today.
12. Irene sits in the third seat of this row.
13. Carl hit a home run in the ninth inning.
14. Ted pitched nine innings.
15. There are twelve houses on this block.
16. Frank lives in the twelfth house down this block.

3. UNDERSTANDING OUR DECIMAL NUMERAL SYSTEM

In our ***decimal system*** for writing numerals, we can write the numeral for any number by using only ten number symbols. These ten symbols, called ***digits,*** or ***figures,*** are:

0, 1, 2, 3, 4, 5, 6, 7, 8, 9

USING DIGITS TO REPRESENT WHOLE NUMBERS

Three, which means three ones, may be written 3.

Ten may be written 10.

Twenty-eight, which means 2 tens and 8 ones, may be written 28.

One hundred may be written 100.

Three hundred sixty-five, which means 3 hundreds, 6 tens, and 5 ones, may be written 365.

Nine hundred eight, which means 9 hundreds, no tens, and 8 ones, may be written 908.

Notice that in 70, which means 7 tens and no ones, the 0 (zero) is a ***place-holder*** because it "holds" the ***ones,*** or ***units,*** place. In 908, which means 9 hundreds, no tens, and 8 ones, the 0 holds the ***tens*** place. The 0 in 9088 holds the ***hundreds*** place, and the 0 in 90,888 holds the ***thousands*** place.

The *place* that a digit occupies in a numeral tells us its *value* in the numeral.

In the numeral 6,666, which is read "six thousand, six hundred sixty-six," the ***place values*** of the digits are:

The first 6 at the right means 6 ones	6
The second 6 from the right means 6 tens	60
The third 6 from the right means 6 hundreds	600
The fourth 6 from the right means 6 thousands	6,000
	6,666

MODEL PROBLEM

Write in figures the numeral that means two hundreds, eight tens, and nine ones. *Answer:* 289

EXERCISES

1. Give the meaning of each figure in each of the following numerals:

a. 49 *b.* 843 *c.* 780 *d.* 1,985 *e.* 6,009

2. Give the value of the 8 in each of the following numerals:

a. 84 *b.* 308 *c.* 6,821 *d.* 85,234 *e.* 98,764

Read the following:

3. *a.* 12 *b.* 94 *c.* 50 *d.* 527 *e.* 609 *f.* 840 *g.* 300
4. *a.* 4,576 *b.* 8,910 *c.* 7,406 *d.* 1,012 *e.* 5,600 *f.* 6,004 *g.* 9,000

5. Write the numeral that means:

a. 8 tens and 4 ones
b. 7 hundreds, 3 tens, and 6 ones
c. 9 thousands, 4 hundreds, and 8 tens
d. 4 thousands, 9 hundreds, and 7 ones

Write as numerals:

6. *a.* fifty-eight *b.* three hundred twenty-five
7. *a.* five hundred ten *b.* two thousand, fifty-seven
8. *a.* four thousand, forty *b.* five thousand, five
9. Write the numeral that represents the largest number in the decimal system that can be written using the digits 8, 6, 9, and 5.
10. Write the numeral that represents the smallest number in the decimal system that can be written using the digits 4, 2, 7, and 8.
11. What position does the 0 hold in the numeral 40?
12. What position does the 0 hold in the numeral 807?

In each of the following, the first four numbers follow a pattern. Using digits, express the missing numbers so that they will follow the same pattern.

	(*a*)	(*b*)	(*c*)
13.	27, 28, 29, 30, __, __	17, 22, 27, 32, __, __	12, 60, 300, 1500, __, __
14.	56, 53, 50, 47, __, __	128, 64, 32, 16, __, __	12, 25, 51, 103, __, __

4. A NEW LOOK AT THE DECIMAL NUMERAL SYSTEM

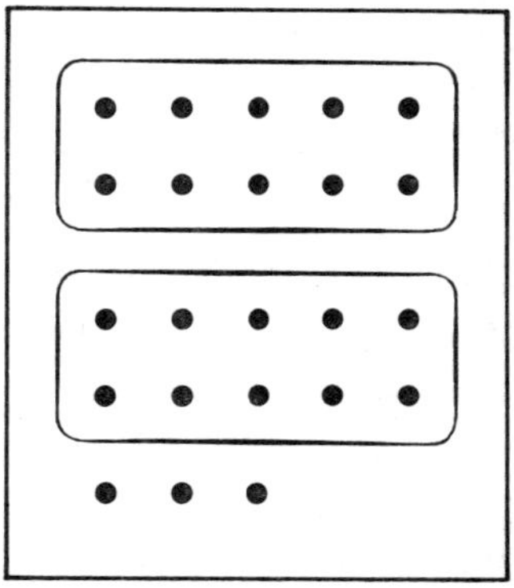

When we count objects, we often group them in tens. At the right, lines have been drawn around groups of ten dots. There are two groups of ten dots and three more dots. Therefore, we say that altogether there are

$$(2 \times 10) + (3 \times 1),$$

or 23 dots.

To count the number of dots in the following diagram, we group them in tens. A dark line has been drawn around one group of ten tens, or one hundred. There are also three groups of tens and seven more dots. Therefore, we say that altogether there are:

$$1 \times (10 \times 10) + (3 \times 10) + (7 \times 1), \text{ or } 137 \text{ dots.}$$

This might also be written as: $(1 \times 100) + (3 \times 10) + (7 \times 1)$, or 137.

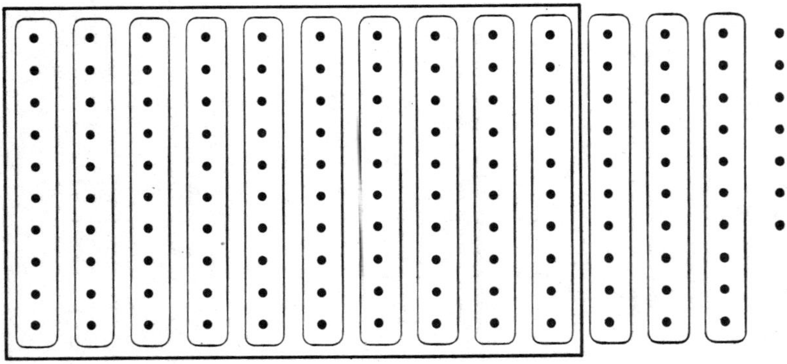

BASES AND EXPONENTS

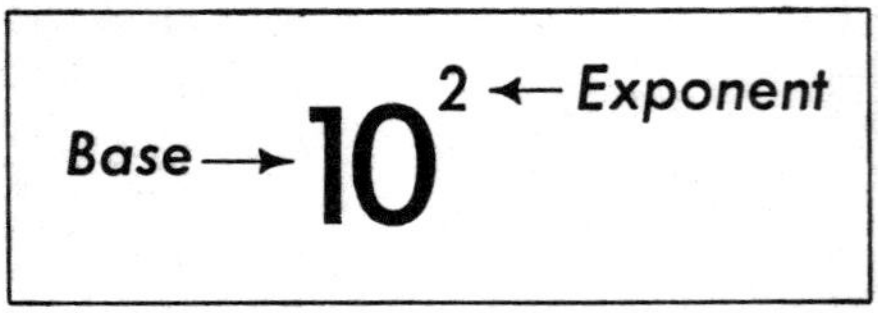

The number 100 may be expressed as a product involving only the number 10. Thus, $100 = 10 \times 10$. A convenient way of writing 10×10 is 10^2. In 10^2, the 10 that is to be repeated when multiplying is called the ***base***. The small ^{2} above and at the right of the 10 is called the ***exponent***. The exponent tells the number of times the base is to be used when multiplying. In 10^3, which means $10 \times 10 \times 10$, the base is 10 and the exponent is 3.

Numbers other than 10 may also be used as the base. For example:

8^1 means 8 and is read "eight to the first power"

5^2 means 5×5, or 25, and is read "five to the second power," or "five squared"

4^3 means $4 \times 4 \times 4$, or 64, and is read "four to the third power," or "four cubed"

2^4 means $2 \times 2 \times 2 \times 2$, or 16, and is read "two to the fourth power"

POWERS

The number represented by an expression like 5^2 is called a ***power***. Since $5^2 = 25$, we say that 25 expressed as a power of 5 is 5^2. Similarly, since $1000 = 10^3$, 10^3 expresses 1000 as a power of 10.

Become familiar with the powers of ten shown at the right because they play an important role in the decimal system. Notice that the number of zeros appearing at the right of the 1 in any numeral that expresses a power of 10 is the same as the exponent in that power of 10. For example, in $100{,}000 = 10^5$, there are 5 zeros at the right of the 1 in 100,000, and the exponent in 10^5 is 5. Therefore, we can see that 10^9 may be written as 1 followed by 9 zeros, or 1,000,000,000 (one billion).

POWERS OF TEN
$10 = 10^1$
$100 = 10^2$
$1000 = 10^3$
$10{,}000 = 10^4$
$100{,}000 = 10^5$
$1{,}000{,}000 = 10^6$

USING POWERS OF TEN TO EXPRESS NUMBERS IN THE DECIMAL SYSTEM

Study the following examples to learn how to use powers of 10 to express numbers in the decimal system:

$20 = 2 \times 10 = 2 \times 10^1$

$300 = 3 \times 100 = 3 \times 10^2$

$4000 = 4 \times 1000 = 4 \times 10^3$

$50{,}000 = 5 \times 10{,}000 = 5 \times 10^4$

$600{,}000 = 6 \times 100{,}000 = 6 \times 10^5$

$7{,}000{,}000 = 7 \times 1{,}000{,}000 = 7 \times 10^6$

$453 = (4 \times 100) + (5 \times 10) + (3 \times 1) = (4 \times 10^2) + (5 \times 10^1) + (3 \times 1)$

$$7{,}984 = (7 \times 1000) + (9 \times 100) + (8 \times 10) + (4 \times 1)$$
$$= (7 \times 10^3) + (9 \times 10^2) + (8 \times 10^1) + (4 \times 1)$$

NOTE

The decimal system of writing numerals has the base ten. Starting at the ones place, each place to the left is given a value 10 times as large as the value of the place before.

EXERCISES

1. Draw eighteen dots and draw a line around a group of ten to show what 18 means.

2. Draw fifty-six dots and draw lines around groups of ten to show what 56 means.

3. Read the following: (*a*) 8^2 (*b*) 9^3 (*c*) 7^4 (*d*) 2^5 (*e*) 6^9 (*f*) 10^{15}

Express in exponent form:

4. 9×9 **5.** $5 \times 5 \times 5$ **6.** $2 \times 2 \times 2 \times 2 \times 2 \times 2$

7. Give the meaning of: (*a*) 2^3 (*b*) 4^2 (*c*) 10^6 (*d*) 1^8

8. Find the number represented by: (*a*) 7^2 (*b*) 5^3 (*c*) 10^5 (*d*) 1^6

9. Which numeral represents the larger number, 2^5 or 5^2?

10. Express 49 as a power of 7.
11. Express 32 as a power of 2.
12. Express 81 as a power of 9.
13. Express 81 as a power of 3.

Express as a power of ten:

14. 10,000,000 **15.** 100,000,000 **16.** 100,000,000,000

17. When each of the following is written as 1 followed by zeros, how many zeros appear at the right of the 1? (*a*) 10^3 (*b*) 10^8 (*c*) 10^{12} (*d*) 10^{35}

Find the number represented by:

18. 3×10^2 **19.** 5×10^2 **20.** 7×10^2 **21.** 2×10^3 **22.** 4×10^3
23. 6×10^3 **24.** 6×10^4 **25.** 8×10^4 **26.** 3×10^5 **27.** 7×10^6

Find the number that can replace the question mark and make the completed sentence true:

28. $400 = ? \times 10^2$ **29.** $600 = ? \times 10^2$ **30.** $7000 = ? \times 10^3$
31. $20{,}000 = ? \times 10^4$ **32.** $400{,}000 = ? \times 10^5$ **33.** $8{,}000{,}000 = ? \times 10^6$

Write as the product of one digit and a power of ten:

34. 500 **35.** 4000 **36.** 30,000 **37.** 600,000
38. 3,000,000 **39.** 70,000,000 **40.** 400,000,000 **41.** 7,000,000,000

42. Express the value of each digit, except the digit in the units place, as the product of the digit and a power of ten: (*a*) 84,732 (*b*) 90,651

8124 may be expressed: $(8 \times 10^3) + (1 \times 10^2) + (2 \times 10^1) + (4 \times 1)$.

43. Express in a similar manner: (*a*) 927 (*b*) 8049 (*c*) 72,650

5. READING AND REPRESENTING LARGE NUMBERS

The following chart will help you read the number 272,750,813,473.

BILLIONS	MILLIONS	THOUSANDS	UNITS
hundred billions, ten billions, billions	hundred millions, ten millions, millions	hundred thousands, ten thousands, thousands	hundreds, tens, ones, or units
2 7 2,	7 5 0,	8 1 3,	4 7 3

The numeral 272,750,813,473 is read as follows: two hundred seventy-two billion, seven hundred fifty million, eight hundred thirteen thousand, four hundred seventy-three.

To help us in understanding and reading a large number, it is marked off with commas into groups of three digits, called ***periods,*** starting from the right and moving toward the left. The first group on the left may have only one or two, instead of three, digits.

MODEL PROBLEM

Stephen reported to his class that three billion, one hundred eighty-four million, eight hundred thirty-six thousand bushels of corn were grown in the United States one year. Using figures, how did he express this number on the blackboard? *Answer:* 3,184,836,000

EXERCISES

Read the following:

1. *a.* 523 *b.* 609 *c.* 8,973 *d.* 6,480 *e.* 4,005
2. *a.* 78,625 *b.* 40,507 *c.* 85,008 *d.* 863,514 *e.* 606,009
3. *a.* 5,000,000 *b.* 125,750,000 *c.* 36,583,194
4. *a.* 942,005,255 *b.* 17,500,000,000 *c.* 250,700,000,000

Group the digits; then read the following:

5. *a.* 1364 *b.* 9850 *c.* 56807 *d.* 706629
6. *a.* 8432175 *b.* 689500000 *c.* 25896300550 *d.* 359700450175

7. Express the following in figures:

a. twenty-five thousand, six hundred seventy-five
b. ten thousand, twenty
c. one hundred twenty-five million, ten thousand
d. seven hundred million
e. nine billion, five hundred million
f. eighteen and one-half billion

8. How many thousands are there in 5,000,000?
9. Which of the following is the same as one-quarter of a billion: 25,000; 250,000; 250,000,000?
10. Which is larger, 42,000,000 or 42 billion?
11. Select the numbers that are greater than two million:

a. 257,625 *b.* 28,629 *c.* 2,693,517 *d.* 13,517,126 *e.* 2,987

12. Read the following statements, which have appeared in the newspapers; then write the numerals in words:

a. Congress is planning a $2,000,000,000 public works program.
b. About 122,000 people marched in the St. Patrick's Day Parade.
c. The New York Central Railroad paid a tax bill of $1,324,000 in Westchester County.
d. Last year there were 52,806,000 telephones in use throughout the world.
e. The total World War I debt that other countries owe the United States is $17,881,503,275.

6. ROUNDING OFF WHOLE NUMBERS

In some situations, it is not convenient or necessary to deal with exact numbers. These numbers may be expressed as approximate numbers by rounding them off to the nearest ten, hundred, thousand, and so on. The approximate number is called a ***rounded number,*** or ***round number.***

ROUNDING OFF NUMBERS TO THE NEAREST TEN

32 rounded off to the *nearest ten* is...... 30
658 rounded off to the *nearest ten* is...... 660

Procedure. **To round off a number to the *nearest ten:***

1. Drop the digit to the right of the tens digit and replace it with a zero.

2. If the units digit that is dropped is less than 5, the number obtained in step 1 is the round number; if it is 5 or greater than 5, increase by 1 the tens digit in the number obtained in step 1 to find the round number.

ROUNDING OFF NUMBERS TO THE NEAREST HUNDRED

8,544 rounded off to the *nearest hundred* is...... 8,500
14,759 rounded off to the *nearest hundred* is...... 14,800

Procedure. **To round off a number to the *nearest hundred:***

1. Drop the digits to the right of the hundreds digit and replace them with zeros.

2. If the tens digit that is dropped is less than 5, the number obtained in step 1 is the round number; if it is 5 or greater than 5, increase by 1 the hundreds digit in the number obtained in step 1 to find the round number.

By similar procedures, numbers may be rounded off to the nearest thousand, ten thousand, hundred thousand, million, and so on.

MODEL PROBLEM

In an election in the state of Florida, 643,849 votes were cast for a Presidential candidate. Round off this number to the nearest (*a*) ten (*b*) hundred (*c*) thousand (*d*) ten thousand (*e*) hundred thousand votes

Answer: (*a*) 643,850 (*b*) 643,800 (*c*) 644,000 (*d*) 640,000 (*e*) 600,000

EXERCISES

1. Is 524 closer to 520 or to 530?

2. Is 6,778 closer to 6,700 or to 6,800?

3. Supply the missing numbers: 3,486 rounded off to the nearest hundred is 3,500 because 3,486 is nearer to ______ than to ______.

4. Round off to the nearest *ten:* (*a*) 78 (*b*) 621 (*c*) 8597 (*d*) 156,835

5. Round off to the nearest *hundred:* (*a*) 342 (*b*) 10,792 (*c*) 3,450,650

6. Round off to the nearest *thousand:* (*a*) 8654 (*b*) 35,482 (*c*) 250,502

7. Round off to the nearest *ten thousand:* (*a*) 985,025 (*b*) 1,850,675

8. Round off to the nearest *hundred thousand:* (*a*) 9,645,000 (*b*) 973,250

9. Round off to the nearest *million:* (*a*) 120,359,700,000 (*b*) 19,500,000

10. On an automobile trip, a man drove 458 miles. Round off this number to the nearest ten miles.

11. A plane was flying at an altitude of 17,425 feet. Round off this number to the nearest thousand feet.

12. Mount Everest, the highest peak in the world, has an altitude of 29,028 feet. What is its altitude rounded off to the nearest ten thousand feet?

13. In one year, the total water-borne commerce of the United States was 923,547,693 tons. Round off this number to the nearest million tons.

7. UNDERSTANDING, READING, AND WRITING ROMAN NUMERALS

In the Roman system, the capital letters I, V, X, L, C, D, and M are used as numerals. Their values in Hindu-Arabic numerals are as follows:

I = 1 **V** = 5 **X** = 10 **L** = 50 **C** = 100 **D** = 500 **M** = 1,000

To read Roman numerals, we proceed from left to right. The following rules will help us understand the meaning of Roman numerals.

Rule 1. When numerals are repeated, their values are added.

II means 1 + 1, or 2 XXX means 10 + 10 + 10, or 30

Only the numerals I, X, C, and M may be repeated. However, no numeral should be written more than three times in succession. We do not write 4 as IIII. We will soon learn that 4 is written IV.

Rule 2. When a numeral is written to the right of a numeral of larger value, its value is added to the larger value.

XV means 10 + 5, or 15 LXI means 50 + 10 + 1, or 61
CCXXX means 100 + 100 + 10 + 10 + 10, or 230

Rule 3. When a numeral is written to the left of a numeral of larger value, its value is subtracted from the larger value.

IV means 5 − 1, or 4 XL means 50 − 10, or 40
CMLXIV means (1000 − 100) + 50 + 10 + (5 − 1),
or 900 + 50 + 10 + 4, or 964

The only numerals that may be written to the left of numerals of higher value are I, X, and C. Each of these numerals may be written to the left of the next two numerals above it in value.

Rule 4. A bar written over a numeral shows multiplication by 1,000.

$\overline{\text{VII}}$ means 7 × 1,000, or 7,000 $\overline{\text{IX}}$ means 9 × 1,000, or 9,000

NUMBERS REPRESENTED BY HINDU-ARABIC AND ROMAN NUMERALS

Hindu-Arabic	*Roman*	*Hindu-Arabic*	*Roman*	*Hindu-Arabic*	*Roman*
1	I	11	XI	30	XXX
2	II	12	XII	40	XL
3	III	13	XIII	50	L
4	IV	14	XIV	60	LX
5	V	15	XV	70	LXX
6	VI	16	XVI	80	LXXX
7	VII	17	XVII	90	XC
8	VIII	18	XVIII	100	C
9	IX	19	XIX	500	D
10	X	20	XX	1000	M

EXERCISES

1. Mr. Talbot bought envelopes at the price of $18.50 per M. How many envelopes did he get for $18.50?

2. Which Roman numeral represents the larger number, XC or CX?

Write in the decimal system:

	(a)	(b)	(c)	(d)	(e)
3.	XVI	XXV	LXVII	CL	MCCLV
4.	XIX	XCVI	XLIX	CDIV	CMXLIX
5.	MDCXXIV	MDCCCXLVIII	MCMXIV	MDCCLXXXI	MCMXXIX
6.	$\overline{\text{III}}$	$\overline{\text{XIV}}$	$\overline{\text{XLII}}$	$\overline{\text{XCV}}$	$\overline{\text{M}}$

Write in Roman numerals:

7.	13	29	58	76	87
8.	42	95	143	654	997
9.	1,863	1,887	1,898	1,958	1,944
10.	8,000	10,000	500,000	400,000	2,000,000

11. The cornerstone of the Westfield Central School is engraved MCMXLII. In what year was the school built?

12. Use Roman numerals to represent the sum of the numbers represented by the following Roman numerals: (*a*) VI and LXI (*b*) IX and XXI.

8. ADDING WHOLE NUMBERS

In an addition problem, the numbers that are being added are called the ***addends.*** The result of the addition is called the ***sum.*** The symbol for addition, which is +, is read ***plus.*** For example, in $23 + 45 = 68$, 23 and 45 are addends, and 68 is the sum.

Add

$$\begin{array}{r} 23 \\ +45 \\ \hline 68 \end{array}$$

Find the sum of 426; 7573; 85; 209.

$$\begin{array}{r} 426 \\ 7573 \\ 85 \\ 209 \\ \hline 8293 \end{array}$$

Add downward. We think: 6 and 3 are 9; and 5, 14; and 9, 23. We write 3 in the ones column and carry 2 to the tens column.

We think: 2 and 2 are 4; and 7, 11; and 8, 19; and 0, 19. We write 9 in the tens column and carry 1 to the hundreds column.

We think: 1 and 4 are 5; and 5, 10; and 2, 12. We write 2 in the hundreds column and carry 1 to the thousands column.

We think: 1 and 7 are 8. We write 8 in the thousands column. To check, add once again going upward.

EXERCISES

Add and check:

	(a)	(b)	(c)	(d)	(e)	(f)
1.	45 32	57 61	519 265	867 489	8058 5724	7596 7865
2.	4 5 8 2	65 80 94 72	246 128 517 200	695 879 798 157	178 37 942 8	842 90 8 636
3.	8143 7106 2015 9350 1678	58,154 69,073 42,860 93,500 12,088	98,753 290 45 6,008 78,694	4,520,637 1,465,733 2,891,008 5,700,529 8,132,680	3,409,872 9,856,670 4,791,008 1,487,633 6,000,775	1,592,603 784,128 8,000 56,878 5,906,894

4. Add and check:

a. 729; 803; 651
b. 736; 48; 402; 6389
c. 4326; 5093; 9846; 6410; 3864
d. 13,812; 21,305; 12,429; 16,496; 19,763

5. Find the sum and check:

a. 426; 7573; 85; 359
b. 6498; 825; 57; 907; 6870; 8781; 968

6. Find the sum and check:

a. \$94,725; \$125,634; \$890,000; \$5,689
b. \$32,655,850; \$8,941,750; \$879,409; \$5,680,584

7. In Eastport High School, there are 658 freshmen, 479 sophomores, 408 juniors, and 379 seniors. Find the total number of pupils in the school.

8. During one week, an automobile company produced 4837 passenger cars, 8952 trucks, and 848 station wagons. How many vehicles did the company produce that week?

9. In Hilltop College, the receipts from sports events during a school year were as follows: football, \$106,525; baseball, \$3928; track, \$1456; basketball, \$15,638; swimming, \$825. Find the total receipts for the year.

10. In a Presidential election, 35,387,015 people voted for candidate *A*; 25,875,408 voted for candidate *B*; 167,826 voted for candidate *C*; and 353,299 voted for other candidates. Find the total number of people who voted in the election.

11. During a five-year period, the appropriations by the Federal government were: \$52,867,672,466; \$67,966,083,088; \$127,788,153,263; \$94,916,821,232; \$74,744,844,305.

a. By rounding off each number to the nearest billion, estimate the total amount appropriated during these years.
b. Find the total amount appropriated during these years.

9. UNDERSTANDING AND USING PRINCIPLES FOR ADDITION

USING SETS TO PICTURE ADDITION

Below we see pictured a collection of objects, a ***set,*** whose number is 4, and a set whose number is 3. The objects of these sets are combined to make a new set. Notice that the number for this new set is 7. Therefore, $4 + 3 = 7$.

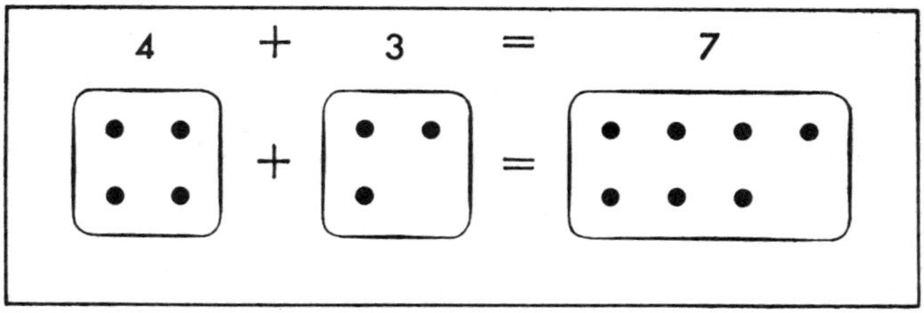

THE COMMUTATIVE PRINCIPLE

If we wish to find the sum of 3 and 4, it makes no difference whether we think "$3 + 4$" or "$4 + 3$." In each case, the answer is 7.

The above example illustrates the ***commutative principle for addition.*** This means that *when two numbers are added, changing the order in which they are added does not change their sum.*

Since $3 + 4 = 7$ and $4 + 3 = 7$, the following mathematical sentence is true:

$$\mathbf{3 + 4 = 4 + 3}$$

The commutative principle for addition explains why we check addition by adding again in the "opposite direction." If we do the problem by "adding upward," we can check by "adding downward."

THE ASSOCIATIVE PRINCIPLE

Marion spent \$2 in one store, \$5 in a second store, and \$8 in a third store. To find out how many dollars she spent, she found the sum of 2, 5, and 8.

Marion could find the sum of 2, 5, and 8 by thinking 2 and 5 are 7, and then 7 and 8 are 15, written $(2 + 5) + 8 = 15$. Or, she might think 2 and the sum of 5 and 8, $(5 + 8)$, gives $2 + 13$, or 15. This may be written $2 + (5 + 8) = 15$. In each case, her answer would be 15.

The above example illustrates the ***associative principle for addition.*** This means that *when we are finding the sum of three or more numbers, changing the way in which we group the numbers does not change their sum.*

Since $(2 + 5) + 8 = 15$ and $2 + (5 + 8) = 15$, the following mathematical sentence is true:

$$\mathbf{(2 + 5) + 8 = 2 + (5 + 8)}$$

The commutative and associative principles for addition make it possible for us to add several numbers in any order we please and to group them in any way that we wish. These principles can help us simplify addition. For example, if we wish to add $78 + 97 + 22$, we can think $78 + 22 = 100$, and $100 + 97 = 197$.

EXERCISES

1. Use pictures of sets to find the number that can replace the question mark and make the resulting sentence true: (*a*) $5 + 2 = ?$ (*b*) $6 + 4 = ?$

2. Does the mathematical sentence illustrate the associative or commutative principle for addition? (*a*) $5 + 9 = 9 + 5$ (*b*) $(5 + 2) + 3 = 5 + (2 + 3)$

In each of the following: (*a*) Replace the question mark with a number that will make the completed sentence true. (*b*) Tell whether the completed sentence illustrates the associative or commutative principle for addition.

3. $7 + 8 = 8 + ?$

4. $(6 + 5) + 8 = 6 + (5 + ?)$

5. $(9 + ?) + 3 = 9 + (4 + 3)$

6. $12 + ? = 9 + 12$

Find the sum. Simplify the addition by using the associative and commutative principles for addition.

7. $79 + 63 + 21$

8. $498 + 224 + 202 + 476$

9. $375 + 97 + 225$

10. $141 + 49 + 259 + 26$

Find the sum. Use the commutative principle to check.

11. $\begin{array}{r} 436 \\ \underline{126} \end{array}$ **12.** $\begin{array}{r} 306 \\ \underline{873} \end{array}$ **13.** $\begin{array}{r} 684 \\ \underline{789} \end{array}$ **14.** $\begin{array}{r} 805 \\ \underline{897} \end{array}$

10. SUBTRACTING WHOLE NUMBERS

In subtraction, the larger number, the one from which we subtract, is called the ***minuend.*** The smaller number, the one that we subtract, is called the ***subtrahend.*** The result of the subtraction is called the ***difference,*** or ***remainder.*** The symbol for subtraction, which is $-$, is read ***minus.*** Thus, in the problem at the right, 75 is the minuend, 32 is the subtrahend, and 43 is the difference, or remainder.

Subtract	*Check*
$\begin{array}{r} 75 \\ \underline{-32} \\ 43 \end{array}$	$\begin{array}{r} 32 \\ \underline{+43} \\ 75 \end{array}$

To check the answer, we add the subtrahend and the difference. The sum must be the same as the minuend.

Mr. Parker has saved \$493. To find how much more he must save in order to buy an E bond, which will cost him \$750, he subtracted, arranging the numbers as follows:

$$\begin{array}{r} \$750 \\ -\ 493 \\ \hline \end{array}$$

Since he could not subtract 3 ones from 0 ones, he exchanged 1 of the 5 tens for 10 ones. 3 ones subtracted from 10 ones gave him 7 ones.

Since he could not subtract 9 tens from the remaining 4 tens, he exchanged 1 of the 7 hundreds for 10 tens, obtaining 14 tens. 9 tens subtracted from 14 tens gave him 5 tens. 4 hundreds subtracted from the remaining 6 hundreds gave him 2 hundreds.

To check his answer, Mr. Parker added \$493 and \$257 and got \$750 as the sum.

Think

$$\begin{array}{lcc} \$7 & 4 & (10) \\ 4 & 9 & 3 \\ \hline \$ & & 7 \end{array}$$

$$\begin{array}{lcc} \$6 & (14) & (10) \\ 4 & 9 & 3 \\ \hline \$2 & 5 & 7 \end{array}$$

Write

$$\begin{array}{r} \$750 \\ -\ 493 \\ \hline \$257 \end{array} \quad \textit{Ans.}$$

EXERCISES

Subtract and check:

	(a)	(b)	(c)	(d)	(e)	(f)
1.	98 72	875 313	3692 322	8697 5342	26,679 2,637	75,681 3,481
2.	692 375	1953 426	4378 149	4995 2926	37,523 2,517	68,417 8,195
3.	534 247	6529 2876	58,623 1,973	7624 509	7690 2004	98,024 4,090
4.	75,000 4,723	39,620 937	89,100 75	652,000 38,750	823,175 65,308	2,576,108 1,829,580

5. Subtract and check:

a. 197 − 68 *b.* 4925 − 1870 *c.* 40,586 − 37,978 *d.* 500,000 − 62,500

6. Subtract 550 from 1300.

7. Subtract 3654 from 8912.

8. Subtract 731,969 from 940,614.

9. From 250,000 take 125,000.

10. Subtract 59,789,500 from 65,410,370.
11. From 9228 take 4875.
12. Find the difference between 3012 and 976.
13. Find the difference between 34,165 and 29,602.
14. A pig weighs 158 pounds and a dog weighs 69 pounds. By how many pounds is the pig heavier than the dog?
15. A dealer bought 8125 dozen shirts. If he sold 4950 dozen, how many dozen does he still have to sell?
16. Mr. Wright wishes to buy a house for $25,300. He can sell his present home for $19,750. How much more expensive than the old house is the new house?
17. A dealer took an order for 750,000 tires. If he has already shipped 575,250 tires, how many does he still have to ship?
18. At one time, the army had 5,629,535 men, and the navy had 986,430 men. How many more men were in the army than in the navy?
19. In 1950, the population of the United States was 151,325,798, and in 1960 it had risen to 179,323,175.

a. By rounding off each number to the nearest million, estimate how much the population increased in the 10-year period.

b. How much did the population increase during the 10-year period?

20. One year, people in the United States bought $45,446,000,000 worth of life insurance. The next year, people bought $48,427,000,000 worth of life insurance. How much was the increase in life insurance that people bought?

11. USING SETS TO PICTURE SUBTRACTION

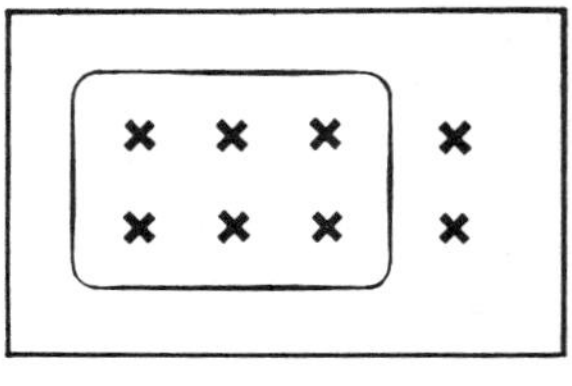

At the right is pictured a collection of objects, a set, whose number is 8. From this group of 8, we have separated a group of 6 by drawing a line around 6 objects. Notice that the number for the group of objects that is left is 2. Therefore, $8 - 6 = 2$. Notice, too, from the picture that $2 + 6 = 8$. Similarly, if $15 - 6 = 9$, then $9 + 6 = 15$.

We say that addition and subtraction are opposite operations, or ***inverse operations***.

Now we can understand why we can check subtraction by adding the difference and the subtrahend. The sum must equal the minuend.

Subtraction	*Check by Addition*
15 minuend	9 difference
− 6 subtrahend	+ 6 subtrahend
9 difference	15 minuend

EXERCISES

Use pictures of sets to find the number that can replace the question mark and make the resulting sentence true:

1. $7 - 5 = ?$ **2.** $9 - 3 = ?$ **3.** $15 - 12 = ?$ **4.** $19 - 14 = ?$

Find the number that can replace the question marks in both part *a* and part *b* and make the resulting sentences true:

5. *a.* $7 - 5 = ?$ *b.* $? + 5 = 7$ **6.** *a.* $17 - ? = 10$ *b.* $10 + ? = 17$
7. *a.* $? - 30 = 20$ *b.* $20 + 30 = ?$ **8.** *a.* $35 + 40 = ?$ *b.* $? - 40 = 35$

Add the two numbers and check the result by subtraction:

9. 213 and 426 **10.** 654 and 185 **11.** 897 and 645 **12.** 592 and 714
13. *a.* Does $8 - 6 = 6 - 8$? *b.* Is subtraction commutative?
14. *a.* Does $8 - (5 - 3) = (8 - 5) - 3$? *b.* Is subtraction associative?

12. MULTIPLYING WHOLE NUMBERS

At the right, we see how $4 \times 6 = 24$ may be written vertically. The number 6 is called the ***multiplicand.*** The number 4 is called the ***multiplier.*** The result of multiplication is called the ***product.*** The multiplication symbol $\times$ is read ***times.***

$$\begin{array}{r} 6 \\ \times\ 4 \\ \hline 24 \end{array}$$

Since $4 \times 6 = 24$ and $6 \times 4 = 24$, $4 \times 6 = 6 \times 4$. Hence:

Principle 1. If we change the order of the numbers that are being multiplied, we do not change the product.

Principle 2. If we multiply a number by 1, the product is that same number.

For example, $1 \times 7 = 7$; also, $8 \times 1 = 8$.

Principle 3. If we multiply a number by 0, the product is 0.

For example, $0 \times 6 = 0$; also, $9 \times 0 = 0$.

MULTIPLYING BY A ONE-DIGIT NUMBER

Let us multiply 2×23.

Think	*Write*
Since 23 means 2 tens and 3 ones, the multiplication 2×23 may be done as follows: 2×3 ones = 6 ones. Write the 6 in the ones place. 2×2 tens = 4 tens. Write the 4 in the tens place.	$\begin{array}{r} 23 \\ \times\ 2 \\ \hline 46 \end{array}$ *Ans.*

Often it becomes necessary ***to carry*** when we are multiplying. Let us multiply 479 by 6.

Think	*Write*
Since 479 means 4 hundreds, 7 tens, 9 ones, the multiplication 6×479 may be done as follows:	479 $\times 6$ 2874 *Ans.*
6×9 ones $= 6 \times 9 = 54$ 6×7 tens $= 6 \times 70 = 420$ 6×4 hundreds $= 6 \times 400 = 2400$ $6 \times 479 = 2874$	

MULTIPLYING BY A TWO-DIGIT NUMBER

Now let us multiply 26×148.

The products we get when we multiply by one digit at a time are called ***partial products.***

In method 2, we do not write the 0 of the number 2960, which is the product of 148×20. Instead, we multiply 148×2 tens, giving 296 tens, and write the 6 in the tens place.

Method 1	*Method* 2
148	148
× 26	× 26
888 (6 × 148)	888 partial product
2960 (20 × 148)	296 partial product
3848 (26 × 148)	3848 product

MULTIPLYING BY A THREE-DIGIT NUMBER

A dealer bought 275 television sets and paid $168 for each one. To find how much he paid for all of them, we multiply 168 by 275.

Method 1	*Method* 2	*Check*
$168	$168	275
× 275	× 275	× 168
840 (5 × 168)	840	2 200
11 760 (70 × 168)	11 76	16 50
33 600 (200 × 168)	33 6	27 5
$46,200 (275 × 168)	$46,200	$46,200

To check a multiplication example, we interchange the multiplier and multiplicand and multiply again, since we know that when we change the order of the numbers being multiplied, we do not change the product.

MULTIPLYING NUMBERS CONTAINING ZEROS

Observe how we arrange our work when we multiply numbers that have zeros in the multiplicand and multiplier. Let us multiply 708 by 670.

Method 1	*Method* 2	
708 × 670 49 560 (70 × 708) 424 800 (600 × 708) 474,360 (670 × 708)	708 × 670 49 560 424 8 474,360	Since 0 × 708 is 0, we place 0 in the units column. When we multiply by 7, we place the first digit obtained for the partial product in the tens column. When we multiply by 6, we place the first digit obtained for the partial product in the hundreds column.

EXERCISES

Multiply and check:

	(*a*)	(*b*)	(*c*)	(*d*)	(*e*)	(*f*)
1.	534 2	5132 3	691 8	6811 9	578 6	79863 5
2.	34 32	47 83	97 69	513 98	743 585	576 985
3.	60 8	8054 6	12,005 7	900 8	940 86	608 589
4.	1803 465	5684 920	3700 800	7059 5860	2050 4003	5880 9300

Find the product of:

5. *a.* 97 and 50 *b.* 532 and 64 *c.* 807 and 940

6. *a.* 500 and 800 *b.* 92,005 and 570 *c.* 125,000 and 500

Find:

7. *a.* 87 × 65 *b.* 529 × 70 *c.* 280 × 709 *d.* 6750 × 300

Multiply:

8. *a.* 70 by 93 *b.* 710 by 350 *c.* 604 by 702 *d.* 8500 by 987

9. Multiply by 10: (*a*) 23 (*b*) 90 (*c*) 708 (*d*) 65,000

10. Multiply by 100: (*a*) 6 (*b*) 40 (*c*) 205 (*d*) 50,000

11. Multiply by 1000: (*a*) 89 (*b*) 692 (*c*) 540 (*d*) 400,000

12. In Southside School, there are 27 classes with 34 pupils in each class. What is the total number of pupils in these classes?

13. A dealer bought 325 radios, each costing $39. How much did he pay for all of them?

14. A ship that averaged 304 miles a day made a trip lasting 28 days.

a. By rounding off the numbers to the nearest ten, estimate the number of miles the ship traveled.

b. Find the number of miles the ship traveled.

15. A dealer sold 180 cars at $2640 each. Find his total sales.

16. A freight car holds 295 cases, each of which weighs 90 pounds. Find the total weight of the cases in the car.

17. A squadron of 105 planes flew a distance of 250 miles. What was the total number of miles flown by the entire squadron?

18. How many pieces of candy are needed to pack 600 boxes, if each box has 8 rows and each row contains 15 pieces?

19. If sound travels 1080 feet in one second, how far does it travel in 3600 seconds?

20. If light travels 186,000 miles in one second, how far does it travel in 1200 seconds?

13. UNDERSTANDING AND USING PRINCIPLES FOR MULTIPLICATION

USING SETS TO PICTURE MULTIPLICATION

Below we see pictured two sets, each having the number 4. A new set is formed consisting of these two sets. Notice that the number for this new set is 8. Therefore, $2 \times 4 = 8$.

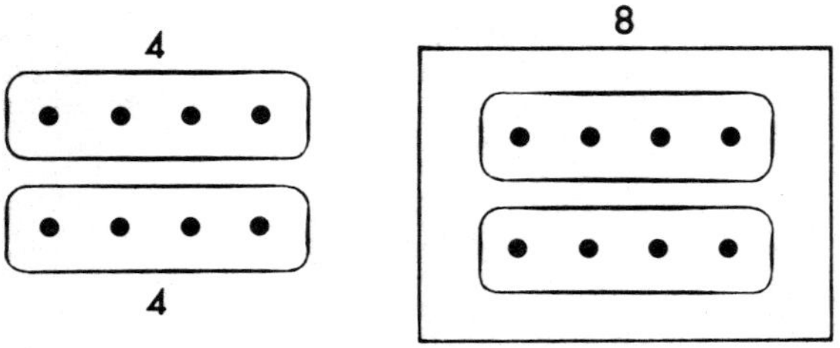

THE COMMUTATIVE PRINCIPLE

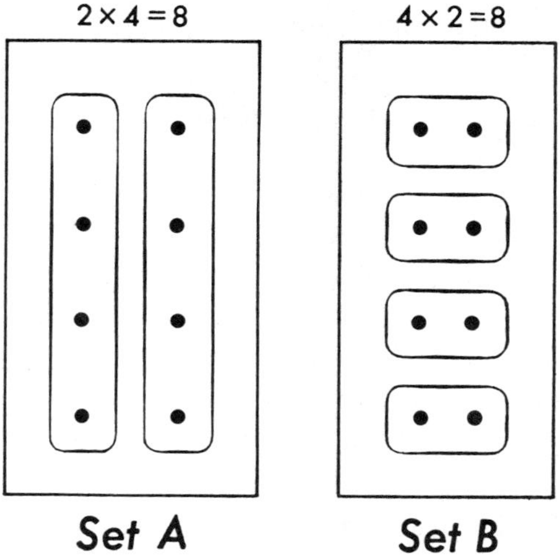

At the right, we see pictured set A, consisting of two sets, each with four members. We also see pictured set B, consisting of four sets, each with two members. Notice that the number for both set A and set B is 8.

For set A, $2 \times 4 = 8$. For set B, $4 \times 2 = 8$. Since 2×4 and 4×2 are both other names for 8, the following mathematical sentence is true:

$$\mathbf{2 \times 4 = 4 \times 2}$$

The above example illustrates the ***commutative principle for multiplication.*** This means that when two numbers are multiplied, changing the order in which they are multiplied does not change their product.

The commutative principle for multiplication explains why we can check multiplication by interchanging the multiplier and multiplicand and multiplying again. The product in the check must be the same as the product in the original solution.

THE ASSOCIATIVE PRINCIPLE

What is the meaning of $2 \times 3 \times 4$? How can we compute its value?

We can first think that 2 times 3 is 6 and then think that 6 times 4 is 24. This can be written $(2 \times 3) \times 4 = 24$. Or, we can first think that 3 times 4 is 12 and then think that 2 times 12 is 24. This can be written $2 \times (3 \times 4) = 24$. In each case, the product is 24. Therefore, either method may be used to compute $2 \times 3 \times 4$.

The above example illustrates the ***associative principle for multiplication.*** This means that when we are finding the product of three or more numbers, changing the way in which we group the numbers does not change the product.

Since $(2 \times 3) \times 4 = 24$ and $2 \times (3 \times 4) = 24$, the following mathematical sentence is true:

$$\mathbf{(2 \times 3) \times 4 = 2 \times (3 \times 4)}$$

THE DISTRIBUTIVE PRINCIPLE

If we wish to multiply the sum of 20 and 3 by 2, we can do the computation in either of the following two ways:

We can think that the sum of 20 and 3 is 23 and then think that 2 times 23 is 46. This can be written $2 \times (20 + 3) = 2 \times 23 = 46$. Or, we can think that 2 times 20 is 40, 2 times 3 is 6, and the sum of 40 and 6 is 46. This can be written $(2 \times 20) + (2 \times 3) = 40 + 6 = 46$. In each case, the result is 46. Therefore, either method may be used to do the computation.

The above example illustrates the ***distributive principle for multiplication.*** This means that when the sum of two or more addends is to be multiplied by a number, the product is the same as the result obtained when each of the addends is multiplied by the number and these products are added.

Since $2 \times (20 + 3) = 46$ and $(2 \times 20) + (2 \times 3) = 46$, the following mathematical sentence is true:

$$\mathbf{2 \times (20 + 3) = (2 \times 20) + (2 \times 3)}$$

EXERCISES

1. Use pictures of sets to find the number that can replace the question mark and make the resulting sentence true: (*a*) $2 \times 3 = ?$ (*b*) $4 \times 5 = ?$

Does the mathematical sentence illustrate the associative, commutative, or distributive principle for multiplication?

2. $7 \times (2 \times 3) = (7 \times 2) \times 3$ **3.** $9 \times 5 = 5 \times 9$
4. $3 \times (4 + 6) = (3 \times 4) + (3 \times 6)$

In each of the following: (*a*) Replace the question mark with a number that will make the completed sentence true. (*b*) Name the principle for multiplication that the completed sentence illustrates.

5. $10 \times 5 = 5 \times ?$ **6.** $8 \times (7 \times 4) = (8 \times 7) \times ?$
7. $5 \times (6 \times 2) = (? \times 6) \times 2$ **8.** $9 \times (6 + 2) = (9 \times 6) + (9 \times ?)$

Multiply and check the product by using the commutative principle for multiplication:

9.	10.	11.	12.	13.
79	596	870	973	640
43	85	264	408	709

14. Find the product in three ways: (*a*) $8 \times 5 \times 6$ (*b*) $4 \times 3 \times 2 \times 5$

Find the product. Simplify the multiplication by using the associative and commutative principles for multiplication.

15. $5 \times 137 \times 2$ **16.** $25 \times 957 \times 40$ **17.** $5 \times 5 \times 834 \times 4$

18. Express the product of 15 and the sum of 16 and 9 in two ways.

Use the distributive principle to rewrite each of the following:

19. $7 \times (9 + 6)$ **20.** $8 \times (5 + 4)$

21. $(9 \times 3) + (9 \times 4)$ **22.** $(4 \times 7) + (4 \times 5)$

23. Find the product of 7×53, first representing 53 as $50 + 3$. Check the product by multiplying in the usual way.

24. To find the product of 9×674, using the distributive principle, we can think that since $674 = 600 + 70 + 4$, then:

$$9 \times 674 = 9 \times (600 + 70 + 4) = (9 \times 600) + (9 \times 70) + (9 \times 4)$$

a. Use the last result to find the product of 9×674.

b. Check the product by multiplying in the usual way.

Use the method outlined in exercise 24 to multiply in each of the following. Check by multiplying in the usual way.

25. 4×38 **26.** 9×79 **27.** 7×264 **28.** 6×862 **29.** 6×940

14. DIVIDING WHOLE NUMBERS

Walter had 18 empty soda bottles that he wished to return to the store. In order to find out how many cases, each holding 6 bottles, he would need, he ***divided*** 18 by 6. We can write 18 divided by 6 equals 3 in two ways.

$$18 \div 6 = 3 \quad \textit{or} \quad \text{divisor } 6\overline{)18} \text{ dividend} \quad (3 \text{ quotient})$$

The number 18, which is to be divided, is called the ***dividend.*** The number 6, by which we divide, is called the ***divisor.*** The result of the division, 3, is called the ***quotient.*** The symbols ÷ and $\overline{)\quad}$ are read "divided by."

If we wish to divide 16 by 3, the division may be written in the following ways:

$$16 \div 3 = 5 \text{ r1} \quad \textit{or} \quad \begin{array}{r} 5 \text{ r1} \\ 3\overline{)16} \end{array}$$

r1 means that 1 is left. The number that is left is called the ***remainder.***

When we divide 728 by 8, we can think: 728, which is 72 tens and 8 ones, is to be divided by 8.

Be careful to place the 9 above the 2 in the tens column and the 1 above the 8 in the ones column.

Think	*Write*
9 tens and 1 one 8)72 tens and 8 ones	91 *Ans.* 8)728

DIVIDING BY A TWO-DIGIT NUMBER

Let us divide 768 by 32.

To estimate the first figure in the quotient, round off the first two figures of the dividend, 76, to the nearest ten, 80. Also round off 32, the divisor, to the nearest ten, 30. Estimate: 80 ÷ 30 is about 2. Place the 2 in the quotient above the 6 in the dividend. Multiply: 2 × 32 = 64. Place the 64 under the 76 and subtract, giving 12. Be sure that this difference is less than the divisor. (If the difference is equal to or greater than the divisor, the estimated figure in the quotient is too small.) Bring down 8, the next figure in the dividend.

```
              24 quotient
divisor 32)768 dividend
           64
           128 partial dividend
           128
```

Check:

```
   24
 × 32
   48
  72
  768
```

Divide 128 by 32. Think: 130 ÷ 30 is about 4. Place the 4 in the quotient above the 8 in the dividend. Multiply: 4 × 32 = 128. Place 128 below the partial dividend, 128. Subtract 128 from 128, giving a remainder of 0.

To check, multiply the quotient by the divisor. The product should be the same as the dividend.

Let us divide 23,873 by 78.

```
      306
78)23,873
   23 4
      473
      468
       r5
```

The answer may be written as

306 r5, or $306\frac{5}{78}$.

Check:

```
   306
 ×  78
  2 448
 21 42
 +    5
 23,873
```

Since the divisor 78 is greater than 23, the number represented by the first *two* figures of the dividend, to estimate the first figure in the quotient, round off the first *three* figures of the dividend, 238, to the nearest ten, 240. Also round off the divisor, 78, to the nearest ten, 80. Think: 240 ÷ 80 is 3. Place the 3 in the quotient above the 8 in the dividend. Multiply: 3 × 78 = 234. Subtract: 238 − 234 = 4. Bring down 7, the next figure in the dividend.

Since 78 does not divide into 47, write 0 in the quotient above the 7 in the dividend. Bring down the next figure, 3. Think: 470 ÷ 80 is about 6. Multiply: 6 × 78 = 468. Subtract: 473 − 468 = 5. The remainder is 5. (Be sure that the final remainder is less than the divisor.)

To check, multiply the quotient by the divisor and add the remainder. The result should be the same as the dividend.

DIVIDING BY A THREE-DIGIT NUMBER

Let us divide 5883 by 147.

$$\begin{array}{r} 40 \\ 147\overline{)5883} \\ \underline{588} \\ 3 \end{array}$$

The answer may be written as

40 r3, or $40\frac{3}{147}$.

To estimate the first figure in the quotient, round off the first three figures of the dividend, 588, to the nearest hundred, 600. Also round off the divisor, 147, to the nearest hundred, 100. Think: 600 ÷ 100 is about 6. Multiply: 6 × 147 = 882. Since 882 is greater than 588, 6 is too large. Try 5 and then 4 until the resulting product is equal to or less than 588. 4 × 147 = 588. Place the 4 in the quotient above the last 8 of 588 in the dividend. Write down the product, 588. Subtract: 588 − 588 = 0. Bring down the 3, the next figure in the dividend. Since 147 does not divide into 3, place 0 in the quotient. The remainder is 3. Check as in the previous problem.

MODEL PROBLEM

A carpenter has a piece of lumber 96 inches long. How many pieces, each 9 inches in length, can he cut from it?

Solution: Divide 96 by 9 to discover how many 9's are contained in 96.

$$\begin{array}{r} 10 \text{ r}6 \\ 9\overline{)96} \end{array}$$

Answer: He can cut 10 pieces.

EXERCISES

Estimate the first digit in the quotient:

1. $43\overline{)903}$ **2.** $38\overline{)2,850}$ **3.** $615\overline{)432,345}$ **4.** $987\overline{)590,226}$

5. Is 3625 ÷ 311 more than 10 and less than 100?

6. 396,198 ÷ 198 is approximately:

a. 20 *b.* 200 *c.* 2,000 *d.* 20,000 *e.* 200,000

7. If the quotient is 32 and the dividend is 1024, find the divisor.

8. In a division example, the quotient is 17, the divisor is 32, and the remainder is 12. Find the dividend.

Divide and check:

	(*a*)	(*b*)	(*c*)	(*d*)	(*e*)
9.	$3\overline{)159}$	$9\overline{)639}$	$3\overline{)65}$	$6\overline{)429}$	$7\overline{)358}$
10.	$4\overline{)928}$	$5\overline{)5755}$	$8\overline{)78,496}$	$4\overline{)3675}$	$9\overline{)85,763}$

	(a)	(b)	(c)	(d)	(e)
11.	4)80,560	7)71,421	21)693	43)645	47)893
12.	34)727	78)986	42)378	68)476	84)515
13.	20)8900	90)8460	80)10,523	41)1517	65)3120
14.	26)1504	19)1907	67)33,801	48)10,031	39)19,520

	(a)	(b)	(c)	(d)
15.	415)11,620	708)34,692	630)44,100	800)52,000
16.	109)77,172	694)120,062	307)184,200	129)645,516
17.	257)6150	109)3273	985)56,173	502)95,380
18.	841)195,526	507)410,374	930)418,539	849)599,394
19.	35)9623	647)93,124	501)45,928	399)158,620

Divide:

20. *a.* 1900 by 95 *b.* 49,040 by 70 *c.* 110,894 by 623

21. *a.* 4008 ÷ 76 *b.* 15,080 ÷ 903 *c.* 639,830 ÷ 800

22. In a certain school, there are 108 boys. How many intramural baseball teams (nine boys on a team) can be formed using all the boys?

23. A farmer sells his eggs by the dozen, 12 in a box. How many boxes will he need for 336 eggs?

24. A carpenter has a piece of lumber 288 inches long. How many pieces, each 17 inches long, can he cut from it?

25. An electrician needs 6 feet of electric wire to wire a lamp. How many lamps can he wire if he has 125 feet of wire?

26. Sally uses 27 inches of cloth to make an apron. If she has 528 inches of cloth, how many aprons will she be able to make?

27. At a bond sale, $50,125 worth of $25 bonds were sold. Find the number of bonds sold.

28. A truck is carrying a load of 11,340 pounds made up of cases weighing 105 pounds each. Find the number of cases in the truck.

29. A machine produces 45,600 articles in 48 hours. How many articles does it produce in 1 hour?

30. Mr. Tobias earned $11,700 during one year. How much did he earn each week?

31. A company that employed 197 workers paid them $1,066,755 during one year. If all workers received the same salary, find, correct to the nearest $100, how much each worker was paid.

15. USING SETS TO PICTURE DIVISION

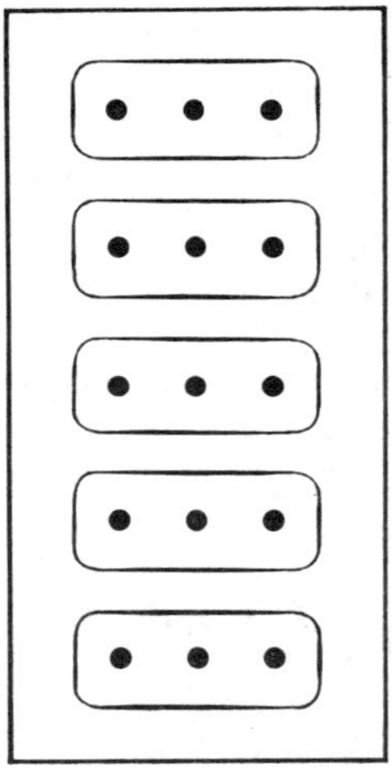

At the right is pictured a set whose number is 15. From the members of this set, we have made sets, each having 3 members. Notice that we were able to form 5 such sets. Therefore, $15 \div 3 = 5$. Notice that since we have 5 sets with 3 members each, together they make a set that has $5 \times 3 = 15$ members. Similarly, if $28 \div 4 = 7$, then $7 \times 4 = 28$.

We say that multiplication and division are opposite operations, or ***inverse operations.***

This relationship makes it possible to check division where there is no remainder by multiplying the divisor by the quotient and obtaining the dividend.

DIVISION INVOLVING ZERO

When zero is divided by any number other than zero, the quotient is zero. For example, $0 \div 5 = 0$. To verify this result, we multiply the quotient, 0, by the divisor, 5, and obtain the dividend, 0.

It is impossible to find a quotient if we attempt to divide any number other than zero by zero. For example, $5 \div 0$ is meaningless because there is no possible quotient. This is so because there is no number that can be multiplied by the divisor, 0, that will give the dividend, 5. Therefore, we say that *a number other than zero cannot be divided by zero.*

EXERCISES

1. Use pictures of sets to find the number that can replace the question mark and make the resulting sentence true: (*a*) $16 \div 8 = ?$ (*b*) $20 \div 5 = ?$

Find the number that can replace the question marks in both part *a* and part *b* and make the resulting sentences true:

2. *a.* $48 \div 12 = ?$ *b.* $? \times 12 = 48$ **3.** *a.* $100 \div 4 = ?$ *b.* $4 \times ? = 100$

Divide and then check the result by two different methods:

4. $600 \div 25$ **5.** $3072 \div 24$ **6.** $17{,}052 \div 84$

7. *a.* Does $40 \div 20 = 20 \div 40$? *b.* Is division commutative?

8. *a.* Does $(40 \div 20) \div 2 = 40 \div (20 \div 2)$? *b.* Is division associative?

16. SOLVING PROBLEMS STEP BY STEP

We approach the solution of a problem by means of a series of related steps:

1. Read the problem carefully to make sure that you understand its wording.
2. Study the information in the problem to discover the given facts and to decide what is to be found.
3. Plan the steps you will follow to arrive at the solution.
4. Decide upon the operations you will use in solving the problem.
5. Estimate a sensible answer for the problem by using round numbers.
6. Do the arithmetic necessary to find the answer.
7. Check your answer.

EXERCISES

1. Tom bought 3 pounds of potatoes at 7 cents per pound and 4 pounds of grapes at 19 cents per pound. How much did he spend?

2. An employer paid each of 12 workers $48 per week and each of 9 workers $75 per week. What is the total amount he paid these people each week?

3. Mr. Wilson bought 6 shirts at $5 per shirt, 3 ties at $3 per tie, and 2 pairs of shoes at $19 per pair. How much change did he receive from a $100 bill?

4. One man earns $5600 per year. A second man earns $375 per month. How much more does the first man earn in a year than the second?

5. If there are 7 frankfurters in 1 pound, how many pounds would you buy for a picnic to supply 21 pupils with 2 frankfurters each?

6. Mr. Preston could purchase a new radio for $75 cash, or he could buy the same radio by paying $10 down and $5 a month for 14 months.
a. Which is the better offer? *b.* How much better?

7. A company produced 1 million staples. It sold 1475 boxes, each containing 500 staples. How many staples were not sold?

17. PROBLEMS WITHOUT NUMBERS

Would you use addition, subtraction, multiplication, or division to solve the problem in the shortest way?

1. If you know the cost of one plane, find the cost of several such planes.

2. Find out how much higher one mountain is than another.

3. If a man earns the same amount of money on each working day, find his salary for the week.

4. If you know how much you paid for several eggs, find the cost of one egg.

5. If you know the length of a cruiser and know how much longer a battleship is, find the length of the battleship.

6. A sum of money is to be shared equally by several children of a family. Find how much each child is to receive.

7. If you know the weight of one can, find the weight of several such cans.
8. If you know how fast a plane is traveling and the number of hours it flies, find the distance it travels.
9. If you know how much you earned in one week and how many hours you worked, find the amount you earned each hour.
10. If you know a man's age and the number of years he is older than his brother, find the brother's age.
11. If you know the sum of two numbers and also one of the numbers, find the other number.
12. If you know the cost of 3 oranges, what is the cost, at the same rate, of a dozen of these oranges?

18. FINDING WHAT IS MISSING

In each of the following questions, one necessary fact has been left out. What fact would you have to know to answer the question?

1. A man bought 4 shirts. How much did he pay for them?
2. If Charles saves $2 per week, how many weeks will it take him to save enough money to buy a portable radio?
3. If you buy a bat and 2 balls for a total of $8, what do you pay for the bat?
4. Wallace bought a bag of potatoes for 35 cents. How much did he pay for a pound of potatoes?
5. Sarah paid $3 for butter. What was the cost of a pound of butter?
6. At $23 per ton, what is the cost of a load of coal?
7. Ed spent 3 minutes reading a page in a book. How many words can Ed read in one minute?
8. If you buy 3 ties, how much change should you receive from a $10 bill?

19. FINDING WHAT IS UNNECESSARY

In each of the following problems, there is one more fact than is necessary for solving the problem. State the unnecessary fact and then solve the problem.

1. How many boxes are needed to package 100 pounds of rice that sells for 23 cents per pound if each box is to contain 2 pounds of rice?
2. Alan needs 4 ft. of lumber to make a shelf. How many such shelves can he make from a piece of lumber 13 ft. long and costing 34 cents per ft.?
3. How much change should Anne receive from a $10 bill if she bought 6 pairs of hose at $1 per pair and 3 handkerchiefs for which she paid $3?
4. Find the cost of 4 tires, each weighing 21 lb., if each tire costs $27.
5. Find the cost of 3 tablecloths, each 108 in. long, if one tablecloth costs $9.

Chapter II

ARITHMETIC OF COMMON FRACTIONS AND MIXED NUMBERS

I. UNDERSTANDING THE MEANINGS OF FRACTIONS

Mrs. Wilson had dinner with her two sons. For dessert, she cut a small pie into 3 equal pieces. Each person received one-third, or $\frac{1}{3}$, of the pie. The 3 parts together are the whole pie. Therefore, the whole pie is $\frac{3}{3}$, or 1.

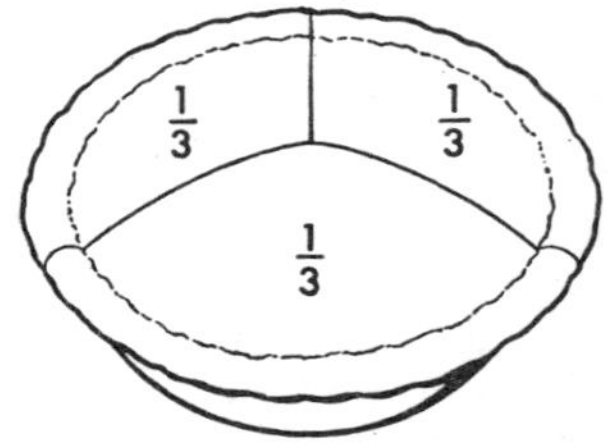

After Mrs. Wilson served herself, 2 of the 3 equal parts remained for the boys. Therefore, two-thirds, or $\frac{2}{3}$, of the pie was served to the boys.

The whole rectangle is divided into 5 equal parts. Since 3 of these parts are shaded, we say that the fraction of the rectangle that is shaded is $\frac{3}{5}$.

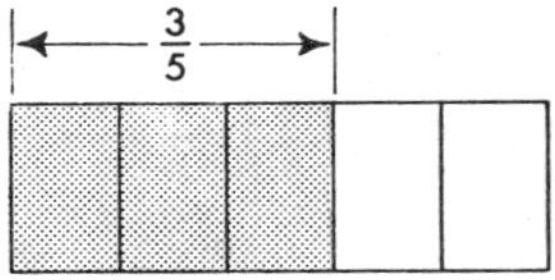

A fraction may be used to represent one or more of the equal parts into which a whole thing has been divided.

$\frac{3}{5}$ numerator / denominator

Every fraction has two parts: the number above the fraction line, the ***numerator,*** and the number below the fraction line, the ***denominator***. These numbers are called the ***terms*** of the fraction.

The *denominator* of the fraction tells the number of equal parts or the kind of equal parts (denomination) into which a whole thing is divided. The *numerator* of the fraction counts (numerates) or tells the number of these equal parts that the fraction represents. In the fraction $\frac{3}{5}$, the denominator, 5, indicates that the whole thing has been divided into 5 equal parts, and the numerator, 3, shows that we are dealing with 3 of these parts.

A fraction whose numerator is 1 and whose denominator is a whole number is called a ***unit fraction.*** Thus, $\frac{1}{2}$, $\frac{1}{3}$, and $\frac{1}{8}$ are unit fractions.

A fraction may be used to represent the division of one number by another number.

If 3 apples are to be divided equally among 4 boys, each boy would receive $\frac{3}{4}$ of an apple.

In a family, there are 2 boys and 3 girls. If we wish to compare the number of boys, 2, with the total number of children, 5, we write the fraction $\frac{2}{5}$. This means that $\frac{2}{5}$ of the children are boys.

We may also wish to compare the number of boys with the number of girls. We then write the fraction $\frac{2}{3}$. This means that there are 2 boys for every 3 girls, or $\frac{2}{3}$ as many boys as girls.

A fraction may be used to express a comparison between the number of members in one group of objects and the number of members in a second group of objects.

KINDS OF FRACTIONS

A ***proper fraction*** is a fraction in which the numerator is less than the denominator, as in $\frac{3}{4}$ and $\frac{9}{10}$. Such a fraction is always less than a whole thing, that is, it is less than 1.

An ***improper fraction*** is a fraction in which the numerator is equal to or greater than the denominator, as in $\frac{4}{4}$ or $\frac{5}{4}$. Such a fraction is either equal to the whole thing, 1, as for example, $\frac{4}{4} = 1$; or is greater than the whole thing, 1, as for example, $\frac{5}{4}$ is greater than 1.

Fractions such as $\frac{3}{4}$, $\frac{9}{10}$, $\frac{4}{4}$, and $\frac{5}{4}$ are examples of ***common fractions.***

EXERCISES

In 1–6, answer the following questions:

a. Into how many equal parts has the figure been divided?
b. What fraction represents each of the equal parts?
c. What fractional part of the whole figure has been shaded?

1. 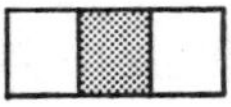

2. 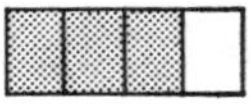

3. 

4. 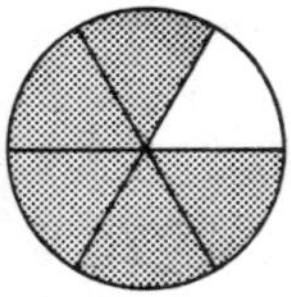

5. 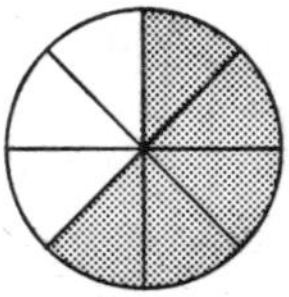

6. 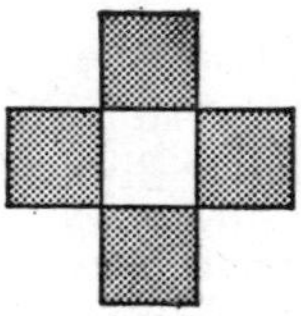

7. A baseball player has 5 balls and 2 bats. Write a fraction that compares:
a. the number of balls with the total number of bats and balls.
b. the number of bats with the total number of bats and balls.
c. the number of balls with the number of bats.

8. Name the numerator and the denominator of each fraction:
a. $\frac{1}{2}$ *b.* $\frac{7}{8}$ *c.* $\frac{3}{10}$ *d.* $\frac{5}{3}$ *e.* $\frac{6}{6}$

9. Tell whether each fraction is a proper or an improper fraction:
a. $\frac{3}{2}$ *b.* $\frac{5}{9}$ *c.* $\frac{7}{10}$ *d.* $\frac{4}{4}$ *e.* $\frac{13}{8}$

10. A whole orange contains $\frac{?}{6}$.

11. A pie that has been divided into 8 equal parts contains $\frac{?}{8}$.

12. A boy has marked off 2 of the 3 equal divisions of a yardstick. What fractional part of the yardstick did he mark off?

13. Mrs. Kelly divided 4 bars of chocolate equally among her 5 children. What part of a bar did each child receive?

14. If 8 boys shared 6 buns equally, what part of a bun did each boy get?

15. Robert earns \$8 a week and saves \$5 of this amount. Write a fraction that compares his savings with his earnings.

16. In a class of 29 pupils, 17 are boys. What fraction of the class is boys? What fraction of the class is girls?

17. Rudy solved 9 of his 10 homework problems correctly. Write a fraction that compares the number he got right with the number he got wrong.

2. REDUCING FRACTIONS TO LOWEST TERMS

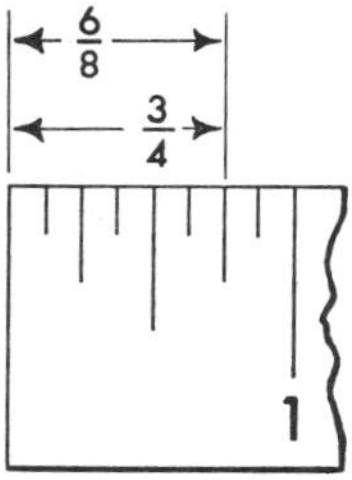

From the piece of ruler pictured at the left, it can be seen that $\frac{6}{8}$ of an inch is the same in length as $\frac{3}{4}$ of an inch: $\frac{6}{8} = \frac{3}{4}$. Observe that we can get $\frac{3}{4}$ by dividing the numerator and the denominator, or both terms, of the fraction $\frac{6}{8}$ by 2. Thus, $\frac{6}{8} = \frac{6 \div 2}{8 \div 2} = \frac{3}{4}$.

Principle. When both the numerator and the denominator of a fraction are divided by the same number (not zero), the value of the fraction does not change.

When we changed $\frac{6}{8}$ to $\frac{3}{4}$, we reduced the fraction $\frac{6}{8}$ to its lowest terms. A fraction is ***reduced to its lowest terms*** when no number other than 1 is exactly divisible into both its numerator and its denominator.

Procedure. To reduce a fraction to its lowest terms, divide both the numerator and the denominator by the largest number that divides evenly into both of them.

MODEL PROBLEM

A ferryboat sailed for 45 minutes. What fraction of an hour did the trip last?

Solution:

1. Change 45 minutes to a fraction of an hour. Since there are 60 minutes in an hour, 45 minutes = $\frac{45}{60}$ of an hour.

2. Reduce the fraction $\frac{45}{60}$ to lowest terms by dividing both the numerator and the denominator by 15.

$$\frac{45}{60} = \frac{45 \div 15}{60 \div 15} = \frac{3}{4} \quad \textit{Answer:}\ \tfrac{3}{4}\text{ of an hour}$$

EXERCISES

Reduce each fraction to lowest terms:

	(a)	(b)	(c)	(d)	(e)
1.	$\frac{2}{4}$	$\frac{15}{60}$	$\frac{60}{180}$	$\frac{200}{1000}$	$\frac{120}{360}$
2.	$\frac{45}{60}$	$\frac{40}{100}$	$\frac{80}{180}$	$\frac{6}{16}$	$\frac{60}{500}$
3.	$\frac{6}{12}$ foot	$\frac{2}{8}$ inch	$\frac{12}{36}$ yard	$\frac{12}{16}$ pound	$\frac{600}{2000}$ ton

Select the fractions with the same value as the fraction in parentheses:

4. $(\frac{1}{2})$ $\frac{5}{10}$, $\frac{8}{4}$, $\frac{1}{4}$, $\frac{7}{14}$, $\frac{15}{20}$

5. $(\frac{2}{3})$ $\frac{6}{8}$, $\frac{8}{12}$, $\frac{10}{15}$, $\frac{24}{16}$, $\frac{4}{5}$

Select the numerals that name the same number as the fraction in parentheses:

6. $(\frac{3}{4})$ $\frac{5}{6}$, $\frac{6}{8}$, $\frac{20}{15}$, $\frac{12}{16}$, $\frac{27}{28}$

7. $(\frac{2}{5})$ $\frac{5}{8}$, $\frac{8}{10}$, $\frac{20}{50}$, $\frac{30}{12}$, $\frac{4}{25}$

Write two fractions that tell what part of the figure has been shaded:

8. 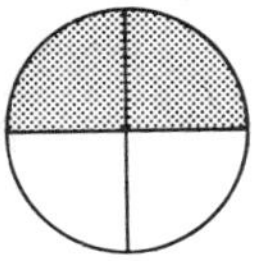

9. 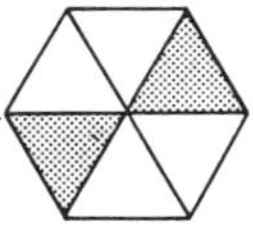

10. 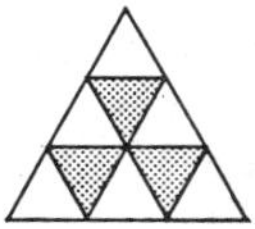

11. Which is the greater fraction?

a. $\frac{8}{10}$ or $\frac{6}{10}$ *b.* $\frac{15}{32}$ or $\frac{30}{32}$ *c.* $\frac{5}{12}$ or $\frac{21}{36}$ *d.* $\frac{3}{4}$ or $\frac{7}{28}$

Reduce each answer to lowest terms:

12. In a class there are 12 boys and 24 girls. What fractional part of the class is boys?

13. Tom used 18 inches of lumber. What fraction of a yard did he use?

14. A man swam for 15 minutes. What fraction of an hour did he swim?

15. Terry worked 8 hours. What part of a day did he work?

16. Ira bought 500 pounds of coal. What fraction of a ton did he buy?

17. Saul grew 3 inches in one year. What fraction of a foot did he grow that year?

3. CHANGING WHOLE OR MIXED NUMBERS TO IMPROPER FRACTIONS

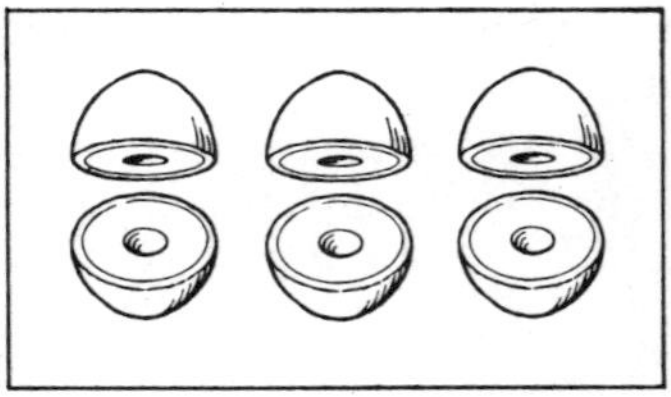

To discover how many halves there are in 3 cantaloupes, we can cut each cantaloupe into 2 halves, getting 6 halves. Therefore, $3 = \frac{6}{2}$.

We can also think out the answer. Since 1 whole = 2 halves, 3 wholes would be 3×2 halves, or 6 halves. This could be written as follows:

$$3 = 3 \times \frac{2}{2} = \frac{3 \times 2}{2} = \frac{6}{2}$$

Note that when we find how many halves there are in 3, we are changing a whole number, 3, to an improper fraction whose denominator is 2.

Procedure. To change a whole number to an improper fraction with a given number as the denominator, multiply the whole number by the given denominator to find the numerator of the improper fraction.

MODEL PROBLEM

How many eighths of an inch are there in 7 inches?

$$7 = 7 \times \frac{8}{8} = \frac{7 \times 8}{8} = \frac{56}{8} \quad \textit{Answer: } \frac{56}{8} \text{ in.}$$

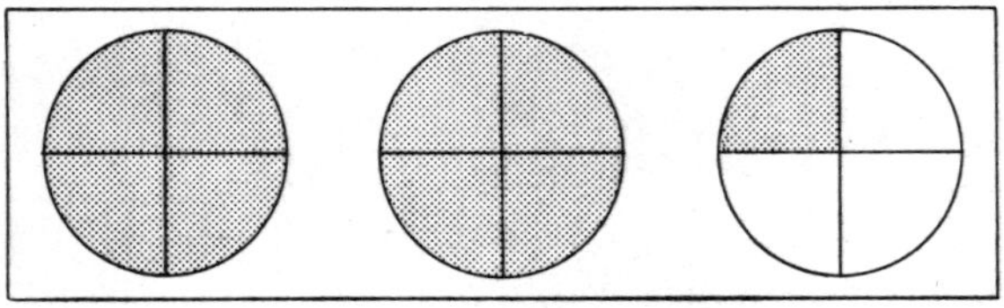

In the above diagram, $2\frac{1}{4}$ circles are shaded. The number $2\frac{1}{4}$ is called a mixed number. A ***mixed number*** is a number that represents the sum of a whole number and a proper fraction. By counting, we can discover that there are $\frac{9}{4}$ circles in $2\frac{1}{4}$ circles.

We can also think out the answer by realizing that 2 whole circles = $\frac{8}{4}$ circles, and $\frac{1}{4}$ circle more gives $\frac{9}{4}$ circles. Therefore, $2\frac{1}{4} = \frac{9}{4}$. Observe that we get $\frac{9}{4}$ when we multiply the 2 by 4, add 1 to get 9, and place the 9 over the denominator, 4.

Procedure. To change a mixed number to an improper fraction, multiply the whole number by the denominator, add the numerator to this product, and write the sum over the denominator.

MODEL PROBLEM

Change $14\frac{3}{8}$ to an improper fraction.

$$14\frac{3}{8} = \frac{8 \times 14 + 3}{8} = \frac{115}{8}$$

$$14\frac{3}{8} = \frac{115}{8} \quad \textit{Ans.}$$

Multiply the whole number, 14, by the denominator, 8. The product is 112.

Add the numerator of the fraction, 3, to this product. The sum is 115.

Write the sum, 115, over the denominator, 8.

EXERCISES

Select the fractions with the same value as the number in parentheses:

1. (3) $\frac{9}{3}$, $\frac{5}{15}$, $\frac{12}{4}$, $\frac{15}{3}$, $\frac{81}{9}$ **2.** ($1\frac{1}{2}$) $\frac{6}{4}$, $\frac{2}{3}$, $\frac{10}{15}$, $\frac{3}{1}$, $\frac{3}{2}$

Select the numerals that name the same number as the numeral in parentheses:

3. (4) $\frac{4}{8}$, $\frac{12}{3}$, $\frac{4}{16}$, $\frac{20}{4}$, $\frac{40}{8}$ **4.** ($1\frac{2}{3}$) $\frac{3}{5}$, $\frac{10}{9}$, $\frac{20}{15}$, $\frac{5}{3}$, $\frac{20}{12}$

5. Change 5 to an improper fraction whose denominator is (*a*) 3 (*b*) 10 (*c*) 12.

6. Change 10 to an improper fraction whose denominator is (*a*) 8 (*b*) 24 (*c*) 60.

Change each whole number to an improper fraction whose denominator is indicated:

7. *a.* $1 = \frac{?}{3}$ *b.* $6 = \frac{?}{2}$ *c.* $7 = \frac{?}{10}$ *d.* $20 = \frac{?}{8}$ *e.* $34 = \frac{?}{6}$

Change to an improper fraction:

	(*a*)	(*b*)	(*c*)	(*d*)	(*e*)	(*f*)	(*g*)
8.	$1\frac{1}{4}$	$2\frac{1}{3}$	$6\frac{1}{2}$	$10\frac{1}{12}$	$15\frac{1}{32}$	$4\frac{3}{5}$	$5\frac{7}{8}$
9.	$8\frac{5}{12}$	$12\frac{7}{32}$	$20\frac{11}{64}$	$8\frac{4}{9}$	$9\frac{4}{11}$	$17\frac{3}{10}$	$35\frac{11}{20}$

10. How many quarters of an inch are there in 8 inches?
11. How many thirds of a yard are there in 12 yards?
12. Mary used $7\frac{1}{8}$ inches of ribbon to trim a dress. How many eighths of an inch of ribbon did she use?
13. If Harold spends $5\frac{2}{3}$ hours on a trip, how many thirds of an hour does he travel?
14. A restaurant served $23\frac{5}{6}$ pies during a day. How many sixths of a pie did it serve?
15. A grocer has $65\frac{3}{4}$ pounds of butter. How many customers can he supply if each customer receives $\frac{1}{4}$ of a pound?
16. A lumber dealer has $124\frac{2}{3}$ yards of molding. How many strips of molding, each $\frac{1}{3}$ of a yard in length, can he cut?

4. CHANGING IMPROPER FRACTIONS TO WHOLE OR MIXED NUMBERS

The improper fraction $\frac{8}{4}$ means $8 \div 4$, which equals 2. Thus, the improper fraction $\frac{8}{4}$ was changed to the whole number 2.

The improper fraction $\frac{9}{2}$ means $9 \div 2$. When 9 is divided by 2, the quotient is 4 and the remainder is 1. This answer may be written as $4\frac{1}{2}$. Therefore, $\frac{9}{2} = 4\frac{1}{2}$.

Procedure. To change an improper fraction to a whole or mixed number, divide the numerator by the denominator. If there is a remainder, write it as the numerator of a proper fraction having the same denominator as the improper fraction. Reduce the proper fraction to lowest terms.

MODEL PROBLEMS

1. Change $\frac{84}{12}$ to a whole number.

$\frac{84}{12} = 7$

Divide the numerator, 84, by the denominator, 12. The result is 7.

2. Change $\frac{56}{6}$ to a mixed number.

$\frac{56}{6} = 9\frac{2}{6} = 9\frac{1}{3}$

Divide the numerator, 56, by the denominator, 6. The quotient is 9 and the remainder is 2.

Write the remainder, 2, over the denominator, 6. Reduce the fraction: $\frac{2}{6} = \frac{1}{3}$.

EXERCISES

Change to a whole or mixed number in lowest terms:

	(a)	(b)	(c)	(d)	(e)
1.	$\frac{8}{4}$	$\frac{27}{4}$	$\frac{84}{12}$	$\frac{120}{20}$	$\frac{360}{60}$
2.	$\frac{7}{2}$	$\frac{37}{5}$	$\frac{153}{8}$	$\frac{89}{16}$	$\frac{67}{12}$
3.	$\frac{154}{10}$	$\frac{542}{64}$	$\frac{168}{24}$	$\frac{250}{60}$	$\frac{320}{20}$

4. Change to a whole or mixed number in lowest terms:
   a. $\frac{36}{12}$ feet *b.* $\frac{120}{60}$ hours *c.* $\frac{6000}{2000}$ tons *d.* $\frac{60}{24}$ days *e.* $\frac{19}{12}$ years
5. Which is greater? (*a*) $\frac{67}{3}$ or 22 (*b*) $\frac{151}{12}$ or $12\frac{5}{12}$ (*c*) $\frac{200}{32}$ or $6\frac{5}{32}$
6. How many pound jars can be filled from $\frac{3200}{16}$ pounds of coffee?
7. Jane needed $\frac{17}{3}$ feet of ribbon to trim a lamp shade. She ordered 6 feet of ribbon. Did she order enough? Explain.
8. A plane flew for 150 minutes. How many hours did it travel?
9. Mr. Thomas has 540 shirts in stock. How many dozen shirts does he have?

5. CHANGING FRACTIONS TO HIGHER TERMS

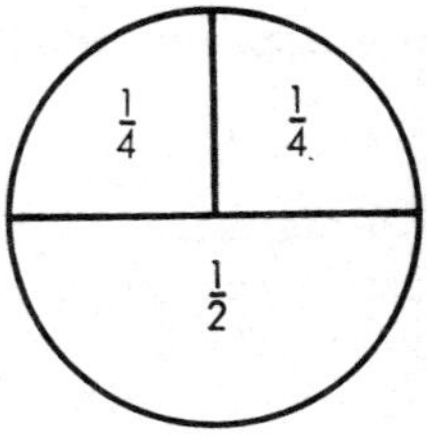

From the figure, we see that $\frac{1}{2}$ of a circle is the same as $\frac{2}{4}$ of the circle. Therefore, we can say that $\frac{1}{2} = \frac{2}{4}$.

Notice that in the fraction $\frac{1}{2}$ when both the numerator, 1, and the denominator, 2, are multiplied by 2, the result is $\frac{2}{4}$.

$$\frac{1}{2} = \frac{1 \times 2}{2 \times 2} = \frac{2}{4}$$

The fractions $\frac{1}{2}$ and $\frac{2}{4}$ are equal in value, or are equivalent. The fraction $\frac{2}{4}$ expresses the fraction $\frac{1}{2}$ in higher terms.

Principle. When both the numerator and the denominator of a fraction are multiplied by the same number (not zero), the value of the fraction does not change.

MODEL PROBLEMS

1. $\frac{3}{4} = \frac{?}{20}$

$$\frac{3}{4} = \frac{3 \times 5}{4 \times 5} = \frac{15}{20}$$

Since the original denominator, 4, must be multiplied by 5 to get the new denominator, 20, then the original numerator, 3, must also be multiplied by 5 in order not to change the value of the fraction.

2. Change $\frac{2}{3}$ and $\frac{3}{4}$ to fractions that have the same denominator.

$$\frac{2}{3} = \frac{2 \times 4}{3 \times 4} = \frac{8}{12}$$

$$\frac{3}{4} = \frac{3 \times 3}{4 \times 3} = \frac{9}{12}$$

Find a number to which the given denominators, 3 and 4, can be changed. To do this, multiply 3 by 4, getting 12. Change each of the given fractions to an equal fraction that has 12 as its denominator.

EXERCISES

Find the missing numerator:

	(*a*)	(*b*)	(*c*)	(*d*)	(*e*)
1.	$\frac{1}{2} = \frac{?}{8}$	$\frac{1}{6} = \frac{?}{12}$	$\frac{1}{10} = \frac{?}{100}$	$\frac{2}{3} = \frac{?}{12}$	$\frac{4}{5} = \frac{?}{10}$
2.	$\frac{7}{12} = \frac{?}{60}$	$\frac{17}{30} = \frac{?}{180}$	$\frac{14}{25} = \frac{?}{100}$	$\frac{27}{90} = \frac{?}{360}$	$\frac{13}{25} = \frac{?}{1000}$

3. Change: (*a*) $\frac{1}{2}$ to 36ths (*b*) $\frac{1}{25}$ to 100ths (*c*) $\frac{5}{8}$ to 32nds (*d*) $\frac{5}{12}$ to 36ths

4. Change to 60ths: $\frac{1}{2}$, $\frac{1}{3}$, $\frac{3}{4}$, $\frac{2}{5}$, $\frac{7}{10}$, $\frac{11}{12}$, $\frac{13}{20}$.

5. Change to 100ths: $\frac{1}{4}$, $\frac{2}{5}$, $\frac{7}{10}$, $\frac{9}{20}$, $\frac{17}{25}$, $\frac{33}{50}$.

6. Change $\frac{7}{4}$ of an inch to sixteenths of an inch.

7. How many 60ths of an hour are there in $\frac{3}{5}$ of an hour?

8. Change the two fractions to equal fractions that have the same denominator: (*a*) $\frac{1}{3}$ and $\frac{1}{4}$ (*b*) $\frac{1}{2}$ and $\frac{1}{3}$ (*c*) $\frac{1}{8}$ and $\frac{1}{3}$ (*d*) $\frac{2}{3}$ and $\frac{4}{5}$ (*e*) $\frac{7}{9}$ and $\frac{3}{4}$

9. *a.* If 4 is added to both the numerator and the denominator of the fraction $\frac{2}{4}$, what is the resulting fraction?

b. Is the resulting fraction obtained in part *a* equal to the original fraction $\frac{2}{4}$?

10. If the same number is added to both the numerator and the denominator of a fraction, is the resulting fraction always equal to the original fraction?

11. *a.* If 5 is subtracted from both the numerator and the denominator of the fraction $\frac{10}{15}$, what is the value of the resulting fraction?

b. Is the resulting fraction obtained in part *a* equal to the original fraction?

12. If the same number is subtracted from both the numerator and the denominator of a fraction, is the resulting fraction always equal to the original fraction?

6. COMPARING FRACTIONS

Principle 1. If fractions have the same numerator, the fraction with the smallest denominator has the largest value.

Thus, among the fractions $\frac{2}{3}$, $\frac{2}{4}$, and $\frac{2}{8}$, the fraction $\frac{2}{3}$ is the largest. Also, $\frac{5}{8}$ is larger than $\frac{5}{32}$, and $\frac{13}{36}$ is smaller than $\frac{13}{18}$.

Principle 2. If fractions have the same denominator, the fraction with the largest numerator has the largest value.

Thus, among the fractions $\frac{1}{4}$, $\frac{2}{4}$, and $\frac{3}{4}$, the fraction $\frac{3}{4}$ is the largest. Also, $\frac{5}{12}$ is larger than $\frac{3}{12}$, and $\frac{13}{36}$ is smaller than $\frac{27}{36}$.

Procedure. To compare the values of fractions, change all the fractions to equal fractions having the same denominator. This denominator should be a number into which all the denominators will divide evenly. One possible denominator can always be obtained by finding the product of the denominators of the given fractions.

MODEL PROBLEM

Is $\frac{2}{5}$ less than, equal to, or greater than $\frac{3}{4}$?

$$\frac{2}{5} = \frac{2 \times 4}{5 \times 4} = \frac{8}{20} \text{ and } \frac{3}{4} = \frac{3 \times 5}{4 \times 5} = \frac{15}{20}$$

Since $\frac{8}{20}$ is less than $\frac{15}{20}$, $\frac{2}{5}$ is less than $\frac{3}{4}$.

Change each fraction to a fraction whose denominator is 20.

EXERCISES

1. Which is the larger fraction?

a. $\frac{7}{12}$ or $\frac{15}{12}$ *b.* $\frac{23}{8}$ or $\frac{18}{8}$ *c.* $\frac{5}{2}$ or $\frac{5}{4}$ *d.* $\frac{10}{12}$ or $\frac{10}{6}$

e. $\frac{125}{120}$ or $\frac{125}{360}$ *f.* $\frac{1}{4}$ or $\frac{1}{3}$ *g.* $\frac{2}{5}$ or $\frac{5}{2}$ *h.* $\frac{7}{4}$ or $\frac{15}{6}$

2. Which is the smaller fraction?

a. $\frac{17}{4}$ or $\frac{27}{4}$ *b.* $\frac{15}{16}$ or $\frac{11}{16}$ *c.* $\frac{9}{4}$ or $\frac{9}{8}$ *d.* $\frac{30}{16}$ or $\frac{30}{32}$

e. $\frac{13}{12}$ or $\frac{4}{3}$ *f.* $\frac{13}{5}$ or $\frac{4}{3}$ *g.* $\frac{7}{2}$ or $\frac{10}{3}$ *h.* $\frac{5}{6}$ or $\frac{8}{9}$

3. Which is the largest fraction? (*a*) $\frac{1}{3}$, $\frac{1}{2}$, or $\frac{1}{4}$ (*b*) $\frac{3}{2}$, $\frac{10}{8}$, or $\frac{7}{4}$
4. Which is the smallest fraction? (*a*) $\frac{1}{6}$, $\frac{1}{3}$, or $\frac{1}{2}$ (*b*) $\frac{3}{10}$, $\frac{7}{25}$, or $\frac{9}{50}$
5. Among the fractions $\frac{1}{2}$, $\frac{7}{25}$, and $\frac{39}{50}$, which fraction is larger than $\frac{3}{5}$?
6. When the denominator of a fraction is decreased and the numerator does not change, is the value of the fraction decreased or increased?
7. If you double the denominator of a fraction and the numerator does not change, do you increase or decrease the value of the fraction?
8. Tom and Barry are painting a fence. Tom has finished $\frac{5}{12}$ of the job and Barry has finished $\frac{1}{3}$ of the job. Who has finished the larger part of the job?
9. Myra and Beatrice are reading the same book. Myra has finished $\frac{7}{8}$ of the book and Beatrice has finished $\frac{11}{12}$ of the book. Who has read fewer pages?

7. FINDING THE LEAST COMMON DENOMINATOR

Fractions with the same denominator are said to have a ***common denominator.*** Such fractions are called ***like fractions.*** Thus, $\frac{7}{8}$ and $\frac{3}{8}$ are like fractions. The fractions $\frac{1}{2}$ and $\frac{1}{3}$ are ***unlike fractions*** because they have different denominators.

The pair of fractions $\frac{1}{2}$ and $\frac{3}{4}$ can be changed to $\frac{2}{4}$ and $\frac{3}{4}$, $\frac{4}{8}$ and $\frac{6}{8}$, $\frac{6}{12}$ and $\frac{9}{12}$, and many other pairs of like fractions. We see that it is possible to find many common denominators.

The smallest possible common denominator is called the ***least common denominator.*** Such a denominator is always the smallest number into which the denominators of the original fractions will divide evenly. For example: The least common denominator for the fractions $\frac{3}{4}$ and $\frac{7}{10}$ is 20, since 20 is the smallest number into which 4 and 10 are exactly divisible.

Procedure. To find the least common denominator of several denominators:

Divide the largest denominator by each of the other denominators.

If the result of each division is a whole number, the largest denominator is the least common denominator.

If each of the other denominators does not divide evenly into the largest denominator, multiply the largest denominator by 2, 3, 4, etc., until the product is a number into which every denominator divides evenly.

MODEL PROBLEM

Change the fractions $\frac{3}{4}$ and $\frac{5}{6}$ into equal fractions that have as their denominators the least common denominator.

$$\frac{3}{4} = \frac{3 \times 3}{4 \times 3} = \frac{9}{12}$$

$$\frac{5}{6} = \frac{5 \times 2}{6 \times 2} = \frac{10}{12}$$

The larger denominator, 6, is not the least common denominator because the smaller denominator, 4, does not divide into 6 evenly. However, if we multiply the larger denominator by 2, the result, 12, turns out to be the least common denominator.

Change both fractions to equal fractions whose denominators are 12.

EXERCISES

1. Find the lowest common denominator for two fractions whose denominators are: (*a*) 5 and 10 (*b*) 16 and 4 (*c*) 6 and 9 (*d*) 3 and 5

Change the given fractions to equal fractions that have the least common denominator:

	(*a*)	(*b*)	(*c*)	(*d*)
2.	$\frac{1}{2}$ and $\frac{5}{8}$	$\frac{2}{3}$ and $\frac{7}{6}$	$\frac{9}{16}$ and $\frac{3}{8}$	$\frac{3}{10}$ and $\frac{37}{100}$
3.	$\frac{1}{2}$ and $\frac{1}{3}$	$\frac{2}{3}$ and $\frac{5}{4}$	$\frac{1}{5}$ and $\frac{4}{3}$	$\frac{13}{4}$ and $\frac{5}{9}$
4.	$\frac{1}{4}$ and $\frac{7}{6}$	$\frac{13}{6}$ and $\frac{5}{8}$	$\frac{7}{10}$ and $\frac{23}{25}$	$\frac{5}{8}$ and $\frac{6}{20}$
5.	$\frac{3}{4}$, $\frac{1}{2}$, and $\frac{7}{8}$	$\frac{4}{5}$, $\frac{17}{20}$, and $\frac{57}{100}$	$\frac{1}{3}$, $\frac{2}{5}$, and $\frac{1}{2}$	$\frac{5}{6}$, $\frac{11}{16}$, and $\frac{7}{12}$

8. ADDING LIKE FRACTIONS AND MIXED NUMBERS THAT CONTAIN LIKE FRACTIONS

ADDING LIKE FRACTIONS

From the figure at the right, it can be seen that $\frac{4}{4} + \frac{3}{4} = \frac{7}{4}$.

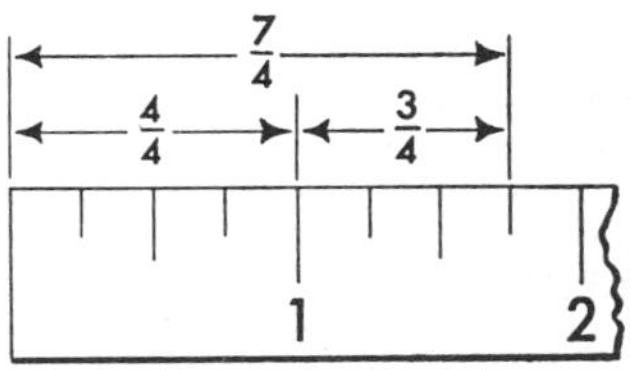

Procedure. To add two or more fractions that have the same denominator (like fractions), find the sum of the numerators and write the sum over the common denominator. Reduce the answer.

MODEL PROBLEMS

Add:

1. $\begin{array}{r}\frac{1}{5}\\ \underline{\frac{1}{5}}\\ \frac{2}{5}\end{array}$

2. $\begin{array}{r}\frac{5}{12}\\ \underline{\frac{3}{12}}\\ \frac{8}{12}\end{array} = \frac{2}{3}$

3. $\begin{array}{r}\frac{3}{8}\\ \underline{\frac{5}{8}}\\ \frac{8}{8}\end{array} = 1$

4. $\begin{array}{r}\frac{3}{4}\\ \underline{\frac{2}{4}}\\ \frac{5}{4}\end{array} = 1\frac{1}{4}$

5. $\begin{array}{r}\frac{5}{6}\\ \underline{\frac{4}{6}}\\ \frac{9}{6}\end{array} = 1\frac{3}{6} = 1\frac{1}{2}$

ADDING MIXED NUMBERS THAT CONTAIN LIKE FRACTIONS

From this figure, we can see that the sum of $2\frac{1}{4}$ circles and $1\frac{2}{4}$ circles is $3\frac{3}{4}$ circles.

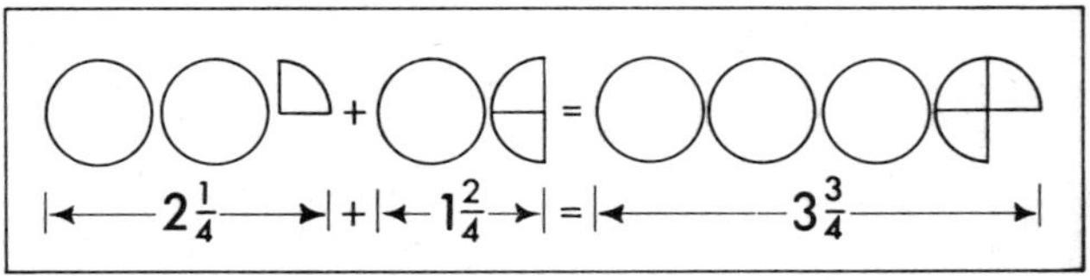

Procedure. To add two or more mixed numbers, find the sum of the fractions and add this sum to the total of the whole numbers. Reduce the answer.

MODEL PROBLEMS

Add:

1. $\begin{array}{r}5\frac{1}{8}\\ \underline{12\frac{3}{8}}\\ 17\frac{4}{8}\end{array} = 17\frac{1}{2}$

2. $\begin{array}{r}4\frac{1}{6}\\ \underline{3\frac{5}{6}}\\ 7\frac{6}{6}\end{array} = 8$

3. $\begin{array}{r}8\frac{5}{9}\\ \underline{12\frac{7}{9}}\\ 20\frac{12}{9}\end{array} = 21\frac{3}{9} = 21\frac{1}{3}$

EXERCISES

Add:

	(a)	(b)	(c)	(d)	(e)	(f)
1.	$\frac{1}{4}$, $\frac{2}{4}$	$\frac{13}{100}$, $\frac{14}{100}$	$\frac{9}{16}$, $\frac{7}{16}$	$\frac{1}{6}$, $\frac{3}{6}$	$\frac{5}{12}$, $\frac{4}{12}$	$\frac{6}{10}$, $\frac{8}{10}$
2.	$\frac{1}{2}$, $\frac{3}{2}$, $\frac{7}{2}$	$\frac{2}{3}$, $\frac{4}{3}$, $\frac{3}{3}$	$\frac{3}{4}$, $\frac{7}{4}$, $\frac{2}{4}$	$\frac{9}{12}$, $\frac{3}{12}$, $\frac{5}{12}$	$\frac{2}{25}$, $\frac{14}{25}$, $\frac{19}{25}$	$\frac{73}{100}$, $\frac{25}{100}$, $\frac{12}{100}$
3.	8, $\frac{3}{5}$	7, $\frac{3}{4}$	$\frac{11}{25}$, 30	$18\frac{3}{32}$, 12	20, $7\frac{13}{64}$	$100\frac{3}{10}$, 200

	(a)	(b)	(c)	(d)	(e)	(f)
4.	$2\frac{1}{4}$	$11\frac{3}{6}$	$5\frac{1}{10}$	$16\frac{13}{32}$	$3\frac{1}{5}$	$10\frac{7}{16}$
	$5\frac{2}{4}$	$9\frac{2}{6}$	$7\frac{4}{10}$	$24\frac{9}{32}$	$8\frac{4}{5}$	$22\frac{11}{16}$
5.	$8\frac{3}{8}$	$6\frac{5}{6}$	$10\frac{5}{9}$	$\frac{4}{10}$	$17\frac{3}{4}$	$\frac{9}{12}$
	$\frac{4}{8}$	$3\frac{1}{6}$	$\frac{4}{9}$	$8\frac{2}{10}$	$\frac{2}{4}$	$9\frac{6}{12}$

6. One day, Will worked $2\frac{1}{4}$ hours before lunch and $1\frac{2}{4}$ hours after lunch. How many hours did Will work that day?

7. The Samuels family traveled $18\frac{7}{10}$ miles going to the beach and $20\frac{5}{10}$ miles returning. What is the total number of miles they traveled?

8. Mrs. Ward needed $3\frac{5}{8}$ yd., $3\frac{4}{8}$ yd., and $2\frac{7}{8}$ yd. of material to make three dresses for the family. In all, how much material did she need?

9. An electrician spent $8\frac{1}{4}$ hours, 9 hours, and $6\frac{3}{4}$ hours doing a job. How long did it take him to do the job?

9. ADDING UNLIKE FRACTIONS AND MIXED NUMBERS THAT CONTAIN UNLIKE FRACTIONS

ADDING UNLIKE FRACTIONS

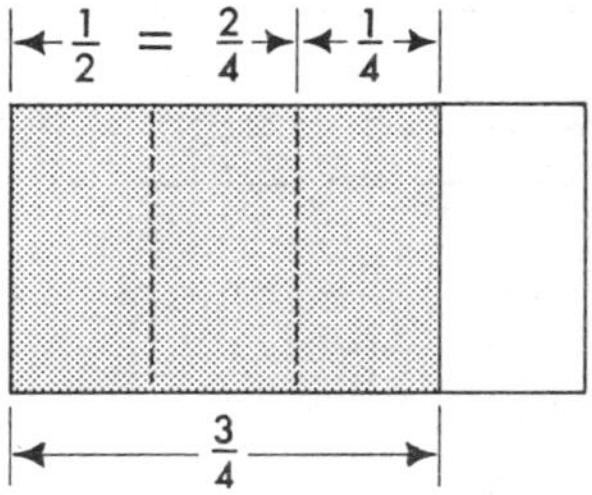

The figure at the right helps us understand why $\frac{1}{2} + \frac{1}{4} = \frac{3}{4}$. Since $\frac{1}{2}$ of the rectangle is the same as $\frac{2}{4}$ of the rectangle, we have $\frac{1}{2} + \frac{1}{4} = \frac{2}{4} + \frac{1}{4} = \frac{3}{4}$. We can write:

$$\begin{array}{r} \frac{1}{2} = \frac{2}{4} \\ \underline{\frac{1}{4} = \frac{1}{4}} \\ \frac{3}{4} \end{array}$$

Procedure. **To add two or more fractions whose denominators are different, change the fractions to equal fractions whose denominators are the least common denominator. Now use the procedure for adding like fractions.**

COMMUTATIVE PRINCIPLE FOR ADDITION

Since $\frac{1}{2} + \frac{1}{4} = \frac{3}{4}$ and $\frac{1}{4} + \frac{1}{2} = \frac{3}{4}$, then $\mathbf{\frac{1}{2} + \frac{1}{4} = \frac{1}{4} + \frac{1}{2}}$.

This example illustrates that *addition is also commutative* when we are dealing with fractions.

ASSOCIATIVE PRINCIPLE FOR ADDITION

Since $(\frac{1}{8} + \frac{3}{8}) + \frac{2}{8} = \frac{4}{8} + \frac{2}{8} = \frac{6}{8}$ and $\frac{1}{8} + (\frac{3}{8} + \frac{2}{8}) = \frac{1}{8} + \frac{5}{8} = \frac{6}{8}$,

then $\mathbf{(\frac{1}{8} + \frac{3}{8}) + \frac{2}{8} = \frac{1}{8} + (\frac{3}{8} + \frac{2}{8})}$.

This example illustrates that *addition is also associative* when we are dealing with fractions.

ADDING MIXED NUMBERS THAT CONTAIN UNLIKE FRACTIONS

If we wish to add $1\frac{1}{4}$ and $1\frac{7}{8}$, we can do so with the help of a ruler. From the markings on the ruler, we see that $1\frac{1}{4} + 1\frac{7}{8} = 3\frac{1}{8}$.

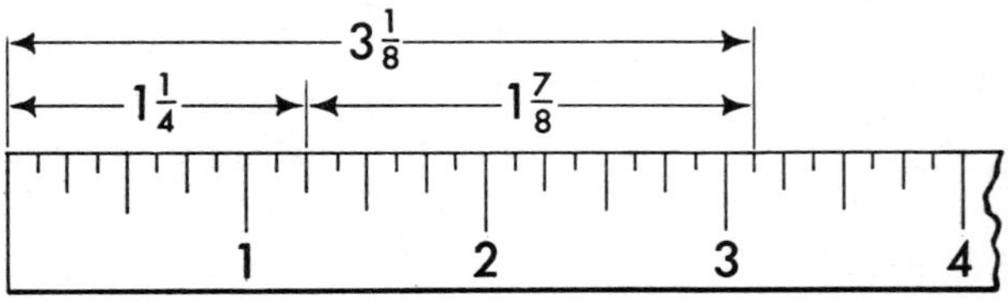

To add without a ruler, we must change the mixed numbers $1\frac{1}{4}$ and $1\frac{7}{8}$ to equal mixed numbers in which the fractions are like fractions. The least common denominator is 8.

$$\begin{aligned} 1\tfrac{1}{4} &= 1\tfrac{2}{8} \\ 1\tfrac{7}{8} &= 1\tfrac{7}{8} \\ \hline &\ 2\tfrac{9}{8} = 2 + 1\tfrac{1}{8} = 3\tfrac{1}{8} \end{aligned}$$

MODEL PROBLEMS

1. Add: $\frac{3}{4} + \frac{2}{3}$.

$$\begin{aligned} \tfrac{3}{4} &= \tfrac{9}{12} \\ \tfrac{2}{3} &= \tfrac{8}{12} \\ \hline &\ \tfrac{17}{12} = 1\tfrac{5}{12} \end{aligned}$$

Find the least common denominator, 12.

Change the fractions $\frac{3}{4}$ and $\frac{2}{3}$ to equal fractions whose denominators are 12.

Add these fractions: $\frac{9}{12} + \frac{8}{12} = \frac{17}{12}$.

Since $\frac{12}{12} = 1$, $\frac{17}{12} = 1\frac{5}{12}$.

2. Add: $3\frac{1}{4}$, $4\frac{3}{8}$, $5\frac{1}{2}$.

$$\begin{aligned} 3\tfrac{1}{4} &= 3\tfrac{2}{8} \\ 4\tfrac{3}{8} &= 4\tfrac{3}{8} \\ 5\tfrac{1}{2} &= 5\tfrac{4}{8} \\ \hline &\ 12\tfrac{9}{8} = 13\tfrac{1}{8} \end{aligned}$$

Change the fractions in the mixed numbers to equal fractions that have the least common denominator, 8. Add the fractions and the whole numbers separately.

NOTE

Fractions must have the same denominator before they can be added.

EXERCISES

Add:

	(a)	(b)	(c)	(d)	(e)	(f)
1.	$\frac{1}{4}$ $\frac{1}{8}$	$\frac{1}{2}$ $\frac{1}{8}$	$\frac{1}{2}$ $\frac{3}{4}$	$\frac{3}{4}$ $\frac{7}{8}$	$\frac{1}{2}$ $\frac{1}{6}$	$\frac{2}{3}$ $\frac{1}{6}$
2.	$\frac{5}{6}$ $\frac{11}{12}$	$\frac{1}{2}$ $\frac{5}{16}$	$\frac{5}{12}$ $\frac{17}{36}$	$\frac{5}{6}$ $\frac{35}{36}$	$\frac{3}{8}$ $\frac{5}{32}$	$\frac{3}{4}$ $\frac{5}{100}$
3.	$\frac{2}{3}$ $\frac{1}{4}$	$\frac{1}{2}$ $\frac{4}{7}$	$\frac{3}{5}$ $\frac{1}{2}$	$\frac{3}{4}$ $\frac{5}{6}$	$\frac{1}{6}$ $\frac{7}{10}$	$\frac{7}{15}$ $\frac{11}{20}$
4.	$\frac{1}{2}$ $\frac{2}{3}$ $\frac{5}{6}$	$\frac{1}{2}$ $\frac{1}{4}$ $\frac{5}{8}$	$\frac{7}{8}$ $\frac{1}{2}$ $\frac{3}{4}$	$\frac{4}{5}$ $\frac{5}{6}$ $\frac{7}{10}$	$\frac{5}{8}$ $\frac{1}{3}$ $\frac{11}{12}$	$\frac{3}{4}$ $\frac{5}{12}$ $\frac{7}{9}$
5.	$2\frac{1}{4}$ $3\frac{3}{8}$	$8\frac{2}{3}$ $9\frac{1}{12}$	$12\frac{3}{5}$ $7\frac{10}{100}$	$8\frac{2}{3}$ $4\frac{1}{4}$	$8\frac{1}{6}$ $9\frac{3}{5}$	$3\frac{5}{9}$ $6\frac{1}{4}$
6.	$2\frac{5}{12}$ $3\frac{1}{3}$	$7\frac{1}{2}$ $2\frac{3}{10}$	$12\frac{3}{25}$ $10\frac{13}{100}$	$2\frac{2}{3}$ $8\frac{5}{6}$	$8\frac{1}{2}$ $7\frac{3}{4}$	$7\frac{3}{4}$ $2\frac{5}{8}$
7.	$2\frac{1}{4}$ $3\frac{1}{8}$ $5\frac{1}{2}$	4 $2\frac{1}{2}$ $6\frac{7}{8}$	8 $2\frac{1}{3}$ $5\frac{3}{4}$	3 $\frac{5}{8}$ $2\frac{2}{3}$	$6\frac{1}{5}$ $\frac{1}{2}$ $5\frac{7}{10}$	13 $12\frac{5}{6}$ $\frac{7}{10}$

Add:

	(a)	(b)	(c)
8.	$\frac{1}{2}+\frac{1}{8}+\frac{1}{4}$	$\frac{1}{2}+\frac{1}{3}+\frac{1}{12}$	$\frac{1}{4}+\frac{1}{8}+\frac{1}{16}$
9.	$\frac{1}{2}+\frac{3}{4}+\frac{5}{8}$	$\frac{2}{3}+\frac{1}{2}+\frac{7}{12}$	$\frac{2}{3}+\frac{3}{5}+\frac{7}{15}$
10.	$\frac{1}{2}+\frac{1}{3}+\frac{1}{4}$	$\frac{3}{4}+\frac{2}{5}+\frac{5}{6}$	$\frac{1}{3}+\frac{7}{12}+\frac{5}{8}$
11.	$3\frac{1}{2}+2\frac{7}{8}+1\frac{3}{4}$	$6\frac{1}{2}+8\frac{3}{4}+5\frac{3}{8}$	$3\frac{3}{4}+1\frac{7}{8}+3\frac{5}{16}$
12.	$2\frac{5}{6}+4\frac{3}{4}+7\frac{1}{2}$	$4\frac{1}{4}+2\frac{3}{5}+10\frac{1}{2}$	$6\frac{1}{2}+4\frac{2}{3}+9\frac{3}{5}$

Find the sum of:

13.	$6\frac{3}{4}$; $2\frac{1}{8}$; $9\frac{1}{2}$	$6\frac{1}{4}$; $3\frac{2}{3}$; $\frac{1}{2}$	$2\frac{2}{3}$; $15\frac{3}{4}$; $11\frac{1}{6}$; $5\frac{7}{12}$
14.	$3\frac{1}{4}$; $8\frac{2}{5}$; $4\frac{1}{2}$	$5\frac{3}{8}$; $4\frac{1}{3}$; $1\frac{1}{2}$; $8\frac{5}{6}$	$12\frac{1}{2}$; $5\frac{1}{3}$; $2\frac{1}{16}$
15.	$3\frac{1}{5}$; $\frac{2}{3}$; $6\frac{4}{15}$	7; $8\frac{3}{4}$; $9\frac{2}{7}$	$\frac{3}{4}$; $1\frac{3}{10}$; 2; $7\frac{2}{5}$

Find the sum. Simplify the arithmetic by using the commutative and associative principles for addition.

	(a)	(b)	(c)
16.	$\frac{3}{4}+\frac{2}{3}+\frac{1}{4}$	$\frac{5}{32}+\frac{7}{9}+\frac{27}{32}$	$\frac{5}{12}+\frac{5}{16}+\frac{7}{12}+\frac{3}{16}$
17.	$3\frac{1}{6}+2\frac{1}{2}+4\frac{5}{6}$	$9\frac{13}{36}+1\frac{1}{5}+3\frac{23}{36}$	$2\frac{7}{8}+5\frac{4}{9}+3\frac{5}{9}+6\frac{1}{8}$

18. Marion bought $\frac{1}{2}$ pound of bread and $\frac{3}{4}$ pound of butter. What was the total weight of her purchase?

19. John weighs $52\frac{1}{2}$ pounds, Mary weighs $33\frac{1}{4}$ pounds, and Helen weighs $36\frac{1}{4}$ pounds. Find the total weight of the three children.

20. Bob is $38\frac{1}{2}$ inches tall. Al is $17\frac{5}{8}$ inches taller. Find Al's height.

21. A plumber estimates that he will need the following lengths of pipe: $8\frac{5}{8}$ in., $10\frac{7}{32}$ in., and $32\frac{3}{4}$ in. Find the total length of pipe he will need.

22. In a hosiery store, the salesmen sold $7\frac{3}{4}$, $10\frac{5}{6}$, $8\frac{1}{2}$, and 12 dozen pairs of stockings on a certain day. Find the total number of dozen pairs of stockings sold that day.

23. Find the total length of the line segment at the right.

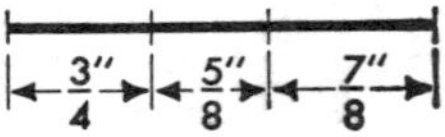

24. In the sketch at the right, find the length of (*a*) *AC* (*b*) *BD* (*c*) *AD*.

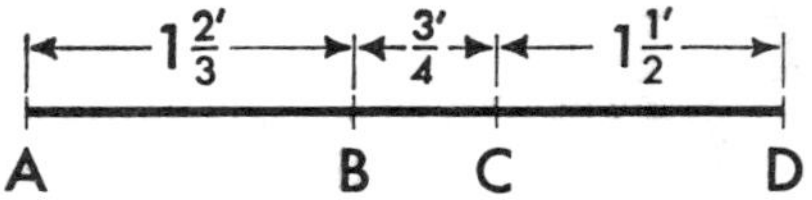

10. SUBTRACTING LIKE FRACTIONS AND MIXED NUMBERS THAT CONTAIN LIKE FRACTIONS

SUBTRACTING LIKE FRACTIONS

If from $\frac{3}{4}$ of an hour we subtract $\frac{2}{4}$ of an hour, the remainder is $\frac{1}{4}$ of an hour. We can write: $\frac{3}{4}-\frac{2}{4}=\frac{1}{4}$.

Procedure. To subtract two fractions that have the same denominator, subtract the numerator of the subtrahend from the numerator of the minuend and write the difference over the common denominator. Reduce the answer.

MODEL PROBLEMS

Subtract:

1. $\frac{7}{8}-\frac{4}{8}=\frac{3}{8}$

2. $\frac{9}{12}-\frac{3}{12}=\frac{6}{12}=\frac{1}{2}$

3. $\frac{9}{4}-\frac{5}{4}=\frac{4}{4}=1$

4. $\frac{17}{8}-\frac{3}{8}=\frac{14}{8}=1\frac{6}{8}=1\frac{3}{4}$

5. $\frac{7}{2}-\frac{7}{2}=0$

SUBTRACTING MIXED NUMBERS THAT CONTAIN LIKE FRACTIONS

The difference between $3\frac{5}{8}$ circles and $2\frac{3}{8}$ circles is $1\frac{2}{8}$, or $1\frac{1}{4}$ circles. We can write: $3\frac{5}{8} - 2\frac{3}{8} = 1\frac{2}{8}$, or $1\frac{1}{4}$.

Procedure. To subtract two mixed numbers in which the fraction in the minuend is larger than the fraction in the subtrahend, find the difference between the fractions and add this difference to the difference between the whole numbers. Reduce the answer.

MODEL PROBLEMS

1. Subtract: $9\frac{7}{8} - 6\frac{3}{8}$.

$$\begin{array}{l} 9\frac{7}{8} \\ \underline{6\frac{3}{8}} \\ 3\frac{4}{8} = 3\frac{1}{2} \end{array}$$

Find the difference between the fractions: $\frac{7}{8} - \frac{3}{8} = \frac{4}{8}$.
Find the difference between the whole numbers: $9 - 6 = 3$.
Add the two results: $3 + \frac{4}{8} = 3\frac{4}{8}$.
Reduce the answer: $3\frac{4}{8} = 3\frac{1}{2}$.

2. From 8 subtract $1\frac{3}{4}$.

$$\begin{array}{lcr} 8 & = & 7\frac{4}{4} \\ 1\frac{3}{4} & = & \underline{1\frac{3}{4}} \\ & & 6\frac{1}{4} \end{array}$$

Since there is no fraction in the minuend, 8, change 1 whole to $\frac{4}{4}$ and write 8 as $7\frac{4}{4}$. Then subtract.

3. Subtract: $20\frac{7}{16} - 13\frac{15}{16}$.

$$\begin{array}{lcr} 20\frac{7}{16} & = & 19\frac{23}{16} \\ 13\frac{15}{16} & = & \underline{13\frac{15}{16}} \\ & & 6\frac{8}{16} = 6\frac{1}{2} \end{array}$$

Since the fraction in the minuend is smaller than the fraction in the subtrahend, change 1 whole to $\frac{16}{16}$ and write $20\frac{7}{16}$ as $19\frac{23}{16}$. Then subtract.

EXERCISES

Subtract:

	(a)	(b)	(c)	(d)	(e)	(f)
1.	$\frac{5}{8}$ $\frac{2}{8}$	$\frac{5}{6}$ $\frac{4}{6}$	$\frac{8}{12}$ $\frac{3}{12}$	$\frac{8}{9}$ $\frac{5}{9}$	$\frac{9}{10}$ $\frac{6}{10}$	$\frac{9}{16}$ $\frac{9}{16}$
2.	$\frac{5}{4}$ $\frac{3}{4}$	$\frac{9}{16}$ $\frac{3}{16}$	$\frac{27}{36}$ $\frac{18}{36}$	$\frac{6}{4}$ $\frac{2}{4}$	$\frac{13}{10}$ $\frac{3}{10}$	$\frac{20}{8}$ $\frac{4}{8}$

	(a)	(b)	(c)	(d)	(e)	(f)
3.	$\frac{7}{4}$ $\frac{1}{4}$	$\frac{19}{16}$ $\frac{7}{16}$	$\frac{17}{10}$ $\frac{5}{10}$	$\frac{63}{32}$ $\frac{13}{32}$	$\frac{95}{60}$ $\frac{15}{60}$	$\frac{123}{100}$ $\frac{3}{100}$
4.	$9\frac{5}{8}$ $4\frac{2}{8}$	$24\frac{3}{4}$ $13\frac{2}{4}$	$7\frac{3}{6}$ $5\frac{2}{6}$	$10\frac{11}{32}$ $8\frac{3}{32}$	$23\frac{9}{25}$ $13\frac{4}{25}$	$100\frac{7}{20}$ $80\frac{7}{20}$
5.	$35\frac{1}{4}$ 35	$21\frac{5}{8}$ 17	6 $1\frac{3}{4}$	32 $8\frac{1}{2}$	13 $5\frac{2}{3}$	30 $29\frac{5}{8}$
6.	$4\frac{1}{4}$ $2\frac{3}{4}$	$8\frac{1}{3}$ $4\frac{2}{3}$	$13\frac{5}{12}$ $9\frac{7}{12}$	$18\frac{2}{5}$ $16\frac{4}{5}$	$45\frac{3}{32}$ $19\frac{5}{32}$	$10\frac{5}{64}$ $9\frac{13}{64}$

7. *a.* Subtract: $\frac{11}{16} - \frac{3}{16}$. *b.* Subtract: $\frac{18}{8} - \frac{6}{8}$.

8. *a.* Subtract $\frac{3}{20}$ from $\frac{11}{20}$. *b.* From $\frac{14}{8}$ subtract $\frac{4}{8}$.

9. How much greater than $\frac{4}{32}$ is $\frac{16}{32}$?

10. *a.* Find: $\frac{9}{4} + \frac{3}{4} - \frac{1}{4}$. *b.* Find: $8\frac{5}{6} + 4\frac{1}{6} - 7\frac{5}{6}$.

11 *a.* Subtract: $32\frac{5}{8} - 21\frac{3}{8}$. *b.* Subtract: $40\frac{1}{4} - 39\frac{3}{4}$.

12. *a.* Subtract $5\frac{1}{2}$ from 10. *b.* From $32\frac{7}{16}$ subtract $15\frac{11}{16}$.

13 How much greater than $3\frac{1}{4}$ is 5?

14. Sam had $\frac{34}{36}$ yd. of wire. If he used $\frac{13}{36}$ yd., how much wire was left?

15. Bill spent $1\frac{3}{4}$ hours on a walk going from his house to town and returning. If it took him $\frac{2}{4}$ of an hour to walk to town, how long did it take him to return to his house?

16. Mr. Archer's tennis racket and press weigh $2\ \frac{2}{16}$ lb. together. If the racket weighs $\frac{14}{16}$ lb., how much does the press weigh?

17. Mrs. Sheridan bought two turkeys that weighed 36 lb. together. If one turkey weighed $15\frac{7}{8}$ lb., what did the second turkey weigh?

11. SUBTRACTING UNLIKE FRACTIONS AND MIXED NUMBERS THAT CONTAIN UNLIKE FRACTIONS

SUBTRACTING UNLIKE FRACTIONS

To find the difference between $\frac{3}{4}$ of a mile and $\frac{1}{8}$ of a mile, we have to subtract: $\frac{3}{4} - \frac{1}{8}$. This problem can be solved as follows:

$$\begin{array}{r} \frac{3}{4} = \frac{6}{8} \\ \frac{1}{8} = \frac{1}{8} \\ \hline \frac{5}{8} \end{array}$$

$\frac{3}{4}$ can be changed to an equal fraction whose denominator is 8: $\frac{3}{4} = \frac{6}{8}$. Then $\frac{6}{8} - \frac{1}{8} = \frac{5}{8}$.

Procedure. To subtract fractions whose denominators are different, change the fractions to equal fractions whose denominators are the least common denominator. Now use the procedure for subtracting fractions having the same denominator.

SUBTRACTING MIXED NUMBERS THAT CONTAIN UNLIKE FRACTIONS

To find the difference between $14\frac{1}{2}$ hours and $8\frac{1}{4}$ hours, we have to subtract: $14\frac{1}{2} - 8\frac{1}{4}$. See how this is done.

Both fractions must have the same denominator. Therefore, we change $14\frac{1}{2}$ to $14\frac{2}{4}$. Then we subtract $8\frac{1}{4}$ from $14\frac{2}{4}$ and get $6\frac{1}{4}$ hours.

$$\begin{array}{rcr} 14\frac{1}{2} & = & 14\frac{2}{4} \\ \underline{8\frac{1}{4}} & = & \underline{8\frac{1}{4}} \\ & & 6\frac{1}{4} \end{array}$$

MODEL PROBLEMS

1. Subtract: $\frac{11}{12} - \frac{1}{4}$.

$$\begin{array}{rcrcl} \frac{11}{12} & = & \frac{11}{12} & & \\ \underline{\frac{1}{4}} & = & \underline{\frac{3}{12}} & & \\ & & \frac{8}{12} & = & \frac{2}{3} \end{array}$$

Find the least common denominator, 12. Change the fraction $\frac{1}{4}$ to an equal fraction, $\frac{3}{12}$. Then subtract: $\frac{11}{12} - \frac{3}{12} = \frac{8}{12} = \frac{2}{3}$.

2. From $3\frac{3}{4}$ subtract $2\frac{1}{3}$.

$$\begin{array}{rcr} 3\frac{3}{4} & = & 3\frac{9}{12} \\ \underline{2\frac{1}{3}} & = & \underline{2\frac{4}{12}} \\ & & 1\frac{5}{12} \end{array}$$

Change the fractions in the mixed numbers to equal fractions that have the least common denominator, 12. Subtract the whole numbers and fractions separately.

3. Subtract $8\frac{2}{3}$ from $16\frac{1}{2}$.

$$\begin{array}{rcrcr} 16\frac{1}{2} & = & 16\frac{3}{6} & = & 15\frac{9}{6} \\ \underline{8\frac{2}{3}} & = & \underline{8\frac{4}{6}} & = & \underline{8\frac{4}{6}} \\ & & & & 7\frac{5}{6} \end{array}$$

Change the fractions in the mixed numbers to equal fractions that have as their denominator the least common denominator, 6. Since the fraction in the subtrahend, $\frac{4}{6}$, is greater than the fraction in the minuend, $\frac{3}{6}$, exchange 1 whole for $\frac{6}{6}$ and write the minuend, $16\frac{3}{6}$, as $15\frac{9}{6}$. Now subtract the fractions and whole numbers separately.

EXERCISES

Subtract:

	(a)	(b)	(c)	(d)	(e)	(f)
1.	$\frac{1}{2}$	$\frac{7}{8}$	$\frac{30}{32}$	$\frac{15}{24}$	$\frac{75}{100}$	$\frac{19}{24}$
	$\frac{1}{4}$	$\frac{5}{16}$	$\frac{3}{4}$	$\frac{1}{3}$	$\frac{2}{5}$	$\frac{2}{3}$
2.	$\frac{1}{2}$	$\frac{2}{3}$	$\frac{1}{4}$	$\frac{5}{6}$	$\frac{9}{10}$	$\frac{7}{8}$
	$\frac{1}{3}$	$\frac{2}{4}$	$\frac{1}{9}$	$\frac{1}{4}$	$\frac{5}{6}$	$\frac{5}{12}$
3.	$3\frac{1}{2}$	$23\frac{5}{8}$	$14\frac{2}{5}$	$8\frac{2}{3}$	$34\frac{3}{5}$	$17\frac{7}{8}$
	$1\frac{1}{4}$	$12\frac{1}{2}$	$8\frac{1}{10}$	$5\frac{1}{2}$	$29\frac{1}{2}$	$6\frac{2}{3}$
4.	$12\frac{5}{8}$	$25\frac{7}{9}$	$17\frac{39}{50}$	$9\frac{1}{2}$	$5\frac{1}{8}$	$15\frac{3}{16}$
	$9\frac{1}{6}$	$16\frac{7}{12}$	$17\frac{3}{4}$	$6\frac{3}{4}$	$4\frac{3}{4}$	$14\frac{7}{8}$
5.	$7\frac{1}{5}$	$10\frac{1}{3}$	$15\frac{3}{8}$	$14\frac{1}{6}$	$9\frac{1}{10}$	$20\frac{1}{20}$
	$5\frac{1}{4}$	$8\frac{3}{4}$	$14\frac{4}{5}$	$8\frac{1}{4}$	$5\frac{3}{4}$	$10\frac{13}{50}$

6. *a.* Subtract: $\frac{9}{16} - \frac{1}{2}$. *b.* From $21\frac{1}{2}$ subtract $9\frac{2}{3}$.

7. Find the difference between $62\frac{1}{6}$ and $27\frac{1}{3}$.

8. *a.* Find: $4\frac{2}{3} + 1\frac{1}{2} - 3\frac{1}{4}$. *b.* Find: $6\frac{1}{3} - 2\frac{1}{4} + 4\frac{1}{12}$.

9. From a roll of ribbon $37\frac{7}{8}$ yd. long, a piece $15\frac{1}{2}$ yd. long is cut. How much ribbon is left on the roll?

10. John weighs $97\frac{5}{8}$ lb. His sister Helen weighs $80\frac{3}{4}$ lb. By how many pounds is John heavier than Helen?

11. If a trip from New York to Boston requires 5 hours by train and only $1\frac{1}{3}$ hours by plane, how many hours can be saved in traveling by plane?

12. Walter is $13\frac{5}{6}$ years old. James is $10\frac{1}{2}$ years old. Find the difference between their ages.

13. A carpenter had a piece of lumber 16 ft. long. He used pieces that measured $5\frac{1}{4}$ ft., $7\frac{3}{4}$ ft., $\frac{1}{2}$ ft., and 2 ft. How much of the original piece of lumber has he used? How much of the original piece of lumber did he have left?

12. MULTIPLYING FRACTIONS AND WHOLE NUMBERS

Joe bought 7 boxes of nuts, each box weighing $\frac{1}{2}$ of a pound. In order to find the total weight of his purchase, he added seven fractions

$$\frac{1}{2} + \frac{1}{2} + \frac{1}{2} + \frac{1}{2} + \frac{1}{2} + \frac{1}{2} + \frac{1}{2}.$$

This gave the result $\frac{7}{2}$, or $3\frac{1}{2}$ pounds.

Since all the boxes were the same weight, Joe might also have found the answer by multiplying $\frac{1}{2}$ by 7. We see that $7 \times \frac{1}{2} = \frac{7}{2}$, or $3\frac{1}{2}$.

The answer can also be found by figuring $7 \times \frac{1}{2} = \frac{7 \times 1}{2} = \frac{7}{2} = 3\frac{1}{2}$.

Procedure. To multiply a whole number and a fraction, multiply the whole number by the numerator of the fraction and write the result over the denominator of the fraction. Reduce the resulting fraction.

Since $7 \times \frac{1}{2} = 3\frac{1}{2}$ and $\frac{1}{2} \times 7 = 3\frac{1}{2}$, we see that **when we multiply a fraction and a whole number, the result remains the same when we change the order of the multipliers, that is, the multiplication is commutative.**

MODEL PROBLEMS

1. Find $\frac{3}{4} \times 8$.

$$\frac{3}{4} \times 8 = \frac{3 \times 8}{4} = \frac{24}{4} = 6 \quad \textit{Ans.}$$

2. Find $\frac{2}{3}$ of 27.

$$\frac{2}{3} \text{ of 27 means } \frac{2}{3} \times 27$$

$$\frac{2}{3} \times 27 = \frac{2 \times 27}{3} = \frac{54}{3} = 18 \quad \textit{Ans.}$$

3. Sue needed $\frac{7}{8}$ of a yard of material to make an apron. How much material did she need to make 12 aprons?

Solution: The answer can be found by multiplying $\frac{7}{8}$ by 12.

$$12 \times \frac{7}{8} = \frac{12 \times 7}{8} = \frac{84}{8} = 10\frac{4}{8} = 10\frac{1}{2} \quad \textit{Answer: } 10\tfrac{1}{2} \text{ yards}$$

EXERCISES

Find:

	(*a*)	(*b*)	(*c*)	(*d*)	(*e*)	(*f*)
1.	$\frac{1}{2}$ of 24	$\frac{1}{3}$ of 39	$\frac{1}{4}$ of 56	$\frac{1}{2} \times 28$	$\frac{1}{4} \times 35$	$\frac{1}{6} \times 50$
2.	$\frac{2}{3}$ of 30	$\frac{3}{4}$ of 44	$\frac{5}{8}$ of 48	$\frac{2}{3} \times 39$	$\frac{5}{6} \times 36$	$\frac{3}{4} \times 88$

	(a)	(b)	(c)	(d)	(e)	(f)
3.	$\frac{1}{2}$ of 13	$\frac{1}{3}$ of 21	$\frac{1}{5}$ of 45	$\frac{1}{6}$ of 42	$\frac{1}{8}$ of 7	$\frac{1}{10}$ of 8
4.	$\frac{2}{3}$ of 17	$\frac{4}{5}$ of 23	$\frac{3}{4}$ of 22	$\frac{5}{8}$ of 85	$\frac{3}{4} \times 13$	$\frac{2}{3} \times 22$
5.	$40 \times \frac{1}{2}$	$18 \times \frac{1}{3}$	$28 \times \frac{2}{7}$	$72 \times \frac{5}{8}$	$68 \times \frac{3}{4}$	$37 \times \frac{2}{3}$
6.	$82 \times \frac{3}{5}$	$96 \times \frac{7}{8}$	$41 \times \frac{2}{5}$	$13 \times \frac{6}{7}$	$46 \times \frac{3}{8}$	$37 \times \frac{7}{10}$

7. Mr. James earned $72 a week. One week he saved $\frac{1}{6}$ of his salary. How much did he save that week?

8. In a class of 36 pupils, $\frac{8}{9}$ of the pupils were present. How many pupils were present?

9. If there are 5280 feet in a mile, how many feet are there in $\frac{7}{8}$ of a mile?

10. Find the weight of 40 boxes of crackers if each box weighs $\frac{3}{4}$ of a pound.

11. A recipe calls for $\frac{2}{3}$ of a cup of water for each cup of cocoa. How much water must be used to make 9 cups of cocoa?

12. Each day Barry walks $\frac{9}{10}$ of a mile in traveling to and from school. How far does he walk in a week if he goes to school on 5 days?

13. MULTIPLYING WHOLE NUMBERS AND MIXED NUMBERS

From the ruler below, we can see that when we have 3 measurements each of which is $1\frac{1}{4}$ in., their total is $1\frac{1}{4}$ in. + $1\frac{1}{4}$ in. + $1\frac{1}{4}$ in. = $3\frac{3}{4}$ in. We can get the same result by multiplying $1\frac{1}{4}$ by 3. Thus, $3 \times 1\frac{1}{4} = 3\frac{3}{4}$.

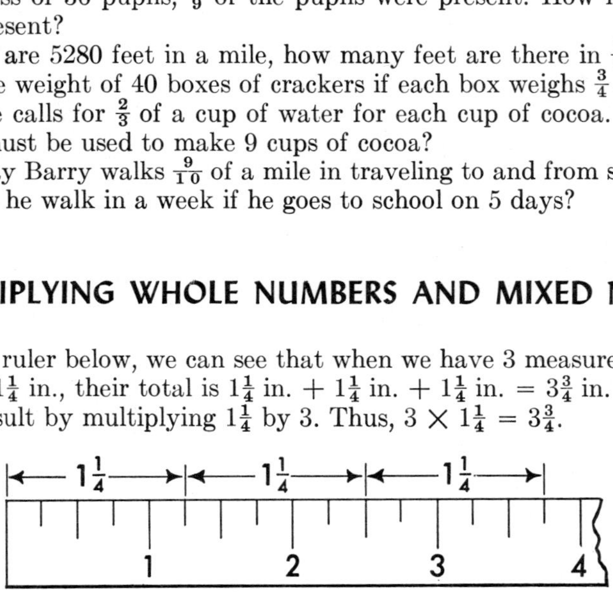

Since $1\frac{1}{4} = 1 + \frac{1}{4}$, we can figure $3 \times 1\frac{1}{4}$ as follows:

Think	*Write*
$3 \times 1 = 3$ Multiply the whole number 1 by 3. $3 \times \frac{1}{4} = \frac{3}{4}$ Multiply the fraction $\frac{1}{4}$ by 3. $3 \times 1\frac{1}{4} = 3\frac{3}{4}$ Add the two products.	$1\frac{1}{4}$ $\times 3$ $3\frac{3}{4}$ *Ans.*

Notice that since $3 \times (1 + \frac{1}{4}) = 3\frac{3}{4}$ and $(3 \times 1) + (3 \times \frac{1}{4}) = 3\frac{3}{4}$, then

$$\mathbf{3 \times (1 + \tfrac{1}{4}) = (3 \times 1) + (3 \times \tfrac{1}{4})}$$

This example illustrates that *multiplication is also distributive* when we are dealing with fractions.

Procedure 1. To multiply a whole number and a mixed number, multiply each part of the mixed number separately by the whole number and add the results.

There is another way of finding the product of $1\frac{1}{4}$ and 3. Since the mixed number $1\frac{1}{4}$ can be changed to the improper fraction $\frac{5}{4}$, we can write:

$$3 \times 1\tfrac{1}{4} = 3 \times \frac{5}{4} = \frac{3 \times 5}{4} = \frac{15}{4} = 3\tfrac{3}{4}$$

Procedure 2. To multiply a whole number and a mixed number, change the mixed number to an improper fraction and then multiply the improper fraction by the whole number.

MODEL PROBLEM

What is the cost of $4\frac{3}{4}$ yards of cloth at \$6 per yard?
Solution: To find the answer, we must multiply \$6 by $4\frac{3}{4}$.

Method 1	*Method 2*
$\begin{array}{rl} 6 & \\ \times\, 4\frac{3}{4} & \\ \hline 4\frac{1}{2} & (\frac{3}{4} \times 6) \\ 24 & (4 \times 6) \\ \hline 28\frac{1}{2} & (4\frac{3}{4} \times 6) \end{array}$ Multiply 6 by 3; then divide the product by 4. *Answer:* \$$28\frac{1}{2}$ or \$28.50	$4\frac{3}{4} \times 6 = \frac{19}{4} \times 6$ $= \frac{19 \times 6}{4}$ $= \frac{114}{4}$ $= 28\frac{1}{2}$

EXERCISES

Multiply:

	(*a*)	(*b*)	(*c*)	(*d*)
1.	$6 \times 2\frac{1}{2}$	$24 \times 2\frac{1}{8}$	$40 \times 6\frac{3}{4}$	$16 \times 7\frac{3}{8}$
2.	$13 \times 5\frac{1}{3}$	$28 \times 8\frac{2}{5}$	$17 \times 4\frac{3}{4}$	$23 \times 21\frac{3}{8}$
3.	$2\frac{1}{2} \times 14$	$12\frac{1}{8} \times 32$	$4\frac{2}{3} \times 24$	$7\frac{2}{5} \times 60$
4.	$5\frac{1}{2} \times 15$	$33\frac{1}{3} \times 5$	$87\frac{1}{2} \times 500$	$126 \times 66\frac{2}{3}$

5. Multiply: (*a*) 125 by $2\frac{3}{5}$ (*b*) 128 by $6\frac{3}{4}$ (*c*) 123 by $12\frac{3}{5}$

6. Find the product of: (*a*) 56 and $8\frac{3}{4}$ (*b*) 48 and $12\frac{5}{6}$ (*c*) 54 and $33\frac{1}{3}$

7. A box of candy weighs $2\frac{1}{2}$ pounds. How much do 10 boxes weigh?

8. What is the cost of $3\frac{2}{3}$ yards of cloth at $9 per yard?

9. What is the cost of $8\frac{1}{2}$ tons of coal at $21 a ton?

10. One barrel of flour can be made from $4\frac{3}{4}$ bushels of wheat. How many bushels of wheat will be needed for 380 barrels of flour?

11. If a mile has 1760 yards, how many yards are there in $8\frac{7}{8}$ miles?

12. If a dress factory uses $3\frac{1}{8}$ yards of cloth to make a dress, how many yards will it take to make 12 dresses?

13. Sally uses $1\frac{3}{4}$ teaspoons of sugar in making a cup of hot chocolate. How much sugar should she use in making 10 cups of hot chocolate?

14. MULTIPLYING FRACTIONS

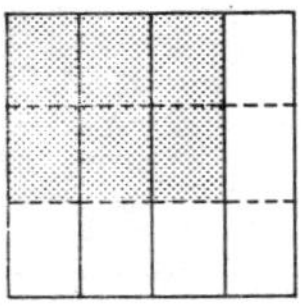

In the picture at the left, the solid lines divide the square into quarters. The dotted lines divide each quarter into three equal parts. From the picture, we see that $\frac{2}{3}$ of $\frac{3}{4}$ of the square is $\frac{6}{12}$ or $\frac{1}{2}$ of the square.

As shown below, the same answer, $\frac{1}{2}$, can be obtained by multiplying the fractions $\frac{2}{3}$ and $\frac{3}{4}$ because $\frac{2}{3}$ of $\frac{3}{4}$ means $\frac{2}{3} \times \frac{3}{4}$.

$$\frac{2}{3} \times \frac{3}{4} = \frac{2 \times 3}{3 \times 4} = \frac{6}{12} = \frac{1}{2}$$

We therefore see that we can multiply fractions by the following method:

Procedure 1. To multiply two or more fractions, multiply the numerators of the fractions to get the numerator of the answer and multiply the denominators to get the denominator of the answer. Reduce the answer.

MODEL PROBLEMS

1. Multiply: $\frac{3}{8} \times \frac{1}{2}$.

$$\frac{3}{8} \times \frac{1}{2} = \frac{3 \times 1}{8 \times 2} = \frac{3}{16}$$

2. $\frac{1}{3} \times \frac{3}{4} \times \frac{2}{5}$.

$$\frac{1}{3} \times \frac{3}{4} \times \frac{2}{5} = \frac{1 \times 3 \times 2}{3 \times 4 \times 5} = \frac{6}{60} = \frac{1}{10}$$

COMMUTATIVE PRINCIPLE FOR MULTIPLICATION

Since $\frac{3}{8} \times \frac{1}{2} = \frac{3}{16}$ and $\frac{1}{2} \times \frac{3}{8} = \frac{3}{16}$, then $\boldsymbol{\frac{3}{8} \times \frac{1}{2} = \frac{1}{2} \times \frac{3}{8}}$.

This example illustrates that *multiplication is also commutative* when we are dealing with fractions.

ASSOCIATIVE PRINCIPLE FOR MULTIPLICATION

Since $(\frac{1}{2} \times \frac{3}{5}) \times \frac{7}{2} = \frac{3}{10} \times \frac{7}{2} = \frac{21}{20}$ and $\frac{1}{2} \times (\frac{3}{5} \times \frac{7}{2}) = \frac{1}{2} \times \frac{21}{10} = \frac{21}{20}$, then $\mathbf{(\frac{1}{2} \times \frac{3}{5}) \times \frac{7}{2} = \frac{1}{2} \times (\frac{3}{5} \times \frac{7}{2})}$.

This example illustrates that *multiplication is also associative* when we are dealing with fractions.

DISTRIBUTIVE PRINCIPLE FOR MULTIPLICATION

Previously, on page 54, we saw that *multiplication is also distributive* when we are dealing with fractions.

USING CANCELLATION

When we multiply the fractions $\frac{2}{5}$ and $\frac{3}{2}$, we have:

$$\frac{2}{5} \times \frac{3}{2} = \frac{2 \times 3}{5 \times 2} = \frac{6}{10} = \frac{6 \div 2}{10 \div 2} = \frac{3}{5}$$

When we reduced the fraction $\frac{6}{10}$, we divided both the numerator and the denominator by 2. This might be written as shown at the right. This process of reducing the fraction by using division is called ***cancellation.***

$$\frac{\overset{3}{\cancel{6}}}{\underset{5}{\cancel{10}}} = \frac{3}{5}$$

We could have reduced or cancelled sooner. When we had $\frac{2}{5} \times \frac{3}{2}$, we could immediately have divided the numerator of any fraction and also the denominator of any fraction by the same number, 2, as shown at the right. Hence, when we multiply fractions, we can shorten the work by using the method of cancellation.

$$\frac{\overset{1}{\cancel{2}}}{5} \times \frac{3}{\underset{1}{\cancel{2}}} = \frac{3}{5}$$

Procedure 2. To multiply two or more fractions, divide any numerator and any denominator by the largest number that divides into them evenly. Multiply the resulting numerators to get the numerator of the answer and multiply the denominators to get the denominator of the answer. Reduce the answer.

MODEL PROBLEMS

1. Multiply: $\frac{4}{9} \times \frac{3}{2}$.

$$\frac{\overset{2}{\cancel{4}}}{\underset{3}{\cancel{9}}} \times \frac{\overset{1}{\cancel{3}}}{\underset{1}{\cancel{2}}} = \frac{2 \times 1}{3 \times 1} = \frac{2}{3}$$

2. Multiply: $\frac{9}{8} \times \frac{6}{5} \times \frac{7}{9}$.

$$\frac{\overset{1}{\cancel{9}}}{\underset{4}{\cancel{8}}} \times \frac{\overset{3}{\cancel{6}}}{5} \times \frac{7}{\underset{1}{\cancel{9}}} = \frac{1 \times 3 \times 7}{4 \times 5 \times 1} = \frac{21}{20} = 1\frac{1}{20}$$

3. Multiply: $\frac{7}{12} \times 6$.

$$\frac{7}{12} \times 6 = \frac{7}{\cancel{12}_{2}} \times \frac{\cancel{6}^{1}}{1} = \frac{7 \times 1}{2 \times 1} = \frac{7}{2} = 3\tfrac{1}{2}$$

Write the whole number 6 as a fraction, $\frac{6}{1}$.

EXERCISES

Find:

	(*a*)	(*b*)	(*c*)	(*d*)	(*e*)
1.	$\frac{1}{2}$ of $\frac{1}{3}$	$\frac{1}{3}$ of $\frac{1}{4}$	$\frac{2}{3}$ of $\frac{4}{5}$	$\frac{3}{5}$ of $\frac{3}{8}$	$\frac{5}{9}$ of $\frac{2}{3}$
2.	$\frac{1}{2} \times \frac{1}{6}$	$\frac{1}{2} \times \frac{1}{5}$	$\frac{3}{5} \times \frac{4}{7}$	$\frac{7}{8} \times \frac{5}{6}$	$\frac{3}{4} \times \frac{7}{10}$
3.	$\frac{1}{2}$ of $\frac{4}{5}$	$\frac{2}{3}$ of $\frac{9}{16}$	$\frac{7}{8}$ of $\frac{5}{14}$	$\frac{4}{5}$ of $\frac{5}{8}$	$\frac{15}{6}$ of $\frac{8}{25}$
4.	$\frac{3}{7} \times \frac{21}{4}$	$\frac{5}{16} \times \frac{8}{3}$	$\frac{4}{6} \times \frac{9}{2}$	$\frac{5}{10} \times \frac{4}{15}$	$\frac{25}{12} \times \frac{8}{15}$
5.	$\frac{2}{3}$ of 18	$\frac{3}{4}$ of 36	$\frac{7}{8}$ of 24	$60 \times \frac{4}{15}$	$180 \times \frac{5}{12}$
6.	$\frac{1}{5} \times 7$	$50 \times \frac{5}{6}$	$\frac{3}{20} \times 5$	$25 \times \frac{7}{30}$	$180 \times \frac{7}{120}$

	(*a*)	(*b*)	(*c*)	(*d*)
7.	$\frac{1}{3} \times \frac{1}{2} \times \frac{1}{4}$	$\frac{1}{4} \times \frac{4}{5} \times \frac{2}{3}$	$\frac{7}{4} \times \frac{3}{2} \times \frac{8}{21}$	$\frac{5}{12} \times \frac{3}{4} \times 16$
8.	$600 \times \frac{4}{100} \times \frac{1}{2}$	$200 \times \frac{3}{100} \times \frac{1}{4}$	$500 \times \frac{6}{100} \times \frac{1}{6}$	$1000 \times \frac{4}{100} \times \frac{1}{4}$

9. Find the product of: (*a*) $\frac{4}{25}$ and $\frac{5}{8}$ (*b*) $\frac{5}{12}$ and 48 (*c*) 600 and $\frac{5}{100}$

10. Sally's glass is $\frac{3}{4}$ full of water. If she drank $\frac{2}{3}$ of the water, what part of a glass of water did she drink?

11. Ted is on a walk that is to cover $\frac{7}{8}$ of a mile. If he has already walked $\frac{2}{3}$ of that distance, what part of a mile has he walked?

12. Ronald has $\frac{15}{16}$ lb. of candy. If he eats $\frac{2}{5}$ of it, what part of a pound does he eat?

13. Robert was to receive $8 for working a full day. How much did he earn for working $\frac{5}{8}$ of a day?

14. A man owns $\frac{5}{3}$ acres of land. If $\frac{3}{10}$ of his land is planted with grass, what part of an acre is planted with grass?

15. MULTIPLYING FRACTIONS AND MIXED NUMBERS

Procedure. To multiply a fraction or a mixed number by a mixed number, change the mixed numbers to improper fractions. Then multiply the fractions.

MODEL PROBLEMS

1. Multiply: $2\frac{1}{8} \times \frac{3}{4}$.

$$2\tfrac{1}{8} \times \frac{3}{4} = \frac{17}{8} \times \frac{3}{4} = \frac{17 \times 3}{8 \times 4} = \frac{51}{32} = 1\tfrac{19}{32}$$

2. Find the product of $3\frac{2}{3}$ and $1\frac{1}{4}$.

$$3\tfrac{2}{3} \times 1\tfrac{1}{4} = \frac{11}{3} \times \frac{5}{4} = \frac{11 \times 5}{3 \times 4} = \frac{55}{12} = 4\tfrac{7}{12}$$

3. Find the value of $8\frac{1}{3} \times \frac{1}{15} \times 6$.

$$8\tfrac{1}{3} \times \frac{1}{15} \times 6 = \frac{\overset{5}{\cancel{25}}}{\underset{1}{\cancel{3}}} \times \frac{1}{\underset{3}{\cancel{15}}} \times \frac{\overset{2}{\cancel{6}}}{1} = \frac{5 \times 1 \times 2}{1 \times 3 \times 1} = \frac{10}{3} = 3\tfrac{1}{3}$$

EXERCISES

Multiply:

	(a)	(b)	(c)	(d)	(e)
1.	$\frac{3}{4} \times 2\frac{1}{2}$	$6\frac{1}{3} \times \frac{4}{3}$	$2\frac{1}{4} \times \frac{2}{3}$	$\frac{4}{5} \times 8\frac{1}{3}$	$\frac{7}{8} \times 3\frac{1}{5}$
2.	$1\frac{1}{2} \times 3\frac{2}{5}$	$2\frac{1}{2} \times 2\frac{1}{2}$	$4\frac{2}{3} \times 3\frac{1}{2}$	$3\frac{1}{8} \times 6\frac{2}{5}$	$23\frac{3}{4} \times 13\frac{3}{5}$
3.	$3\frac{1}{7} \times 14$	$8 \times 5\frac{1}{4}$	$5 \times 2\frac{1}{6}$	$72 \times 3\frac{5}{12}$	$36 \times 12\frac{1}{4}$

	(a)	(b)	(c)	(d)
4.	$1\frac{1}{2} \times 2 \times \frac{1}{3}$	$1\frac{1}{2} \times 2\frac{1}{4} \times \frac{2}{3}$	$1\frac{1}{3} \times \frac{3}{10} \times 3\frac{3}{4}$	$4\frac{4}{5} \times \frac{7}{8} \times 4$

5. Find: (*a*) $\frac{2}{5}$ of $7\frac{1}{2}$ (*b*) $\frac{2}{3}$ of $2\frac{1}{7}$ (*c*) $\frac{5}{6}$ of $1\frac{4}{5}$ (*d*) $\frac{3}{4}$ of $2\frac{2}{3}$

6. Find the product of: (*a*) $\frac{5}{8}$ and $2\frac{2}{3}$ (*b*) $4\frac{4}{5}$ and $\frac{7}{8}$ (*c*) $2\frac{3}{4}$ and $3\frac{1}{11}$ (*d*) $3\frac{1}{7}$ and $4\frac{2}{3}$

7. How much will $4\frac{1}{2}$ lb. of pears cost at 18 cents per pound?

8. In a certain concrete mixture, $\frac{3}{4}$ of the mixture is sand. How much sand is there in $3\frac{1}{3}$ cubic yards of this concrete mixture?

9. What is the cost of $4\frac{2}{3}$ yards of material at \$$1\frac{1}{2}$ per yard?

10. To seed his lawn, Bert used $5\frac{1}{3}$ boxes of seed, each box weighing $4\frac{1}{2}$ pounds. How many pounds of seed did he use?

11. Find the weight of 80 cans of paint if each can weighs $5\frac{1}{4}$ pounds.

12. Arnold walked at the rate of $2\frac{1}{4}$ miles per hour. How far did he walk in $2\frac{2}{3}$ hours?

13. A cubic foot holds about $7\frac{1}{2}$ gallons. How many gallons of milk will $5\frac{3}{4}$ cubic feet hold?

16. DIVIDING WHOLE NUMBERS BY FRACTIONS AND MIXED NUMBERS

DIVIDING A WHOLE NUMBER BY A FRACTION

The picture at the right shows us that $\frac{3}{4}$ is contained in 3 wholes 4 times. This means that

$$3 \div \frac{3}{4} = 4$$

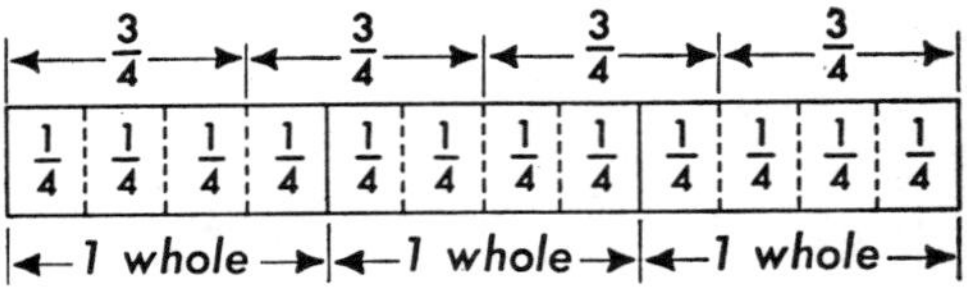

Notice that we get the same result when we ***invert*** the divisor, $\frac{3}{4}$, so that it becomes $\frac{4}{3}$ and then multiply the inverted divisor, $\frac{4}{3}$, by 3.

$$3 \div \frac{3}{4} = 3 \times \frac{4}{3} = \frac{12}{3} = 4 \quad \text{or} \quad 3 \div \frac{3}{4} = \frac{\overset{1}{\cancel{3}}}{1} \times \frac{4}{\underset{1}{\cancel{3}}} = \frac{4 \times 1}{1 \times 1} = \frac{4}{1} = 4$$

Procedure. To divide a whole number by a fraction, invert the divisor and multiply the inverted divisor by the whole number.

DIVIDING A WHOLE NUMBER BY A MIXED NUMBER

Procedure. To divide a whole number by a mixed number, change the mixed number to an improper fraction. Then divide the whole number by the improper fraction.

MODEL PROBLEMS

1. Divide 5 by $\frac{3}{4}$.

$$5 \div \frac{3}{4} = \frac{5}{1} \times \frac{4}{3} = \frac{5 \times 4}{1 \times 3} = \frac{20}{3} = 6\frac{2}{3}$$

2. How many pieces of ribbon $2\frac{1}{4}$ inches in length can be cut from a piece 36 inches long?

Solution: Find how many times $2\frac{1}{4}$ is contained in 36 by dividing 36 by $2\frac{1}{4}$.

$$36 \div 2\frac{1}{4} = 36 \div \frac{9}{4} = \frac{\overset{4}{\cancel{36}}}{1} \times \frac{4}{\underset{1}{\cancel{9}}} = \frac{4 \times 4}{1 \times 1} = \frac{16}{1} = 16$$

Answer: 16 pieces

EXERCISES

Divide:

	(a)	(b)	(c)	(d)	(e)
1.	$2 \div \frac{1}{3}$	$4 \div \frac{1}{5}$	$6 \div \frac{2}{3}$	$8 \div \frac{3}{4}$	$4 \div \frac{2}{5}$
2.	$36 \div \frac{4}{5}$	$21 \div \frac{3}{8}$	$60 \div \frac{5}{12}$	$13 \div \frac{3}{4}$	$15 \div \frac{7}{12}$
3.	$12 \div \frac{3}{2}$	$8 \div \frac{4}{3}$	$44 \div \frac{22}{7}$	$9 \div \frac{6}{5}$	$8 \div \frac{7}{6}$
4.	$16 \div 1\frac{1}{3}$	$25 \div 8\frac{1}{3}$	$9 \div 1\frac{1}{5}$	$25 \div 7\frac{1}{2}$	$36 \div 7\frac{1}{5}$

5. When a whole number is divided by a proper fraction, is the answer always smaller or larger than the whole number? Give an example.

6. When a whole number is divided by a mixed number, is the answer always smaller or larger than the whole number? Give an example.

7. How many (*a*) $\frac{1}{8}$'s are there in 8? (*b*) $\frac{2}{3}$'s are contained in 18?

8. How many (*a*) $\frac{3}{4}$'s are contained in 24? (*b*) $1\frac{1}{4}$'s are contained in 10?

9. How many packages of rice, each containing $\frac{1}{2}$ of a pound, can be filled from 100 pounds of rice?

10. How many pieces of metal, each $\frac{3}{8}$ of an inch long, can be cut from a metal strip 72 inches long?

11. A machine produces a piece of work in $\frac{5}{6}$ of an hour. How many pieces of work does it produce in 10 hours?

12. How many badges $3\frac{1}{2}$ inches long can be cut from 7 yards of ribbon?

13. A man bought $8\frac{1}{3}$ yards of cloth for $50. How much did he pay for 1 yard?

14. A bus driver spends $1\frac{1}{3}$ hours in making a round trip on his route. How many round trips does he make on a day when he works 8 hours?

17. DIVIDING FRACTIONS AND MIXED NUMBERS BY FRACTIONS

DIVIDING A FRACTION BY A FRACTION

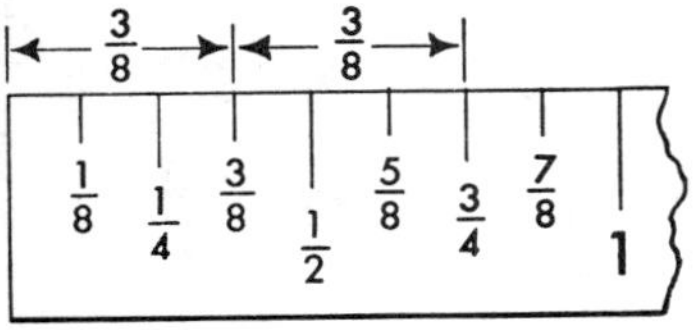

Using the ruler at the left, we see that $\frac{3}{8}$ of an inch is contained in $\frac{3}{4}$ of an inch 2 times. This means that $\frac{3}{4} \div \frac{3}{8} = 2$. We can get the same result by using division, as follows:

$$\frac{3}{4} \div \frac{3}{8} = \frac{\overset{1}{\cancel{3}}}{\underset{1}{\cancel{4}}} \times \frac{\overset{2}{\cancel{8}}}{\underset{1}{\cancel{3}}} = \frac{1 \times 2}{1 \times 1} = \frac{2}{1} = 2$$

Procedure. **To divide a fraction by a fraction, invert the divisor and then multiply the inverted divisor by the dividend.**

MODEL PROBLEM

Divide: $\frac{8}{3} \div \frac{4}{5}$.

$$\frac{8}{3} \div \frac{4}{5} = \frac{\overset{2}{\cancel{8}}}{3} \times \frac{5}{\underset{1}{\cancel{4}}} = \frac{2 \times 5}{3 \times 1} = \frac{10}{3} = 3\tfrac{1}{3}$$

DIVIDING A MIXED NUMBER BY A FRACTION

Procedure. **To divide a mixed number by a fraction, change the mixed number to an improper fraction and then divide the fractions.**

MODEL PROBLEM

How many crates, each weighing $\frac{1}{20}$ of a ton, are there on a truck whose load is $1\frac{1}{2}$ tons?

Solution: Divide the total load, $1\frac{1}{2}$ tons, by the weight of each crate, $\frac{1}{20}$ of a ton.

$$1\tfrac{1}{2} \div \frac{1}{20} = \frac{3}{2} \div \frac{1}{20} = \frac{3}{\underset{1}{\cancel{2}}} \times \frac{\overset{10}{\cancel{20}}}{1} = \frac{3 \times 10}{1 \times 1} = \frac{30}{1} = 30$$

Answer: 30 crates

EXERCISES

Divide:

	(a)	(b)	(c)	(d)	(e)
1.	$\frac{3}{4} \div \frac{1}{8}$	$\frac{4}{5} \div \frac{1}{10}$	$\frac{5}{9} \div \frac{2}{3}$	$\frac{12}{25} \div \frac{4}{5}$	$\frac{3}{5} \div \frac{4}{9}$
2.	$\frac{2}{3} \div \frac{3}{10}$	$\frac{7}{16} \div \frac{5}{8}$	$\frac{3}{4} \div \frac{3}{4}$	$\frac{9}{4} \div \frac{2}{3}$	$\frac{36}{25} \div \frac{18}{5}$
3.	$2\frac{1}{2} \div \frac{1}{8}$	$1\frac{1}{8} \div \frac{9}{16}$	$6\frac{1}{4} \div \frac{7}{10}$	$1\frac{5}{8} \div \frac{3}{4}$	$4\frac{1}{5} \div \frac{7}{20}$
4.	$3\frac{1}{4} \div \frac{5}{6}$	$6\frac{3}{5} \div \frac{11}{10}$	$3\frac{1}{9} \div \frac{4}{3}$	$1\frac{1}{4} \div \frac{5}{4}$	$4\frac{1}{2} \div \frac{9}{2}$

5. Find the quotient: (*a*) $\frac{9}{64} \div \frac{3}{8}$ (*b*) $\frac{10}{9} \div \frac{5}{6}$ (*c*) $7\frac{1}{2} \div \frac{5}{8}$

6. How many bags of candy, each containing $\frac{1}{8}$ of a pound, can be filled from a box of candy that contains $\frac{3}{4}$ of a pound?

7. How many stakes, each $\frac{3}{4}$ of a foot in length, can be cut from a piece of lumber that is $4\frac{1}{2}$ feet long?

8. How many pieces of ribbon, each $\frac{5}{12}$ of a yard in length, can be cut from a piece of ribbon that is $6\frac{1}{4}$ yards long?

9. A truck is loaded with machines, each of which weighs $\frac{4}{5}$ of a ton. If the total load is $4\frac{4}{5}$ tons, how many machines are on the truck?

10. Mr. Tracy can make a poster in $\frac{3}{4}$ of an hour. How many posters can he make in $7\frac{1}{2}$ hours?

18. DIVIDING FRACTIONS AND MIXED NUMBERS BY WHOLE NUMBERS

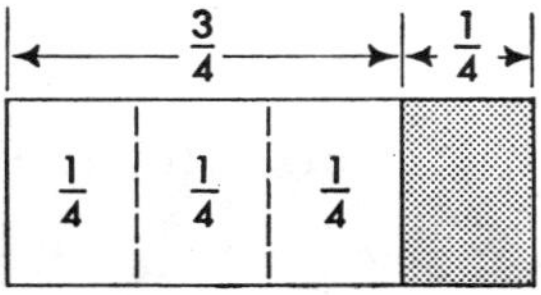

The picture shows us that if $\frac{3}{4}$ of a candy bar is divided into 3 equal pieces, each piece is $\frac{1}{4}$ of the original candy bar. This means that $\frac{3}{4} \div 3 = \frac{1}{4}$.

We can also find the answer to the problem $\frac{3}{4} \div 3$ as follows:

$$\frac{3}{4} \div 3 = \frac{3}{4} \div \frac{3}{1} = \frac{\overset{1}{\cancel{3}}}{4} \times \frac{1}{\underset{1}{\cancel{3}}} = \frac{1 \times 1}{4 \times 1} = \frac{1}{4}$$

We can find the answer to the problem $2\frac{1}{2} \div 3$ as follows:

$$2\tfrac{1}{2} \div 3 = 2\tfrac{1}{2} \div \frac{3}{1} = \frac{5}{2} \div \frac{3}{1} = \frac{5}{2} \times \frac{1}{3} = \frac{5 \times 1}{2 \times 3} = \frac{5}{6}$$

Procedure. To divide a fraction or a mixed number by a whole number, change the whole number to an improper fraction by placing the whole number over 1. Then divide by this improper fraction.

MODEL PROBLEMS

1. Divide $\frac{3}{4}$ by 2.

$$\frac{3}{4} \div 2 = \frac{3}{4} \div \frac{2}{1} = \frac{3}{4} \times \frac{1}{2} = \frac{3 \times 1}{4 \times 2} = \frac{3}{8}$$

2. Divide $1\frac{1}{3}$ by 4.

$$1\tfrac{1}{3} \div 4 = 1\tfrac{1}{3} \div \frac{4}{1} = \frac{4}{3} \div \frac{4}{1} = \frac{\overset{1}{\cancel{4}}}{3} \times \frac{1}{\underset{1}{\cancel{4}}} = \frac{1 \times 1}{3 \times 1} = \frac{1}{3}$$

EXERCISES

Divide:

	(a)	(b)	(c)	(d)	(e)
1.	$\frac{1}{2} \div 2$	$\frac{1}{4} \div 3$	$\frac{7}{8} \div 3$	$\frac{5}{6} \div 2$	$\frac{3}{4} \div 5$
2.	$\frac{9}{10} \div 6$	$\frac{5}{4} \div 20$	$\frac{3}{4} \div 9$	$\frac{3}{8} \div 12$	$\frac{12}{5} \div 8$
3.	$1\frac{1}{2} \div 3$	$6\frac{2}{3} \div 2$	$5\frac{5}{8} \div 5$	$3\frac{1}{2} \div 7$	$6\frac{1}{3} \div 4$

4. A piece of wire $\frac{25}{36}$ of a yard long is to be cut into 5 equal pieces. Find the length of each piece.

5. Sol travels $\frac{7}{8}$ of a mile in going back and forth between his home and his school. How far is it from his home to his school?

6. Ten boys are to share $3\frac{3}{4}$ pounds of candy equally. How much candy does each boy get?

7. A board $6\frac{3}{4}$ feet in length is cut into 3 equal pieces. How long is each piece?

19. DIVIDING MIXED NUMBERS BY MIXED NUMBERS

Procedure. To divide a mixed number by a mixed number, change the mixed numbers to improper fractions and then divide the fractions.

MODEL PROBLEM

Divide $3\frac{3}{8}$ by $2\frac{1}{4}$.

$$3\frac{3}{8} \div 2\frac{1}{4} = \frac{27}{8} \div \frac{9}{4} = \frac{\overset{3}{\cancel{27}}}{\underset{2}{\cancel{8}}} \times \frac{\overset{1}{\cancel{4}}}{\underset{1}{\cancel{9}}} = \frac{3 \times 1}{2 \times 1} = \frac{3}{2} = 1\frac{1}{2}$$

EXERCISES

Divide:

	(a)	(b)	(c)	(d)	(e)
1.	$3\frac{3}{4} \div 3\frac{1}{8}$	$2\frac{1}{4} \div 7\frac{3}{4}$	$16\frac{2}{3} \div 6\frac{2}{3}$	$1\frac{2}{5} \div 3\frac{1}{4}$	$1\frac{3}{16} \div 7\frac{1}{3}$
2.	$6\frac{1}{4} \div 3\frac{1}{3}$	$1\frac{3}{5} \div 3\frac{1}{5}$	$12\frac{3}{4} \div 1\frac{8}{9}$	$1\frac{7}{9} \div 5\frac{1}{3}$	$37\frac{1}{2} \div 6\frac{1}{4}$
3.	$8\frac{1}{4} \div 8\frac{1}{4}$	$19\frac{1}{2} \div 6\frac{1}{4}$	$9\frac{3}{4} \div 9\frac{3}{4}$	$50\frac{2}{3} \div 2\frac{2}{3}$	$12\frac{1}{2} \div 12\frac{1}{2}$

4. A dealer sold $6\frac{1}{4}$ yards of silk for \$$62\frac{1}{2}$. What was the cost of 1 yard of silk?

5. How many boxes of candy, each weighing $1\frac{1}{2}$ pounds, can be filled from a bag containing $67\frac{1}{2}$ pounds of candy?

6. How many pieces of ribbon, each $1\frac{3}{4}$ feet long, can be cut from a roll that is $40\frac{1}{4}$ feet long?

7. How many shelves, each $2\frac{3}{4}$ feet in length, can be cut from a piece of lumber that is $15\frac{2}{3}$ feet long?

20. USING RECIPROCALS IN DIVIDING BY A FRACTION

When the product of two numbers is equal to 1, the numbers are called ***reciprocals*** of each other.

Since $\frac{2}{3} \times \frac{3}{2} = 1$, $\frac{2}{3}$ and $\frac{3}{2}$ are reciprocals of each other.

Notice that the reciprocal of a fraction can be found by interchanging its numerator and its denominator, that is, by *inverting* the fraction. Thus, the reciprocal of $\frac{2}{3}$ is $\frac{3}{2}$; the reciprocal of $\frac{5}{6}$ is $\frac{6}{5}$.

If we wish to find the reciprocal of 4, we can think of 4 as the improper fraction $\frac{4}{1}$ and then invert the fraction $\frac{4}{1}$ to get $\frac{1}{4}$ as the result.

We have learned that to divide a number by a fraction, we invert the divisor and multiply the inverted divisor by the number. Since, when we invert the divisor we are finding the reciprocal of the divisor, we may restate the procedure for dividing a number by a fraction as follows:

Procedure. To divide a number by a fraction, multiply the reciprocal of the divisor by the number.

Thus, $9 \div \frac{3}{4} = 9 \times \frac{4}{3} = \frac{36}{3} = 12$. (The reciprocal of $\frac{3}{4}$ is $\frac{4}{3}$.)

EXERCISES

Find the reciprocal of:

1. $\frac{3}{4}$ **2.** $\frac{7}{8}$ **3.** $\frac{9}{10}$ **4.** $\frac{13}{16}$ **5.** 2 **6.** 6

7. 10 **8.** 12 **9.** $1\frac{1}{2}$ **10.** $2\frac{1}{3}$ **11.** $4\frac{3}{4}$ **12.** $3\frac{4}{5}$

Find the quotient by using the reciprocal of the divisor:

13. $\frac{5}{8} \div \frac{1}{2}$ **14.** $\frac{12}{16} \div \frac{4}{8}$ **15.** $\frac{5}{12} \div \frac{7}{36}$ **16.** $\frac{8}{9} \div 2$ **17.** $\frac{20}{36} \div 4$

18. $3\frac{1}{5} \div 2$ **19.** $6\frac{1}{4} \div 5$ **20.** $3\frac{1}{3} \div \frac{5}{9}$ **21.** $4\frac{1}{6} \div 6\frac{2}{3}$ **22.** $9\frac{1}{3} \div 1\frac{1}{6}$

21. FINDING WHAT FRACTIONAL PART OF A NUMBER ANOTHER NUMBER IS

What fractional part of all 4 circles are the 2 darkened circles? We have learned that a fraction may be used to compare the numbers in two groups of things. We may therefore write the fraction whose numerator is 2 and whose denominator is 4. The result is $\frac{2}{4}$, which is equal to $\frac{1}{2}$.

Notice that we get the same result, $\frac{2}{4}$, if we consider 4 as the number representing the whole group of circles (the denominator) and 2 as the number representing the part of the group that is darkened (the numerator).

Procedure. To find what fractional part of a number that represents a whole (to be called the first number) another number (to be called the second number) is, write a fraction whose numerator is the second number and whose denominator is the first number. Reduce the fraction.

MODEL PROBLEMS

1. What part of 36 is 9?

$$\frac{9}{36} = \frac{1}{4}$$

The numerator is 9, the number that is being compared with the whole, 36. The denominator is 36, the number that represents the whole.

2. 18 is what part of 24?

$\frac{18}{24} = \frac{3}{4}$

3. What part of 8 is 8?

$\frac{8}{8} = 1$, or the whole thing

EXERCISES

1. *a.* What part of 10 is 5? *b.* 4 is what part of 12?

2. *a.* What part of 27 is 18? *b.* What part of 50 is 35?

3. *a.* 25 is what part of 60? *b.* 1500 is what part of 2000?

4. *a.* 8 is what part of 35? *b.* What part of 16 is 16?

5. *a.* What part of 16 is 4? *b.* What part of 120 is 45?

6. *a.* What part of 4 is $\frac{1}{3}$? *b.* What part of 100 is $16\frac{2}{3}$?

7. At a meeting of a club that has 24 members, 2 members were absent. What part of the membership was absent?

8. In a month of 28 days, 21 days were fair. What part of the month was fair?

9. A newsboy bought 100 papers and sold 96. What part of his newspapers did he sell?

10. In a class, there are 21 boys and 14 girls. What part of the class is boys? What part of the class is girls?

11. In 120 pounds of concrete mixture, which consists of cement and sand, there are 90 pounds of sand. What part of the mixture is cement?

22. FINDING A WHOLE WHEN A FRACTIONAL PART OF IT IS KNOWN

When Marion came home from shopping, her mother noticed that she had paid 72 cents for $\frac{3}{4}$ of a dozen oranges. She asked Marion for the cost of a dozen oranges. Marion, not remembering, figured it out as follows:

$\frac{3}{4}$ dozen oranges cost 72 cents.

$\frac{1}{4}$ dozen, which is $\frac{1}{3}$ of $\frac{3}{4}$ dozen, would cost $\frac{1}{\cancel{3}_1} \times \overset{24}{\cancel{72}}$, or 24 cents.

A whole dozen, or $\frac{4}{4}$ dozen, would cost 4×24 cents, or 96 cents.

Procedure 1. To find a whole when a fractional part of it is known, first find what number a unit fraction of the whole is equal to. Then multiply that number by the denominator of the unit fraction.

In the previous problem, Marion divided 72 cents by 3 to find the cost of $\frac{1}{4}$ dozen. She then multiplied that answer, 24, by 4, to find the cost of a whole dozen. Notice that if Marion had divided 72 cents by $\frac{3}{4}$, she would also have found the correct answer, 96 cents. Thus:

$$72 \div \frac{3}{4} = \frac{72}{1} \div \frac{3}{4} = \frac{\overset{24}{\cancel{72}}}{1} \times \frac{4}{\underset{1}{\cancel{3}}} = \frac{24 \times 4}{1 \times 1} = \frac{96}{1} = 96$$

Procedure 2. To find a whole when a fractional part of it is known, divide the number that represents the part of the whole by the fraction which that part represents.

MODEL PROBLEMS

1. 16 is $\frac{1}{4}$ of what number?

Method 1	*Method* 2
$\frac{1}{4}$ of the whole number $= 16$	$16 \div \frac{1}{4} = \frac{16}{1} \times \frac{4}{1}$
$\frac{4}{4}$ or the whole number $= 4 \times 16 = 64$	$= \frac{16 \times 4}{1 \times 1} = 64$
Answer: 64	

2. $\frac{3}{5}$ of what number is 30?

Method 1	*Method* 2
$\frac{3}{5}$ of the whole number $= 30$ $\frac{1}{5}$ of the whole number $= \frac{1}{3}$ of $30 = 10$ $\frac{5}{5}$ or the whole number $= 5 \times 10 = 50$ *Answer:* 50	$30 \div \frac{3}{5} = \frac{30}{1} \div \frac{3}{5}$ $= \frac{\overset{10}{\cancel{30}}}{1} \times \frac{5}{\underset{1}{\cancel{3}}} = \frac{10 \times 5}{1 \times 1} = 50$

EXERCISES

1. *a.* 5 is $\frac{1}{4}$ of what number? *b.* 25 is $\frac{1}{5}$ of what number?

2. *a.* $\frac{1}{2}$ of what number is 4? *b.* $\frac{1}{8}$ of what number is 8?

3. *a.* 18 is $\frac{2}{3}$ of what number? *b.* 45 is $\frac{3}{4}$ of what number?

4. *a.* $\frac{3}{8}$ of what number is 27? *b.* $\frac{7}{10}$ of what number is 56?

5. *a.* 42 is $\frac{6}{7}$ of what number? *b.* $\frac{5}{9}$ of what number is 40?

6. If $\frac{1}{3}$ of a yard of cloth costs \$2, what is the cost of 1 yard of cloth?

7. If there are 440 yards in $\frac{1}{4}$ of a mile, find the number of yards in 1 mile.

8. If there are 200 pounds in $\frac{1}{10}$ of a ton, how many pounds are there in a ton?

9. If $\frac{3}{4}$ of a pound of meat costs 48 cents, what is the price of a pound of meat?

10. During a sale, Mr. Lorey paid \$64 for a record player. If this was $\frac{2}{3}$ of the original price, what was the original price of the record player?

11. A carpenter used $\frac{7}{8}$ of the number of feet of lumber he had ordered for a job. If he used 350 feet of lumber, how many feet had he ordered?

12. Only $\frac{5}{6}$ of the membership of a club went on its annual boat ride. If 245 members went, what was the number of the entire membership?

Chapter III

ARITHMETIC OF DECIMAL FRACTIONS

I. UNDERSTANDING THE MEANING OF DECIMAL FRACTIONS

Common fractions whose denominators are 10, 100, or 1,000, etc.—like $\frac{8}{10}$, $\frac{75}{100}$, $\frac{125}{1000}$—are called ***decimal fractions.*** With the use of a ***decimal point,*** we can write a decimal fraction without a denominator.

$\frac{8}{10}$ = .8 (*8 tenths*) $\frac{75}{100}$ = .75 (*75 hundredths*)

$\frac{125}{1000}$ = .125 (*125 thousandths*)

When a road sign states "To Smithtown—1.5 miles," it means that the distance to Smithtown is $1\frac{5}{10}$ or 1 and $\frac{5}{10}$ miles. The number 1.5 is called a ***mixed decimal.*** It means *1 and 5 tenths.*

A *mixed decimal* is a number that contains a whole number and a decimal fraction. For example, 1.5 (1 and 5 tenths, $1\frac{5}{10}$); 2.75 (2 and 75 hundredths, $2\frac{75}{100}$); and 12.375 (12 and 375 thousandths, $12\frac{375}{1000}$) are mixed decimals.

Both decimal fractions and mixed decimals are usually called simply ***decimals.***

PLACE VALUES OF THE DIGITS IN A DECIMAL

In a decimal, the place which a digit occupies to the right of the decimal point tells us the value of the digit.

In the decimal .666, the *place values* of the digits are as follows. Notice that as we move to the right, the value of each place is $\frac{1}{10}$ the value of the place to the left of it.

The first 6 to the right of the decimal point means 6 tenths. . . . . $\frac{6}{10}$

The second 6 to the right of the decimal point means 6 hundredths. . . . . $\frac{6}{100}$

The third 6 to the right of the decimal point means 6 thousandths. . . . . . $\frac{6}{1000}$

EXERCISES

1. Write these common fractions as decimals:

a. $\frac{3}{10}$ *b.* $\frac{19}{100}$ *c.* $\frac{375}{1000}$ *d.* $\frac{4}{100}$ *e.* $\frac{25}{1000}$ *f.* $\frac{9}{1000}$

2. Write these decimals as common fractions:

a. .2 *b.* .45 *c.* .625 *d.* .06 *e.* .045 *f.* .008

3. Write these mixed numbers as decimals:

a. $1\frac{5}{10}$ *b.* $2\frac{12}{100}$ *c.* $5\frac{137}{1000}$ *d.* $8\frac{9}{100}$ *e.* $6\frac{75}{1000}$ *f.* $4\frac{2}{1000}$

4. Write these decimals as mixed numbers:

a. 4.1 *b.* 6.32 *c.* 7.145 *d.* 5.09 *e.* 4.055 *f.* 3.008

5. The ruler on the right is marked so that each inch is divided into ten equal parts. Give each of the readings *a*, *b*, *c*, *d*, and *e* as a decimal.

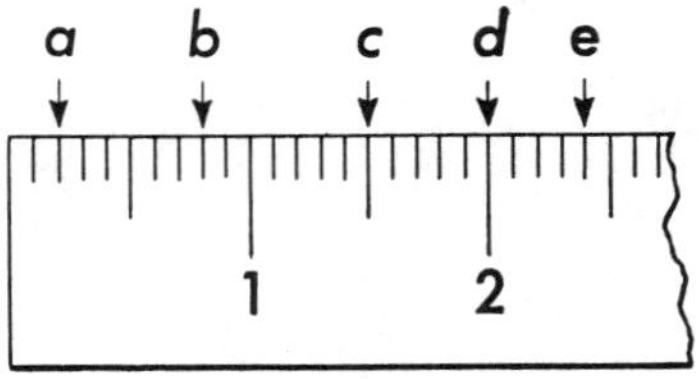

6. Sue bought a sheet of 100 stamps. If she used 95 stamps to mail letters, represent as a decimal the fractional part of the stamps she used.

7. Mr. Turner earned \$100 a week. He spent \$80 and saved the rest. Represent as a decimal the fractional part of his salary that he saved.

8. In the decimal .77, does the first or second 7 to the right of the decimal point have the larger value? How many times as much?

9. In the decimal .333, does the first or the third 3 to the right of the decimal point have the larger value? How many times as much?

2. READING AND WRITING DECIMAL FRACTIONS

The number of figures to the right of the decimal point tells us whether the decimal is to be read as tenths, hundredths, thousandths, etc.

When a decimal has 1 figure to the right of the decimal point, the decimal represents tenths. For example, .7 means $\frac{7}{10}$ and is read "7 *tenths*."

When a decimal has 2 figures to the right of the decimal point, the decimal represents hundredths. For example, .87 means $\frac{87}{100}$ and is read "eighty-seven hundredths."

When a decimal has 3 figures to the right of the decimal point, the decimal represents thousandths. For example, .563 means $\frac{563}{1000}$ and is read "five hundred sixty-three thousandths."

Notice in the previous examples that the number of figures to the right of the decimal point is always the same as the number of zeros in the denominator of the fraction that the decimal represents.

Procedure. To read a decimal fraction, read the number to the right of the decimal point and use as a denominator the name of the position of the last number on the right.

See the chart below for the names that are given to the different positions to the right of the decimal point.

Sometimes decimals contain fractions. For example, the decimal $.5\frac{1}{2}$ means $\frac{5\frac{1}{2}}{10}$ and is read "$5\frac{1}{2}$ tenths."

The mixed decimal 1.5 means $1\frac{5}{10}$ and is read "one and five tenths."

The mixed decimal 12.75 means $12\frac{75}{100}$ and is read "twelve and seventy-five hundredths."

Procedure. When reading a mixed decimal, the decimal point is read "and." First read the whole number, then read the word "and" to show the position of the decimal point, and then read the decimal fraction.

thousands	hundreds	tens	ones, or units	**and**	tenths	hundredths	thousandths	ten-thousandths	hundred-thousandths
8	9	0	2	.	5	4	9	8	7

The number at the bottom of the chart is read: eight thousand, nine hundred two and fifty-four thousand, nine hundred eighty-seven hundred-thousandths.

Note that in reading a number the only time "and" is used is to show the position of the decimal point.

To write the mixed decimal six and twenty-five thousandths using figures, we use a decimal point to represent the word "and." We write 6.025.

Procedure. To write a mixed decimal, write the whole number, place the decimal point after the whole number, and then write the decimal fraction.

MODEL PROBLEMS

1. Read the number .0085. *Ans.* eighty-five ten-thousandths

2. Write 425.23 in words. *Ans.* four hundred twenty-five and twenty-three hundredths

3. Write in figures: one hundred twenty-seven thousandths. *Ans.* .127

4. Write in figures: thirty-seven and five-hundredths. *Ans.* 37.05

EXERCISES

Write in words:

	(a)	(b)	(c)	(d)	(e)	(f)
1.	.7	.9	.69	.01	.40	$.37\frac{1}{2}$
2.	.535	.609	.057	.086	.010	$.002\frac{1}{2}$
3.	.6153	.0063	.85126	.09784	.00009	.00078
4.	8.6	100.5	9.48	12.07	7.135	9.054
5.	18.003	13.703	4.1575	8.0056	17.0524	894.0003

Write in figures:

6. eight tenths

7. fifty-nine hundredths

8. five hundredths

9. nine hundred twenty-five thousandths

10. fifty-seven thousandths

11. five thousand eight hundred twenty-five ten-thousandths

12. one hundred twenty-eight and two tenths

13. thirty-seven and twenty-five hundredths

14. three thousand and eight hundredths

15. four hundred twenty-seven and fifteen thousandths

16. two thousand seven and forty-five thousandths

17. three hundred twenty-five and forty-five thousandths

18. two hundred four and four thousandths

19. one hundred forty and thirty-eight ten-thousandths

20. twenty and four thousand one hundred twenty-five hundred-thousandths

Read the sentences and write in words the decimals that appear in them:

21. Last month we had 3.8 inches of rainfall in our city.

22. The thickness of this sheet of paper is .025 inch.

23. The value of π is 3.14.

24. A centimeter is equal to .39 of an inch.

25. A kilogram is equal to 2.2 pounds.

26. In a race, the winning car traveled at a rate of 110.87 miles per hour.

27. The circumference of a wheel on Jim's wagon is 8.625 inches.

28. The area of Helen's garden is 150.75 square feet.

3. COMPARING DECIMALS

We can compare the decimals .7 and .68 as follows:

Since $.7 = .70$ and .70 is more than .68, then .7 is more than .68.

Procedure. To compare decimals, change them, by annexing zeros on the right, into equal decimals all of which have the same number of decimal places. Then compare the resulting numbers.

MODEL PROBLEM

Which number has the smallest value: .8; .88; .08; .808; .088?
Solution: Change all the decimals to three-place decimals.

$$.8 = .800; \quad .88 = .880; \quad .08 = .080; \quad .808 = .808; \quad .088 = .088$$

Answer: Since .080 is the smallest decimal, .08 has the smallest value.

EXERCISES

1. Select the decimals having the same value:

a. .2; .02; .20; .200 *b.* 1.8; 1.80; 1.08; 1.800

2. Select the larger number:

a. .8; .82 *b.* .7; .68 *c.* .9; 1.0 *d.* .004; .04
e. .0017; .002 *f.* 3.2; 3.18 *g.* 6.9; 7.001 *h.* .3; .033

3. Arrange the numbers in order of size, with the smallest number first:

a. .04; .004; .0004 *b.* .0025; .005; .0505 *c.* 2.6; 2.606; 2.66

4. Which of the following decimals is greater than $4\frac{1}{2}$: 4.09; 4.6; 4.492?

4. ROUNDING OFF DECIMALS

The number .87 rounded off to the nearest tenth is .9, because .87 is closer to .90 (.9) than it is to .80 (.8).

The number .042 rounded off to the nearest hundredth is .04, because .042 is closer to .040 (.04) than it is to .050 (.05).

The number .1509 rounded off to the nearest thousandth is .151, because .1509 is closer to .1510 (.151) than it is to .1500 (.150).

The number 7.4 rounded off to the nearest whole number is 7, because 7.4 is closer to 7 than it is to 8.

Procedure. To round off a decimal to a given number of places:

1. Write down as many figures of the given decimal as are required by the problem and drop the rest of the figures.

2. Reading from the left, if the first of the figures dropped is less than 5, the number obtained in step 1 is the round number. If the first of the figures dropped is 5 or greater than 5, increase by 1 the last figure in the number obtained in step 1 to find the round number.

MODEL PROBLEMS

1. Round off 3.14 to the nearest tenth. *Ans.* 3.1

2. Round off 11.975 to the nearest tenth. *Ans.* 12.0

3. Round off 10.396 to the nearest hundredth. *Ans.* 10.40

4. Round off 2.0563 to the nearest thousandth. *Ans.* 2.056

5. Round off $35.62 to the nearest dollar. *Ans.* $36

EXERCISES

1. Round off to the nearest *tenth:*
a. .24 *b.* .87 *c.* .09 *d.* 4.81 *e.* 3.97 *f.* 15.625

2. Round off to the nearest *hundredth:*
a. .891 *b.* 2.046 *c.* 14.897 *d.* 4.5103 *e.* 2.0249

3. Round off to the nearest *thousandth:*
a. .5147 *b.* .0503 *c.* 1.1204 *d.* 6.0095 *e.* 13.5608

4. Round off to the nearest *ten-thousandth:*
a. .51247 *b.* .08912 *c.* 1.00764 *d.* 5.00098

5. Round off to the nearest *hundred-thousandth:*
a. .689352 *b.* .000557 *c.* 2.405295 *d.* 8.087504

6. Round off to the nearest *whole number:*
a. 5.6 *b.* 11.32 *c.* 29.75 *d.* 3.49 *e.* 16.553 *f.* 25.485

7. Round off to the nearest *dollar:*
a. $8.65 *b.* $4.32 *c.* $23.49 *d.* $17.59 *e.* $12.09 *f.* $19.90

8. The width of a house is 34.88 feet. Round off the width correct to the nearest tenth of a foot.

9. The height of a box is 4.375 inches. Round off the height correct to the nearest hundredth of an inch.

10. A value of π is 3.14159. Round this value off correct to the nearest thousandth.

5. ADDING DECIMALS

Since decimals are really fractions, the principles learned for the addition of fractions help us in adding decimals.

Let us add .15 and .72.

Think	*Write*
Since $.15 = \frac{15}{100}$	.15
and $.72 = \frac{72}{100}$	.72
then $.15 + .72 = \frac{87}{100} = .87$	.87

Let us add 1.3 and .9.

Think	*Write*
Since $1.3 = 1\frac{3}{10} = \frac{13}{10}$	1.3
and $.9 = \frac{9}{10} = \frac{9}{10}$	.9
then $1.3 + .9 = \frac{22}{10} = 2\frac{2}{10} = 2.2$	2.2

Let us add .7 and 1.05.

Think	*Write*		
Since $.7 = \frac{7}{10} = \frac{70}{100}$	.7		.70
and $1.05 = 1\frac{5}{100} = \frac{105}{100}$	1.05	*or*	1.05
then $.7 + 1.05 = \frac{175}{100} = 1\frac{75}{100} = 1.75$	1.75		1.75

Procedure. To add decimals, write them in a vertical column so that the decimal points are directly underneath each other. Add the numbers in the same way that whole numbers are added. Place the decimal point in the sum directly under the decimal points in the addends. Check as in whole numbers.

MODEL PROBLEMS

1. Add: \$3.25, \$6.82, and \$13.55.

```
 $ 3.25
   6.82
  13.55
 ------
 $23.62
```

2. Find the sum of 5.807, 48.4, .89, and 3.

```
  5.807              5.807
 48.4               48.400
   .89      or        .890
  3.                 3.000
 ------             ------
 58.097             58.097
```

Since decimals are fractions, *addition is commutative* and *addition is associative* when we are dealing with decimals.

EXERCISES

Add and check:

	(a)	(b)	(c)	(d)	(e)	(f)
1.	.5 .4	.8 .4	3.58 2.17	.87 .09	3.06 6.58	12.78 24.14
2.	1.42 4.56 3.81 9.21	8.43 6.08 2.90 5.76	8.6 14.9 32.1 8.4	8.90 25.75 18.63 257.32	\$337.58 192.24 684.97 132.12	\$ 87.95 126.54 8.98 350.50
3.	1.4 5. .8 9.2	8.48 12. 5.8 6.05	33.5 .85 9. 125.48	8.92 .03 5. .003	9.035 32. 127.6 8.25	406.3 6.03 80. 3789.

4. Add:

a. 95.6; .008; 4; 578
b. 4.5; 8.163; 10.01; 36
c. 4.98; 62.5; 3.865; 587
d. 3.1415; 41.05; 371.4; 362; .45

5. Find the sum of:

a. 1.762 + 25 + .98 + 1.5
b. .375; 16; 22.25; .5; 1.125

6. Add:

a. \$.87; \$.92; \$.05; \$.50
b. \$25.98; \$.85; \$4.49; \$8
c. \$125; \$325.50; \$20.89; \$.07
d. \$43.50; \$9.95; \$7.98; \$67.33

7. Find the sum. Simplify the arithmetic by using the commutative and associative principles for addition.

a. .98 + .79 + .11
b. \$1.98 + \$6.42 + \$9.02 + \$3.58

8. In the box at the right, we see the meaning of each of the digits of the decimal .4125. Notice that the total of the four decimals is .4125.

.4
.01
.002
.0005
.4125

Arrange each of the following decimals in the manner shown at the right.

a. .625 *b.* .9823 *c.* .5555 *d.* .7058 *e.* .0605

9. Mary was promised a new summer outfit. This was what she planned to buy: a dress, \$14.75; shoes, \$8.00; hat, \$6.50; stockings, \$.98; gloves, \$3.25. How much money did she need for the entire outfit?

10. Mr. Johnson issued the following checks in one day: \$28.98, \$150, \$300.41, \$2.50, and \$14.05. Find the total amount of the checks issued that day.

11. During a week, a salesman traveled the following distances: Mon., 135.8 miles; Tues., 55 miles; Wed., 72.9 miles; Thurs., 200.7 miles; Fri., 8.2 miles. Find the total number of miles he traveled during the week.

12. During a series of flights, a pilot flew a plane for the following distances: 327.8 miles, 154 miles, 624.2 miles, 458.4 miles, and 249.6 miles. What was the total number of miles the pilot flew?

13. Find the total length of the piece of steel shown at the right.

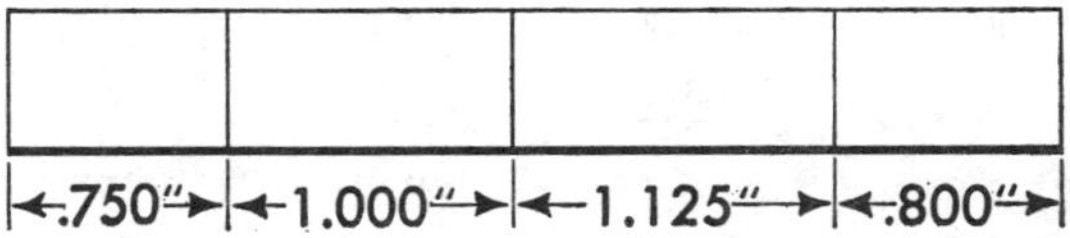

6. SUBTRACTING DECIMALS

Let us subtract 1.75 from 3.24.

Think	*Write*
Since $3.24 = 3\frac{24}{100} = \frac{324}{100}$	3.24
and $1.75 = 1\frac{75}{100} = \frac{175}{100}$	1.75
then $3.24 - 1.75 = \frac{149}{100} = 1\frac{49}{100} = 1.49$	1.49

Procedure. To subtract decimals, write the subtrahend under the minuend so that the decimal points are directly underneath each other. Subtract the numbers in the same way that whole numbers are subtracted. Place the decimal point in the difference directly under the decimal points above. Check as in whole numbers.

MODEL PROBLEMS

1. From \$8.75 subtract \$3.98.

$$\begin{array}{r} \$8.75 \\ 3.98 \\ \hline \$4.77 \end{array}$$

2. Subtract .7 from 5.2.

$$\begin{array}{r} 5.2 \\ .7 \\ \hline 4.5 \end{array}$$

3. Subtract \$2.49 from \$4.

$$\begin{array}{r} \$4.00 \\ 2.49 \\ \hline \$1.51 \end{array}$$

Annex zeros in the minuend so that it will have the same number of decimal places as the subtrahend.

4. 8.17 − 7.5.

$$\begin{array}{r} 8.17 \\ 7.5 \\ \hline .67 \end{array} \quad \textit{or} \quad \begin{array}{r} 8.17 \\ 7.50 \\ \hline .67 \end{array}$$

EXERCISES

Subtract and check:

	(a)	(b)	(c)	(d)	(e)	(f)
1.	.9	.08	.31	.56	.64	.73
	.2	.04	.19	.08	.09	.34
2.	5.8	6.7	84.2	57.6	78.8	196.5
	2.4	1.3	42.4	34.7	46.9	73.8
3.	79.5	6.13	44.32	81.64	36.27	413.25
	29.8	2.47	29.75	38.95	18.89	298.75
4.	5.5	29.6	23.7	38.52	49.98	869.74
	3.5	19.6	12.9	27.82	39.98	563.79
5.	4.508	38.06	400.6	9.005	10.54	28.09
	2.798	29.08	205.8	7.052	9.49	27.90
6.	\$529.38	\$75.08	\$300.50	\$4625.24	\$500.42	\$610.00
	129.18	64.92	200.48	932.78	127.32	410.98
7.	3.	6.	15.	30.	17.	300.
	1.2	.48	.22	29.53	.705	50.75
8.	8.5	9.37	6.035	56.98	4.003	8.506
	.7	2.5	1.52	2.	3.8	4.09
9.	.5	.6	.8	.03	.83	.09
	.28	.57	.005	.012	.5675	.005

10. Subtract:

a. .06 − .045 *b.* 3.06 − 1.88 *c.* 5 − 4.7 *d.* 36 − .03 *e.* 27.8 − 15.35

11. *a.* From \$8.29 subtract \$7.84. *b.* From .005 subtract .0045.
c. From \$10 subtract \$6.42. *d.* From 81.75 subtract 73.5.

12. *a.* Subtract .375 from .8. *b.* Subtract \$300.50 from \$600.

13. Helen spent \$36.50 for a coat and Jane spent \$28.75 for a coat. How much more did Helen spend for her coat than Jane spent for hers?

14. Walter ran 100 yards in 10.8 seconds. Gerald ran the same distance in 12.2 seconds. By how many seconds was Walter faster than Gerald?

15. Robert went to the store with a \$10 bill. If he spent \$5.43, how much change should he have received?

16. When Peter left home, the speedometer of his car read 16,214.9 miles. When he reached Buffalo, it read 16,405.2 miles. How far had he traveled?

17. A piece of metal 41.54 feet long became .005 feet shorter during cold weather. What was its new length?
18. Sidney weighs 100.2 pounds. If Julius weighs 14.3 pounds less than Sidney, how much does Julius weigh?

7. MULTIPLYING DECIMALS AND WHOLE NUMBERS

Notice that in each of the examples at the right the product has the same number of decimal places as the decimal multiplier.

.5	\$ 1.75	2
× 7	× 6	× .3
3.5	\$10.50	.6

Procedure. To multiply a whole number and a decimal, multiply the numbers in the same way that whole numbers are multiplied. The number of decimal places in the product is the same as the number of decimal places in the decimal multiplier. Check as in whole numbers.

MODEL PROBLEMS

1. Multiply: 1.5 × 36.

$$\begin{array}{r} 1.5 \\ \times\ 36 \\ \hline 9\ 0 \\ 45\ \ \\ \hline 54.0 \end{array}$$

The answer may also be written as 54.

Alternate Method Using Fractions

Since $1.5 = 1\frac{5}{10} = 1\frac{1}{2} = \frac{3}{2}$,

$$1.5 \times 36 = \frac{3}{\cancel{2}_1} \times \cancel{36}^{18} = 54$$

2. Find the product of 17 and .005.

$$\begin{array}{r} 17 \\ \times\ .005 \\ \hline .085 \end{array}$$

Since the answer must be a *three*-place decimal, or thousandths, place a zero to the left of 85 as a placeholder.

Since decimals are fractions, *multiplication is commutative, associative, and distributive* when we are dealing with decimals.

EXERCISES

1. How many decimal places are there in each product?

a.	*b.*	*c.*	*d.*	*e.*	*f.*
.2	1.72	.945	6	5	28
× 3	× 8	× 23	× 1.8	× .375	× 1.46

2. Place the decimal point correctly in each product:

a.	*b.*	*c.*	*d.*	*e.*
23.1	2.31	.231	105	105
× 105	× 105	× 105	× 2.31	× .231
24255	24255	24255	24255	24255

Multiply:

	(*a*)	(*b*)	(*c*)	(*d*)	(*e*)	(*f*)
3.	.2	1.8	12.5	3.08	1.123	.003
	3	7	50	27	4	23
4.	28	800	14	$8.98	$5.50	$4.75
	.32	.025	.002	32	60	700

5. Find the product. Simplify the arithmetic by using the commutative and associative principles for multiplication.

a. .25 × 1.9 × 4 *b.* .5 × 8.7 × 20 *c.* 4 × 9.79 × .75

6. Use the distributive principle for multiplication to simplify the arithmetic for obtaining the result:

a. 93 × .95 + 93 × .05 *b.* 57 × 4.6 + 57 × 3.1 + 57 × 2.3

7. In each of the following: (*a*) Estimate the answer by using round numbers. (*b*) Find the product.

a.	*b.*	*c.*	*d.*	*e.*	*f.*
1.95	12.01	144	1.48	125	3.487
× 4	× 6	× .52	× 28	× .997	× 120

8. Mr. Bell uses on the average .7 of a ton of coal per week to heat his house. How much coal does he use in 4 weeks?

9. A dealer buys 72 dresses, each of which costs $3.75. How much does he pay for the dresses?

10. A piece of paper is .008 inch thick. If a package contains 24 of these sheets, how thick is the package?

11. A boy has 125 marbles. If he loses .6 of them in a game, how many marbles does he lose?

12. Mr. Smith has found that his truck can average 13 miles per gallon of gasoline. How far can he go on 7.5 gallons of gasoline?

13. Mr. Frank had $400. If he spent .12 of his money for a suit, how much did he pay for the suit?

14. A merchant sold 150 pairs of shoes at $3.98 a pair. (*a*) Use round numbers to estimate the amount he received for the shoes. (*b*) Find how much he actually received for the shoes.

8. MULTIPLYING A DECIMAL BY A DECIMAL

Let us multiply .9 by .7.

Think	*Write*
Since $.9 = \frac{9}{10}$	.9
and $.7 = \frac{7}{10}$	$\times$ .7
then $.9 \times .7 = \frac{9}{10} \times \frac{7}{10} = \frac{9 \times 7}{10 \times 10} = \frac{63}{100} = .63$	.63
Answer: .63	

Notice that the number of decimal places in the answer is the same as the total number of decimal places in the multiplicand and multiplier.

Procedure. To multiply two decimals, multiply the numbers in the same way that whole numbers are multiplied. The number of decimal places in the product is the same as the total number of decimal places in the multiplicand and multiplier. Check as in whole numbers.

MODEL PROBLEMS

1. Multiply 4.08 by .5.

$$\begin{array}{r} 4.08 \\ .5 \\ \hline 2.040 \end{array} = 2.04$$

2. Find the product of 12.5 and .75.

$$\begin{array}{r} 12.5 \\ .75 \\ \hline 625 \\ 8\,75 \\ \hline 9.375 \end{array}$$

3. Multiply 1.2 by .002.

$$\begin{array}{r} 1.2 \\ .002 \\ \hline .0024 \end{array}$$

Since there should be 4 decimal places (ten-thousandths) in the answer, place two zeros to the left of 24 as placeholders.

4. How much did Mrs. Green pay for 3.5 yards of material at $.59 per yard?
Solution: We must multiply $.59 by 3.5.

$$\begin{array}{r} \$\ .59 \\ 3.5 \\ \hline 295 \\ 1\,77 \\ \hline \$2.065 \end{array}$$

Answer: Mrs. Green paid $2.07. (In business, it is the practice for the customer to pay an extra cent if the bill involves a part of a cent.)

EXERCISES

1. Find the number of decimal places in each product:

a. .5 × .7 *b.* .2 × .64 *c.* .25 × .38 *d.* .008 × .01

2. Place the decimal point in the product:

a. 3.75 × .13 = 4875 *b.* 375 × .013 = 4875
c. 6742 × 1.24 = 836008 *d.* .6742 × 12.4 = 836008

Multiply and check:

	(*a*)	(*b*)	(*c*)	(*d*)	(*e*)	(*f*)
3.	.5 .9	.2 .1	15.7 .9	1.8 2.5	16.9 7.4	60.3 50.7
4.	.95 .5	8.75 1.4	6.08 13.7	.28 .52	80.65 1.03	65.84 5.06
5.	.625 5.4	84.7 .065	4.63 .008	2.04 60.5	800 1.05	.075 .8

Add an extra cent if the answer involves a part of a cent.

6.	$2.50 × 3.5	$4.75 × 13.5	$10.85 × 9.75	$8.36 × .25	$75.20 × .05	$708.55 × .06

7. Find the product of:

a. .5 and .8 *b.* .18 and .75 *c.* .5 and .01 *d.* .001 and .01

8. Multiply:

a. 20.6 by 1.38 *b.* 25.6 by .075 *c.* .006 by .05 *d.* .003 by 40

9. What is the cost of 10.7 gallons of gasoline at $.32 per gallon?

10. A plane averaged 357.8 miles per hour on a flight. How far did the plane fly in 3.5 hours? Round off the answer to the nearest mile.

11. One gallon of milk weighs about 8.7 pounds. How much do 12.75 gallons weigh? Round off the answer to the nearest tenth of a pound.

12. The average amount of rainfall in a city is 2.78 inches per month. Find the average rainfall during .25 of a year.

13. A girl bought 8.75 yards of curtain material at $2.98 a yard. How much did she pay?

14. Mary bought 4.5 yards of dress material at $2.15 per yard. How much change should she get from a $20 bill?

9. MULTIPLYING DECIMALS BY 10, 100, 1000, ETC.

In solving problems, there are times when we have to multiply a decimal by 10, 100, 1000, etc. Let us learn a short way of doing this.

Notice how the decimal point changes position as we multiply .125 by 10, 100, and 1000. In each problem at the right, study the position of the decimal point in the multiplicand and in the product.

.125	.125	.125
10	100	1000
1.250	12.500	125.000

Procedure. To multiply a decimal by 10, 100, 1000, etc., move the decimal point as many places to the right as there are zeros in the multiplier.

MODEL PROBLEMS

1. Multiply: 15.37 × 10.
15.37 × 10 = 153.7

2. Find the product of 12.75 and 100.
12.75 × 100 = 1275.

3. Multiply: 9.36 × 1000.

9.36 × 1000 = 9360.

Notice that we annex a zero to the right of the 6 so that we can move the decimal point three places to the right. We can also think of 9.36 as 9.360 and then multiply by 1000.

4. Multiply: $2.49 × 10.

$2.49 × 10 = $24.9 = $24.90

In order to express the answer $24.9 in dollars and cents, we write $24.90. We can do this because $.9 and $.90 are equal.

EXERCISES

1. Multiply by 10: (*a*) .875 (*b*) .7 (*c*) 3.05 (*d*) .025 (*e*) .0075
2. Multiply by 100: (*a*) 7.62 (*b*) .3 (*c*) .825 (*d*) 1.025 (*e*) .0003
3. Multiply by 1000: (*a*) .157 (*b*) 20.86 (*c*) .4 (*d*) 9.0048 (*e*) .009
4. A man bought 100 pounds of apples at \$.08 a pound. How much did the apples cost him?
5. If a shirt costs \$4.98, what is the cost of 10 shirts?
6. A man bought 1000 yards of cloth at \$1.625 a yard. How much did the cloth cost him?
7. If it costs \$.0425 to ride 1 mile on a train, what must a man pay for a 1000-mile trip?
8. A baker charges \$.0625 for a cupcake. What is the cost of 100 cupcakes?

10. DIVIDING A DECIMAL BY A WHOLE NUMBER

Study the following examples in division:

Think	*Write*
$1.6 \div 2 = 1\frac{6}{10} \div 2 = \frac{16}{10} \div \frac{2}{1} = \frac{\overset{8}{\cancel{16}}}{10} \times \frac{1}{\underset{1}{\cancel{2}}} = \frac{8 \times 1}{10 \times 1} = \frac{8}{10} = .8$	$2\overline{)1.6}$ quotient .8
$.24 \div 6 = \frac{24}{100} \div \frac{6}{1} = \frac{\overset{4}{\cancel{24}}}{100} \times \frac{1}{\underset{1}{\cancel{6}}} = \frac{4 \times 1}{100 \times 1} = \frac{4}{100} = .04$	$6\overline{).24}$ quotient .04

Procedure. To divide a decimal by a whole number, place the decimal point in the quotient directly above the decimal point in the dividend. Then divide as in whole numbers.

MODEL PROBLEMS

1. Divide 69.3 by 3.

```
  23.1
3)69.3
```

2. Divide \$16.24 by 8.

```
 $ 2.03
8)$16.24
```

3. A wire 40.2 feet long is cut into 12 equal pieces. What is the length of each piece?

Solution: To find the length of each piece, we divide 40.2 by 12.

$$\begin{array}{r} 3.35 \\ 12\overline{)40.20} \\ \underline{36} \\ 4\,2 \\ \underline{3\,6} \\ 60 \\ \underline{60} \end{array}$$

Ans. 3.35 ft.

Notice that since there was a remainder of 6, we annexed a zero in the dividend, which did not change the value of the dividend, and continued the division until there was no remainder. Sometimes it is necessary to annex more than one zero in order to divide until there is no remainder.

4. Find, correct to the nearest tenth, .5 ÷ 3.

$$\begin{array}{r} .16 \\ 3\overline{).50} \\ \underline{3} \\ 20 \\ \underline{18} \\ 2 \end{array}$$

Ans. .2

When .5 is divided by 3, no matter how many zeros we annex in the dividend, the quotient will not come out evenly. We therefore annex enough zeros in the dividend so that we can find the quotient to 1 decimal place more than the question requires. We then round off the answer as required in the question.

When .16 is rounded off to the nearest tenth, the result is .2.

EXERCISES

Divide and check:

	(a)	(b)	(c)	(d)	(e)
1.	$3\overline{)9.3}$	$8\overline{)40.80}$	$4\overline{)\$.16}$	$7\overline{)\$14.21}$	$12\overline{)\$60.72}$
2.	$4\overline{).024}$	$8\overline{).616}$	$12\overline{).084}$	$7\overline{)52.605}$	$9\overline{)73.809}$
3.	$23\overline{)82.8}$	$64\overline{)67.20}$	$36\overline{)15.228}$	$25\overline{)1.525}$	$72\overline{).7272}$
4.	$2\overline{).7}$	$4\overline{)8.6}$	$8\overline{)8.6}$	$5\overline{).32}$	$2\overline{)5.47}$

Find to the nearest tenth:

5.	$2\overline{).3}$	$8\overline{)8.32}$	$3\overline{).7}$	$7\overline{)29.3}$	$12\overline{)1.9}$

Find to the nearest hundredth:

6.	$3\overline{).963}$	$4\overline{).87}$	$6\overline{).8}$	$8\overline{)24.23}$	$9\overline{)49.4}$

7. A board 12.8 feet long was divided into 4 equal parts. What was the length of each part?

8. Mrs. Blake paid $5.25 for 5 pounds of steak. How much did she pay for a pound of steak?

9. A dealer purchased 8 boxes of candy for $8.16. If each box contained 12 candy bars, what was the cost of each bar?

10. In a city, the rainfall in a year was 26 inches. Find, to the nearest tenth of an inch, the average rainfall during one month.

11. DIVIDING A WHOLE NUMBER BY A LARGER WHOLE NUMBER

Let us find the quotient of 1 ÷ 4. We can write 1 ÷ 4 as $\frac{1}{4}$. Since $\frac{1}{4} = \frac{1.00}{4}$, we can solve the problem as shown at the right.

$$4\overline{)1.00}\quad .25$$

Ans. .25 of a mile

Let us find the quotient of 2 ÷ 3 to the nearest tenth.

```
   .66
 3)2.00
   1 8
    20
    18
     2
```

Ans. .7

Since the question requires that the quotient be rounded off to the nearest tenth, we annex two zeros in order to find the quotient as a two-place decimal. We then round off.

When we round off .66 to tenths, the result is .7.

Procedure. To divide a whole number by a larger whole number: Place a decimal point after the dividend and annex zeros to the right of the decimal point until there is no remainder in the division.

If the quotient is to be found to a given number of decimal places, find the quotient to one decimal place more than the question requires. Then round off the quotient to the required number of decimal places.

EXERCISES

1. Divide:

a. 2 ÷ 5 *b.* 4 ÷ 5 *c.* $12\overline{)9}$ *d.* $36\overline{)9}$ *e.* $64\overline{)32}$

2. Find to the nearest tenth:

a. 1 ÷ 6 *b.* 3 ÷ 8 *c.* $6\overline{)5}$ *d.* 60 ÷ 90 *e.* $200\overline{)13}$

3. Find to the nearest hundredth: (*a*) $4 \div 12$ (*b*) $18\overline{)10}$ (*c*) $24\overline{)20}$

4. Find to the nearest cent: (*a*) $5\overline{)\$4.98}$ (*b*) $8\overline{)\$7.75}$ (*c*) $12\overline{)\$1.49}$

5. If a piece of lumber 10 feet long is cut into 12 equal pieces, find the length of each piece, correct to the nearest tenth of a foot.

6. If 36 gumdrops weigh 11 ounces, find the weight of a gumdrop to the nearest hundredth of an ounce.

12. DIVIDING A WHOLE NUMBER OR A DECIMAL BY A DECIMAL

If we wish to divide 1.68 by .8, we can write $\frac{1.68}{.8}$. Since we already know how to divide a decimal by a whole number, let us find a fraction equivalent to $\frac{1.68}{.8}$ whose denominator is a whole number. In the fraction $\frac{1.68}{.8}$ we can multiply both the numerator 1.68 and the denominator .8 by 10, or move the decimal point one place to the right, without changing the value of the fraction. The result $\frac{16.8}{8}$ has the same value as $\frac{1.68}{.8}$. Now we divide 16.8 by 8 and get the quotient 2.1. We can arrange the decimal division as shown at the right. Note that we use a caret (∧) to show the new position of the decimal point in the divisor and the dividend after the divisor has been changed to a whole number.

$$\begin{array}{r} 2.1 \\ .8_{\wedge}\overline{)1.6_{\wedge}8} \end{array}$$

If we wish to divide .156 by .12, we can write $\frac{.156}{.12}$. To make the denominator a whole number, we multiply the numerator and the denominator by 100, or move the decimal point two places to the right in both the dividend and the divisor. We then get the fraction $\frac{15.6}{12}$. The result of this division is 1.3. We can arrange the decimal division as shown at the right.

$$\begin{array}{r} 1.3 \\ .12_{\wedge}\overline{).15_{\wedge}6} \\ \underline{12} \\ 36 \\ \underline{36} \end{array}$$

Procedure. To divide a whole number or a decimal by a decimal:

1. Make the divisor a whole number by multiplying it by 10, or 100, etc., which means moving the position of the decimal point the necessary number of places to the right. Use a caret (∧) to show the new position of the decimal point.

2. In order not to change the value of the quotient, multiply the dividend by the same multiplier used in step 1, which means moving the decimal point in the dividend the same number of places to the right as in step 1. Annex zeros if necessary. Use a caret (∧) to show the new position of the decimal point.

3. Place a decimal point in the quotient directly above the caret in the dividend.

4. Divide the resulting numbers as in division by a whole number.

5. Check as in whole numbers.

MODEL PROBLEMS

1. Divide .008 by .2.

$$\begin{array}{r} .04 \\ .2_\wedge\overline{).0_\wedge08} \end{array}$$

2. Divide: \$6.40 ÷ \$.04.

$$\begin{array}{r} 1\,60. \\ .04_\wedge\overline{)6.40_\wedge} \end{array}$$

3. Divide .7 by .008.

$$\begin{array}{r} 87.5 \\ .008_\wedge\overline{).700_\wedge0} \end{array}$$

4. A car uses 1.3 gallons of gasoline to go 17 miles. Find, to the nearest tenth of a mile, the distance it can go using 1 gallon of gasoline.

Solution: Divide the distance traveled, 17 miles, by the number of gallons of gasoline used.

Find the quotient to one decimal place more than the number of places required by the question. Then round off the answer to the number of places asked for.

When 13.07 is rounded off to the nearest tenth, the result is 13.1.

$$\begin{array}{r} 13.07 \\ 1.3_\wedge\overline{)17.0_\wedge00} \\ \underline{13} \\ 40 \\ \underline{39} \\ 100 \\ \underline{91} \\ 9 \end{array}$$

Ans. 13.1 miles

EXERCISES

Divide and check:

	(*a*)	(*b*)	(*c*)	(*d*)	(*e*)
1.	$.3\overline{)63}$	$.5\overline{)125}$	$.6\overline{)360}$	$2.5\overline{)625}$	$4.3\overline{)8600}$
2.	$.4\overline{)14.4}$	$.8\overline{)164.8}$	$.9\overline{)9.81}$	$.5\overline{)137.15}$	$.8\overline{)640.64}$
3.	$.8\overline{).0840}$	$.4\overline{).001}$	$2.5\overline{)22.25}$	$7.2\overline{)17.28}$	$29.1\overline{)6896.7}$

	(a)	(b)	(c)	(d)	(e)
4.	$.03\overline{).69}$	$.15\overline{)1.2}$	$.09\overline{)8.01}$	$.57\overline{)68.97}$	$.64\overline{)13.12}$
5.	$.6\overline{).0012}$	$.09\overline{).00981}$	$.015\overline{)18.6}$	$.024\overline{)19.8}$	$.052\overline{)6.5}$
6.	$.07\overline{)56}$	$.09\overline{)27}$	$.11\overline{)88}$	$.32\overline{)128}$	$.72\overline{)360}$

Find to the nearest tenth:

7.	$.3\overline{)2}$	$.9\overline{).4}$	$.8\overline{)7}$	$1.2\overline{)7.6}$	$9.5\overline{)983}$
8.	$54.9\overline{)64.1}$	$.32\overline{)54.3}$	$.81\overline{)59.8}$	$.73\overline{)9.83}$	$.84\overline{)1.78}$

Find to the nearest hundredth:

9.	$.3\overline{)3.2}$	$.6\overline{)40}$	$1.8\overline{)7.46}$	$61.2\overline{)7.15}$	$.87\overline{)8.64}$

Divide:

	(a)	(b)	(c)	(d)
10.	84.64 by .4	150 by .25	1.596 by .12	2.088 by .24
11.	9.18 by 2.7	5.672 by .08	20.25 by .045	6.5224 by 2.48
12.	53,238 by .76	106.75 by 2.5	14.762 by 1.22	1.95489 by .321
13.	217.6 ÷ .32	5000 ÷ .02	86.45 ÷ .035	19,224 ÷ .48

14. How many ball point pens can be bought for $5 if each pen costs $.25?

15. How many streamers, each 1.5 feet long, can be cut from a roll of crepe paper 36 feet long?

16. Anthony wishes to buy a sport jacket that costs $22.50. If he can save $2.50 per week, how many weeks does he have to save to buy the jacket?

17. At 25 cents each, how many school lunches can be bought for $6.00?

18. Henry can save 5 quarters each week. How many weeks must he save to purchase an E bond that costs $18.75?

19. A ship sailed 370 miles in 18.5 hours. What was the ship's average rate of speed?

20. The water in a tank weighs 668.75 pounds. If one cubic foot of water weighs 62.5 pounds, how many cubic feet of water does the tank contain?

21. An airplane uses 53.8 gallons of gasoline every hour it flies. If the plane carries 215.2 gallons of gasoline, how many hours can it fly?

22. A butcher paid $115.75 for 55 chickens. What was the average price he paid for each chicken? Find the answer correct to the nearest cent.

23. If potatoes sell at 5 pounds for $.37, how many pounds can be bought for $2.22?

13. DIVIDING WHOLE NUMBERS AND DECIMALS BY 10, 100, 1000, ETC.

Notice how the position of the decimal point changes as we divide 25, which may be written 25., by 10, 100, and 1000. In each of the following examples, study the position of the decimal point in the dividend and the quotient.

$$10\overline{)25.0} = 2.5 \qquad 100\overline{)25.00} = .25 \qquad 1000\overline{)25.000} = .025$$

$$10\overline{)2.50} = .25 \qquad 100\overline{)2.500} = .025 \qquad 1000\overline{)2.5000} = .0025$$

Procedure. To divide a whole number or a decimal by 10, 100, or 1000, etc., move the decimal point as many places to the left as there are zeros in the divisor.

MODEL PROBLEMS

1. 300 ÷ 100
300 ÷ 100 = 3.00

2. 72.6 ÷ 10
72.6 ÷ 10 = 7.26

3. 1.6 ÷ 1000

1.6 ÷ 1000 = .0016

Since we must move the decimal point three places to the left, we place two zeros to the left of the 1.

EXERCISES

Divide by 10:

	(a)	(b)	(c)	(d)	(e)
1.	80	900	250	83	7
2.	2.65	.875	.8	.07	.045

Divide by 100:

3.	400	6000	50,000	930	25
4.	12.6	6	.1	.07	.001

Divide by 1000:

5.	9000	85	3	.02	.05

Replace the blank with a number that will make the completed sentence true:

6. To divide a number by 30, first divide the number by 10 and then divide the result by ______.

7. To divide 600 by 200, first divide 600 by 100 and then divide the result by ______.

8. To divide a number by 6000, first divide the number by ______ and then divide the result by 6.

9. A man paid \$6 for a 100-pound bag of potatoes. How much did a pound of potatoes cost him?

10. If 1000 feet of electric wire cost \$90, what is the cost of 1 foot of wire?

11. A builder paid \$9000 for 10 lots. What was the average cost of 1 lot?

12. If 10 pounds of melon cost \$.65, what is the cost of 1 pound of melon?

14. FIGURING WITH FRACTIONS AND DECIMALS

Problem 1. Fred bought $\frac{3}{4}$ of a dozen colored electric light bulbs that sold for \$1.20 per dozen. To determine the cost of the purchase, Fred had to find $\frac{3}{4}$ of \$1.20. He therefore multiplied \$1.20 by $\frac{3}{4}$.

$$\frac{3}{\cancel{4}_{\,1}} \times \overset{\$.30}{\cancel{\$1.20}} = \$.90$$

Answer: \$.90

Problem 2. Mrs. Robinson passed a store that was having a sale. A sign in the window said that the customer would pay only $.87\frac{1}{2}$ of the price of an article. If Mrs. Robinson bought a dress whose original price was \$39.50, how much did she pay for the dress?

To find the answer, we multiply \$39.50 by $.87\frac{1}{2}$.

$$\begin{array}{r} \$39.50 \\ \underline{.87\frac{1}{2}} \\ 1975 \\ 2\,7650 \\ \underline{31\,600} \\ \$34.5625 \end{array} \qquad \left(\frac{1}{\cancel{2}_{\,1}} \times \overset{1975}{\cancel{3950}} = 1975\right)$$

Answer: \$34.57

Problem 3. Arlene spent 2.5 hours doing her homework. If she spent .5 of an hour doing her spelling, what fractional part of the entire time did she spend on her spelling?

To find the answer, we can write a fraction whose numerator is .5, the number that represents the part of the whole, and whose denominator is 2.5, the number that represents the whole.

$$\frac{.5}{2.5} = \frac{.5 \times 10}{2.5 \times 10} = \frac{5}{25} = \frac{1}{5}$$

Answer: $\frac{1}{5}$

Problem 4. Richard correctly spelled .9 of the words on his spelling test. If he spelled 45 words correctly, how many words were there on the test?

$45 \div .9$
$.9_\wedge\overline{)45.0_\wedge}$ quotient $50.$
Answer: 50 words

We can find the answer by dividing the number that represents the decimal part of the whole we are trying to find by the decimal fraction which that part represents.

Problem 5. Find the cost of 8 bars of candy at 2 bars for \$.09.

METHOD 1

Since 8 bars are 4 times as much as 2 bars ($8 \div 2 = 4$), 8 bars will cost 4 times as much as 2 bars. Therefore, 8 bars cost $4 \times \$.09 = \$.36$. *Answer:* \$.36

METHOD 2

First find the cost of 1 bar. $\$.09 \div 2 = \$.04\frac{1}{2}$. Then multiply the cost of 1 bar by 8. $8 \times \$.04\frac{1}{2} = \$.36$ *Answer:* \$.36

EXERCISES

Find:

	(*a*)	(*b*)	(*c*)	(*d*)	(*e*)
1.	$\frac{1}{2} \times \$.46$	$\frac{1}{3} \times \$3.60$	$\frac{3}{5} \times \$6.80$	$\frac{7}{8} \times \$9.60$	$\frac{9}{10} \times \$75$
2.	$\frac{1}{3} \times \$2.25$	$\frac{1}{5} \times \$4.98$	$\frac{3}{4} \times \$.59$	$\frac{2}{3} \times \$1.19$	$1\frac{2}{3} \times \$10.20$

Find:

3.	$\frac{1}{4}$ of \$4.80	$\frac{3}{4}$ of \$4.56	$\frac{5}{8}$ of \$8.40	$\frac{1}{4}$ of \$8.25	$\frac{2}{3}$ of \$7.60
4.	\$100 $\times .37\frac{1}{2}$	\$5.40 $\times .33\frac{1}{3}$	\$480 $\times .06\frac{1}{4}$	\$395.10 $\times .16\frac{2}{3}$	\$425.25 $\times .83\frac{1}{3}$
5.	.04 of \$120	.06 of \$50.95	$.01\frac{1}{2}$ of \$800	.95 of \$150	$.33\frac{1}{3}$ of \$6.90
6.	.7 of 85	.4 of 6.8	.02 of 1.8	.03 of .45	.15 of .012

7. *a.* What part of 8 is .5? *b.* What part of .15 is .03?

8. *a.* .6 is what part of 12? *b.* .02 is what part of 1.2?

9. *a.* .1 of what number is 3? *b.* .02 of what number is 4?
c. .05 of what number is 25? *d.* .25 of what number is 7.5?

10. Find the cost of $\frac{3}{4}$ of a pound of peas at \$.24 per pound.

11. What is the cost of $\frac{2}{3}$ of a yard of woolen cloth at \$6.75 per yard?

12. Amy bought $5\frac{1}{3}$ yards of curtain material at \$1.68 a yard. How much did the material cost her?

13. What is the cost of 240 feet of lumber at $\$.08\frac{1}{2}$ a foot?

14. Mr. Black bought some shirts at a sale. If his bill was \$18.75 and he had to pay only $.66\frac{2}{3}$ of the bill, how much did he actually pay?
15. If Sue paid \$.35 for .5 of a dozen eggs, what was the price of a dozen eggs?
16. Anthony saved \$7.50 in a week. If this represented .15 of his weekly salary, how much did he earn that week?
17. One day 6 players of the Southridge High School football team failed to appear for practice. If this represented .08 of the squad, how many players were there on the squad?
18. What would be the cost of 40 pencils at the rate of 2 for 11 cents?
19. At 2 for 25 cents, how many cans of soup can be bought for one dollar?
20. Find the cost of a dozen cans of tomato sauce if 3 cans cost \$.23.
21. Sarah bought $3\frac{1}{2}$ pounds of grapes at \$.29 a pound and $2\frac{3}{4}$ pounds of pears at \$.25 a pound. How much change should she get from a \$10 bill?
22. Mr. Bailey has planned a 240-mile trip. Before lunch he traveled $4\frac{1}{3}$ hours at an average rate of 42.3 miles per hour. How many miles does he still have to travel?

15. CHANGING A COMMON FRACTION TO A DECIMAL FRACTION

When a common fraction is changed to an equal fraction whose denominator is 10, 100, or 1000, etc., the common fraction is being changed to a decimal fraction. Study the following examples:

$$\frac{3}{5} = \frac{3 \times 2}{5 \times 2} = \frac{6}{10} = .6 \qquad \frac{12}{25} = \frac{12 \times 4}{25 \times 4} = \frac{48}{100} = .48$$

$\frac{7}{8} = 7 \div 8$

$$8\overline{)7.000}\quad \text{quotient } .875$$

or

$$8\overline{)7.00}\quad \text{quotient } .87\tfrac{1}{2}$$

$\frac{7}{8} = .875$ *or* $.87\frac{1}{2}$

There are many fractions, such as $\frac{7}{8}$, that cannot easily be changed to an equal fraction whose denominator is 10, 100, or 1000. In such cases, we use division. For example, if we wish to change $\frac{7}{8}$ to a decimal, we divide 7 by 8 because $\frac{7}{8}$ means 7 divided by 8. See how the division is done at the left.

Procedure. To change a common fraction to a decimal fraction:
If it can be done easily, change the fraction to an equal fraction whose denominator is 10, 100, or 1000. Then write this fraction as a decimal. ***or***
If the change is not easy, divide the numerator by the denominator, finding the quotient to as many decimal places as are necessary.

MODEL PROBLEMS

1. Change $\frac{5}{4}$ to a decimal fraction.

$\frac{5}{4} = 5 \div 4$

$$4\overline{)5.00}\quad 1.25$$

$\frac{5}{4} = 1.25$

2. Change $\frac{1}{6}$ to a decimal correct to the nearest hundredth.

Method 1	*Method* 2
$\frac{1}{6} = 1 \div 6$	$\frac{1}{6} = 1 \div 6$
$6\overline{)1.00}\quad .16\frac{4}{6}$	$6\overline{)1.000}\quad .166 +$ remainder
$\frac{1}{6} = .16\frac{4}{6} = .16\frac{2}{3}$	$\frac{1}{6} = .166 +$ remainder
$.16\frac{2}{3}$ rounded off to hundredths $= .17$	$.166$ rounded off to hundredths $= .17$

3. Which is larger, $\frac{3}{8}$ or $\frac{4}{9}$?
Solution: Change the fractions to decimals. Then compare the decimals.

$\frac{3}{8} = 3 \div 8$ *or* $8\overline{)3.000}\quad .375$ $\qquad$ $\frac{4}{9} = 4 \div 9$ *or* $9\overline{)4.000}\quad .444 +$ remainder

Answer: $\frac{4}{9}$ is larger than $\frac{3}{8}$.

4. Multiply $10\frac{1}{4}$ by $14\frac{1}{2}$.

Method 1 (*Using Decimals*)	*Method* 2 (*Using Fractions*)
$10.25\ (10\frac{1}{4} = 10.25)$	$14\frac{1}{2} \times 10\frac{1}{4} = \frac{29}{2} \times \frac{41}{4} = \frac{29 \times 41}{2 \times 4}$
$\times\ 14.5\ (14\frac{1}{2} = 14.5)$	$= \frac{1189}{8} = 148\frac{5}{8}$
5 125	
41 00	
102 5	
148.625	

Answer: 148.625 *or* $148\frac{5}{8}$

EXERCISES

Change to decimals:

	(*a*)	(*b*)	(*c*)	(*d*)	(*e*)	(*f*)	(*g*)	(*h*)
1.	$\frac{1}{2}$	$\frac{1}{5}$	$\frac{1}{25}$	$\frac{3}{5}$	$\frac{2}{5}$	$\frac{9}{10}$	$\frac{1}{8}$	$\frac{9}{12}$
2.	$\frac{7}{10}$	$\frac{3}{20}$	$\frac{5}{200}$	$\frac{475}{1000}$	$\frac{12}{25}$	$\frac{8}{16}$	$\frac{17}{100}$	$\frac{7}{50}$
3.	$\frac{7}{4}$	$\frac{8}{5}$	$\frac{12}{8}$	$2\frac{1}{2}$	$1\frac{3}{4}$	$1\frac{2}{5}$	$2\frac{3}{5}$	$5\frac{5}{8}$

Change to decimals correct to the nearest hundredth:

	(a)	(b)	(c)	(d)	(e)	(f)	(g)	(h)
4.	$\frac{1}{3}$	$\frac{2}{3}$	$\frac{5}{6}$	$\frac{4}{7}$	$\frac{1}{9}$	$\frac{1}{12}$	$\frac{7}{12}$	$\frac{5}{32}$
5.	$\frac{1}{64}$	$\frac{5}{64}$	$\frac{13}{64}$	$\frac{10}{3}$	$\frac{11}{9}$	$3\frac{2}{3}$	$1\frac{5}{6}$	$2\frac{2}{7}$

Which is the largest decimal?

	(a)	(b)	(c)
6.	.5, .9, .2	.56, .65, .48	.25, .2, .098
7.	.532, .476, .8	.05, .005, .15	.7, .693, .701

Which is larger?

	(a)	(b)	(c)	(d)
8.	$\frac{3}{4}$ or .7	.38 or $\frac{3}{8}$	.67 or $\frac{2}{3}$	$\frac{1}{16}$ or .06

Which is the smallest fraction?

9. $\frac{1}{5}, \frac{3}{10}, \frac{1}{4}$ $\quad$ $\frac{1}{2}, \frac{4}{5}, \frac{5}{8}$ $\quad$ $\frac{3}{4}, \frac{7}{10}, \frac{8}{9}$ $\quad$ $\frac{5}{6}, \frac{14}{16}, \frac{32}{36}$

Change the common fractions to equivalent decimals and then multiply:

10. $20\frac{7}{10} \times 32\frac{3}{10}$ $\quad$ $12\frac{17}{100} \times 4\frac{3}{100}$ $\quad$ $12\frac{1}{4} \times 13\frac{1}{2}$ $\quad$ $15\frac{3}{4} \times 18\frac{1}{2}$

Change the mixed numbers to improper fractions and then multiply:

11. $12\frac{1}{5}$ by $10\frac{1}{4}$ $\quad$ $62\frac{1}{2}$ by $5\frac{9}{10}$ $\quad$ $18\frac{3}{25}$ by $12\frac{1}{3}$ $\quad$ $10\frac{1}{8}$ by $84\frac{4}{5}$

12. Mrs. Thomas served $\frac{4}{6}$ of a pie at lunch. Represent as a decimal, correct to the nearest hundredth, the fraction of the pie that she served.

13. Represent as a decimal, correct to the nearest hundredth, the fractional part of a week that 1 day is.

14. During a season, the Tigers won .6 of the games they played. In the same league, the Panthers won 5 games out of the 8 they played. Which team had a better standing?

15. At the end of a baseball season, the American League standing was:

	Team	*Won*	*Lost*		*Team*	*Won*	*Lost*
(*a*)	New York	97	57	(*b*)	Detroit	82	72
(*c*)	Cleveland	88	66	(*d*)	Baltimore	69	85
(*e*)	Chicago	85	69	(*f*)	Washington	59	95
(*g*)	Boston	84	70	(*h*)	Kansas City	52	102

Represent as a decimal, correct to the nearest thousandth, the fractional part of the games played that each team won.

16. Find the cost of $25\frac{3}{4}$ yards of material at \$$11\frac{1}{2}$ per yard.

17. Find the cost of $18\frac{1}{2}$ dozen sheets at \$$35\frac{3}{4}$ per dozen.

18. What is the cost of $20\frac{3}{4}$ dozen pairs of stockings at \$$10\frac{1}{2}$ per dozen?

CHART OF COMMON FRACTIONS AND EQUIVALENT DECIMALS

Common Fraction	*Equivalent Decimal*	*Common Fraction*	*Equivalent Decimal*	*Common Fraction*	*Equivalent Decimal*
$\frac{1}{2}$	.50	$\frac{1}{10}$	.10	$\frac{1}{6}$	$.16\frac{2}{3}$
$\frac{1}{4}$	.25	$\frac{3}{10}$	.30	$\frac{5}{6}$	$.83\frac{1}{3}$
$\frac{3}{4}$	.75	$\frac{7}{10}$	.70	$\frac{1}{8}$	$.12\frac{1}{2}$
$\frac{1}{5}$	.20	$\frac{9}{10}$	.90	$\frac{3}{8}$	$.37\frac{1}{2}$
$\frac{2}{5}$	.40	$\frac{1}{3}$	$.33\frac{1}{3}$	$\frac{5}{8}$	$.62\frac{1}{2}$
$\frac{3}{5}$	.60	$\frac{2}{3}$	$.66\frac{2}{3}$	$\frac{7}{8}$	$.87\frac{1}{2}$
$\frac{4}{5}$	.80				

16. CHANGING A DECIMAL FRACTION TO A COMMON FRACTION

We know that a decimal represents a fraction whose denominator is 10, 100, or 1000, etc. Thus,

$$.7 = \frac{7}{10} \qquad .25 = \frac{25}{100} = \frac{5}{20} = \frac{1}{4} \qquad .225 = \frac{225}{1000} = \frac{45}{200} = \frac{9}{40}$$

Procedure. To change a decimal fraction to a common fraction, write a fraction whose numerator is the given decimal without the decimal point and whose denominator is 10, 100, or 1000, etc., depending upon the number of decimal places in the given decimal. Reduce the fraction if possible.

MODEL PROBLEMS

1. Change .045 to a common fraction. $.045 = \frac{45}{1000} = \frac{9}{200}$

2. Change $.37\frac{1}{2}$ to a common fraction.

$$.37\frac{1}{2} = .375 = \frac{375}{1000} = \frac{75}{200} = \frac{15}{40} = \frac{3}{8}$$

3. Find the cost of 16 feet of lumber if each foot costs $\$.12\frac{1}{2}$.
Solution: Since $\$.12\frac{1}{2} = \frac{1}{8}$ of a dollar, the cost of 16 feet of lumber is $16 \times \$\frac{1}{8} = \$2$.

EXERCISES

Change to common fractions:

	(a)	(b)	(c)	(d)	(e)	(f)
1.	.3	.6	.45	.79	.04	.60
2.	.125	.045	.0625	.0025	$.66\frac{2}{3}$	$.87\frac{1}{2}$
3.	1.5	2.25	3.40	$1.33\frac{1}{3}$	$2.12\frac{1}{2}$	$1.37\frac{1}{2}$

Do the following questions mentally by using equivalent fractions for the decimals:

	(a)	(b)	(c)	(d)	(e)
4.	$.10 \times 60$	$.50 \times 12$	$28 \times .25$	$48 \times .75$	$90 \times .80$
5.	$.12\frac{1}{2} \times 56$	$.37\frac{1}{2} \times 40$	$32 \times .87\frac{1}{2}$	$33\frac{1}{3} \times 24$	$18 \times .66\frac{2}{3}$
6.	$.125 \times 80$	$.375 \times 32$	$64 \times .875$	$72 \times .625$	44×1.25

7. A baseball team won .75 of the 24 games it played. How many games did it win?

8. How much do 32 yards of ribbon cost at \$$.62\frac{1}{2}$ a yard?

9. If a towel costs \$$.87\frac{1}{2}$, find the cost of 2 dozen towels.

10. If $.33\frac{1}{3}$ of the 690 pupils in a school were absent on a certain day, how many pupils were absent?

11. Sarah saved $.66\frac{2}{3}$ of the $9.30 that she earned. How much did she save?

Chapter IV

ARITHMETIC OF PER CENTS

1. UNDERSTANDING THE MEANING OF PER CENTS

Gerald took a test of 100 questions. He correctly answered 93 questions. The teacher wrote on the top of his paper 93%. When he arrived at home, Gerald told his mother that he had received 93 per cent on the test.

Since Gerald correctly answered 93 out of 100 questions, he correctly answered $\frac{93}{100}$ or 93 *hundredths* of the questions. Instead of writing $\frac{93}{100}$, we may write 93 *per cent.* ***Per cent*** **means hundredths.**

The symbol % means per cent. Thus, 93 per cent may be written 93%.

1% means	$\frac{1}{100}$, or 1 out of 100
75% means	$\frac{75}{100}$, or 75 out of 100
$87\frac{1}{2}$% means	$\frac{87\frac{1}{2}}{100}$, or $87\frac{1}{2}$ out of 100
100% means	$\frac{100}{100}$, or 100 out of 100

A certain number of per cent means a fraction whose numerator is the given number and whose denominator is 100. 100% means the whole of anything, or 1.

MODEL PROBLEMS

1. Write five-hundredths as a per cent. *Ans.* 5%

2. Write $\frac{13}{100}$ as a per cent. *Ans.* 13%

3. Write 73% as a common fraction. *Ans.* $\frac{73}{100}$

4. Write $3\frac{1}{2}$% as a fraction whose denominator is 100. *Ans.* $\frac{3\frac{1}{2}}{100}$

5. If a man spends 95% of his salary, what per cent does he have left?

Ans. From his whole salary, which is represented by 100%, subtract the part he spends, 95%. The result, 5%, is the part he has left.

EXERCISES

1. Write as a common fraction whose denominator is 100 and as a per cent:
   a. 98 out of 100 *b.* 50 out of 100
2. Write as a fraction whose denominator is 100:
   a. 4% *b.* 12% *c.* 78% *d.* $37\frac{1}{2}\%$ *e.* $6\frac{1}{4}\%$
3. Write as a per cent:
   a. 3 hundredths *b.* 37 hundredths *c.* $\frac{5}{100}$ *d.* $\frac{15}{100}$ *e.* $\frac{65}{100}$
4. Robert received $100 in Christmas presents. He spent $15 for books, $45 for a bicycle, $20 for a camera, and saved the rest.
   a. What per cent of his present did he spend for books?
   b. What per cent did he spend for the bicycle?
   c. What per cent did he spend for the camera?
   d. What per cent did he save?
5. Among 100 lamps, 3 were defective. What per cent were imperfect?
6. Explain the statement, "This blouse is 100% silk."
7. A scout troop had 14 members. If 100% of the troop went on a hike, how many scouts went?
8. If a man pays 30% of his salary as income taxes, what per cent of his salary does he have left after taxes?
9. If $62\frac{1}{2}\%$ of a class is present in school, what per cent of the class is absent?
10. A boy spends 30% of his allowance for candy, 30% for a notebook, and saves the balance. What per cent of his allowance does he save?
11. In his will, Mr. Hanson left 50% of his possessions to his wife, 30% to his son, and the remainder to his daughter. What per cent of his possessions did the daughter receive?

2. CHANGING PER CENTS TO DECIMALS

Let us learn how to change per cents to decimals quickly.

23% means $\frac{23}{100} = .23$

2% means $\frac{2}{100} = .02$

1.5% means $\frac{1.5}{100} = .015$

$2\frac{1}{2}\%$ means $\frac{2\frac{1}{2}}{100} = .02\frac{1}{2}$, or $2\frac{1}{2}\% = 2.5\% = .025$

Procedure. To change a per cent to a decimal, move the decimal point two places to the left and omit the per cent sign.

MODEL PROBLEMS

Write each per cent as a decimal:

1. $55\% = .55$ **2.** $6\% = .06$ **3.** $2.3\% = .023$

4. $12\frac{1}{2}\% = .12\frac{1}{2}$ or $.125$ **5.** $\frac{1}{4}\% = .00\frac{1}{4}$ or $.0025$ **6.** $.1\% = .001$

EXERCISES

Write as decimals:

	(*a*)	(*b*)	(*c*)	(*d*)	(*e*)	(*f*)
1.	27%	50%	3%	$37\frac{1}{2}\%$	$18\frac{3}{4}\%$	$3\frac{1}{4}\%$
2.	15.5%	20.8%	1.25%	$33\frac{1}{3}\%$	$8\frac{1}{3}\%$	$12\frac{1}{3}\%$
3.	$\frac{1}{2}\%$	$\frac{3}{8}\%$	$\frac{2}{3}\%$	$.2\%$	$.6\%$	$.01\%$

4. Write one-half of one per cent in figures.

5. Select the greatest quantity: (*a*) $\frac{4}{5}$; 75%; $.66$ (*b*) $\frac{1}{3}$; $\frac{2}{3}\%$; $.3$

6. Select the quantities which are the same: $37\frac{1}{2}\%$; 37.5; $.375$; $\frac{3}{4}$

7. If a farmer sold 95% of his corn crop, what decimal fraction of his crop did he sell?

8. If 80% of the boys in a school can swim, what decimal fraction of the boys can swim?

9. A bank pays $3\frac{1}{4}\%$ interest. Write this rate of interest as a decimal.

10. A baseball player made hits 22.5% of the times he was at bat. Express this per cent, which is his batting average, as a decimal.

3. CHANGING DECIMALS TO PER CENTS

To change a decimal to a per cent, we must find the number of hundredths to which the decimal is equal.

$$.17 = \tfrac{17}{100} = 17\%$$

$$.05 = \tfrac{5}{100} = 5\%$$

$$.025 = \frac{25}{1000} = \frac{25 \div 10}{1000 \div 10} = \frac{2.5}{100} = 2.5\%$$

$$.5 = \frac{5}{10} = \frac{5 \times 10}{10 \times 10} = \frac{50}{100} = 50\%$$

Procedure. To change a decimal to a per cent, move the decimal point two places to the right and write the per cent sign, %.

MODEL PROBLEMS

Write each decimal as a per cent:

1. $.35 = 35\%$ **2.** $.08 = 8\%$ **3.** $.0625 = 6.25\%$
4. $.02\frac{1}{2} = 2\frac{1}{2}\%$ **5.** $.002 = .2\%$ **6.** $.00\frac{1}{3} = \frac{1}{3}\%$

NOTE

Per cent means hundredths.

EXERCISES

Write as per cents:

	(a)	(b)	(c)	(d)	(e)	(f)
1.	.38	.40	.09	.06	.1	.9
2.	.375	.0325	.008	$.85\frac{1}{3}$	$.01\frac{3}{4}$	$.00\frac{1}{2}$

3. A fruit dealer found that .05 of his peaches were spoiled. What per cent of his peaches were spoiled?

4. A family spent .225 of its income for rent. What per cent of its income was used for rent?

5. The leading team in the National League won .614 of the games it played. What per cent of the games played did it win?

6. A baseball player's batting average is .250. What per cent of his official times at bat did he get a hit?

7. If .875 of the weight of a melon is water, what per cent of the weight of the melon is water?

4. CHANGING PER CENTS TO COMMON FRACTIONS

Maria read that 75% of an orange crop had been destroyed by a snowstorm. To find the fractional part of the crop that had been destroyed, Maria changed 75% to the common fraction $\frac{75}{100}$, which reduces to $\frac{3}{4}$.

Therefore, Maria realized that $\frac{3}{4}$ of the crop had been destroyed.

Procedure. To change a given per cent to a common fraction, write a fraction whose numerator is the given number and whose denominator is 100. Reduce the fraction to lowest terms.

MODEL PROBLEMS

1. Change 40% to a common fraction.

$$40\% = \tfrac{40}{100} = \tfrac{2}{5}$$

2. Change $8\frac{1}{3}\%$ to a common fraction.

$$8\tfrac{1}{3}\% = \frac{8\frac{1}{3}}{100} = 8\tfrac{1}{3} \div 100 = \frac{25}{3} \div \frac{100}{1} = \frac{\overset{1}{\cancel{25}}}{3} \times \frac{1}{\underset{4}{\cancel{100}}} = \frac{1}{12}$$

3. Change 2.5% to a common fraction.

$$2.5\% = \frac{2.5}{100} = \frac{2.5 \times 10}{100 \times 10} = \frac{25}{1000} = \frac{5}{200} = \frac{1}{40}$$

EXERCISES

Change to common fractions:

	(a)	(b)	(c)	(d)	(e)	(f)
1.	10%	30%	60%	80%	65%	95%
2.	24%	36%	2%	4%	6%	9%
3.	$33\frac{1}{3}\%$	$12\frac{1}{2}\%$	$62\frac{1}{2}\%$	$87\frac{1}{2}\%$	$6\frac{1}{4}\%$	$83\frac{1}{3}\%$
4.	7.5%	18.5%	23.6%	5.4%	3.2%	1.25%

5. To which of the following is $33\frac{1}{3}\%$ equal: $.033\frac{1}{3}$; $1.3\frac{1}{3}$; $\frac{1}{3}$?

6. To which of the following is 60% equal: $\frac{6}{100}$; .06; $\frac{3}{5}$?

7. If 30% of a circle is to be colored, what fractional part is to be colored?

8. If in cutting a piece of material to make a dress 20% is wasted, what fractional part of the material is wasted?

9. If John lost 15% of his weight during an illness, what fractional part of his weight did he lose?

10. If a mixture of water and alcohol in a car radiator contains 40% alcohol, what fractional part of the mixture is alcohol?

11. During a dry spell, Mr. White lost $37\frac{1}{2}\%$ of his tomato plants. What fractional part of his plants did he lose?

12. During a meat shortage, a butcher received 62.5% of his order. What fractional part of his order did he receive?

13. In a sale, a store advertises "30% off." If you buy an article at the sale, what fractional part of the original price must you pay for the article?

14. Sol spent $66\frac{2}{3}\%$ of his allowance and saved the rest. What fractional part of his allowance did he save?

Summarized below are some per cents with their equivalent values written as common fractions. Try to memorize them because they are very frequently used.

TABLE OF PER CENTS AND
EQUIVALENT COMMON FRACTIONS

Per cent	*Equivalent Common Fraction*	*Per cent*	*Equivalent Common Fraction*	*Per cent*	*Equivalent Common Fraction*
10	$\frac{1}{10}$	80	$\frac{4}{5}$	75	$\frac{3}{4}$
20	$\frac{1}{5}$	90	$\frac{9}{10}$	$87\frac{1}{2}$	$\frac{7}{8}$
30	$\frac{3}{10}$	100	$\frac{1}{1}$ or 1	$16\frac{2}{3}$	$\frac{1}{6}$
40	$\frac{2}{5}$	$12\frac{1}{2}$	$\frac{1}{8}$	$33\frac{1}{3}$	$\frac{1}{3}$
50	$\frac{1}{2}$	25	$\frac{1}{4}$	$66\frac{2}{3}$	$\frac{2}{3}$
60	$\frac{3}{5}$	$37\frac{1}{2}$	$\frac{3}{8}$	$83\frac{1}{3}$	$\frac{5}{6}$
70	$\frac{7}{10}$	$62\frac{1}{2}$	$\frac{5}{8}$		

5. CHANGING COMMON FRACTIONS TO PER CENTS

If we wish to find the number of per cent to which $\frac{21}{25}$ is equal, we change $\frac{21}{25}$ to an equal fraction whose denominator is 100.

$$\frac{21}{25} = \frac{21 \times 4}{25 \times 4} = \frac{84}{100} = 84\%$$

Procedure 1. To change a common fraction to an equivalent per cent, change the common fraction to an equal fraction whose denominator is 100. The numerator of this fraction represents the number of per cent equivalent to the given fraction.

There is another way to find the number of per cent to which $\frac{21}{25}$ is equal. The fraction $\frac{21}{25}$ can mean $\frac{21}{25}$ of a whole thing. Remember that 1 whole is equal to 100%. Therefore,

$$\frac{21}{25} \text{ of a whole} = \frac{21}{25} \text{ of } 100\% = \frac{21}{\cancelto{1}{25}} \times \cancelto{4\%}{100\%} = 84\%$$

Procedure 2. To change a common fraction to an equivalent per cent, find that fraction of 100%, that is, multiply 100% by the fraction.

$\frac{21}{25}$ means $21 \div 25$

$$\begin{array}{r} .84 \\ 25\overline{)21.00} \\ \underline{20\ 0} \\ 1\ 00 \\ \underline{1\ 00} \end{array}$$

$\frac{21}{25} = .84 = 84\%$

There is still a third method of changing $\frac{21}{25}$ to a per cent. We can divide 21 by 25 and express the quotient as a two-place decimal, hundredths. The number of hundredths in the quotient tells us the number of per cent to which $\frac{21}{25}$ is equal.

Procedure 3. To change a common fraction to an equivalent per cent, divide the numerator by the denominator and find the quotient as a two-place decimal, hundredths. Write this decimal as a per cent.

MODEL PROBLEMS

1. Express $\frac{1}{4}$ as a per cent.

Method 1	*Method* 2	*Method* 3
$\frac{1}{4} = \frac{1 \times 25}{4 \times 25} = \frac{25}{100} = 25\%$	$\frac{1}{4} = \frac{1}{\cancel{4}_1} \times \cancel{100\%}^{25\%} = 25\%$	$\frac{1}{4}$ means $1 \div 4$ $\begin{array}{r} .25 \\ 4\overline{)1.00} \end{array}$ $\frac{1}{4} = .25 = 25\%$

2. Change $\frac{18}{27}$ to a per cent. $\frac{18}{27} = \frac{2}{3} = \frac{2}{3} \times 100\% = \frac{200}{3}\% = 66\frac{2}{3}\%$

3. Mr. Jackson used $\frac{4}{15}$ of his monthly salary for rent. Find what per cent of his salary was used for rent. Give the answer correct to the nearest whole per cent.

$\frac{4}{15} = 4 \div 15$

$$\begin{array}{r} .266 \\ 15\overline{)4.000} \\ \underline{3\ 0} \\ 1\ 00 \\ \underline{90} \\ 100 \\ \underline{90} \\ 10 \end{array}$$

$\frac{4}{15} = .27 = 27\%$

We find the quotient as a three-place decimal in order to round off to hundredths.

When we round off .266 to hundredths, we get .27.

EXERCISES

Change to equivalent per cents:

	(a)	(b)	(c)	(d)	(e)	(f)
1.	$\frac{1}{2}$	$\frac{3}{4}$	$\frac{2}{5}$	$\frac{3}{10}$	$\frac{12}{25}$	$\frac{7}{50}$
2.	$\frac{1}{3}$	$\frac{1}{6}$	$\frac{3}{8}$	$\frac{5}{8}$	$\frac{7}{8}$	$\frac{2}{3}$
3.	$\frac{14}{28}$	$\frac{12}{15}$	$\frac{3}{9}$	$\frac{35}{56}$	$\frac{21}{35}$	$\frac{25}{30}$

4. Change to the nearest whole per cent:

a. $\frac{1}{9}$ *b.* $\frac{7}{9}$ *c.* $\frac{1}{16}$ *d.* $\frac{3}{16}$ *e.* $\frac{1}{12}$ *f.* $\frac{7}{12}$ *g.* $\frac{5}{32}$ *h.* $\frac{13}{36}$

5. Mr. Rawlings has $\frac{3}{50}$ of his salary deducted for his pension. What per cent of his salary goes toward his pension?

6. If the school baseball team has won $\frac{7}{8}$ of its games, what per cent of its games has it won?

7. Mr. Reilly used $\frac{5}{6}$ of his coal during the winter. What per cent of his coal did Mr. Reilly use?

8. By buying his overcoat for cash, Harry received a $\frac{1}{20}$ reduction. What was the per cent of his reduction?

9. If $\frac{3}{35}$ of a class was absent, what per cent of the class was absent? Give the answer to the nearest whole per cent.

10. Mr. Wellington bought a house and made a down payment of $\frac{1}{7}$ of the price. What per cent of the price was the down payment? Give the answer to the nearest whole per cent.

11. What per cent of the entire figure is shaded?

(a)

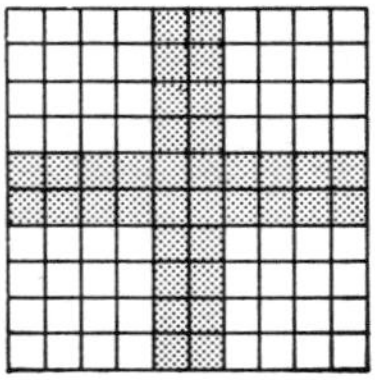

(b)

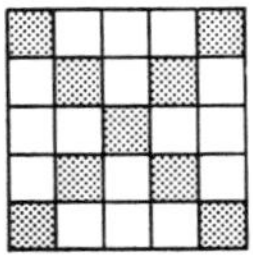

(c)

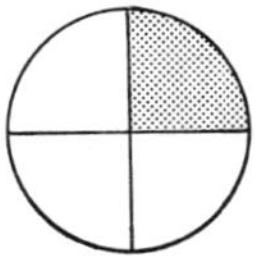

(d)

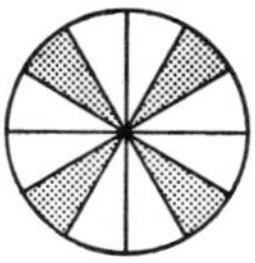

(e)

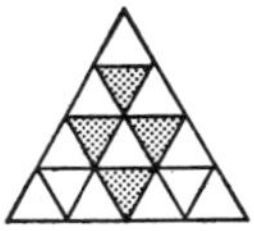

6. FINDING A PER CENT OF A NUMBER

If we wish to find 25% of \$36, let us remember that 25% may be written as a decimal, .25. Now, to find .25 of \$36, we multiply: .25 × \$36.

$$\begin{array}{r} \$36 \\ \times .25 \\ \hline 1\ 80 \\ 7\ 2 \\ \hline \$9.00 \end{array}$$

We can shorten our work if we know the common fraction to which the given per cent is equivalent. Let us find $87\frac{1}{2}\%$ of 48. Since $87\frac{1}{2}\%$ is equivalent to the fraction $\frac{7}{8}$, we can say:

$$\frac{7}{8} \text{ of } 48 = \frac{7}{\cancel{8}_1} \times \overset{6}{\cancel{48}} = 42$$

Procedure. To find a per cent of a number:

Change the per cent to a decimal and multiply the number by the decimal.

or

Change the per cent to an equivalent common fraction and multiply the number by the fraction.

MODEL PROBLEMS

1. Find 18% of 39.

18% = .18

$$\begin{array}{r} 39 \\ \times .18 \\ \hline 3\ 12 \\ 3\ 9 \\ \hline 7.02 \end{array}$$

2. Find $33\frac{1}{3}\%$ of \$24.

$33\frac{1}{3}\% = \frac{1}{3}$

$$\frac{1}{\cancel{3}_1} \times \overset{\$8}{\cancel{\$24}} = \$8$$

3. Find $\frac{1}{2}\%$ of \$10,000.

$\frac{1}{2}\% = .00\frac{1}{2}$ or .005

$$\begin{array}{r} \$10{,}000 \\ \times .00\frac{1}{2} \\ \hline \$50.00 \end{array} \quad \textit{or} \quad \begin{array}{r} \$10{,}000 \\ \times .005 \\ \hline \$50.000 \end{array} \text{ or } \$50.00$$

4. Find 2% of \$8.98 to the nearest cent.

2% = .02

$$\begin{array}{r} \$8.98 \\ \times .02 \\ \hline \$.1796 \end{array} \text{ or } \$.18$$

EXERCISES

Find:

	(a)	(b)	(c)	(d)
1.	2% of 48	6% of 120	27% of 92	79% of 453
2.	2.9% of 60	13.6% of 123	$4\frac{1}{2}$% of 180	$3\frac{1}{3}$% of 360
3.	20% of 25	80% of 55	$33\frac{1}{3}$% of 36	$12\frac{1}{2}$% of 32
4.	75% of 744	$66\frac{2}{3}$% of 24	$8\frac{1}{3}$% of 60	$62\frac{1}{2}$% of 64
5.	25% of 240	90% of 45	75% of 250	30% of 75
6.	$33\frac{1}{3}$% of $99	$33\frac{1}{3}$% of $8.10	$33\frac{1}{3}$% of $10.08	$12\frac{1}{2}$% of $48
7.	$12\frac{1}{2}$% of $9.04	$8\frac{1}{3}$% of $60	$8\frac{1}{3}$% of $13.20	$37\frac{1}{2}$% of $560
8.	$66\frac{2}{3}$% of $402	$87\frac{1}{2}$% of $800	$83\frac{1}{3}$% of $600	$62\frac{1}{2}$% of $400
9.	$\frac{1}{2}$% of $4000	$\frac{1}{4}$% of $1200	$\frac{1}{5}$% of $15,000	$\frac{1}{10}$% of $1000

10. Find to the nearest cent: (If the answer has a fraction of a cent, drop the fraction and add another cent.)

a. $33\frac{1}{3}$% of $95.70 *b.* $87\frac{1}{2}$% of $16.96 *c.* $66\frac{2}{3}$% of $17
d. $37\frac{1}{2}$% of $49 *e.* 75% of $18.50 *f.* 3% of $127.13

11. There were 80 planes on an airfield. If 75% of the planes took off for a flight, how many planes took off?
12. A dealer bought $400 worth of fruit. If 15% of the fruit spoiled before he could sell it, what was the amount of his loss?
13. If Sue buys a $25 dress marked "10% off," how much does she save?
14. Mary's standing on a spelling test was 88%. If there were 50 words in the test, how many words did Mary misspell?
15. The price of a new car is $3450. Mr. Saltz made a down payment of 15% when he bought the car. How much did he pay at that time?
16. Inez earns $47.20 per week. If she receives an increase of 12%, find, correct to the nearest cent, the amount of her increase.
17. A team that played 152 games during a season won $62\frac{1}{2}$% of its games. How many games did it win?
18. Mr. Hobson left $36,000 to his family when he died. If he left $66\frac{2}{3}$% to his wife, how much did his wife receive?
19. Mr. Rogers, who earns $3600 a year, spends $8\frac{1}{3}$% of his salary for clothes. How much does he spend for clothes?
20. Mr. Sawyer's salary is $8400 per year. He pays 21% of his salary for rent. (*a*) What is his rent for the year? (*b*) What is his rent each month?

7. FINDING WHAT PER CENT ONE NUMBER IS OF ANOTHER

In a class of 40 pupils, 36 pupils contributed to the Red Cross drive. Bruce, the class president, wished to report to the class the per cent of pupils who gave to the Red Cross. He figured that since 36 out of 40 pupils gave to the Red Cross, then $\frac{36}{40}$ or $\frac{9}{10}$ of the class gave. $\frac{9}{10} = \frac{90}{100} = .90 = 90\%$.

Notice that 36, the number that represented the part of the class, became the numerator of the fraction $\frac{36}{40}$; and 40, the number that represented the whole class, became the denominator of the fraction $\frac{36}{40}$.

Procedure. To find what per cent of a number that represents a whole (to be called the first number) another number (to be called the second number) is, write a fraction whose numerator is the second number and whose denominator is the first number. Then change this fraction to an equivalent per cent.

MODEL PROBLEMS

1. 18 is what per cent of 24?

$$\frac{18}{24} = \frac{3}{4} = .75 = 75\%$$

The numerator is 18, the number that is being compared with the whole, 24. The denominator is 24, the number that represents the whole.

2. What per cent of 80 is 50?

$$\frac{50}{80} = \frac{5}{8} = .62\tfrac{1}{2} = 62\tfrac{1}{2}\%$$

The numerator is 50, the number that is being compared with the whole, 80. The denominator is 80, the number that represents the whole.

3. 4 is ?% of 4?
$\frac{4}{4} = 1 = 100\%$

4. Marjorie bought a soiled dress for \$2.40 that formerly sold for \$4.80. What per cent of the original price did she pay?
Solution: We must find what per cent of \$4.80 is the payment of \$2.40.

$$\frac{\$2.40}{\$4.80} = \frac{\$2.40 \times 100}{\$4.80 \times 100} = \frac{\$240}{\$480} = \frac{1}{2} = .50 = 50\% \textit{ Ans.}$$

EXERCISES

1. *a.* 6 is what per cent of 12? *b.* 16 is what per cent of 40?
c. What per cent of 20 is 14? *d.* What per cent of 35 is 28?

2. *a.* 3 is what per cent of 9? *b.* 3 is ?% of 18?
c. What per cent of 15 is 10? *d.* What per cent of 80 is 30?

3. *a.* 10 is what per cent of 10? *b.* What per cent of 50 is 50?

4. *a.* What per cent of \$6.60 is \$3.30? *b.* What per cent of \$15 is \$.75?
c. \$1.20 is what per cent of \$3.60? *d.* \$2.50 is ?% of \$3?

5. Find, correct to the nearest per cent:
a. What per cent of 14 is 2? *b.* What per cent of 45 is 25?
c. 9 is what per cent of 33? *d.* What per cent of 125 is 87?
e. ?% of 32 is 17? *f.* \$7 is ?% of \$39?

6. Henry did 8 problems for homework. If 2 were wrong, what per cent of the problems were wrong?

7. Mary receives \$1.00 each week as an allowance. If she saves 35 cents of it, what per cent does she save?

8. If, on the average, one potato out of every 50 in a bin spoils, what per cent of the potatoes spoil?

9. In a group of 21 girls, there were 14 Girl Scouts. What per cent of the 21 were Scouts?

10. A class raised \$54 of the school's \$200 contribution to the Junior Red Cross. What per cent of the contribution was raised by this class?

11. Marvin has read 50 pages of a book that has 250 pages. What per cent of the book has he read?

12. Mr. Roll has traveled 250 miles of a 1000-mile trip. What per cent of the total distance has he already traveled?

13. Our school team has won 9 games of the 15 it has played. What per cent of its games has our team won?

14. Ronald has to put up 18 storm windows. He has already put up 12 of them. What per cent of the total number of windows does he still have to put up?

15. What per cent of a pound of butter is 12 ounces?

16. In the first half of the school year, there was a total of 90 school days. If Joe was absent 9 days, on what per cent of the school days was he present?

17. A girl earned \$24 a week. She received an increase of \$3.60. What per cent of her original salary was the increase?

18. If a manufacturer pays an excise tax of \$2 on a \$20 article, what per cent of the value of the article is the excise tax?

19. A man bought a diamond ring whose value was \$400. In addition, he had to pay an \$80 tax. What per cent of the value of the ring was the tax?

20. Out of 350 men who were examined, 137 were found to have poor eyesight. Find, correct to the nearest whole per cent, the per cent of the men who had poor eyesight.

21. During a baseball season, the New York Yankees won 97 games and lost 57 games. What per cent of the games that they played did the Yankees win? Give the answer correct to the nearest whole per cent.

8. FINDING A NUMBER WHEN A PER CENT OF IT IS KNOWN

By purchasing a fur coat at a clearance sale, Mrs. Kay saved \$150, which represented a saving of 30% of the original price. Let us find the original price of the coat.

30% of the original price is \$150
1% of the original price is $\frac{1}{30}$ of \$150, or \$5
100% or the whole original price is 100 × \$5, or \$500

There is a second method of finding the original price of the coat. \$150 represents 30% or $\frac{3}{10}$ of the original cost of the coat. To find the original price, we have to find the whole when we know that $\frac{3}{10}$ of it is \$150. Therefore, we divide \$150 by $\frac{3}{10}$.

$$\$150 \div \frac{3}{10} = \$150 \times \frac{10}{3} = \frac{\$1500}{3} = \$500$$

Procedure. To find a number when a per cent of it is known:
Find what number 1% is equal to and then multiply that number by 100.

or

Divide the known part by the fractional equivalent of the per cent.

MODEL PROBLEMS

1. 25% of what number is 200?

Method 1	*Method* 2
25% of the number = 200 1% of the number = $\frac{1}{25}$ of 200 = 8 100% of the number = 100 × 8 = 800 *Ans.*	25% = .25 = $\frac{25}{100}$ = $\frac{1}{4}$ 200 ÷ $\frac{1}{4}$ = $\frac{200}{1} \times \frac{4}{1}$ = 800 *Ans.*

2. In the first week of a Red Cross drive, a class collected \$14.49. If this amount represented $87\frac{1}{2}\%$ of the total amount the class collected, how much did the class collect?

Solution: We have to find a number $87\frac{1}{2}\%$ of which is \$14.49.

$$87\tfrac{1}{2}\% = \frac{7}{8} \qquad \$14.49 \div \frac{7}{8} = \frac{\overset{\$2.07}{\cancel{\$14.49}}}{1} \times \frac{8}{\underset{1}{\cancel{7}}} = \$16.56 \text{ } Ans.$$

EXERCISES

1. 20 is 10% of what number?
2. If 12 is 6% of a number, what is the number?
3. If 24 is 8% of a number, find the number.
4. If 50 is 25% of a number, find the number.
5. *a.* 8 is 50% of what number? *b.* 64 is 80% of what number?
   c. 4 is 20% of what number? *d.* 6% of what number is 18?
6. *a.* 2 is $12\frac{1}{2}$% of what number? *b.* 24 is $66\frac{2}{3}$% of what number?
   c. 33 is $37\frac{1}{2}$% of what number? *d.* $33\frac{1}{3}$% of what number is 36?
7. A boy spent $2, which was 25% of his money. How much did he have before he spent the two dollars?
8. A baseball team lost 2 games, which was 20% of the total number of games it played. How many games did the team play?
9. In a school, there are 60 pupils who participate in a certain activity. If this is 12% of the student body, how many pupils are there in the school?
10. Julia saved $18 when she bought a suit at a sale. If this represented a $33\frac{1}{3}$% reduction, what was the original cost of the suit?
11. Harry earned $36.40 during one week selling magazines. If this represented 40% of his total sales, what were his total sales for the week?
12. Perry saved $1.25, which was $62\frac{1}{2}$% of his allowance for the week. What was his weekly allowance?

9. FINDING THE AMOUNT OF INCREASE OR DECREASE

Mr. London bought a record player for $64. After using it for several months, he sold it for $37\frac{1}{2}$% less than the amount he paid for it. How much did Mr. London receive for his record player?

Solution: Step 1. We find the amount he lost by finding $37\frac{1}{2}$% of $64.
Step 2. We subtract this amount from the amount he paid to find the amount he received when he sold the record player.

Step 1

$$37\tfrac{1}{2}\% \text{ of } \$64 = \frac{3}{\cancel{8}_{1}} \times \cancel{\$64}^{\$8} = \$24$$

Step 2

$64	(amount he paid)
− 24	(amount he lost)
$40	(amount he received) *Ans.*

EXERCISES

1. A dealer bought a painting for \$120. He sold it for 40% more than he paid for it. How much did he gain when he sold the painting?
2. Marilyn bought a camera for \$15 and later sold it for $33\frac{1}{3}$% less than she paid for it. How much less than the amount paid for the camera did she receive when she sold it?
3. Mr. Brown's weekly salary is \$76.50. If he receives a 6% increase, what will his new salary be?
4. Coal was selling for \$22.50 per ton. If the price of coal dropped 8%, what is the new price of coal?
5. The population of a city of 720,000 increased $37\frac{1}{2}$% during a five-year period. What was the population at the end of the five-year period?
6. Last year a company sold \$40,000,000 worth of its product. This year it had an increase of 15% over last year's sales. What is the amount of its sales this year?
7. In a city, there were 500,000 people employed last year. If employment decreased 8% during the year, how many people are now employed?

10. FINDING THE PER CENT OF INCREASE OR DECREASE

The number of members in a club increased from 20 to 24. What was the per cent of increase in the membership?

Solution: Step 1. We find the increase in membership. *Step* 2. We find what fractional part of the original membership the increase represents. *Step* 3. We change this fraction to a per cent.

Step 1	*Step* 2	*Step* 3
24 (new membership) − 20 (old membership) 4 (increase)	$\frac{\text{increase}}{\text{original number}} = \frac{4}{20} = \frac{1}{5}$	$\frac{1}{5} = 20\%$

Answer: The increase in the membership was 20%.

Procedure. **To find the per cent of increase (or decrease), write a fraction whose numerator is the amount of the increase (or decrease) and whose denominator is the original amount. Then change this fraction to an equivalent per cent.**

EXERCISES

1. Harry had 60 marbles and won 15 from Fred. This represented an increase of how many per cent of the original number of marbles he had?
2. The price of milk rose from 20 cents a quart to 22 cents a quart. What was the per cent of increase in price?
3. The price of tomatoes dropped from 40 cents a pound to 28 cents a pound. What was the per cent of decrease in price?
4. Before an illness, a boy weighed 110 lb. On the day he left the hospital, he weighed 99 lb. What per cent of his former weight did he lose?
5. Mr. Brown's income was \$42 per week. He received an increase of \$6.30. What is the per cent of increase?
6. During a sale, a dress was reduced from \$14 to \$12. What is the per cent of the reduction to the nearest whole per cent?
7. Sid bought a car for \$3650. Now his car is worth \$2800. Find, to the nearest whole per cent, the per cent of decrease in the value of his car.
8. The population of a village has increased from 1500 to 1900. Find the per cent of increase in the population correct to the nearest whole per cent.

11. WORKING WITH PER CENTS LARGER THAN 100%

Dr. Ward announced in the assembly that the number of pupils in the school now is 200% of the number of pupils attending when the school was opened. Let us see what 200% of a number means.

We have learned that the whole, or all, of any number is represented by 100%. Therefore, 200% (2 times 100%) of a number is two times, or twice, that number.

Now we can understand that Dr. Ward meant there are at present twice as many pupils in the school as there were when the school opened.

Whenever we have to work with per cents larger than 100%, we use the same procedures as those we learned for per cents less than 100%.

MODEL PROBLEMS

1. Write $\frac{150}{100}$ as a per cent. *Ans.* 150%
2. Write 175% as a decimal. *Ans.* 1.75
3. Write 3 as a per cent. *Ans.* 300%
4. Write 4.75 as a per cent. *Ans.* 475%
5. Change 150% to a common fraction.

   $150\% = \frac{150}{100} = \frac{3}{2}$

6. Change $\frac{7}{2}$ to a per cent.

   $$\frac{7}{2} = \frac{7}{2} \times 100\% = \frac{700\%}{2} = 350\%$$

7. Find 120% of \$30.60.

$120\% = \frac{6}{5}$

$\frac{6}{\not 5_1} \times \not{\$30.60}^{\$6.12} = \$36.72$

8. What per cent of 16 is 20?

$\frac{20}{16} = \frac{5}{4} = 1.25 = 125\%$

EXERCISES

1. Write as a decimal or a whole number:
a. 125% *b.* 150% *c.* 275% *d.* 400% *e.* 500%

2. Write as a per cent: (*a*) 1.75 (*b*) 2.25 (*c*) 3.2 (*d*) 2 (*e*) 10

3. Change to common fractions:
a. 120% *b.* 175% *c.* 225% *d.* $112\frac{1}{2}\%$ *e.* $166\frac{2}{3}\%$

4. Change to equivalent per cents:
a. $1\frac{1}{4}$ *b.* $1\frac{2}{3}$ *c.* 7 *d.* 6 *e.* $\frac{15}{12}$ *f.* $\frac{36}{27}$ *g.* $\frac{50}{25}$

5. Find:
a. 150% of 18 *b.* 175% of 400 *c.* $133\frac{1}{3}\%$ of 36 *d.* 103% of 17

6. *a.* 15 is what per cent of 10? *b.* What per cent of 16 is 18?

7. Find, correct to the nearest whole per cent:
a. What per cent of 27 is 35? *b.* What per cent of 84 is 111?

8. *a.* 125% of what number is 20? *b.* $112\frac{1}{2}\%$ of what sum is \$72?

9. How many times a number is each of the following per cents of the number? (*a*) 200% (*b*) 500% (*c*) 100% (*d*) 150% (*e*) 225%

10. If the population of a town has doubled, then the new population is what per cent of the original population?

11. A merchant now does $1\frac{1}{2}$ times as much business as he did a year ago. What per cent of the business he did a year ago is he now doing?

12. During a period of 10 years, the population of a town increased 200%. The population now is how many times the population 10 years ago?

13. Mr. James got a 10% increase in salary. What per cent of his original salary is he now earning?

14. A company spent \$50,000 to enlarge its business last year. This year it expects to spend 150% of that sum for the same reason. How much will be spent this year to enlarge the business?

15. A man bought a diamond ring for \$150. Many years later, he sold it for \$375. What was the per cent of increase in the value of the ring?

16. Stanley sold an article for 120% of the amount he paid for it. If he sold it for \$30, how much did the article cost him?

Chapter V

ARITHMETIC OF MEASURES

1. UNDERSTANDING THE MEANING OF MEASUREMENT

To measure a distance, we compare it with a *standard unit* of length whose size we know. If the unit of length, a *yard*, is contained in a distance 8 times, we say that the length of the distance is 8 yards. To measure a period of time, we compare it with a standard unit of time whose duration we know. If a period of time is 10 times as long as the unit, a *minute*, we say that the period of time measures 10 minutes.

To measure any type of thing, we compare it with a unit of the same type of thing whose standard size is already known to all.

EXERCISES

1. Why is the yard a standard unit of measure?

2. Why is a boxful not a standard unit of measure?

3. State whether or not each of the following is a standard unit of measure:

a. mile	*b.* second	*c.* hatful	*d.* quart
e. pace	*f.* hour	*g.* glassful	*h.* truckload
i. gallon	*j.* dozen	*k.* pocketful	*l.* yard

Complete each statement by giving the most frequently used standard unit of measure:

4. A man's weight is measured by the ______.

5. The distance between two towns is measured by the ______.

6. Eggs are sold by the ______.

7. Milk is sold by the ______.

8. Coal is measured by the ______.

9. The time required to run 100 yards is measured by the ______.

10. Temperature is measured by the ______.

11. Gasoline is sold by the ______.

12. The altitude of a plane is measured by the ______.

13. The length of a piece of dress material is measured by the ______.

2. USING MEASURES OF LENGTH

Measures of Length

12 inches (in.)	=	1 foot (ft.)
36 inches	=	1 yard (yd.)
3 feet	=	1 yard
$16\frac{1}{2}$ feet	=	1 rod (rd.)
$5\frac{1}{2}$ yards	=	1 rod
5280 feet	=	1 mile (mi.)
1760 yards	=	1 mile
320 rods	=	1 mile

Barnett needed 14 yards of molding for a room. Since molding is sold by the foot, Barnett had to change 14 yards to an equivalent number of feet. He knew that each yard contains 3 feet. Therefore, he multiplied 14 by 3 and found that he had to buy 42 feet of molding.

Procedure. To change a given measure made with a larger unit to an equivalent measure made with a smaller unit, first determine the number of times the smaller unit is contained in the larger unit. Then multiply the given measure by this number.

Sue needed 72 inches of ribbon to trim a dress. Since she knew that ribbon is sold by the yard, Sue had to change 72 inches to an equivalent number of yards. She knew that each yard has 36 inches. Therefore, she found how many 36's are contained in 72 by dividing 72 by 36. Sue discovered that she had to buy 2 yards of ribbon.

Procedure. To change a given measure made with a smaller unit to an equivalent measure made with a larger unit, first determine the number of times the smaller unit is contained in the larger unit. Then divide the given measure by this number.

MODEL PROBLEMS

1. How many inches are there in $4\frac{1}{3}$ yards? [1 yard = 36 inches]

$$\begin{array}{r} 36 \\ \times\ 4\frac{1}{3} \\ \hline 12 \\ 144 \\ \hline 156 \end{array}$$

Multiply the number of inches in a yard, 36, by the number of yards, $4\frac{1}{3}$.

Answer: 156 inches

2. What fractional part of a mile is 440 yards? [1 mile = 1760 yards]

$$\frac{\overset{1}{\cancel{440}}}{\underset{4}{\cancel{1760}}} = \frac{1}{4}$$

Divide the number of yards, 440, by the number of yards in a mile, 1760.

Answer: $\frac{1}{4}$ of a mile

EXERCISES

1. Express in miles the altitude of an airplane flying at 21,120 feet.

2. What part of a yard is 27 inches?

3. Find the number of feet in (*a*) .6 of a mile (*b*) $\frac{7}{8}$ of a mile

Supply the missing numbers:

	(*a*)	(*b*)	(*c*)
4.	4 ft. = _____ in.	$\frac{2}{3}$ ft. = _____ in.	3 yd. = _____ ft.
5.	$5\frac{2}{3}$ yd. = _____ ft.	$\frac{3}{4}$ yd. = _____ in.	8 rd. = _____ ft.
6.	6 mi. = _____ ft.	$\frac{1}{2}$ mi. = _____ yd.	$\frac{1}{8}$ mi. = _____ ft.
7.	24 in. = _____ ft.	18 in. = _____ ft.	30 ft. = _____ yd.
8.	8 ft. = _____ yd.	9 in. = _____ yd.	880 yd. = _____ mi.
9.	66 ft. = _____ rd.	165 ft. = _____ rd.	10,560 ft. = _____ mi.
10.	4 ft. 3 in. = _____ in.	7 yd. 2 ft. = _____ ft.	3 mi. 400 yd. = _____ yd.
11.	8 ft. 4 in. = _____ ft.	3 ft. 9 in. = _____ ft.	2 yd. 27 in. = _____ yd.

12. If 30 inches of leather will make one belt, how many feet of leather would be needed to make 12 belts?

13. A woman bought 12 yards of linen toweling. How many towels, each 27 inches long, can be cut from this length of material?

3. USING MEASURES OF TIME

Jimmy was told when he entered school that he would be late at 8:50 A.M. and would be dismissed at 3:10 P.M. Jimmy knew that A.M. describes time after 12 o'clock midnight and before 12 o'clock noon. P.M. describes time after 12 o'clock noon and before 12 o'clock midnight. He also knew that the colon (:) separates hours from minutes when clock time is written. Thus, he knew that he had to get to school before 50 minutes after 8 o'clock in the morning, and that he would be dismissed at 10 minutes after 3 o'clock in the afternoon.

Measures of Time

60 seconds (sec.)	= 1 minute (min.)
60 minutes	= 1 hour (hr.)
24 hours	= 1 day (da.)
7 days	= 1 week (wk.)
4 weeks (approx.)	= 1 month (mo.)
12 months	= 1 year (yr.)
365 days	= a common year
366 days	= a leap year
10 years	= a decade
100 years	= a century

MODEL PROBLEM

How much would Cary earn if he worked for 4 hours and 20 minutes at \$1.20 an hour?

Solution: Step 1. Find the number of hours in 4 hours and 20 minutes.
Step 2. Find how much Cary will earn in that number of hours.

Step 1

[1 hour = 60 minutes]

20 minutes = $\frac{20}{60}$ of an hour = $\frac{1}{3}$ of an hour

4 hours and 20 minutes = $4\frac{1}{3}$ hours

Step 2

$$\begin{array}{r} \$1.20 \\ \times\ 4\frac{1}{3} \\ \hline 40 \\ 4\ 80 \\ \hline \$5.20 \end{array}$$ *Ans.*

EXERCISES

1. Explain the meaning of (*a*) 10:30 A.M. (*b*) 12:45 P.M. (*c*) 5:15 P.M.
2. How many minutes are there in (*a*) 5 hours? (*b*) a day?
3. What part of a minute is 45 seconds?
4. How many years are there in (*a*) 5 decades? (*b*) 4 centuries?

Supply the missing numbers:

	(*a*)	(*b*)	(*c*)
5.	5 min. = _____ sec.	$8\frac{1}{3}$ min. = _____ sec.	$\frac{1}{5}$ hr. = _____ min.
6.	$4\frac{3}{4}$ hr. = _____ min.	8 wk. = _____ da.	26 wk. = _____ da.
7.	3 yr. = _____ mo.	$\frac{3}{4}$ yr. = _____ mo.	$1\frac{1}{2}$ yr. = _____ mo.
8.	120 sec. = _____ min.	15 sec. = _____ min.	105 min. = _____ hr.
9.	30 min. = _____ hr.	72 hr. = _____ da.	36 hr. = _____ da.
10.	70 da. = _____ wk.	48 mo. = _____ yr.	9 mo. = _____ yr.

	(*a*)	(*b*)
11.	3 min. 40 sec. = _____ sec.	4 hr. 15 min. = _____ min.
12.	4 wk. 3 da. = _____ da.	1 hr. 6 min. 5 sec. = _____ sec.
13.	2 yr. 8 mo. = _____ mo.	2 da. 8 hr. 15 min. = _____ min.

14. Sam ran the 100-yard dash in 12 seconds. In what fractional part of a minute did he run this race?
15. Fifty minutes is what fractional part of an hour?
16. Each week Charles spends 135 minutes in shopwork. How much time in hours does he spend in this work?

17. At \$1.50 an hour, how much will a boy earn in 5 days if he works 6 hours a day?
18. If a man pays a water bill of \$6.75 every three months, how much does he pay in one year?
19. At \$1.68 cents an hour, how much will a boy earn in 5 hours and 15 minutes?

4. USING MEASURES OF WEIGHT

When Helen goes shopping, there are many items of food that she buys by weight. The merchant uses a scale to measure weight.

Measures of Weight
(Avoirdupois)

16 ounces (oz.) = 1 pound (lb.)
100 pounds = 1 hundredweight (cwt.)
2000 pounds = 1 ton (T.)
2240 pounds = 1 long ton

MODEL PROBLEM

One day Helen bought 5 pounds of candy for a class party. How many bags could she fill if she planned to put 2 ounces in each bag?

Solution: Step 1. Find the number of ounces in 5 pounds. *Step* 2. Divide that number by 2 to discover how many times 2 is contained in that number.

Step 1	*Step* 2	
[1 pound = 16 ounces] $5 \times 16 = 80$	$80 \div 2 = 40$	*Ans.* 40 bags

EXERCISES

Supply the missing numbers:

	(*a*)	(*b*)	(*c*)
1.	3 lb. = _____ oz.	8 lb. = _____ oz.	12 lb. = _____ oz.
2.	$\frac{1}{2}$ lb. = _____ oz.	$\frac{3}{4}$ lb. = _____ oz.	$\frac{7}{8}$ lb. = _____ oz.
3.	$2\frac{1}{4}$ lb. = _____ oz.	$5\frac{1}{8}$ lb. = _____ oz.	$4\frac{5}{8}$ lb. = _____ oz.
4.	4 T. = _____ lb.	10 T. = _____ cwt.	20,000 T. = _____ lb.
5.	$\frac{3}{4}$ T. = _____ lb.	$2\frac{1}{2}$ T. = _____ cwt.	$3\frac{1}{8}$ T. = _____ lb.
6.	32 oz. = _____ lb.	4 oz. = _____ lb.	10 oz. = _____ lb.
7.	20 oz. = _____ lb.	36 oz. = _____ lb.	76 oz. = _____ lb.

	(a)	(b)
8.	2 lb. 3 oz. = _____ oz.	6 lb. 8 oz. = _____ oz.
9.	6 T. 400 lb. = _____ lb.	10 T. 500 lb. = _____ lb.
10.	38 oz. = _____ lb. _____ oz.	84 oz. = _____ lb. _____ oz.
11.	9500 lb. = _____ T. _____ lb.	35,750 lb. = _____ T. _____ lb.

12. A truck has a capacity of 10,000 pounds.
   a. How many tons will it hold?
   b. How many hundredweight will it hold?

13. A ship transported 201,600 pounds of coal. How many long tons is that?
14. What fractional part of a pound is (*a*) 12 ounces? (*b*) 10 ounces?
15. How many 4-ounce bags can be filled from 12 pounds of peanuts?
16. How many bags of coal, each weighing 50 pounds, can be filled from one ton of coal?
17. What will be the cost of 3500 pounds of coal at $20 a ton?
18. What will 12 ounces of tea cost at $1.20 a pound?
19. A turkey weighed 16 lb. 12 oz. How much did it cost at $.56 per pound?
20. What is the cost of 6 lb. 14 oz. of pears at $.19 per pound?
21. How much change should you receive from five dollars after paying for 1 pound 4 ounces of sirloin steak at $1.28 a pound?

5. USING LIQUID MEASURES

Liquid Measures

4 fluid ounces (fl. oz.)	= 1 gill (gi.)
16 fluid ounces	= 1 pint (pt.)
2 measuring cups	= 1 pint
2 pints	= 1 quart (qt.)
4 quarts	= 1 gallon (gal.)

Liquids are often measured by the amount of space they occupy. Cream is sold by the pint, milk is sold by the quart, gasoline is sold by the gallon.

MODEL PROBLEM

If a pint of milk fills 2 glasses, how many quarts will Harriet need to serve a glass of milk to each of 12 children?

Solution: Step 1. Find the number of pints needed to supply the children each with 1 glass or $\frac{1}{2}$ of a pint of milk. *Step* 2. Change this number of pints to quarts.

Step 1

$12 \times \frac{1}{2} = 6$

Step 2

[2 pints = 1 quart]

$6 \div 2 = 3$

Ans. 3 quarts

EXERCISES

Supply the missing numbers:

	(a)	(b)	(c)
1.	4 pt. = ____ cups	$1\frac{1}{2}$ pt. = ____ cups	4.5 pt. = ____ cups
2.	5 qt. = ____ pt.	12 qt. = ____ gi.	$6\frac{1}{2}$ qt. = ____ pt.
3.	4 gal. = ____ qt.	$1\frac{3}{4}$ gal. = ____ pt.	$1\frac{1}{2}$ gal. = ____ gi.
4.	12 cups = ____ pt.	10 pt. = ____ qt.	40 qt. = ____ gal.
5.	7 qt. = ____ gal.	20 pt. = ____ gal.	18 qt. = ____ gal.
6.	2 qt. 1 pt. = ____ pt.	3 gal. 2 qt. = ____ qt.	4 gal. 1 qt. = ____ qt.

7. How many quart bottles can be filled from four 10-gallon cans of milk?
8. How many quarts are there in $3\frac{1}{2}$ gallons of milk?
9. What per cent of a gallon is 3 quarts?
10. How many quarts of milk will be required by a school lunchroom to serve one pint of milk to each of 32 pupils?
11. How much would you receive for 1 gallon of orangeade if you sold it for 5 cents per $\frac{1}{2}$-pint glass?

6. USING DRY MEASURES

Dry Measures

2 pints (pt.) = 1 quart (qt.)
8 quarts = 1 peck (pk.)
4 pecks = 1 bushel (bu.)

Although solids are generally sold by weight, some solids are sold by the amount of space they occupy. For example, strawberries are sold by the quart; wheat is sold by the bushel.

EXERCISES

Supply the missing numbers:

	(a)	(b)	(c)
1.	3 qt. = ____ pt.	$\frac{3}{4}$ pk. = ____ qt.	$1\frac{1}{2}$ pk. = ____ qt.
2.	$\frac{1}{8}$ pk. = ____ pt.	$\frac{1}{4}$ bu. = ____ pk.	$2\frac{1}{2}$ bu. = ____ pt.
3.	16 pt. = ____ qt.	24 qt. = ____ pk.	2 qt. = ____ pk.
4.	9 qt. = ____ pk.	16 qt. = ____ bu.	3 pk. = ____ bu.

5. *a.* 4 bu. 2 pk. = ____ pk. *b.* 3 pk. 6 qt. = ____ qt.
6. How many pecks of potatoes are there in 4 bushels of potatoes?
7. At 48 cents a quart, what is the cost of $\frac{1}{2}$ peck of berries?
8. If a farmer has 108 bushels of potatoes, how many 100-pound bags will he be able to fill completely? [Use 1 bu. of potatoes = 60 lb.]
9. If a bushel of coal weighs 80 pounds, how many bushels of coal are there in a ton?
10. A grocer paid $2.40 for a bushel of potatoes and sold them for 80 cents a peck. How much did he gain?

7. USING MEASURES OF COUNTING

Some articles are commonly sold by the dozen, which is a group of 12 of the articles. Items sold by the dozen are eggs, oranges, etc. Other measures of counting appear in the table at the right.

Measures of Counting

12 units = 1 dozen (doz.)
12 dozen = 1 gross (gr.)
144 units = 1 gross
20 units = 1 score

MODEL PROBLEM

A farmer has 96 eggs. If he sells them at \$.60 per dozen, how much will he receive for them?

Solution: Step 1. Find the number of dozens in 96 eggs. *Step* 2. Multiply \$.60 by this number.

Step 1	*Step* 2
[12 units = 1 dozen]	$8 \times \$.60 = \$4.80$ *Ans.*
$96 \div 12 = 8$	

EXERCISES

Supply the missing numbers:

	(*a*)	(*b*)
1.	5 doz. = _____ articles	$\frac{1}{2}$ doz. = _____ articles
2.	$4\frac{1}{4}$ doz. = _____ articles	2 gr. = _____ articles
3.	$\frac{3}{4}$ gr. = _____ articles	$1\frac{1}{2}$ gr. = _____ doz.
4.	36 articles = _____ doz.	120 articles = _____ doz.
5.	9 articles = _____ doz.	42 articles = _____ doz.
6.	24 doz. = _____ gr.	36 articles = _____ gr.

7. Harry bought 4 shirts. What fractional part of a dozen did he buy?
8. How much is "four score and seven years" ago?
9. What is the cost of 6 eggs at \$.70 per dozen?
10. What is the cost of 3 shirts at \$48 per dozen?
11. On three successive days, Mary gathered 56 eggs, 48 eggs, and 40 eggs. How many dozens did she gather altogether?
12. Pencils cost \$8.64 per gross. At the same rate, what would the cost be per dozen?
13. At 3 for 14 cents, what will a dozen bars of candy cost?
14. If four oranges cost 29¢, what will one dozen cost?
15. What will 4 dozen ears of corn cost at the rate of 6 ears of corn for \$.39?

8. USING THE METRIC SYSTEM

In the United States and in Great Britain, the units commonly used to measure length and distance are the inch, foot, yard, and mile. These are units in the ***English system.*** In South America and in most of Europe, however, the units of length are the centimeter, meter, and kilometer. These are units in the ***metric system.***

The metric system is a system of measurement based entirely upon the number 10. It is called a decimal system.

In the metric system the standard unit of length is the *meter*.

1 yard = 36 inches
1 meter = 39.37 inches

Smaller units of length are obtained by using tenths, hundredths, and thousandths of a meter. Notice that:

deci means .1, or $\frac{1}{10}$
centi means .01, or $\frac{1}{100}$
milli means .001, or $\frac{1}{1000}$

Therefore, in a meter there are 10 decimeters, or 100 centimeters, or 1000 millimeters.

Metric System

Measures of Length

1 decimeter (dm.) = .1 meter (m.)
1 centimeter (cm.) = .01 meter
1 millimeter (mm.) = .001 meter

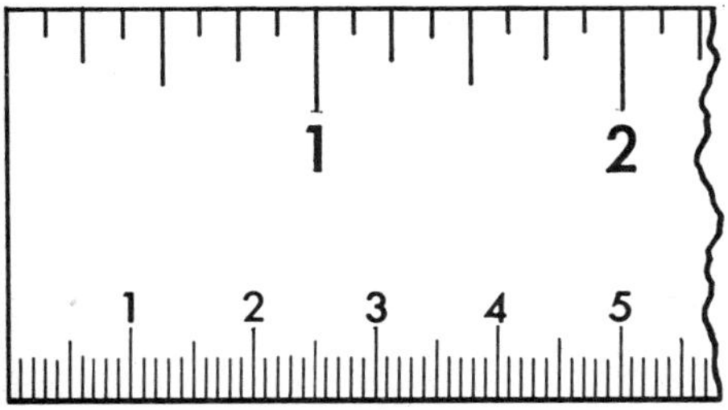

On the ruler at the left, the upper scale is in inches and the lower scale is in centimeters. Each larger unit on the lower scale is 1 centimeter. Each smaller unit is 1 millimeter. Notice that 1 centimeter is less than $\frac{1}{2}$ inch. Also, observe that 1 inch is about 2.5 centimeters, or 25 millimeters.

To help us change units of length from one system to the other, it would be helpful to remember:

1 meter = 39.37 inches	1 meter = 1.1 yards
1 centimeter = .4 inch	1 yard = .9 meter

Larger units of length are obtained by taking multiples of the meter in tens, hundreds, and thousands. The most frequently used large unit is a length of 1000 meters, called the *kilometer* (km.). *Kilo* means 1000.

In the metric system, distances between cities are measured in kilometers. A kilometer is approximately $\frac{5}{8}$ or .62 of a mile. A mile is approximately 1.61 kilometers.

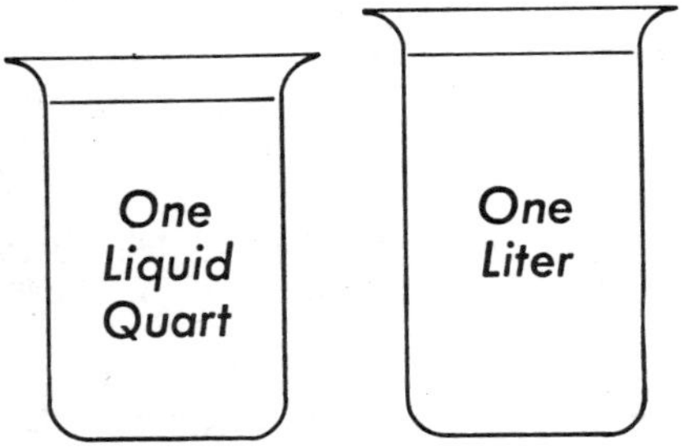

To measure liquids in our system, we use the pint, quart, and gallon. In the metric system, the *liter* is the standard unit of liquid measure. The liter is a little more than a liquid quart, or approximately 1.1 liquid quarts. The liquid quart is approximately .95 liter.

To measure weight in our system, we use the ounce and pound. In the metric system, the *gram* (g.) is used to measure the weight of light objects; the *kilogram* (kg.), which is 1000 grams, is used to measure the weight of heavier objects. The gram is a small weight of about .035 ounce. The kilogram is approximately 2.2 pounds. There are, correct to the nearest 10 grams, 450 grams in a pound and 30 grams in an ounce.

MODEL PROBLEMS

1. The distance between two cities in Europe is 450 kilometers. How many miles apart are these cities?

Solution: Since 1 kilometer = .62 of a mile, multiply .62 by 450.

$$\begin{array}{r} 450 \\ \times\ .62 \\ \hline 9\ 00 \\ 270\ 0 \\ \hline 279.00 \end{array} \quad \textit{Answer: } 279 \text{ miles}$$

2. If a package weighs 44 pounds, how many kilograms does it weigh?

Solution: Since 1 kilogram = 2.2 pounds, to find how many kilograms there are in 44 pounds, divide 44 by 2.2.

$$\begin{array}{r} 2\,0\,. \\ 2.2_{\wedge}\overline{)44.0_{\wedge}} \\ \underline{44}\phantom{.0_{\wedge}} \end{array} \quad \textit{Answer: } 20 \text{ kilograms}$$

EXERCISES

1. Which is longer, a meter stick or a yard stick?
2. Which is a longer race, the 100-yard dash or the 100-meter dash?
3. Which is a longer race, a mile run or a 1500-meter run?
4. How many millimeters are there in a centimeter?

Supply the missing numbers:

5. *a.* 4 cm. = _____ mm. *b.* 3.5 cm. = _____ mm.
c. 80 mm. = _____ cm. *d.* 45 mm. = _____ cm.

6. *a.* 2 m. = _____ cm. *b.* 7.5 m. = _____ cm.
c. 400 cm. = _____ m. *d.* 850 cm. = _____ m.

7. *a.* 8000 m. = _____ km. *b.* 1200 m. = _____ km.
c. 13 km. = _____ m. *d.* 7.6 km. = _____ m.

8. For each of the following sizes of motion picture film, find the width in centimeters and in inches: [Use 1 cm. = .4 in.]
a. 35 mm. wide *b.* 16 mm. wide *c.* 8 mm. wide

9. Find the number of yards in: [Use 1 m. = 1.1 yd.]
a. 100 m. *b.* 200 m. *c.* 400 m. *d.* 1500 m. *e.* 10,000 m.

10. Find the number of meters in: [Use 1 yd. = .9 m.]
a. 100 yd. *b.* 220 yd. *c.* $\frac{1}{2}$ mi. *d.* 1 mi. *e.* 2 mi.

11. Find the number of miles in: [Use 1 km. = .62 mi.]
a. 250 km. *b.* 376 km. *c.* 628 km. *d.* 340 km. *e.* 500 km.

12. Find the number of kilometers in: [Use 1 mi. = 1.6 km.]
a. 140 mi. *b.* 215 mi. *c.* 68 mi. *d.* 1500 mi. *e.* 1092 mi.

13. If the speed limit on a certain road is 80 km. per hour, to how many miles per hour is this equivalent?

14. If a plane flies at the rate of 450 km. per hour, how many miles per hour is the plane flying?

15. How many times as large as a liquid quart is a liter?

16. Are 4 liters more or less than a gallon?

17. A tank holds 32 quarts. How many liters does it contain? [Use 1 qt. = .95 liter.]

18. A gas tank holds 20 gallons of gasoline. How many liters does it hold? [Use 1 qt. = .95 liter.]

19. While traveling in Europe, Mr. Barton had 40 liters of gas put in his gas tank. How many gallons of gas did he buy? [Use 1 liter = 1.1 quarts.]

20. If your gas tank—capacity 16 gallons—is empty, what is the largest whole number of liters you would order to fill the tank? [Use 1 liter = 1.1 quarts.]

Supply the missing numbers using the round values given:
[Use 1 oz. = 30 g.; 1 lb. = 450 g.]

	(*a*)	(*b*)	(*c*)
21.	4 oz. = _____ g.	8 oz. = _____ g.	10 oz. = _____ g.
22.	2 lb. = _____ g.	$1\frac{1}{2}$ lb. = _____ g.	$3\frac{1}{2}$ lb. = _____ g.
23.	90 g. = _____ oz.	360 g. = _____ oz.	330 g. = _____ oz.
24.	1350 g. = _____ lb.	675 g. = _____ lb.	2025 g. = _____ lb.

25. Supply the missing numbers: [Use 1 kg. = 2.2 lb.]

a. 15 kg. = _____ lb. *b.* 84 kg. = _____ lb.
c. 24.5 kg. = _____ lb. *d.* 123.4 kg. = _____ lb.

26. Find, correct to the nearest tenth of a kilogram, each of the following weights: [Use 1 kg. = 2.2 lb.]

a. 66 lb. *b.* 110 lb. *c.* 15.4 lb. *d.* 91.7 lb. *e.* 158.5 lb.

27. If 120 grams of meat are used to make a hamburger, about how many ounces is this?

28. If a dealer bought 10,000 kilograms of butter, how many pounds of butter did he buy?

9. ADDING MEASURES

4 lb. 12 oz.
3 lb. 10 oz.
7 lb. 22 oz.
or 8 lb. 6 oz.

Mrs. Browning bought two chickens. One weighed 4 lb. 12 oz., and the other weighed 3 lb. 10 oz. See how she added these weights to find the total weight. She placed one measure underneath the other, being careful to place pounds in one column and ounces in another column. Then she added each column separately to get the answer, 7 lb. 22 oz. Since 22 oz. is more than a pound and there are 16 oz. in 1 lb., she changed 22 oz. to 1 lb. 6 oz. Therefore, she found that the total weight was 8 lb. 6 oz.

Procedure. To add measures, arrange them in vertical columns. Be careful to place like units of measure in the same column. Add each column separately. If possible, simplify the result by converting smaller units to larger ones.

MODEL PROBLEM

A train that was supposed to arrive at 8:30 A.M. was 1 hour and 45 minutes late. At what time did it arrive?

Solution: 8:30 A.M. means 8 hr. 30 min. after midnight. Add 1 hr. 45 min. to 8 hr. 30 min.

8 hr. 30 min.
1 hr. 45 min.
9 hr. 75 min. or 10 hr. 15 min. *Answer:* 10:15 A.M.

EXERCISES

Add:

	(a)	(b)	(c)
1.	3 ft. 2 in. 5 ft. 8 in.	2 ft. 10 in. 6 ft. 8 in.	10 yd. 2 ft. 9 yd. 1 ft.
2.	15 hr. 25 min. 20 hr. 10 min.	8 hr. 35 min. 2 hr. 45 min.	12 min. 10 sec. 14 min. 50 sec.
3.	4 yr. 3 mo. 7 yr. 8 mo.	14 wk. 5 da. 8 wk. 4 da.	20 wk. 1 da. 5 wk. 6 da.
4.	1 pk. 2 qt. 1 pk. 3 qt.	13 gal. 2 qt. 15 gal. 3 qt.	9 gal. 3 qt. 12 gal. 1 qt.
5.	5 lb. 4 oz. 8 lb. 2 oz. 3 lb. 8 oz.	18 lb. 6 oz. 6 lb. 12 oz. 21 lb. 5 oz.	2 lb. 8 oz. 14 lb. 5 oz. 10 lb. 3 oz.

6. *a.* 4 hr. 20 min. 50 sec.
3 hr. 25 min. 30 sec.

   b. 5 yd. 2 ft. 9 in.
6 yd. 2 ft. 8 in.

7. George and his father caught two fish that weighed 2 lb. 7 oz. and 4 lb. 15 oz. What was the total weight of the two fish?

8. Mrs. Carney bought three pieces of ribbon that measured 5 yd. 18 in., 4 yd. 24 in., and 10 yd. 15 in. What was the total length of the three pieces?

9. Mr. Smith's car is 16 ft. 9 in. long. If Mr. Barkley's car is 1 ft. 6 in. longer, what is the length of Mr. Barkley's car?

10. A friend, who was expected to arrive at 5:30, was 1 hour and 40 minutes late. At what time did he arrive?

11. A train scheduled to arrive in New York at 11:50 A.M. was 25 minutes late. At what time in the afternoon did it reach New York?

12. Mr. Weeks started a trip at 1:50 P.M. He drove for 3 hours and 35 minutes and spent 45 minutes getting some refreshments before arriving at his destination. At what time did he arrive there?

10. SUBTRACTING MEASURES

9 ft. 3 in.	=	8 ft. 15 in.
3 ft. 6 in.	=	3 ft. 6 in.
		5 ft. 9 in.

Helene had 9 ft. 3 in. of ribbon. She used 3 ft. 6 in. to trim a dress. To find how much ribbon she had left, Helene had to subtract. She made 9 ft. 3 in. the minuend and 3 ft. 6 in. the subtrahend. She carefully wrote the subtrahend under the minuend. Now she subtracted each column separately.

Helene began in the column with the smaller unit of measure. Since she could not subtract 6 in. from 3 in., she converted 1 ft. of the 9 ft. in the minuend to its equivalent, 12 in. She then wrote the minuend, 9 ft. 3 in., as 8 ft. 15 in. Then, she subtracted in each column. The difference was 5 ft. 9 in.

Procedure. To subtract measures:

1. Place the subtrahend beneath the minuend. Be careful to place like units of measure in the same column.

2. Starting with the column that has the smallest unit of measure, subtract each column separately.

3. When in any column the number in the subtrahend is larger than the number in the minuend, change 1 of the next larger unit in the minuend to its equivalent number of smaller units.

MODEL PROBLEM

A plane took off from an air base at 9:45 A.M. and returned at 1:43 P.M. Find the flying time in hours and minutes.

Solution: Step 1. Find the number of hours and minutes from 9:45 A.M. until noon, 12:00 o'clock. *Step* 2. To this number, add 1 hour and 43 minutes.

Step 1

12 hr. 00 min.	=	11 hr. 60 min.
− 9 hr. 45 min.	=	9 hr. 45 min.
		2 hr. 15 min.

Step 2

	2 hr. 15 min.
+	1 hr. 43 min.
	3 hr. 58 min. *Ans.*

EXERCISES

Subtract:

	(a)	(b)	(c)
1.	6 ft. 10 in.	13 yd. 1 ft.	10 ft.
	3 ft. 2 in.	10 yd. 2 ft.	4 ft. 6 in.
2.	35 min. 50 sec.	42 min. 18 sec.	6 hr. 55 min.
	18 min. 30 sec.	13 min. 45 sec.	2 hr.
3.	9 wk. 6 da.	15 yr. 2 mo.	75 yr.
	4 wk. 3 da.	8 yr. 4 mo.	64 yr. 10 mo.
4.	3 pk. 6 qt.	2 pk. 5 qt.	24 gal. 2 qt.
	1 pk. 2 qt.	1 pk. 7 qt.	18 gal.
5.	8 lb. 14 oz.	43 lb. 8 oz.	75 lb.
	3 lb. 6 oz.	17 lb. 10 oz.	41 lb. 12 oz.
6.	6 yd. 2 ft. 4 in.	9 yr. 2 mo. 25 da.	8 hr. 10 min. 25 sec.
	3 yd. 1 ft. 8 in.	2 yr. 8 mo. 12 da.	3 hr. 30 min. 45 sec.

7. Marvin ran a quarter of a mile in 1 min. 34 sec. Harry ran the same distance in 1 min. 48 sec. How much faster was Marvin?

8. Mr. Trask has a piece of lumber 8 yd. long. How much must he cut off to have 3 yd. 1 ft. left?

9. A loaded truck weighs 3 T. 800 lb. If the truck when empty weighs 1 T. 1000 lb., what is the weight of the load in the truck?

10. Mr. Service bought 40 gal. of milk. If he sold 28 gal. 3 qt., how much milk did he have left?

11. Phyllis is 8 yr. 10 mo. old. Her brother Leonard is 13 yr. 2 mo. old. How much younger is Phyllis than Leonard?

12. It takes a train 5 hr. 12 min. to travel from New York to Boston. A plane travels the same distance in 1 hr. 40 min. How much time can be saved by traveling by plane?

13. Mildred weighs 108 lb. 8 oz. She should weigh 125 lb. How much is she underweight?

14. Bill won first place in the shot-put with a heave of 37 feet 8 inches, and John won second place with a toss of 36 feet 10 inches. How much farther did Bill heave the shot than John?

15. Jane is 4 ft. 7 in. tall, and her brother Dick is 5 ft. 4 in. tall. How much taller is Dick than Jane?

16. A train leaves New York City at 9:20 A.M. and arrives in Buffalo at 4:55 P.M. on the same day. How long, in hours and minutes, does the train take to make the trip?

17. On a certain day, the sun rises at 7:30 A.M. and sets at 5:15 P.M. How many hours and minutes is it from sunrise to sunset?

18. John worked from 8:00 A.M. until 11:30 A.M. at the rate of \$1.20 per hour. How much did he receive for his work?

19. A boy worked from 10:30 A.M. until 1:00 P.M. at \$1.00 an hour. How much did he earn?

11. MULTIPLYING MEASURES

Seymour's vegetable garden was a plot of land that measured 5 yd. 2 ft. on each of its 4 sides. In order to find how much fencing material he needed to put around the garden, Seymour multiplied 5 yd. 2 ft. by 4. Since he knew that 5 yd. 2 ft. really means 5 yd. *and* 2 ft., he multiplied 5 yd. by 4 and 2 ft. by 4. The result he obtained was 20 yd. 8 ft. Since 8 ft. is equal to 2 yd. 2 ft., his answer was 22 yd. 2 ft.

5 yd. 2 ft.
4
20 yd. 8 ft.
or 22 yd. 2 ft.

Procedure. To multiply a measure that contains two or more units by a number, multiply each of the parts of the measure by the number. If possible, simplify the answer by converting smaller units to larger ones.

MODEL PROBLEM

A record player ready to be shipped weighs 8 lb. 4 oz. What is the weight of 5 of these record players?

Solution: Multiply 8 lb. 4 oz. by 5.

8 lb. 4 oz.
5
40 lb. 20 oz. or 41 lb. 4 oz. *Ans.*

EXERCISES

Multiply:

	(*a*)	(*b*)	(*c*)
1.	3 ft. 2 in. × 4	13 yd. 2 ft. × 10	9 ft. 6 in. × 8
2.	4 hr. 3 min. × 13	22 hr. 47 min. × 2	12 wk. 2 da. × 3

	(a)	(b)	(c)
3.	15 yr. 8 mo. × 7	2 wk. 3 da. × 7	25 qt. 1 pt. × 5
4.	4 gal. 2 qt. × 4	14 lb. 3 oz. × 5	20 lb. 4 oz. × 8

5. *a.* 3 mi. 400 yd. 2 ft. × 2   *b.* 6 hr. 20 min. 50 sec. × 4

6. An electrician needs 10 pieces of wire that measure 7 ft. 8 in. each. What is the total length of wire he needs?
7. A family uses 2 lb. 12 oz. of meat for supper. How much meat does this family use for 4 meals?
8. Mr. May took a trip that lasted 2 wk. 4 da. Mr. Parsons took a trip that lasted 4 times as long. How much time did Mr. Parsons spend on his trip?
9. Mrs. North knows that she can have 7 portions of ice cream from 1 qt. 1 pt. How much ice cream should she order to serve 35 guests?
10. A truck is loaded with 5 doz. wheels. Each wheel weighs 8 lb. 12 oz. What is the total weight of all the wheels?
11. Each day after school, Sanford worked in a store for 2 hr. 45 min. If he received $1.20 an hour, how much did he earn in a 5-day week?

12. DIVIDING MEASURES

	2 lb.	7 oz.	
4)	9 lb.	12 oz.	
	8 lb.		
	1 lb.	12 oz.	= 28 oz.
			28 oz.

Fred, John, Arthur, and Pete had a vegetable garden. One day they picked 9 lb. 12 oz. of peas. In order to find how much each boy was to receive, they had to divide 9 lb. 12 oz. by 4. Fred, who did the figuring, started the division by dividing 9 lb. by 4. This gave a quotient of 2 lb. and a remainder of 1 lb. He now had to divide 1 lb. 12 oz. by 4. Since 1 lb. = 16 oz., 1 lb. 12 oz. = 28 oz. He divided 28 oz. by 4. The quotient was 7 oz. Therefore, Fred's answer was 2 lb. 7 oz.

Procedure. To divide a measure that contains two or more units by a number, start with the largest unit and divide each part of the measure by the number. If in dividing a unit there is a remainder, convert the remainder to the equivalent number of next smaller units and add this result to the given number of those units in the dividend. Then continue the division.

EXERCISES

Divide:

	(*a*)	(*b*)	(*c*)
1.	$5\overline{)15\text{ ft. }10\text{ in.}}$	$5\overline{)21\text{ yd. }2\text{ ft.}}$	$3\overline{)7\text{ ft. }3\text{ in.}}$
2.	$6\overline{)42\text{ min. }30\text{ sec.}}$	$5\overline{)12\text{ hr. }10\text{ min.}}$	$6\overline{)5\text{ hr. }30\text{ min.}}$
3.	$2\overline{)6\text{ wk. }4\text{ da.}}$	$5\overline{)21\text{ yr. }3\text{ mo.}}$	$3\overline{)8\text{ wk. }1\text{ da.}}$
4.	$2\overline{)8\text{ gal. }2\text{ qt.}}$	$3\overline{)10\text{ gal. }2\text{ qt.}}$	$3\overline{)4\text{ bu. }2\text{ pk.}}$
5.	$5\overline{)20\text{ lb. }15\text{ oz.}}$	$8\overline{)17\text{ lb. }8\text{ oz.}}$	$2\overline{)9\text{ lb. }5\text{ oz.}}$

6. *a.* $4\overline{)8\text{ yd. }10\text{ ft. }4\text{ in.}}$ *b.* $3\overline{)10\text{ hr. }5\text{ min. }15\text{ sec.}}$

7. A load of 5 T. 500 lb. is to be divided equally among 3 trucks. How much of the load will each truck carry?

8. Elliot and Harvey are to share equally 5 gal. 2 qt. of berries that they picked. Find each boy's share.

9. Margaret studies four subjects each day. She spends 2 hr. 40 min. doing this. If she divides the time equally among the four subjects, how much time does she spend on each subject?

10. If 5 jars of strawberry jam weigh 17 lb. 8 oz., what is the average weight of one jar?

11. A man had a piece of lumber 10 feet 4 inches long that he wished to cut into four equal lengths. How long would each piece be if he made no allowance for loss in cutting?

Chapter VI

ARITHMETIC IN THE HOME

1. KEEPING A CASH ACCOUNT

On August 1, Ralph Forbes had $4.65, ***cash on hand,*** in his money box. During the month of August, he received several sums of money, called ***receipts,*** and paid out other sums of money, called ***expenditures.*** At the end of the month, he had $3.55 left in the money box. This amount is called the ***balance.*** Below is a detailed record of his monies, called a ***cash account.***

RALPH'S CASH ACCOUNT FOR AUGUST

Date	*Receipts*			*Date*	*Expenditures*		
Aug. 1	Cash on hand	4	65	Aug. 7	Boat model	2	40
Aug. 6	Selling newspapers	8	70	Aug. 15	Movies	1	20
Aug. 12	Mowing lawn	3	00	Aug. 21	Film	1	30
Aug. 20	Working in store	4	50	Aug. 24	Baseball game	1	65
Aug. 25	Running errands	1	25	Aug. 30	Savings bank	15	00
Aug. 29	Mowing lawn	3	00	Aug. 31	Balance	3	55
	Total	25	10		Total	25	10

In a cash account, the total receipts should always equal the total expenditures.

EXERCISES

1. Rule a form for a cash account. Prepare a cash account for Sue Waters for the week of March 7 to March 13. Find the balance at the end of the week.

March 7: cash on hand, $2.30; allowance, $3.00; March 8: bought school supplies, $.65; March 9: ran errands, $.50, and bought a magazine, $.25; March 10: helped with the dishes, $.25; March 11: went to the movies, $.75, and put $2.00 in the savings bank; March 12: did baby sitting, $1.50; March 13: made a church contribution, $.50.

2. Prepare Mrs. Robbin's cash account for the week of January 10. Find the balance at the end of the week.

Jan. 10: cash on hand, \$8.23; weekly household allowance, \$50.00; butcher's bill, \$10.81; Jan. 11: milk bill, \$3.25; laundry bill, \$2.13; Jan. 13: gas bill, \$4.18; electric bill, \$3.72; Jan. 14: grocery bill, \$15.79; carpet sweeper, \$8.50; Jan. 15: miscellaneous, \$2.15.

2. MAKING A BUDGET

A ***budget*** is a plan, prepared in advance, for the wise spending of one's income. A budget may be made for an individual, a family, a business, or a government.

In preparing a ***family budget,*** it is possible to list all the expenditures of a family under the six major headings, with suggested subheadings, shown below.

TYPES OF FAMILY EXPENDITURES

Shelter	*Operating Expenses*	*Food*	*Clothing*	*Miscellaneous*	*Savings*
Rent Payments on a house Fire insurance Repairs Furniture	Car Light Heat Telephone Supplies	Meat Groceries Vegetables Dairy products	New Repair Cleaning	Education Recreation Reading materials Charity Medical care Taxes	Banks Life insurance Investments

EXERCISES

1. Under which of the six major headings in the previous table should each of the following family expenditures be placed?

a. milk *b.* telephone *c.* movie tickets *d.* savings bonds
e. taxes *f.* shoe repairs *g.* car repairs *h.* plumbing repairs
i. meat *j.* doctor's bills *k.* electricity *l.* magazines

2. Mr. Harvey earns \$400 a month. In his budget, he allows 35% for food. How much money does he allow for food?

3. Mr. Jackson, who earns \$300 a month, pays \$75 for rent each month. What per cent of his salary is this?

4. Mr. Simpson saved \$40 a month. If this represented 8% of his earnings, how much did he earn a month?

5. The Degnan family had an income of \$6000 one year. They spent \$1950 for food, \$1200 for shelter, \$1050 for clothing, \$550 for operating expenses, \$650 for miscellaneous items, and saved the rest.

a. How much did they save during the year?
b. What per cent of their income did they save?
c. What per cent of their total income did they spend for each item?

6. A savings bank showed the following budget for a family of four having a yearly income between \$3500 and \$4500: Shelter, 25%; Clothing, 15%; Food, 30%; Savings, 10%; Operating expenses, 8%; Miscellaneous, 12%. If the annual income of a family that follows this budget is \$4000, how much should the family allow for each item?

7. A newspaper suggested the following budget for a family of three with an income of \$400 a month: Shelter, 25%; Clothing, 13%; Food, 25%; Savings, 12%; Operating expenses, 10%; Miscellaneous, 15%. If a family follows this budget, how much should be allowed for each item during the year?

3. MAKING BILLS AND RECEIPTS

Mr. Byron bought two tires and a battery from his garage. He received the bill shown below.

A ***bill,*** or ***sales slip,*** indicates the articles bought, the cost of each article, and the total amount of the purchase. When Mr. Byron paid the bill, the merchant marked it PAID. The paid bill then acts as a ***receipt,*** or proof of payment.

Tel. MAin 1-6847 THE PARK GARAGE
120 PARK AVENUE
TROY, NEW YORK

To: Saul Byron May 20, 1961
405 Main Street
Troy, New York

2	Goodyear tires @ 24.25	48	50
1	Exide battery @ 21.75	21	75
	PAID 5/20/61 Pat Ford Total	70	25

EXERCISES

1. The Triumph Department Store in Yourtown sold Mrs. Harold Sax the following goods on July 18, 1964, and received payment from her on August 1, 1964:

$\frac{3}{4}$ doz. pillow cases @ \$16.00 per doz. $3\frac{5}{8}$ yd. silk @ \$4.98 per yd.
$1\frac{2}{3}$ doz. towels @ \$10.50 per doz. $\frac{1}{2}$ doz. sheets @ \$30.00 per doz.

a. Make out a bill such as the Triumph Department Store might have sent to Mrs. Harold Sax.
b. What is the total amount of the bill? *c.* Receipt the bill.

2. John Doe sold Mary Roe the following goods on June 16, 1965, and Mrs. Roe paid the bill on the next day: 2 dozen oranges at 70¢ per dozen, $1\frac{1}{2}$ dozen eggs at 64¢ per dozen, 2 boxes of cereal at 23¢ per box, $\frac{1}{2}$ lb. of butter at 68¢ per lb., and 3 cans of soup at 29¢ per can.

a. Make out a bill such as Mr. Doe might have sent to Mrs. Roe.
b. What is the total amount of the bill? *c.* Receipt the bill.

3. Mrs. Shelly bought the following items in a department store. What was the total amount of her purchases?

1 luncheon cloth @ \$8.98 each $\frac{1}{2}$ doz. dish cloths @ 3 cloths for \$1.49
8 napkins @ \$5.00 per doz. $6\frac{3}{4}$ yd. toweling @ \$.65 per yd.

4. On May 10, 1965, Sam Sport paid his tennis club membership fee of \$100 to the treasurer, Carl Spark. Fill out the receipt that Mr. Spark issued to Mr. Sport.

____________________19____

Received of______________________________

______________________________Dollars

For______________________________

____

\$__________ ____________________

4. BUYING WISELY

A housewife can be economical in running her home by buying items that are on sale, buying items in large quantities, and buying items that are inexpensive because they are in season.

EXERCISES

1. A can of tuna fish sells regularly for 38 cents. At a sale, 3 cans are sold for $1. How much does Mrs. Bird save by buying 1 dozen cans at the sale price?
2. Mrs. Buhl regularly pays 35¢ for 2 cans of frozen orange juice. How much does she save if she buys 6 cans of frozen orange juice at a sale and pays at the rate of 3 cans for 49¢?
3. What will be a housewife's total saving if she buys the items listed below at the sale prices instead of the regular prices?

Sale Prices	*Regular Prices*
2 boxes of crackers for 39¢	23¢ a box
3 boxes of cleanser for 29¢	12¢ a box
1 gallon of bleach for 79¢	25¢ a quart
1 dozen oranges for 75¢	8¢ apiece
3 packages of frozen string beans for 59¢	24¢ a package

4. At an end-of-summer-season sale, Mr. Roz bought a suit that regularly sold for $90 at a 25% reduction. How much did he save?
5. Potatoes sell for $.06 per pound. How much does Mrs. Thrift save by buying a 100-lb. bag of potatoes for $4.75?
6. A 2-lb. box of sugar costs $.23; a 5-lb. box costs $.49. If Mrs. Price needs 20 lb. of sugar, how much can she save by buying 5-lb. boxes rather than 2-lb. boxes?
7. Mrs. Baker uses a lot of tomatoes in planning her menus during the month of August because tomatoes are then in season. In August, she paid $.29 for 2 lb. of tomatoes. In December, she paid $.45 per lb. How much more did 6 lb. of tomatoes cost in December than in August?

5. BUYING ON THE INSTALLMENT PLAN

Mr. Whitney bought a television set that was marked $190. Instead of paying cash for it, he bought it on the ***installment plan.*** By this plan, he made a ***down payment*** of $30 and agreed to make ten monthly payments, or ***installments,*** of $20 each.

Of course, Mr. Whitney realized that he was paying $40 more in buying on the installment plan. This sum, which is the difference between the amount paid on the installment plan and the amount of the cash price, is called the ***carrying charge,*** or service charge.

MODEL PROBLEM

Mr. Backwin bought on the installment plan a tape recorder that would have cost $150 if bought for cash. He paid $50 down and made 11 monthly payments of $12 each. (*a*) How much did he pay for the tape recorder? (*b*) What was the carrying charge?

Solution:

a. Down payment	$ 50	*b.* Cost on installment plan	$182
11 installments at $12 each	+132	Cost for cash	−150
Cost on installment plan	$182	Carrying charge	$ 32

EXERCISES

1. Mr. Apt bought on the installment plan a refrigerator whose cash price was $325. He paid $80 as a down payment and then made 12 monthly payments of $22 each.

a. How much did Mr. Apt pay for the refrigerator?
b. What was the carrying charge?

2. Mrs. Burrows bought on the installment plan a radio marked $125. She paid $25 down and then made 6 monthly installments of $18.95 each.

a. How much did Mrs. Burrows pay for the radio?
b. What was the service charge?

3. Mr. Drysdale bought on the installment plan a home freezer whose cash price was $420. He paid 25% of the cash price as a down payment and paid 6% of the unpaid balance as a carrying charge. Find, correct to the nearest cent, how much he has to pay in each of 12 equal installments to pay completely for the freezer.

4. Mr. Brown purchased a television set marked $199.50. He made a down payment of $19.50 and signed a contract to pay the balance in 18 installments of $12 each.

a. How much did the television set really cost when all payments were complete?
b. How much could Mr. Brown have saved by paying cash?

6. COMPUTING THE GAS BILL

Gas is used in the home for cooking and heating. At regular periods, the gas company sends a man to read the ***meter.*** The meter is an instrument that measures the amount of gas used. The unit of measure is the ***cubic foot*** (cu. ft.). When the gas company prepares a customer's bill for a certain period of time, the company computes the charge for the number of cubic feet of gas that was consumed during that period.

EXERCISES

A gas company charges the following rates for residential service:

First 200 cu. ft. for	\$1.15
Next 1,200 cu. ft.	@ 34.5¢ per 100 cu. ft.
Next 1,200 cu. ft.	@ 22.0¢ per 100 cu. ft.
Next 6,000 cu. ft.	@ 18.5¢ per 100 cu. ft.
Excess of 8,600 cu. ft.	@ 15.5¢ per 100 cu. ft.

Use the gas rates given above to find the cost of using:

1. 800 cu. ft. **2.** 1,700 cu. ft. **3.** 2,900 cu. ft. **4.** 5,200 cu. ft.

5. On July 29, the reading on Mr. Simpson's gas meter was 44,800 cu. ft. On Aug. 30, the reading was 47,200 cu. ft.

a. For how many cubic feet of gas should the gas company charge Mr. Simpson for the period July 29, to Aug. 30?

b. How much will Mr. Simpson's gas bill be for this period? (Use the rates given above.)

6. On July 26, the reading on Mrs. Chase's gas meter was 12,900 cu. ft. On Aug. 27, the reading was 16,300 cu. ft.

a. How much gas was consumed between July 26, and Aug. 27?

b. How much will Mrs. Chase's gas bill be? (Use the rates given above.)

7. USING ELECTRICITY

Electricity is used in the home mainly for lighting and the running of electrical appliances. Our living is made more comfortable by the use of such appliances as a toaster, an iron, a radio, and a television set, which are run by electrical power.

The unit used to measure electrical power is the ***watt.*** To light an electric bulb marked 60 W (watts), we need 60 units of electrical power. To run an electric heater marked 1,000 watts, we need 1,000 watts of electrical power.

When one watt of power is used for one hour, then one ***watt-hour*** of electricity is consumed. The watt-hour is the unit used to measure the amount of electricity consumed.

Procedure. To find the number of watt-hours of electricity used by an appliance, multiply the number of watts needed to run the appliance by the number of hours that the appliance is in use.

The quantity 1,000 watts is called a ***kilowatt.***

The quantity 1,000 watt-hours is called 1 ***kilowatt-hour.*** Electricity is sold by the kilowatt-hour.

Procedure. To change a number of watt-hours to kilowatt-hours, divide the number of watt-hours by 1,000.

MODEL PROBLEM

What is the cost of using a 100-watt bulb for 40 hours at 5 cents per kilowatt-hour?

Solution: Step 1. Find the number of watt-hours consumed. *Step* 2. Change this number of watt-hours to kilowatt-hours. *Step* 3. Find the cost for this number of kilowatt-hours.

Step 1. $40 \times 100 = 4{,}000$ watt-hours

Step 2. 4,000 watt-hours $\div$ 1,000 = 4 kilowatt-hours

Step 3. 4×5 cents = 20 cents *Ans.*

EXERCISES

In 1–5, find: *a.* the number of watt-hours used
b. the number of kilowatt-hours used
c. the cost of operation at 5 cents per kilowatt-hour

1. A 75-watt electric bulb burns for 40 hours.

2. A 60-watt electric bulb burns for 20 hours.

3. A 50-watt radio is played 4 hours a day for 30 days.

4. A 1,000-watt electric iron is used 3 hours a week for 50 weeks.

5. A 1,000-watt toaster is used 30 minutes a day for 14 days.

6. What fractional part of a kilowatt-hour is used by a 15-watt bulb that burns for 20 hours?

7. Mrs. Simpson keeps a 15-watt night light burning for the children on the average of 8 hours each night. How much does it cost Mrs. Simpson to do this for 50 weeks at the rate of 4 cents per kilowatt-hour?

8. Mrs. Baxter uses a 750-watt sandwich grill on the average of 20 minutes to prepare lunch for the family. If she pays $4\frac{1}{2}$ cents per kilowatt-hour, how much does the electricity cost her to prepare 15 lunches?

8. COMPUTING THE ELECTRIC BILL

At regular periods, the electric company sends a man to read the ***meter,*** which is the instrument that measures the amount of electricity consumed. The unit of measure is the ***kilowatt-hour*** (K.W.H.).

EXERCISES

An electric company charges the following rates for residential service:

First 10 K.W.H. or less......	$.95
Next 42 K.W.H.............	@ 5¢ per K.W.H.
Next 48 K.W.H.............	@ 4.5¢ per K.W.H.
Next 100 K.W.H............	@ 3.2¢ per K.W.H.
Excess of 200 K.W.H........	@ 2.5¢ per K.W.H.

Use the electric rates given above to find the cost of using:

1. 8 K.W.H. **2.** 37 K.W.H. **3.** 79 K.W.H. **4.** 155 K.W.H. **5.** 276 K.W.H.

6. On July 29, Mr. Samuel's meter read 3,494 kilowatt-hours. On Aug. 30, it read 3,572 kilowatt-hours.

a. How many kilowatt-hours were used during this period?

b. Using the rates given above, find the charge for the electricity consumed during this period.

7. On January 31, Mrs. Stanley's meter read 8,625 kilowatt-hours. On February 28, it read 8,803 kilowatt-hours.

a. How many kilowatt-hours were used during February?

b. Using the rates given above, find the charge for the electricity consumed during the month.

9. ADJUSTING RECIPES

Mrs. Hammond's recipe for making enough gingered beets to serve 6 persons calls for $2\frac{1}{2}$ cups of cooked beets. In order to find out how many cups of beets she should use to serve 18 persons, Mrs. Hammond compared 18 with 6. She divided 18 by 6 and learned that she had to use 3 times the amount the recipe called for. She multiplied $2\frac{1}{2}$ by 3 and discovered that she needed $7\frac{1}{2}$ cups of beets.

EXERCISES

1. A cookbook suggests that $\frac{3}{4}$ lb. asparagus should be used to serve 2 persons. How much asparagus should be used to serve 8 persons?
2. Mrs. Swift uses $\frac{1}{4}$ lb. of uncooked rice in preparing a dish to serve 2 persons. How much rice should she use to serve 12 persons?
3. A cake recipe for 10 persons calls for 2 cups of sugar. How many cups of sugar will be needed to make a cake for 25 persons?
4. Mae wishes to make lemonade using a recipe that suggests 2 lemons for every 7 cups of lemonade. How many lemons should she use to make 35 cups of lemonade?
5. Mrs. Weber uses a liquid plant food for her geraniums. The directions on the bottle tell her to use $1\frac{1}{2}$ tablespoonfuls of liquid with a gallon of water. How much liquid should she use with 1 quart of water?
6. Helen is making $\frac{3}{4}$ of her recipe for brownies. How much sugar must she use if the recipe calls for $2\frac{1}{3}$ cups of sugar?
7. Coca-Cola Sparkle is made according to the following recipe: 1 package prepared lemon gelatin, $\frac{1}{2}$ cup boiling water, 2 bottles Coca-Cola. This recipe serves 4 persons. Adjust the recipe so that it will serve 24 persons.
8. Helen's recipe for Date and Nut Bread is: 2 eggs, $\frac{1}{2}$ cup sugar, 1 cup dates, 4 cups flour, 2 cups milk, 1 cup chopped nuts, 8 teaspoons baking powder. How should Helen adjust her recipe so that she will have $\frac{1}{2}$ the amount of cake the recipe would produce?

10. FINDING AVERAGES

Helen's marks in mathematics for 5 consecutive days were: 80, 95, 90, 85, and 100. To find her average, she added the 5 marks and divided by 5. The total of her marks was 450. Dividing 450 by 5, she obtained the result, 90, which was her average.

Procedure. To find the average of several numbers, add them and then divide the sum by the number of addends.

MODEL PROBLEM

The 4 members of the Teller family spent $280.60 on an automobile trip. What was the average cost for each member of the family?

Solution: Divide the total expenses by the number of people on the trip.

$$\begin{array}{r} \$\ 70.15 \quad \textit{Ans.} \\ 4\overline{)\$280.60} \end{array}$$

EXERCISES

1. Harvey received the following marks on his report card: English, 85; French, 92; History, 76; Geometry, 87; Science, 80. What is the average of Harvey's marks?
2. Mr. Jackson has 4 children whose weights are: 150 lb., 132 lb., 144 lb., and 122 lb. What is the children's average weight?
3. Edwin earned the following sums of money working after school: Monday, \$2.25; Tuesday, \$3.50; Wednesday, \$2.75; Thursday, \$3.10; Friday, \$1.90. What was Edwin's average daily earnings?
4. Mr. Pierce's automobile was driven 134.8 miles on 9 gallons of gasoline. Find, to the nearest mile, how many miles per gallon the car averaged.
5. It cost Mr. Sims \$29.40 to drive 560 miles. What was the average cost of driving one mile?
6. Mr. Hart has three sons whose heights are 5 ft. 8 in., 6 ft. 1 in., and 5 ft. 6 in. What is their average height?
7. Ellen spent 3 hours and 45 minutes studying for 5 classes. What was the average time in minutes that she spent in studying for each class?
8. John gained $10\frac{5}{8}$ lb. in 5 months. What was his average monthly gain?
9. One year, the United States produced about 14,400,000 bales of cotton, which sold for \$756,000,000.
   a. At that rate, what was the value of a bale of cotton that year?
   b. There are about 500 pounds in a bale of cotton. What was the average price of a pound of cotton that year?

11. BUYING FIRE INSURANCE

Mr. Thomas bought a new home. To protect himself against loss by fire, he wanted to buy a ***fire insurance policy.*** This is a written agreement, or contract, by which the company agrees to make good to Mr. Thomas for any fire loss up to a fixed sum, or the ***face value,*** stated in the policy.

He bought a policy that would be in force for a ***term*** of one year. The ***rate*** for a one-year term was \$.10 per \$100 of insurance.

To find the cost, or ***premium,*** for a \$20,000 policy for one year, Mr. Thomas proceeded as follows:

First, he discovered how many \$100's are contained in \$20,000 by dividing \$20,000 by \$100. The result was 200. Since each unit of \$100 of insurance costs \$.10, he figured that 200 units would cost 200 × \$.10, or \$20. Therefore, the premium would be \$20.

Procedure. To find the premium for a policy when the rate per \$100 of insurance is known, divide the face value of the policy by 100. Multiply the result by the rate per \$100 of insurance.

MODEL PROBLEM

At the rate of \$.35 per \$100 for a 3-year policy, what is the premium on an insurance policy for \$8000?

Solution: Step 1. Divide the face value of the policy, \$8000, by \$100. *Step* 2. Multiply this result by \$.35.

Step 1

$$\frac{\$8000}{\$100} = 80$$

Step 2

$$\begin{array}{r} \$.35 \\ \times\ 80 \\ \hline \$28.00 \end{array} \quad \textit{Ans.}$$

EXERCISES

1. A house was insured for \$12,000 at a rate of 60 cents per \$100. What was the premium?

2. What is the cost of a \$2500 fire insurance policy on Mr. Bacon's furniture at the rate of \$.18 per \$100?

Find the annual premium:

	Face Value	*Rate per* \$100		*Face Value*	*Rate per* \$100
3.	\$6000	\$.17	**4.**	\$7500	\$.30
5.	\$12,500	\$.32	**6.**	\$18,000	\$.44
7.	\$22,000	\$.42	**8.**	\$30,000	\$.25

9. Mr. Godfrey insured his house for \$17,500 at a rate of \$.17 per \$100, and its contents for \$9000 at a rate of \$.23 per \$100. Find the total premium.

10. Mr. Jones insures his barn and contents valued at \$30,000 for $\frac{4}{5}$ of their value. At 60 cents per \$100 of insurance, what premium does he pay?

11. Mr. Jayne wants to insure his \$15,000 house for 80% of its value. He can buy insurance for 1 year or for a period of 3 years. The rate is \$.40 per \$100 for a 1-year policy or $2\frac{1}{2}$ times the yearly rate for a 3-year policy.

a. For what amount does he want to insure his house?

b. How much would he save in 3 years by taking a 3-year policy rather than three 1-year policies?

12. BUYING AUTOMOBILE INSURANCE

Mr. Stevens bought a new car. To protect himself against claims where he is responsible for an accident, he bought ***personal liability*** and ***property damage*** insurance. Mr. Stevens could also have bought for his protection *fire, theft, collision,* and *comprehensive* insurance.

EXERCISES

1. At the rate of 60 cents per $100, how much does it cost to insure a car against fire and theft for $900?
2. Find the premium for fire and theft insurance on an automobile whose value is $3900 if the rate is $.80 per $100.
3. Mr. Harris owns a car that had a value of $2800 when new. Now he insures his car against fire and theft at 50% of its original value, at $.90 per $100. Find the premium.
4. Mr. Burns drove a car for almost 20 years without an accident. During that time, he paid, on the average, $140 a year for automobile liability insurance. After having paid 20 premiums, he injured a man. As a consequence, the insurance company paid damages amounting to $25,000. How much did Mr. Burns save by having paid insurance for the 20 years?
5. Mr. Walker drove a car for nearly 15 years without an accident. During that time, he paid, on the average, $130 a year for automobile liability insurance. After having paid for 15 years, he injured a man seriously. The damages amounted to $40,000, which the insurance company paid. How much did Mr. Walker save by having insurance for the 15 years?

13. BUYING LIFE INSURANCE

Mr. Black wishes to protect his family in the event of his death. He does this by buying a ***life insurance policy.*** Mr. Black, the person for whom the policy is written, is called the ***insured.*** He makes an annual payment, the ***premium,*** to the company. In return, the company agrees to pay a specified sum of money, the ***face value*** of the policy, in case of Mr. Black's death. The one who receives the benefits of the policy is the ***beneficiary.***

The four common types of life insurance are *term, ordinary life, limited payment life,* and *endowment.* They are designed to meet the varying needs of different people.

EXERCISES

The table below shows some annual premiums for each $1000 of insurance on the different types of policies. Premiums are shown for every 5 years from age 20 to age 40.

ANNUAL PREMIUM FOR $1000 OF LIFE INSURANCE

Age	*10-Year Term*	*Ordinary Life*	*20-Payment Life*	*20-Year Endowment*
20	6.66	17.34	28.56	47.99
25	7.39	19.80	31.30	48.51
30	8.51	22.85	34.45	49.31
35	10.22	26.68	38.13	50.54
40	12.83	31.59	42.51	52.44

In 1–8, use the above table to find the annual premium for each of these policies:

	Face Value	*Age*	*Type*		*Face Value*	*Age*	*Type*
1.	$2000	25	10-year term	**2.**	$3000	30	20-payment life
3.	$15,000	35	10-year term	**4.**	$20,000	35	20-payment life
5.	$5000	20	ordinary life	**6.**	$8000	25	20-year endowment
7.	$10,000	40	ordinary life	**8.**	$25,000	30	20-year endowment

9. What will the premiums paid on a 20-year endowment policy for $5000, issued at age 30, amount to in 20 years?

10. Mr. Thompson took out a 20-year endowment policy for $6000. The rate was $47.50 per $1000.

a. How much was his yearly premium?

b. What will be the total amount paid in premiums if he lives until the policy is paid up?

11. Mr. Day purchases a $1000, 20-year endowment policy at an annual premium of $52.41 per $1000.

a. How much will he pay in premiums during the 20-year period?

b. If Mr. Day dies after making five payments, how much more will the beneficiary receive from the company than Mr. Day paid in premiums?

12. How much must be saved each month to pay the yearly premium on a $5000 life insurance policy if the rate of premium is $28.80 per thousand?

14. BUYING OTHER FORMS OF INSURANCE

EXERCISES

1. Mr. Ross placed $12,000 storm insurance on his home at a rate of $\frac{1}{2}\%$. What was his annual premium?
2. Find the premium for theft insurance on a new automobile whose value is $2900 if the rate is $.75 per $100.
3. Mr. Warren insured his wife's jewelry against theft for $8000 at $1\frac{3}{4}\%$. What was his annual premium?
4. Mr. Reif works in a cafeteria and receives a weekly salary of $54.75. While at work, he broke his arm. Through the Workmen's Compensation Law, he received $\frac{1}{3}$ of his weekly salary for each of the 14 weeks he was disabled. In addition, he received $1800 because the injury lessened his future earning power. What was the total compensation that he received?
5. A working man paid 3% of his salary to the government to be eligible for social security benefits. Find how much he paid each week if his weekly salary was:

   a. $48.00 *b.* $55.00 *c.* $67.00 *d.* $77.50

15. PAYING REAL ESTATE AND SCHOOL TAXES

The community of Bridgetown plans to raise $70,000 for town services by taxing real estate. For tax collection purposes, the estimated value of all the real estate in town, the ***assessed valuation,*** is $2,000,000. In order to prepare each property owner's real estate bill, the town must first determine the ***realty tax rate*** that must be used in order to raise the required $70,000.

To compute the tax rate, it is necessary to find what per cent the amount to be raised, $70,000, is of the total assessed valuation, $2,000,000. This is done by dividing $70,000 by $2,000,000.

$$\frac{\$70{,}000}{\$2{,}000{,}000} = \frac{7}{200} = \frac{3.5}{100} = .035 = 3.5\% \text{ or } 3\tfrac{1}{2}\%$$

Each taxpayer will have to pay $3\frac{1}{2}\%$ of the assessed valuation of his property as his realty tax.

Procedure. To find the realty tax rate as a per cent, divide the amount of money to be raised through realty taxes by the total assessed valuation and express the quotient as a per cent.

A tax rate of $3\frac{1}{2}\%$ can be expressed in any of the following ways:

1. $3\frac{1}{2}\%$ of the assessed valuation
2. 3.5 cents on each dollar of assessed valuation
3. 35 mills on each dollar of assessed valuation (1 mill = $\frac{1}{10}$ cent)
4. \$3.50 on each 100 dollars of assessed valuation
5. \$35 on each 1000 dollars of assessed valuation

Mr. Brody, who lives in Bridgetown, owns a home that is assessed at \$14,500. To find his realty tax, he must find $3\frac{1}{2}\%$, or .035, of the assessed valuation, \$14,500.

Since the tax rate can also be expressed as 3.5 cents per dollar of assessed valuation, the tax can also be found by multiplying 3.5 cents or .035 dollar by 14,500. The answer can then be converted to the usual dollar and cents form.

```
   14,500
  × .035
  -------
   72 500
  435 00
  -------
  507.500
```

Answer: \$507.50

Procedure. To find the amount of taxes, multiply the number of dollars in the assessed valuation by the tax per one dollar of assessed valuation.

MODEL PROBLEM

A school district with an assessed valuation of \$1,443,200 requires \$90,200 for its school. What would the district have to set as its tax rate per \$1000?

Solution: Step 1. Divide the amount required, \$90,200, by the total assessed valuation, \$1,443,200, to find the tax for each dollar of assessed valuation. *Step* 2. Find the tax rate for each \$1000 of assessed valuation.

Step 1

$$\frac{\$90{,}200}{\$1{,}443{,}200} = .0625$$ The tax rate is \$.0625 per \$1 of assessed valuation.

Step 2

Therefore, the tax rate per \$1000 would be \$.0625 × 1000, or \$62.50.

Answer: \$62.50 per \$1000 of assessed valuation

EXERCISES

1. How much per $100 is each of the following tax rates?

   a. $.027 per dollar
   b. $1\frac{3}{4}$ cents per dollar
   c. 2% of the assessed value
   d. $3\frac{1}{2}$% of the assessed value
   e. 7 mills per dollar
   f. $12\frac{1}{2}$ mills per dollar

2. Find the tax rate per $1000 in each of the following problems:

	Assessed Value	*Money to Be Raised*
a.	$2,000,000	$50,000
b.	$3,000,000	$150,000
c.	$15,000,000,000	$360,000,000

3. In a town where the assessed valuation is $750,000, a sum of $45,000 is needed for education. What would be the tax rate per $100?
4. The assessed valuation in a city is $2,580,000. The education budget calls for $72,240. What is the education tax rate expressed as mills per dollar of assessed value?
5. Find the taxes on properties that have the following assessments:

	Assessed Value	*Rate*
a.	$10,000	$2.75 per $100
b.	$12,500	$2.95 per $100
c.	$15,750	$3\frac{1}{4}$%
d.	$125,000	$32\frac{1}{2}$ mills per $1

6. A man owns property assessed at $7000. If the tax rate is $35 per $1000 this year, what is the amount of the tax on his property?
7. A piece of land is assessed at $8200. What is the tax on it at the rate of 46 mills per $1?
8. If the school tax rate in a school district is $.04615 on each dollar of assessed valuation, what is the school tax on a piece of property having an assessed valuation of $1000?
9. A man paid $25,000 for a house. The assessed valuation is $\frac{4}{5}$ of the cost. How much tax must he pay if the rate is $43.59 per $1000?
10. A community with an assessed valuation of $1,862,000 has a budget that amounts to $55,860.

    a. What must be the tax rate per $100?
    b. If a man owns a house that is assessed at $7200, how much tax will he have to pay?

11. The assessed valuation of a certain school district is $14,425,200. It is necessary to raise $461,606.40 by a tax to pay the cost of operating the school system.

    a. What will the school tax rate be?
    b. How much school tax will Mr. Benson pay on his property if it is assessed at $6400?

16. PAYING MANY KINDS OF LOCAL, STATE, AND FEDERAL TAXES

To determine the amount of a particular tax, multiply the amount to be taxed by the tax rate.

EXERCISES

1. If a sales tax is 3%, what would the tax be on a purchase of $8.50?
2. Find the 3% sales tax and the total bill on articles costing:

   a. $24.00   *b.* $250.00   *c.* $36.20   *d.* $99.75   *e.* $243.84
3. A man rents a safe deposit box for $8 a year. If the federal tax on the rental is 20%, how much is the tax?
4. At a rate of 50 cents per hundred pounds or any part of a hundred pounds, find the cost of a state license for a car that weighs:

   a. 3100 lb.   *b.* 3400 lb.   *c.* 3675 lb.   *d.* 3515 lb.   *e.* 4175 lb.
5. Mr. Deacon bought a watch costing $125 that was subject to a 10% federal luxury tax. How much did he pay in all?
6. The price of gasoline that sells at 30 cents a gallon includes a state tax of 6 cents. What per cent of the selling price is the tax?
7. How much will an importer have to pay the federal government if he imports 60,000 tulip bulbs when the federal tax, or import duty, is $4.50 per 1000 bulbs?
8. If Mr. Trevor's weekly social security tax is 3% of his $72 weekly salary, how much is his social security tax for a full year?

17. FINDING DISTANCE, RATE, AND TIME

Vincent was at college, 70 miles away from home. On Saturday, he drove in his car to visit his parents. He traveled at an average rate of 35 miles per hour, taking 2 hours to get there. The number of miles he traveled, 70, is called the ***distance.*** The number of miles he traveled in 1 hour is called his ***rate of speed,*** or rate. Since he traveled 35 miles in 1 hour, his rate of speed was 35 miles per hour, which may be written 35 mph. The number of hours he traveled, 2, is called the ***time.***

Distance = Rate × Time
or
D = R × T

Stephen took a 3-hour auto trip and traveled at an average rate of 35 miles per hour. To find the distance he covered, he multiplied the rate at which he traveled, 35 miles per hour, by the time he traveled, 3 hours. He found that he had traveled 105 miles: $35 \times 3 = 105$.

Rate = Distance ÷ Time
or
R = D ÷ T

A DC-7 flew 1400 miles in 4 hours. To find its average rate of speed, we divide the number of miles the plane flew, 1400, by the number of hours the plane flew, 4. We find that it flew at the rate of 350 miles per hour: 1400 ÷ 4 = 350.

Time = Distance ÷ Rate
or
T = D ÷ R

If we want to find how long it took a train to travel 300 miles at an average rate of 60 miles per hour, we divide the number of miles the train traveled, 300, by the number of miles the train traveled in 1 hour, 60. We find that the time required was 5 hours: 300 ÷ 60 = 5.

EXERCISES

1. Harry took a 5-hour auto trip on which he averaged 40 miles per hour. What distance did he travel?

2. A plane flew 240 miles an hour for $4\frac{3}{4}$ hours. How far did the plane fly?

3. A train left New York at 11:30 A.M. and arrived in Philadelphia at 1:15 P.M. The train averaged 52 miles per hour. What is the distance from New York to Philadelphia?

4. A plane flew 840 miles at an average rate of 160 miles per hour. How long did the trip take?

5. How many minutes are required to drive 30 miles at the rate of 40 miles per hour?

6. A car traveled 228 miles in 6 hours. What was the average rate of speed?

7. A train left New York at 10:00 A.M. and arrived at Washington, D.C., at 1:45 P.M. If the distance from New York to Washington is 225 miles, find the average rate of speed of the train.

8. A man drove 80 miles in 2 hours and then 100 miles in 3 hours. What was his average speed for the 5 hours?

9. An airplane started on a flight at 8:45 A.M. and arrived at its destination at 12:15 P.M. It flew a distance of 630 miles and used 189 gallons of gas.

a. How many hours were required for the flight?
b. What was the average rate of speed of the plane?
c. What was the average number of gallons of gas used per hour?

10. Mr. Verity left his home in his car at 9:00 A.M. and drove to a city 209 miles away. He arrived there at 4:45 P.M. He had stopped $2\frac{1}{4}$ hours on the road for lunch and to rest.

a. How many hours did he spend driving his car?
b. What was the average rate of speed at which he drove, expressed in miles per hour?

Chapter VII

ARITHMETIC IN BUSINESS PRACTICE

1. FINDING SELLING PRICE, MARGIN, PROFIT AND LOSS

Mr. Hamilton, who has a men's shop, bought a dozen shirts for $36. He sold them for $57, an increase of $21 over the amount he paid for them. The amount that he paid for the shirts, $36, is called the ***cost.*** The amount for which he sold the shirts, $57, is called the ***selling price.*** The difference between the selling price and the cost, $21, is called the ***margin.*** Margin is also known as ***markup*** and ***gross profit.***

margin = selling price − cost *or* **selling price = cost + margin**

In running his business, Mr. Hamilton had to spend money for rent, light, heat, salaries, advertising, etc. These are called ***expenses,*** or ***overhead.*** Mr. Hamilton estimated that his overhead in selling the shirts was $12. Therefore, his net gain, or ***net profit,*** was $21 − $12 = $9.

net profit = margin − overhead *or* **margin = overhead + net profit**

We therefore also see that:

selling price = cost + overhead + net profit

or

net profit = selling price − (cost + overhead)

When a dealer has to sell an article for less than the total of the cost and the overhead, he sells the article at a ***loss.***

MODEL PROBLEMS

1. Mr. Bancroft paid $5.75 for a basketball. What must be the selling price if he plans to have a margin of $2.50 in order to meet his expenses?

Solution:

$$\begin{array}{rl} \$5.75 & \text{(cost)} \\ +\ 2.50 & \text{(margin)} \\ \hline \$8.25 & \text{(selling price)} \end{array}$$

Answer: The selling price must be $8.25.

2. Mr. Pryor bought a sofa for \$64 and sold it for \$112. If his overhead was 25% of the cost, how much was his net profit?

Solution: Step 1. Find the amount of overhead. *Step* 2. Add this to the cost. *Step* 3. Subtract this total from the selling price.

Step 1		*Step* 2		*Step* 3	
\$64	(cost)	\$64	(cost)	\$112	(selling price)
× .25	(rate of overhead)	+ 16	(overhead)	− 80	(cost + overhead)
3 20		\$80	(total)	\$32	(net profit)
12 8					
\$16.00	(overhead)				

Answer: The net profit was \$32.

3. A dealer figured that the cost of a chair and the overhead for selling it is \$45. If it was sold at a loss of 20% of this total, what was the selling price?

Solution: Step 1. Find the loss. *Step* 2. Subtract the loss from the total of the cost and overhead.

Step 1		*Step* 2	
\$45	(cost + overhead)	\$45	(cost + overhead)
× .20	(rate of loss)	− 9	(loss)
\$9.00	(loss)	\$36	(selling price)

Answer: The selling price was \$36.

EXERCISES

1. Mr. George bought a piano for \$450 and sold it for \$675. What was the margin?

2. A man bought a house for \$5500. During the first year, he had expenses amounting to \$1200. At the end of the year, he sold the house, making a \$750 profit. Find the selling price.

3. Leon bought a bicycle for \$25 and spent \$8 in repairing it. He then sold the bicycle at a loss of \$3. Find the selling price.

4. A dealer bought a table for \$36 and sold it for \$40. If his overhead was \$7, did he have a net profit or a loss? How much?

5. A dealer bought a television set for \$150. For how much did he sell the set if his margin was 40% of the cost?

6. An antique dealer bought some silverware for \$750. If his margin was 100% of the cost, how much did he receive when he sold the silverware?

7. Find the selling price of a dress bought for \$16 and marked up 75%.

8. In a sporting goods store, a week's sales amounted to \$920. The cost of the merchandise sold was \$570. The overhead was 20% of the amount of sales. What was the gross profit? What was the net profit?

9. The cost and overhead of a gallon of ice cream is $2.50. If a dealer bought 5 gallons of ice cream and sold them at $.90 a quart, what was his total profit?
10. A merchant paid $14 for 4 dozen pairs of hose and sold them at 45 cents a pair. Find the margin.
11. The cost and overhead to a dealer for 400 lb. of spinach total $40. If he sold the spinach at the rate of 2 lb. for $.35, find his profit.
12. At a sale, a department store sold books that cost $1.50 each for 20% less than the cost. The overhead was $.15 per book. If 164 books were sold, how much did the store lose?
13. At the end of a season, a dealer had 4 dozen tennis balls that cost him $4.80 per dozen. He sold them for $33\frac{1}{3}$% of the cost. The overhead was $.96 per dozen. Find his loss.
14. A merchant paid $9000 for merchandise that he sold at a net profit of $33\frac{1}{3}$% of the cost. His overhead was 20% of the cost. How much did he receive for the merchandise?
15. Find the selling price of a lamp that cost $25, if the markup for overhead is 45% of the cost and the markup for profit is 35% of the cost.
16. A merchant who sold tomatoes made the following sales during one week: Monday, 175 lb.; Tuesday, 225 lb.; Wednesday, 150 lb.; Thursday, 250 lb.; Friday, 325 lb.; Saturday, 375 lb. He paid $2.10 for a 25-lb. box and sold the tomatoes for $.19 per pound.
    a. How many pounds of tomatoes did he sell during the week?
    b. How much did the tomatoes cost him?
    c. What were his total sales?
    d. What was his margin for the week?

2. FINDING THE PER CENT OF PROFIT OR LOSS BASED ON THE COST

Mr. Greenwood paid $20 for a table and made a profit of $4 when he sold it. He wanted to find what per cent of the cost his profit was. He compared the profit, $4, with the cost, $20, as follows:

$$\text{rate of profit} = \frac{\text{amount of profit}}{\text{cost}} = \frac{\$4}{\$20} = \frac{1}{5} = 20\%$$

Therefore, the profit was 20% of the cost.

Procedure. To find the per cent of profit or loss based on the cost, divide the amount of profit or loss by the cost and express the result as a per cent.

EXERCISES

1. A piece of jewelry that cost \$200 was sold at a profit of \$150. Find the per cent of profit based on the cost.
2. A crate of oranges that cost \$6 was sold at a profit of \$2. Find the per cent of profit based on the cost.
3. If a boy bought a baseball glove for \$5.50 and sold it for \$4.40, what was the per cent of loss on the original cost?
4. A rug that cost \$95 was sold for \$75. Find, to the nearest per cent, the per cent of loss based on the cost.

In 5–8: *a.* Determine whether there is a net profit or a net loss.
b. Find the per cent of net profit or loss based on the cost.

	Cost	*Overhead*	*Selling Price*		*Cost*	*Overhead*	*Selling Price*
5.	\$50	\$10	\$80	**6.**	\$48	\$8	\$44
7.	\$400	\$60	\$610	**8.**	\$2.40	\$.40	\$3.60

9. A grocer bought soap for \$1.25 per doz. He sold it at 3 cakes for 50 cents. What was his per cent of profit based on the cost?
10. A hardware dealer bought hammers at \$18 per dozen and sold them for \$2.25 each. What was his per cent of profit based on the cost?
11. A furniture dealer bought two chairs for \$24. He sold one for \$22 and the other for \$26. What was his per cent of profit based on the cost?
12. A dress merchant bought 10 dresses at \$25 each. He sold 3 of them for \$40 each, 2 of them for \$25 each, and 5 of them for \$15 each. His average overhead per dress was \$3. Find his per cent of loss based on the cost.
13. Mr. Jones paid \$8,500 for a building. He repaired it for \$1,500, then sold it for \$13,000. What per cent of the total cost was his profit?

3. FINDING THE PER CENT OF PROFIT OR LOSS BASED ON THE SELLING PRICE

When Mr. Sayer sold a hat for \$16, he made a profit of \$6. He wanted to find what per cent of the selling price was his profit. He compared the profit, \$6, with the selling price, \$16, as follows:

$$\text{rate of profit} = \frac{\text{amount of profit}}{\text{selling price}} = \frac{\$6}{\$16} = \frac{3}{8} = 37\tfrac{1}{2}\%$$

Therefore, the profit was $37\frac{1}{2}\%$ of the selling price.

Procedure. To find the per cent of profit or loss based on the selling price, divide the amount of profit or loss by the selling price and express the result as a per cent.

EXERCISES

1. A man bought a lot for $1200 and sold it for $1500. His gain was what per cent of the selling price?
2. A cabinet was sold for $80, at a loss of $10. Find the per cent of loss based on the selling price.
3. Warren paid $2.25 for a stamp. He sold it for $3.00. His profit was what per cent of the selling price?
4. Mrs. Benson paid $95 for a painting. She later sold it for $150. Find, to the nearest per cent, the per cent of profit based on the selling price.

In 5–10:

a. Determine whether there is a net profit or a net loss.
b. Find the per cent of net profit or loss based on the selling price.

	Cost	*Overhead*	*Selling Price*		*Cost*	*Overhead*	*Selling Price*
5.	$44	$5	$56	**6.**	$640	$20	$600
7.	$42	$8	$75	**8.**	$7.90	$.90	$6.40
9.	$4.10	$.40	$5	**10.**	$3.00	$.22	$2.80

Find the margin and the per cent of margin based on the selling price:

	Cost	*Selling Price*		*Cost*	*Selling Price*
11.	$6	$10	**12.**	$65	$100
13.	$3.50	$5	**14.**	$1.08	$1.44

15. Mr. Sackett bought 10 boxes of apples, each box weighing 25 lb. He paid $1.25 per box for the apples and sold them at a price of 2 lb. for $.19. (*a*) What was his margin? (*b*) What was his per cent of margin, to the nearest whole number of per cent, based on the selling price?
16. A storekeeper bought merchandise that cost $1440 and sold it for $2400. His overhead averaged 15% of his sales. What per cent of the selling price was his net profit?

4. BUYING AT A DISCOUNT

At the end of the summer, the Marion Shop had some summer dresses left over. In order to clear them out, it ran a "40% off" sale. At this sale, Mrs. Breyer bought a dress originally marked $20. To find how much the dress would be reduced, Mrs. Breyer found 40% of $20: .40 × $20 = $8. Then, to find how much she would have to pay for the dress, she subtracted the amount she saved, $8, from the original price, $20. She discovered that the dress would cost her $12.

The original price, \$20, is called the ***list price,*** or ***marked price.*** The per cent of reduction, 40%, is called the ***rate of discount.*** The amount of reduction, \$8, is called the ***discount.*** The reduced price at which the article was sold, \$12, is called the ***sale price,*** or ***net price.***

Procedure. To find the discount, multiply the list price by the rate of discount.

Procedure. To find the net price, subtract the discount from the list price.

NOTE

$$\text{discount} = \text{rate of discount} \times \text{list price}$$
$$\text{net price} = \text{list price} - \text{discount}$$
$$\text{rate of discount} = \frac{\text{amount of discount}}{\text{marked price}} \text{ expressed as a per cent}$$

MODEL PROBLEM

At a sale, Bruce paid \$120 for a canoe whose marked price was \$150. What was the rate of discount on the marked price?

Solution: Step 1. Find the amount of the discount. *Step* 2. Compare this amount with the marked price to find the rate of discount on the marked price.

Step 1

\$150	(marked price)
− 120	(sale price)
\$30	(discount)

Step 2

$$\frac{\text{amount of discount}}{\text{marked price}} = \frac{\$30}{\$150} = \frac{1}{5} = 20\%$$

Answer: The rate of discount was 20%.

EXERCISES

1. If a book listed at \$4.50 is sold at a discount of 20%, how much is the discount?
2. How much will the discount be on a coat listed at \$45 and sold at a discount of 40%?
3. A merchant sells a \$60 suit at a 20% reduction. What is the reduced price?
4. A toy wagon listed at \$14.50 was sold at a discount of 10%. What was the sale price?
5. A camera listed at \$72 was sold at a discount of 15%. What was the net price?
6. An old-model electric toaster is marked at \$22.50. If you receive a 40% discount on the marked price, what will you have to pay for the toaster?

Find the discount and the net price:

	List Price	*Rate of Discount*		*List Price*	*Rate of Discount*
7.	$140	15%	**8.**	$265	18%
9.	$856	$12\frac{1}{2}$%	**10.**	$19.48	2%
11.	$56.69	4%	**12.**	$78.32	35%

Find the discount and the net amount of the following bills:

	Face of Bill	*Rate of Discount*		*Face of Bill*	*Rate of Discount*
13.	$525	4%	**14.**	$315	2%
15.	$3250	$1\frac{1}{2}$%	**16.**	$354.32	3%
17.	$856.24	4%	**18.**	$729.48	$2\frac{1}{2}$%

19. If you received a discount of $2 on a bill of $20, what rate of discount did you receive?

20. Bob paid 68¢ for a toy listed at 85¢. What was the rate of discount?

21. A dealer bought a $16 tire for $13.60. What was the rate of discount?

Find the net price:

	List Price	*Successive Discounts*		*List Price*	*Successive Discounts*
22.	$360	20%, 2%	**23.**	$275	10%, 3%
24.	$640	$12\frac{1}{2}$%, 4%	**25.**	$36.45	$33\frac{1}{3}$%, 2%
26.	$25.38	15%, 4%	**27.**	$64.13	24%, 6%

28. A refrigerator listed at $164.80 is sold with discounts of 25% and 5%. Find the net price.

29. Mr. Harvey found that he could buy a radio listed at $160 with a cash discount of 4%, or a similar radio listed at $200 with discounts of 15% and 10%. Which is the better offer? How much better is it?

30. The owner of a women's dress shop bought 15 dresses listed at $12.50 each and 18 suits listed at $21.75 each. She was allowed a discount of 2% for paying cash. Find the net amount she paid.

31. A school purchased 8 desks at $36 each, 10 chairs at $12.75 each, and 3 tables at $17.50 each. There was a trade discount of $33\frac{1}{3}$% and a cash discount of 2%. If the bill was paid at once, find the net amount paid.

5. WORKING ON COMMISSION

Selma decided to sell Christmas cards for a company in order to make some extra money. She did not receive a salary, but was allowed to keep 30% of her sales. Selma's total sales were $320. To find how much she had earned, Selma found 30% of $320: $.30 \times \$320 = \$96$. To find how much she had to send to the company, Selma subtracted the amount she earned, \$96, from the total sales, \$320, and found that she had to send \$224 to the company.

In this job, Selma's pay was a part of her sales. We say that she was ***working on commission.*** The amount that Selma earned, \$96, is called the ***commission.*** The per cent of the sales that she kept, 30%, is called the ***rate of commission.*** The total sales, \$320, is called the ***gross proceeds.*** The amount that was left after the commission was deducted, \$224, is called the ***net proceeds.***

Procedure. To find the commission, multiply the amount of sales by the rate of commission.

Procedure. To find the net proceeds, subtract the commission from the amount of sales.

NOTE

$$\text{commission} = \text{rate of commission} \times \text{total sales}$$

$$\text{net proceeds} = \text{total sales} - \text{commission}$$

$$\text{rate of commission} = \frac{\text{amount of commission}}{\text{total sales}} \text{ expressed as a per cent}$$

MODEL PROBLEMS

1. A salesman receives an 8% commission on all sales above \$10,000. One month, his sales were \$16,000. How much did he earn that month?

Solution: Step 1. Find the amount of sales above \$10,000. *Step* 2. Find the commission on this amount.

Step 1		*Step* 2	
\$16,000	(total sales)	\$6,000	(difference)
− 10,000	(required sales)	× .08	(rate of commission)
\$6,000	(difference)	\$480.00	(commission)

Answer: He earned \$480 that month.

2. Martin sold \$240 worth of magazines. Of this sum, he received \$48 as commission. What was the rate of commission on his sales?

Solution: To find the rate of commission, compare the commission with the total sales.

$$\text{rate of commission} = \frac{\text{amount of commission}}{\text{total sales}} = \frac{\$48}{\$240} = \frac{1}{5} = 20\%$$

Answer: The rate of commission was 20%.

EXERCISES

1. A salesman receives a 5% commission on his sales. If his sales for the past week were \$800, how much did he receive?
2. A salesman sold \$960 worth of merchandise on a $12\frac{1}{2}\%$ commission. How much did he earn?
3. Mr. Tauton sold a \$578 car and received a 12% commission. How much did he earn?
4. Jane sold Christmas cards for a total of \$150. Of this total, she received \$60 as a commission. What was the rate of commission on her sales?
5. Tom sold \$120 worth of magazines during one month. He received \$40 in commission for his work. What was the rate of commission on his sales?
6. A real-estate agent sold a house for \$18,000 and received a $2\frac{1}{2}\%$ commission. How much did he receive?

Find the commission:

	Sales	*Rate*		*Sales*	*Rate*		*Sales*	*Rate*
7.	\$670	35%	**8.**	\$5000	2%	**9.**	\$1325	6%
10.	\$20,000	$1\frac{1}{2}\%$	**11.**	\$25,000	$\frac{1}{2}\%$	**12.**	\$125.50	8%

13. A salesman sold 40 dozen shirts at \$18 a dozen. He received a 5% commission. How much did he earn?
14. A commission merchant sold 1200 boxes of pears at \$4.50 a box. If he received a 3% commission, how much did he earn?
15. A company employs 15 salesmen who work on a commission of 10%. Together, they sold \$1,350,000 worth of merchandise during the year.
    a. Find the average amount of sales for each salesman during the year.
    b. Find each salesman's average earnings during the year.
16. Hilda works in a store where she earns a salary of \$50 a week. She is also given a commission of 2% on her sales. If she sold \$640 worth of merchandise during one week, how much did she earn that week?

Find the commission and net proceeds:

	Sales	*Rate*		*Sales*	*Rate*		*Sales*	*Rate*
17.	\$628	2%	**18.**	\$124	20%	**19.**	\$732	5%
20.	\$24,000	$\frac{3}{8}\%$	**21.**	\$5600	$2\frac{1}{4}\%$	**22.**	\$258.40	$3\frac{1}{2}\%$

23. An agent sold \$64,000 worth of merchandise at a commission of $\frac{3}{8}\%$. What was his commission? What were the net proceeds?
24. A broker sold a house for Mr. Martin for \$18,500. The broker's commission was 5% for the first \$5000 and $2\frac{1}{2}\%$ for the amount above \$5000.
    a. Find the broker's commission. *b.* Find the net return to Mr. Martin.
25. A salesman receives a salary of \$2700 a year and 2% commission on his sales. During the past year, his sales amounted to \$125,000.
    a. What was his total income for the year?
    b. What was his average weekly salary during this year?

Chapter VIII

ARITHMETIC IN BANKING PRACTICE

I. USING THE REGULAR CHECKING ACCOUNT

Jack Burns opened a ***checking account*** in the Citizens National Bank.

To start his checking account, Mr. Burns made a ***deposit*** of $716.46 on September 26, 1961. He filled out the ***deposit slip*** shown below and gave the slip and the money to the bank clerk, or ***teller.*** A carbon copy of the deposit slip serves as his record of the deposit.

DEPOSITED WITH
CITIZENS NATIONAL BANK
Glendale, New York

DATE Sept. 26 1961
FOR ACCOUNT OF
Jack Burns,
152 First Street,
Glendale, N.Y.

	DOLLARS	CENTS
CASH	131	20
CHECKS	375	32
	118	10
	91	84
TOTAL	716	46

To pay out money, the bank gave Mr. Burns a ***checkbook*** containing blank checks. The checkbook is made so that a check may be detached from the book while a part, called the ***stub,*** remains in the book. The check, properly filled out and signed, serves as a written order to the bank to pay out a certain sum of money from Mr. Burns' account. When properly filled out, the stub serves Mr. Burns as a record of the check.

On September 28, 1961, Mr. Burns made another deposit of $115.22 and issued his first check to Fred Lee for a plumbing bill of $37.50. Mr. Burns is called the ***maker,*** and Fred Lee is called the ***payee.***

An essential part of the check is the ***signature.*** Only Mr. Burns can sign checks drawn on his account.

The fully written out check and stub are shown on the next page.

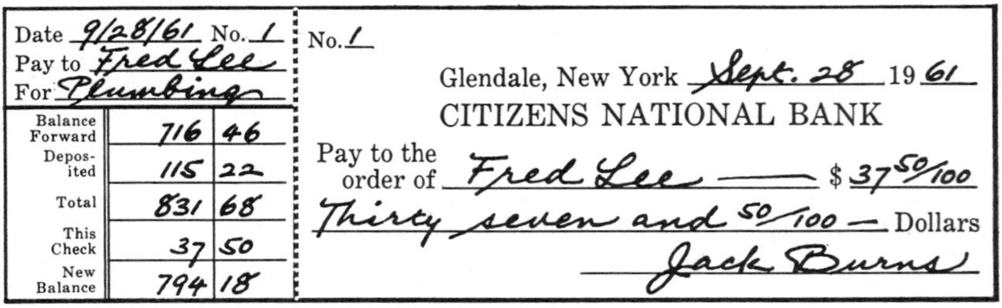

Date 9/28/61 No. 1
Pay to Fred Lee
For Plumbing

Balance Forward	716	46
Deposited	115	22
Total	831	68
This Check	37	50
New Balance	794	18

No. 1

Glendale, New York Sept. 28 1961
CITIZENS NATIONAL BANK
Pay to the order of Fred Lee —— $ 37 50/100
Thirty seven and 50/100 — Dollars
Jack Burns

If Mr. Lee deposits the check in his bank, he ***endorses*** it by signing his name on the back of the check, as shown below under ***limited endorsement.*** His account will then be credited with the amount of the check. When the check is sent to Mr. Burns' bank, it will debit (charge) Mr. Burns' account.

Mr. Lee might not deposit the check in his account. He might cash it with someone. He would then endorse it to the order of that person, as shown below under ***endorsement in full.***

Limited Endorsement

Endorsement in Full

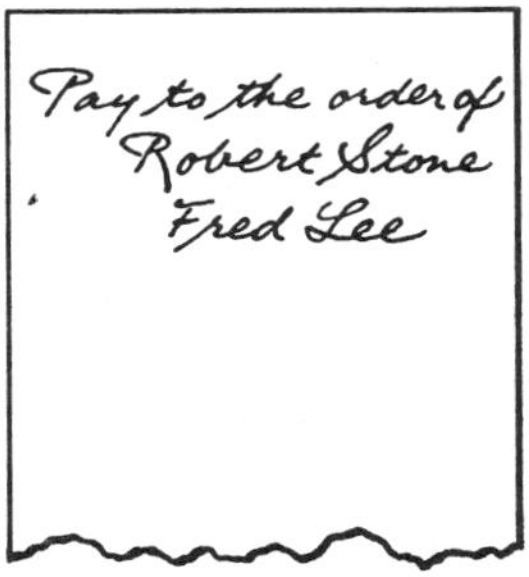

The bank keeps a continuous record of the sums that Mr. Burns deposits and the sums that he withdraws. At the end of each month, the bank sends him a copy of this record, called a ***statement.*** The bank also returns to him all the checks that it paid out for his account.

EXERCISES

1. James Waverly, who worked in the Pick-Quick Market, made daily deposits in the bank for the store. On January 22, 1965, he took to the bank for deposit \$200.00 in bills, \$15.00 in quarters, \$8.00 in dimes, \$10.00 in half-dollars, \$6.00 in nickels, \$4.00 in pennies, and a check for \$24.50. Prepare a deposit slip using the model on page 161.

2. Using the model on page 162, write and sign checks for the following:

a. The sum of \$125 to be paid to Arthur Whitney.

b. The sum of \$8.43 to be paid to the Consolidated Edison Company.

c. The sum of \$112.69 to be paid to the City Collector.

3. The check stubs shown below are from the checkbook of Harry Stuart. Answer the following questions by referring to the stubs.

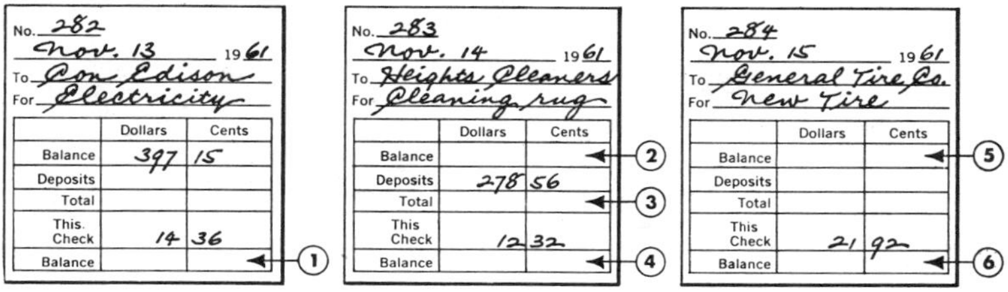

a. Find the amounts that should be written in the spaces indicated by 1, 2, 3, 4, 5, and 6 on the stubs above.

b. Who was the maker of check No. 284?

c. Who was the payee of check No. 284?

d. Write the two ways in which the amount to be paid on check No. 284 was written on the check.

4. Mr. Pat Harrison bought merchandise worth $258 from Barton Bros. He received a 2% discount for paying his bill in less than 10 days.

a. What was the net amount that Mr. Harrison had to pay Barton Bros.?

b. Write a check for this amount as Mr. Harrison would have written it on May 14, 1965. Use a model check.

5. On December 1, Mr. Peterson had a balance of $457.60 in the bank. During the month, he deposited checks and cash as follows: $75.48; $36.92; $145.60; $25.35; $89.13. During the same month, he drew checks as follows: $15.30; $79.80; $42.21; $84.35; $76.09. The bank deducted from the account 5 cents for each deposit and 5 cents for each check. What was Mr. Peterson's bank balance on January 1?

2. USING TRAVELERS CHECKS

Phil Crowley is planning to take a trip. He does not want to carry with him a large amount of cash. What does he do?

Mr. Crowley can go to a bank or to an office of the American Express Company and exchange cash for ***travelers checks,*** which are issued in denominations of $10, $20, $50, and $100. He pays a small charge of 1% of the value of the checks he buys. If he buys $1000 in checks, the charge is $10 and he pays $1010 for them.

In the presence of the person from whom he buys the checks, Mr. Crowley signs each one in the upper left-hand corner. The signature therefore identifies Mr. Crowley as the real owner of those checks. When Mr. Crowley wishes to cash a check, he fills it out and signs it again, this time in the lower left-hand corner and in the presence of the dealer he is paying it to. The dealer, seeing that the new signature is identical with the original signature, feels assured that the person cashing the check is Mr. Crowley—and the rightful owner.

EXERCISES

If the charge is 1% of the value of the checks, how much must be paid when purchasing travelers checks for each of the following amounts?

1. $200 **2.** $350 **3.** $500 **4.** $750 **5.** $1000 **6.** $2500

3. FINDING INTEREST AND AMOUNT

Mr. Story needed $800 to build a rumpus room in the basement of his house. His uncle, Mr. Sweeney, agreed to lend Mr. Story the $800 for a period of 2 years and said that he would charge him 3% of the $800 for each of the 2 years. A charge for the use of borrowed money is called ***interest.*** The sum of money that Mr. Story borrowed, $800, is called the ***principal.*** 3% is the ***annual rate of interest;*** 2 years is the ***time.***

To find what the interest would be for borrowing the money for the first year, Mr. Story found 3% of $800: $.03 \times \$800 = \$24$. Then he multiplied the interest for one year, $24, by the number of years he was borrowing the money, 2, and found that the interest charge would be $2 \times \$24$, or $48.

Procedure. To find simple interest, multiply the principal by the annual rate of interest. Then multiply by the time in years.

To find the amount of money he would have to return to Mr. Sweeney at the end of 2 years, Mr. Story added the principal, \$800, and the interest, \$48. He found that the amount would be \$800 + \$48, or \$848. This sum of money—the sum of the principal and the interest—is called the ***amount.***

Procedure. To find the amount, add the principal and the interest.

NOTE

interest = principal × rate × time, or I = P × R × T

amount = principal + interest

$$\text{annual rate of interest} = \frac{\text{interest for 1 year}}{\text{principal}} \text{ expressed as a per cent}$$

In the first rule above, "rate" means annual rate of interest and "time" means time in years. In business transactions, to simplify calculations, a month is considered 30 days; a year, 360 days.

MODEL PROBLEMS

1. Mr. Bates borrowed \$400 for 3 years. Find the interest he paid if the annual rate of interest was 6%.

Solution:

Method 1	*Method* 2
\$400 (principal)	principal = \$400
×.06 (rate of interest)	rate = 6% = .06
\$24.00 (interest for 1 year)	time = 3 years
×3 (number of years)	interest = principal × rate × time
\$72.00 (interest for 3 years)	interest = \$400 × .06 × 3
	interest = \$24.00 × 3 = \$72.00

Answer: He paid \$72 in interest.

2. Mr. Sweet borrowed \$900 for 60 days at an annual rate of interest of 5%. Find how much interest he paid.

Solution: Since 60 days is $\frac{60}{360}$ or $\frac{1}{6}$ of a year, the time is $\frac{1}{6}$ year.

Method 1	*Method* 2
\$900 (principal) ×.05 (rate of interest) \$45.00 (interest for 1 year) Interest for $\frac{1}{6}$ year is $\frac{1}{6}$ of \$45.00. $6\overline{)\$45.00}$ = \$7.50	P = \$900, R = 5% = $\frac{5}{100}$, T = 60 da. = $\frac{1}{6}$ yr. I = P × R × T $I = \overset{\$9}{\cancel{\$900}} \times \frac{5}{\underset{1}{\cancel{100}}} \times \frac{1}{6} = \frac{\$45}{6}$ I = \$7.50

Answer: He paid \$7.50 in interest.

3. Mr. Crews borrowed \$840 at $4\frac{1}{2}\%$ for 3 months.

a. How much interest will he have to pay?

b. What is the total amount he will have to repay?

Solution: Since 3 months is $\frac{3}{12}$ or $\frac{1}{4}$ of a year, the time is $\frac{1}{4}$ year.

a. *Step* 1	*Step* 2	*b.* *Step* 3
\$840 (principal) ×.045 (rate) 4 200 33 60 \$37.800 (interest for 1 year)	Interest for $\frac{1}{4}$ year is $\frac{1}{4}$ of \$37.80. $4\overline{)\$37.80}$ = \$9.45	\$840.00 (principal) +9.45 (interest) \$849.45 (amount)

Answer: *a.* \$9.45 *b.* \$849.45

EXERCISES

In 1–10, find the interest:

	Principal	*Rate*	*Time*		*Principal*	*Rate*	*Time*
1.	\$700	5%	1 year	**2.**	\$680	4%	2 years
3.	\$925	6%	3 years	**4.**	\$2750	3%	1 year
5.	\$4625	4%	2 years	**6.**	\$1375	5%	3 years
7.	\$825.75	2%	1 year	**8.**	\$650.25	4%	2 years
9.	\$550	$4\frac{1}{2}\%$	1 year	**10.**	\$1700	$3\frac{1}{2}\%$	2 years

In 11–20, find the interest:

	Principal	*Rate*	*Time*		*Principal*	*Rate*	*Time*
11.	\$300	5%	3 months	**12.**	\$450	6%	4 months
13.	\$850	6%	6 months	**14.**	\$2500	4%	8 months
15.	\$200	6%	30 days	**16.**	\$1200	5%	60 days
17.	\$750	4%	90 days	**18.**	\$1500	6%	1 year, 6 months
19.	\$600	$5\frac{1}{2}$%	30 days	**20.**	\$480	$4\frac{1}{2}$%	2 months

In 21–28, find the interest and the amount:

	Principal	*Rate*	*Time*		*Principal*	*Rate*	*Time*
21.	\$725	3%	2 years	**22.**	\$840	5%	9 months
23.	\$1450	6%	60 days	**24.**	\$750	$4\frac{1}{2}$%	8 months
25.	\$3200	$3\frac{1}{2}$%	6 months	**26.**	\$5400	$4\frac{3}{4}$%	2 years
27.	\$600,000	$\frac{1}{2}$%	30 days	**28.**	\$24,000,000	$\frac{3}{4}$%	30 days

In 29–32, find the annual rate of interest:

	Principal	*Interest*	*Time*		*Principal*	*Interest*	*Time*
29.	\$800	\$32	1 year	**30.**	\$600	\$72	2 years
31.	\$720	\$5	60 days	**32.**	\$250	\$2.50	4 months

33. Mr. Smith needed \$4500 to buy a business. He borrowed this sum of money for 2 years at 4%. How much interest did he pay?

34. Mr. Carey needed \$750 to buy a car. He borrowed this sum for 2 years and 6 months at 6% interest. How much interest did he pay?

35. Mr. Carson borrowed \$1500 from a bank for 90 days at 3% interest. What interest did he pay?

36. Mr. Trout wished to start a business. He borrowed \$2500 from a bank at 4% interest for a period of 10 months. What was the cost of the loan?

37. The yearly income from a \$2400 loan is \$120. What is the annual rate of interest?

38. Raymond paid \$3 in interest for a loan of \$225 for a period of 4 months. What was the annual rate of interest he was paying?

39. Reuben borrowed some money for 1 year at a rate of 4%. If he paid \$24 in interest that year, how much did he borrow?

40. Mr. Hahn borrowed \$600 at 3% for 90 days. How much did he pay at the end of 90 days when he repaid the loan and paid the interest that was due?

41. Mr. Purvis bought a business for \$12,000. He paid \$7500 in cash and agreed to pay the balance in 9 months with interest at 6%. How much was the payment that he had to make at the end of that time? How much did the business actually cost Mr. Purvis?

4. PROMISSORY NOTES

Peter Garvey needed $700 to improve his business. His friend, Henry Ross, agreed to lend that sum to him for 6 months at an interest rate of 5% per annum. Upon receiving the money, Mr. Garvey gave Mr. Ross the following ***promissory note.***

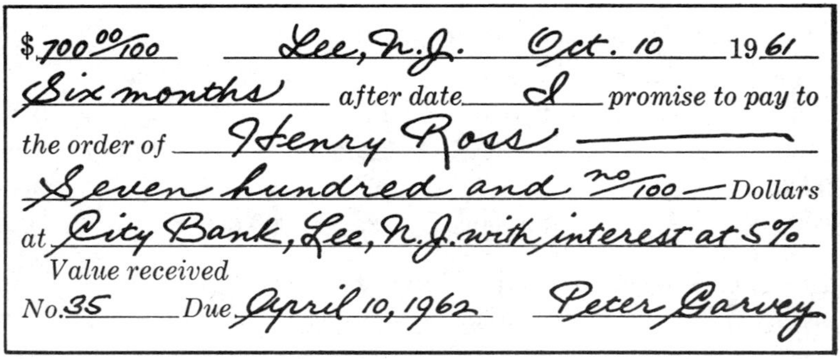

$700 00/100 Lee, N.J. Oct. 10 1961

Six months after date I promise to pay to

the order of Henry Ross

Seven hundred and no/100 Dollars

at City Bank, Lee, N.J. with interest at 5%

Value received

No. 35 Due April 10, 1962 Peter Garvey

The ***maker*** of the note is Peter Garvey, the person who issued the note. The ***payee*** is Henry Ross, the person who will collect the $700 and interest. The ***face,*** or ***principal,*** is $700. The ***time*** is 6 months. The ***rate of interest*** is 5% per annum. The ***date of maturity*** is the date when payment is due. In the above model note, the date of maturity is April 10, 1962.

To find the amount of interest he would have to pay, Mr. Garvey multiplied the principal, $700, by the rate of interest, 5%, and then multiplied the result by $\frac{1}{2}$, since 6 months equal $\frac{1}{2}$ year. Since $\$700 \times .05 \times \frac{1}{2} = \$17.50$, he knew that the interest would be $17.50. He then added the principal, $700, and the interest, $17.50, and found that the amount he would have to repay Mr. Ross would be $717.50.

EXERCISES

1. From the following facts, write promissory notes. (Use the model on page 168.)

	Principal	*Date*	*Time*	*Place*	*Rate*	*Maker*	*Payee*
a.	$600	7/13/61	4 mo.	Troy, N.Y.	3%	M. Hart	I. Sax
b.	$800	2/15/61	6 mo.	Rye, N.Y.	6%	S. Carr	J. Wahl
c.	$450	4/10/61	60 da.	Rome, N.Y.	4%	R. Bang	H. Rowe

Find the interest and amounts for each of the following notes:

	Principal	*Rate*	*Time*		*Principal*	*Rate*	*Time*
2.	\$650	3%	30 da.	**3.**	\$800	6%	90 da.
4.	\$320	4%	75 da.	**5.**	\$535	2%	6 mo.
6.	\$720	5%	4 mo.	**7.**	\$900	$3\frac{1}{2}$%	1 mo.

5. DISCOUNTING NOTES

In order to buy some merchandise, Robert Hamilton borrowed \$3000 from the Citizens National Bank for a period of 4 months at a rate of interest of 6% per year. He gave the bank the following promissory note:

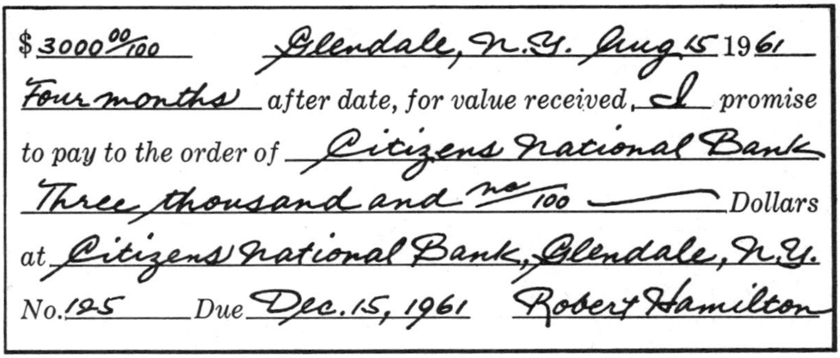

\$3000 00/100 Glendale, N.Y. Aug 15 1961

Four months *after date, for value received,* I *promise*

to pay to the order of Citizens National Bank

Three thousand and no/100 —— *Dollars*

at Citizens National Bank, Glendale, N.Y.

No. 125 *Due* Dec. 15, 1961 Robert Hamilton

The bank asked Mr. Hamilton to pay the interest *in advance.* For this reason, the interest was not mentioned in the promissory note that Mr. Hamilton gave the bank. Interest collected by the bank in advance is called ***bank discount.*** The bank discounted Mr. Hamilton's note.

The interest on the note amounted to $\$3000 \times .06 \times \frac{1}{3}$, or \$60. Since Mr. Hamilton paid the interest, \$60, in advance, he received $\$3000 - \$60$, or \$2940. This amount remaining after the discount was deducted is called the ***proceeds.***

EXERCISES

Find the bank discount and proceeds of the following notes:

	Principal	*Rate*	*Time*		*Principal*	*Rate*	*Time*
1.	\$400	4%	60 da.	**2.**	\$780	6%	90 da.
3.	\$250	5%	30 da.	**4.**	\$580	4%	60 da.
5.	\$375	6%	30 da.	**6.**	\$750	6%	3 mo.
7.	\$940	5%	6 mo.	**8.**	\$1200	6%	2 mo.
9.	\$3000	4%	4 mo.	**10.**	\$5000	6%	10 mo.

6. SAVING MONEY IN SAVINGS BANKS, AND UNITED STATES GOVERNMENT BONDS

SAVINGS BANKS

Mrs. Wooster saved \$100 from the family income. On January 2, she opened an account in a savings bank and deposited the money there. She was given a ***bankbook*** in which a record is kept of her deposits and withdrawals. This book shows the amount to her credit, her ***balance,*** at all times.

The savings bank paid interest at the rate of 4% a year. However, the interest for the first half-year was added to Mrs. Wooster's original principal at the end of 6 months. This total sum became the principal on which the bank figured the interest for the second half-year. Interest paid on the sum of the principal and the interest already earned is called ***compound interest.*** In Mrs. Wooster's account, the interest was *compounded twice a year,* or *semiannually.* If the interest is computed at the end of every 3 months, or four times a year, we say that interest is being *compounded quarterly.*

Let us see how much Mrs. Wooster had to her credit at the end of the year. There are two ways that can be used to find the interest on \$100 at 4% for $\frac{1}{2}$ year. One way is by finding 4% of \$100, which is \$4, and then dividing by 2, which is \$2. Another way is by realizing that if the principal earns 4% as interest during a whole year, then during the first 6 months, or $\frac{1}{2}$ year, it will earn $\frac{1}{2}$ of 4%, or 2% interest.

The amount of her credit at the end of the year can be found by the following two-step method:

Step 1

\$100	(principal for first 6 months, first interest period)
× .02	(rate of interest for 6-month interest period)
\$2.00	(interest for first interest period)
+ 100.00	(principal for first interest period)
\$102.00	(amount at end of first interest period, which becomes the principal for the second interest period)

Step 2

\$102	(principal for second interest period)
× .02	(rate of interest for 6-month interest period)
\$2.04	(interest for second interest period)
+ 102.00	(principal for second interest period)
\$104.04	(amount at end of second interest period)

Instead of having \$104.04 to her credit at the end of the year, Mrs. Wooster would have had only \$104 if the bank had paid her only simple interest of 4%: \$100 × .04 = \$4; adding \$4 to her principal of \$100 gives \$104.

Although Mrs. Wooster gained only the small sum of 4¢ in this example, the difference between compound interest and simple interest becomes more significant as the principal becomes larger and the period of time longer.

To simplify computations, banks use compound interest tables and even mechanical computers. The compound interest table that follows shows the sum of money that $1 will amount to if it is compounded by periods at interest rates of 1, $1\frac{1}{2}$, 2, $2\frac{1}{2}$, and 3% per period.

If money is invested for 5 years at 4% a year compounded quarterly:

1. Each period will be $\frac{1}{4}$ of a year.
2. There will be 4 periods in a year.
3. There will be 5 × 4, or 20 periods in 5 years.
4. The rate of interest for each period will be $\frac{1}{4}$ of 4%, or 1%.

AMOUNT OF $1 AT COMPOUND INTEREST

Periods	1%	$1\frac{1}{2}$%	2%	$2\frac{1}{2}$%	3%
1	1.01000	1.01500	1.02000	1.02500	1.03000
2	1.02010	1.03023	1.04040	1.05063	1.06090
3	1.03030	1.04568	1.06121	1.07689	1.09273
4	1.04060	1.06136	1.08243	1.10381	1.12551
5	1.05101	1.07728	1.10408	1.13141	1.15927
6	1.06152	1.09344	1.12616	1.15969	1.19405
7	1.07214	1.10984	1.14869	1.18869	1.22987
8	1.08286	1.12649	1.17166	1.21840	1.26677
9	1.09369	1.14339	1.19509	1.24886	1.30477
10	1.10462	1.16054	1.21899	1.28008	1.34392
12	1.12683	1.19562	1.26824	1.34489	1.42576
16	1.17258	1.26899	1.37279	1.48451	1.60471
20	1.22019	1.34686	1.48595	1.63862	1.80611

We can use the above table to find the amount to Mrs. Wooster's credit at the end of 1 year as follows: Since her money earns interest semiannually, there are 2 periods in one year. Also, since each period is $\frac{1}{2}$ year, the rate of interest for each period is $\frac{1}{2}$ of 4%, or 2%. Looking down the column headed "2%" and across the line opposite "2 periods," we find that $1 will amount to $1.04040. To find what $100 will amount to, we multiply $1.04040 by 100:

100 × $1.04040 = $104.04 (amount at end of 2 periods, or 1 year)

UNITED STATES GOVERNMENT BONDS

Another way of saving money is to buy *United States Savings Bonds.* By issuing ***bonds,*** the government is borrowing money and promising to pay it back at a specified time and to pay interest to the purchaser at a specified rate.

The bonds most frequently purchased are the United States Savings Bonds—Series E. These bonds, which can be bought in denominations of \$18.75, \$37.50, \$75, \$375, and \$750, mature in 7 years and 9 months, or 7.75 years, and have ***maturity values*** of \$25, \$50, \$100, \$500, and \$1000 respectively. If these bonds are bought and kept until maturity, they will earn an average annual interest of 4.3%.

EXERCISES

1. Mrs. Sikorsky had \$400 in her account in the Central Savings Bank on Jan. 1, 1963. The bank paid 4% interest compounded annually. Mrs. Sikorsky made no deposits or withdrawals until Dec. 31, 1964.
a. How much did Mrs. Sikorsky have to her credit in the bank on Jan. 1, 1965?
b. How much interest did Mrs. Sikorsky receive on her money between Jan. 1, 1963 and Jan. 1, 1965?
c. How much would Mrs. Sikorsky have received over the 2-year period if she had invested her money at an annual rate of 4% simple interest?
d. Find the difference between the amounts found in part *b* and part *c*.

2. Mr. Tepper kept \$800 in a savings bank that paid compound interest of 4% annually and compounded the interest quarterly. What was the amount of interest he earned from July 1, 1964 to June 30, 1965?

In 3–6, use the compound interest table on page 171:

3. If \$800 is kept in a bank that pays interest at the rate of 4% a year compounded quarterly, how much will be in the account at the end of 2 years?

4. \$2500 is invested for 4 years at 5% a year compounded semiannually.
a. What will the \$2500 amount to at the end of the 4-year period?
b. How much interest was earned during this period of time?

5. How much will \$5000 amount to if it is invested for 5 years at 6% per year compounded quarterly?

6. Find the interest earned when \$2000 is invested for 4 years at 6% per year compounded semiannually.

7. Carl saves \$.25 a week. For how many weeks must he save in order to have \$18.75 for the purchase of a \$25 E-bond?

8. Carl paid \$18.75 for a United States Savings Bond—Series E. After 7.75 years, he cashed the bond for \$25.
a. How much interest did he receive?
b. What was the average interest earned per year?
c. Find, to the nearest tenth of a per cent, the average annual rate of interest.

9. Mr. Clark decided to buy government savings bonds through the payroll-savings plan of the company for which he works. During the year, he plans to buy 20 bonds at \$37.50 each. How much should be withheld from his pay check each month to carry out this plan?

Chapter IX

RATIO

1. COMPARING NUMBERS BY USING RATIOS

If Richard is 16 years old and Martha is 4 years old, we can compare their ages in two ways. (1) We can say that Richard is 12 years older than Martha; we subtracted 4 from 16. Or (2) we can say that Richard is 4 times as old as Martha; we divided 16 by 4.

When we compare two numbers by division, we are finding their ***ratio.*** The ratio of Richard's age to Martha's age is $16 \div 4$ or $\frac{16}{4}$ or $\frac{4}{1}$. This means that Richard's age is 4 times as great as Martha's age.

Another way of writing the ratio $\frac{16}{4}$ is 16:4. Both the expressions $\frac{16}{4}$ and 16:4 are read "16 to 4."

The comparison of Richard's age with Martha's age may be expressed: For the 16 years in Richard's age, there are 4 years in Martha's age; or the number of years in Richard's age to the number of years in Martha's age is 16 to 4. This may be written $\frac{16}{4}$ or 16:4.

The comparison of Richard's age with Martha's age may also be expressed: For every 8 years in Richard's age, there are 2 years in Martha's age; or the number of years in Richard's age to the number of years in Martha's age is 8 to 2. This may be written $\frac{8}{2}$ or 8:2.

The comparison of Richard's age with Martha's age may also be expressed: For every 4 years in Richard's age, there is 1 year in Martha's age; or the number of years in Richard's age to the number of years in Martha's age is 4 to 1. This may be written as $\frac{4}{1}$ or 4:1.

In the ratio $\frac{4}{1}$ or 4:1, the numbers 4 and 1 are called the ***terms*** of the ratio. In representing a ratio, it is important to write the terms in their proper order. The ratio of Richard's age to Martha's age may be written $\frac{4}{1}$ or 4:1. This means that Richard is 4 times as old as Martha. The ratio of Martha's age to Richard's age may be written $\frac{1}{4}$ or 1:4. This means that Martha's age is $\frac{1}{4}$ of Richard's age.

Procedure. To find the ratio of two numbers, divide the first number by the second number (not zero).

FINDING THE RATIO OF UNITS OF MEASURE

If we wish to find the ratio of 4 quarters to 2 dimes, we must realize that 4 quarters = 100 cents and 2 dimes = 20 cents. Now we can find the ratio of 100 cents to 20 cents, which may be written $\frac{100}{20}$ [100:20] or $\frac{5}{1}$ [5:1]. This means that 4 quarters is 5 times as much as 2 dimes.

Procedure. To find the ratio of two quantities that are expressed in different units of measure, first express both quantities in the same unit of measure and then divide the first of the resulting quantities by the second.

FINDING EQUAL RATIOS

Principle. When both terms of a ratio are multiplied or divided by the same number (not zero), the results are the terms of another ratio equal to the original ratio.

Thus, $\frac{3}{2} = \frac{3 \times 5}{2 \times 5} = \frac{15}{10}$, *or* $3:2 = 15:10$

and $\frac{16}{4} = \frac{16 \div 2}{4 \div 2} = \frac{8}{2}$, *or* $16:4 = 8:2$.

A ratio is expressed in simplest form when both terms of the ratio are whole numbers with no whole number other than 1 as a common divisor. Thus, to express the ratio $\frac{45}{60}$ in simplest form, we can divide both its terms by their largest common divisor, 15, and obtain the ratio $\frac{3}{4}$ as the result.

USING RATIOS TO EXPRESS RATES

When we say that a car is traveling at a rate of 40 miles per hour, we are comparing 40, the number of miles traveled, with 1, the number of hours traveled. The word *per* indicates that we are comparing two different kinds of quantities. We have already learned that when we compare two quantities, we are finding their ratio. Thus, we can express the rate of 40 miles per hour as the ratio $\frac{40}{1}$.

If apples are sold at the rate of 29 cents per pound, the word *per* indicates the ratio of two different kinds of quantities, 29 cents and 1 pound. The ratio, which may be written $\frac{29}{1}$, expresses the rate 29 cents for 1 pound.

Similarly,

8 apples for 4 boys may be expressed $\frac{8}{4}$ or $\frac{2}{1}$; 2 for 1 (2 apples per boy)

60 miles in 2 hours may be expressed $\frac{60}{2}$ or $\frac{30}{1}$; 30 per 1 (30 miles per hour)

6 cans of juice for 70 cents may be expressed $\frac{6}{70}$ or $\frac{3}{35}$; 3 for 35 (3 cans for 35 cents)

MODEL PROBLEMS

1. Phil's forearm is $1\frac{1}{2}$ ft. long. In a photograph, his forearm is 1 in. long. What is the ratio of the actual length of Phil's forearm to the length of the forearm in the picture?

Solution:

$$\text{ratio} = \frac{\text{actual length of Phil's forearm}}{\text{length of forearm in picture}} = \frac{1\frac{1}{2}\text{ ft.}}{1\text{ in.}} = \frac{18\text{ in.}}{1\text{ in.}} = \frac{18}{1}$$

Answer: 18:1

2. Express the ratio $1\frac{3}{4}$ to $1\frac{1}{2}$ in simplest form.

Solution: ratio $= 1\frac{3}{4}:1\frac{1}{2} = 1\frac{3}{4} \div 1\frac{1}{2} = \frac{7}{4} \div \frac{3}{2} = \frac{7}{4} \times \frac{2}{3} = \frac{14}{12} = \frac{7}{6}$ *Ans.*

3. The ratio of the width of a rectangle to the length of the rectangle is $\frac{1}{2}$. If the length of the rectangle is 20 ft., find the width of the rectangle.

Solution: When the ratio of the width of a rectangle to its length is $\frac{1}{2}$, the width is $\frac{1}{2}$ of its length. Therefore, if the length is 20 ft., the width is $\frac{1}{2}$ of 20 ft., or 10 ft. *Ans.*

EXERCISES

1. Write the following ratios (*a*) as fractions and (*b*) using the colon:

(1) 36 to 12 (2) 48 to 24 (3) 40 to 25 (4) 2 to 3 (5) 5 to 4

2. Reduce the following ratios to simplest form:

a. $\frac{8}{32}$ *b.* $\frac{40}{5}$ *c.* $\frac{12}{28}$ *d.* 20:10 *e.* 15:45 *f.* 15:24

3. In each of the following:

a. Write in simplest form the ratio of the larger to the smaller number.
b. The larger number is how many times the smaller number?

(1) 10 and 5 (2) 12 and 8 (3) 9 and 27 (4) 15 and 25

4. *a.* If the ratio of two numbers is 5:1, find the quotient of the terms of the ratio.
b. The larger of the two numbers is how many times the smaller number?

5. In each of the following:

a. Write in simplest form the ratio of the smaller to the larger number.
b. The smaller number is what part of the larger number?
(1) 8 and 24 (2) 50 and 10 (3) 12 and 18 (4) 36 and 27

6. Which of the following ratios are equal?

a. $\frac{2}{3}$, $\frac{6}{9}$, $\frac{10}{30}$, $\frac{28}{36}$, $\frac{50}{75}$ *b.* 10:8, 20:16, 15:13, 4:5, 50:40

7. Find the missing numbers:

a. $\frac{1}{2} = \frac{?}{24}$ *b.* $\frac{2}{3} = \frac{10}{?}$ *c.* $\frac{4}{5} = \frac{?}{100}$ *d.* $\frac{5}{8} = \frac{25}{?}$

8. Find 4 pairs of numbers whose ratio is (*a*) $\frac{1}{2}$ (*b*) 3:1 (*c*) $\frac{3}{5}$ (*d*) 4:3

9. *a.* Find the quotient of all pairs of numbers whose ratio is 5:2.
b. The larger number in each pair is how many times the smaller number?

10. *a.* Find the quotient of all pairs of numbers whose ratio is 5:8.
b. The smaller number in each pair is what fractional part of the larger number?

11. Express the following ratios in simplest form:

a. $\frac{3}{4}$ to $\frac{1}{4}$ *b.* $1\frac{1}{8}$ to $\frac{3}{8}$ *c.* $2\frac{1}{2}:1\frac{1}{4}$ *d.* $1\frac{2}{3}:2\frac{1}{2}$
e. 1.2 to 2.4 *f.* .75 to .25 *g.* 1.2:4 *h.* .5:.01

12. Find, in simplest form, the ratio of the larger to the smaller quantity:

a. $1\frac{1}{4}$ hr. and $\frac{3}{4}$ hr. *b.* $\frac{2}{3}$ yd. and $\frac{5}{6}$ yd. *c.* .5 ft. and 2.5 ft.

13. Express the following ratios in simplest form:

a. 1 yd. to 1 ft. *b.* 2 ft. to 6 in. *c.* $\frac{1}{2}$ yd. to 8 in. *d.* 4 oz. to $\frac{1}{2}$ lb.
e. 1 hr. to 15 min. *f.* 8 hr. to 2 da. *g.* $\frac{3}{4}$ min. to 10 sec. *h.* $3 to 75¢

14. Express each of the following rates as a ratio in simplest form:

a. 4 apples for 24 cents *b.* 6 bars of candy for 30 cents
c. 500 miles in 5 hours *d.* 24 sandwiches for 12 people

15. A baseball team played 144 games and won 96.
a. What is the ratio of the number of games won to the number of games played?
b. For every 3 games played, how many games were won?

16. A student did 6 out of 10 problems correctly.
a. What is the ratio of the number right to the number wrong?
b. For every two answers that were wrong, how many answers were right?

17. A cake recipe calls for $1\frac{1}{2}$ cups of milk to $1\frac{3}{4}$ cups of flour. What is the ratio of the number of cups of milk to the number of cups of flour in this recipe?

18. The ratio of the number of pupils in the seventh grade of a school to the number of pupils in the eighth grade is $\frac{4}{5}$. If the number of pupils in the eighth grade is 125, find the number of pupils in the seventh grade.

19. The ratio of a father's age to his son's age is $\frac{7}{2}$. If the son's age is 10 years, how old is the father?

20. Mr. Corliss is following a budget in which the ratio of the amount paid for rent to the total monthly income is 1:4. If Mr. Corliss earns $400 a month, what is his monthly rent?

21. Tom and his helper Bill agree to do a job for $60. They also agree to share this sum of money in the ratio 3:1.
a. What fractional part of the $60 should each boy receive?
b. How much money should each boy receive?

22. Divide a profit of $30,500 between two partners, Mr. Samuels and Mr. Deegan, who have agreed to share the profits in the ratio 2:3.

23. If a man earns $8.40 in 4 hours, how much will he earn in 40 hours at the same rate of pay?

24. If 2 pieces of pastry cost 15 cents, find the cost of a dozen pieces.

25. If 2 feet of lumber cost $.49, what is the cost of 4 yards of this lumber?

2. EXPRESSING RATIOS AS DECIMALS AND AS PER CENTS

On a fishing trip, Perry caught 15 fish and Ray caught 10 fish. In order to find what per cent of the entire catch he had caught, Perry compared the number of fish that he had caught with the total number of fish caught. In other words, he found the ratio of 15 to 25. He discovered that he had caught $\frac{15}{25}$ or $\frac{3}{5}$ of the entire catch. He changed the fraction $\frac{3}{5}$ to its equal decimal, .60. Therefore, he knew that he had caught 60% of the entire catch.

We see, therefore, that a ratio can be expressed not only as a fraction, but also as a decimal and as a per cent.

EXERCISES

1. Write the following ratios as decimals:

a. $\frac{19}{100}$ *b.* $\frac{7}{10}$ *c.* $\frac{49}{50}$ *d.* $\frac{18}{72}$ *e.* $\frac{24}{48}$ *f.* $\frac{9}{6}$

2. Write the following ratios as per cents:

a. $\frac{51}{100}$ *b.* $\frac{13}{25}$ *c.* $\frac{6}{24}$ *d.* $\frac{24}{32}$ *e.* $\frac{21}{24}$ *f.* $\frac{15}{12}$

3. Write the following ratios as decimals correct to the nearest hundredth:

a. 20:80 *b.* 47:50 *c.* 37:100 *d.* 9:25 *e.* 17:20
f. 38:200 *g.* 36:30 *h.* 15:45 *i.* 35:8 *j.* 73:96

4. Write the following ratios as per cents correct to the nearest per cent:

a. 9:100 *b.* 7:10 *c.* 5:20 *d.* 18:25 *e.* 20:8
f. 48:64 *g.* 11:33 *h.* 21:36 *i.* 56:79 *j.* 42:23

5. Express the ratio of 6 inches to 1 foot as a per cent.

6. On a test, a pupil got 8 questions right and 2 questions wrong. Express, as a per cent, the ratio of the number of questions he got right to the total number of questions on the test. What per cent did he receive on the test?

7. One foot is what per cent of a yard?

8. One quart is what per cent of a pint?

9. Forty-five minutes is what per cent of an hour?

10. A rectangle is 20 feet wide and 25 feet long. What is the ratio of its width to its length? What per cent of its length is its width?

11. If 70% of the pupils in a school are boys, what is the ratio of the number of boys to the number of girls in the school?

12. Susie got 90% on her science test. If all questions counted the same number of credits, what was the ratio of the number of questions she got wrong to the number she got right?

3. UNDERSTANDING THE USE OF SCALE DRAWINGS

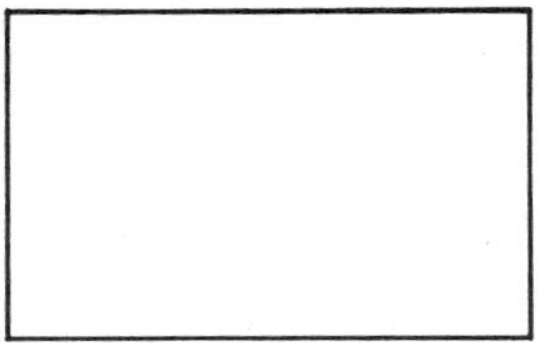

Scale: $\frac{1}{16}$ in. = 1 ft.

Mr. Shriver was discussing with his wife the plans for the outdoor patio that he was about to build. He told her that he was thinking of making it rectangular in shape, 22 ft. long and 14 ft. wide. When she told him that she could not picture what it would look like, he made a ***scale drawing*** of the patio. In making the scale drawing, he used $\frac{1}{16}$ of an inch on the scale drawing to represent a length of 1 foot on the patio. On the scale drawing, the length of 22 ft. was represented by a line $\frac{22}{16}$ in. long, and the 14 ft. width was represented by a line $\frac{14}{16}$ in. long. The shape of the scale drawing that Mr. Shriver made was exactly the same as the shape of the patio he was going to build.

A scale drawing always pictures the exact shape of the object that the drawing represents. When a large object must be pictured in a small space, a scale is chosen that will make the scale drawing smaller than the object—for example, when drawing a road map or the floor plan of a building. When a small object is to be pictured in a large space, a scale is chosen that will make the scale drawing larger than the object—for example, when drawing the detailed blueprint for a very small part of a complicated machine.

The scale used in a scale drawing can be expressed as a ratio. If a length of 1 ft. in a scale drawing represents a length of $\frac{1}{4}$ in. in the actual object, the scale of the drawing can be given as the ratio of 1 ft. to $\frac{1}{4}$ in. Since 1 ft. = 12 in., the scale of the drawing can be given as the ratio $12:\frac{1}{4}$ or $48:1$ $[\frac{48}{1}]$. This means that 48 units of length in the scale drawing represent 1 of the same units of length in the actual object. The scale given as the ratio $48:1$ also means that each length in the drawing is 48 times as great as the length that it represents in the actual object. The scale drawing would therefore be an enlargement of the actual object.

In the scale of a scale drawing, the first term refers to lengths in the scale drawing and the second term refers to corresponding lengths in the actual object. Therefore, it follows that:

1. If the first term of the scale is *less* than the second, the scale drawing is a *reduction* of the actual object—for example, when the scale is 1:4.
2. If the first term of the scale is the *same* as the second, the scale drawing is the *same* size as the actual object—for example, when the scale is 1:1.
3. If the first term of the scale is *greater* than the second, the scale drawing is an *enlargement* of the actual object—for example, when the scale is 3:2.

MODEL PROBLEM

In a scale drawing, the scale that is being used is $\frac{1}{4}$ in. = 1 ft. What is the length of a line whose measurement on the scale drawing is $1\frac{1}{2}$ inches?

Solution: Since $\frac{1}{4}$ in. represents 1 ft., to find the number of feet represented by $1\frac{1}{2}$ in., we find how many times $\frac{1}{4}$ in. is contained in $1\frac{1}{2}$ in. That is, we divide $1\frac{1}{2}$ by $\frac{1}{4}$.

$$1\tfrac{1}{2} \div \tfrac{1}{4} = \tfrac{3}{2} \div \tfrac{1}{4} = \tfrac{3}{2} \times \tfrac{4}{1} = \tfrac{12}{2} = 6 \quad \textit{Answer: } 6 \text{ feet}$$

EXERCISES

1. Tell whether the scale drawing will be a reduction, an enlargement, or the same size as the actual object if the scale that is being used is:

a. $\frac{1}{4}$ in. = 5 ft. *b.* 1 in. = $\frac{1}{16}$ in. *c.* 5:2 *d.* $\frac{1}{8}$ *e.* 4:4 *f.* $1\frac{1}{2}$:1

In 2–3, refer to the diagram at the right:

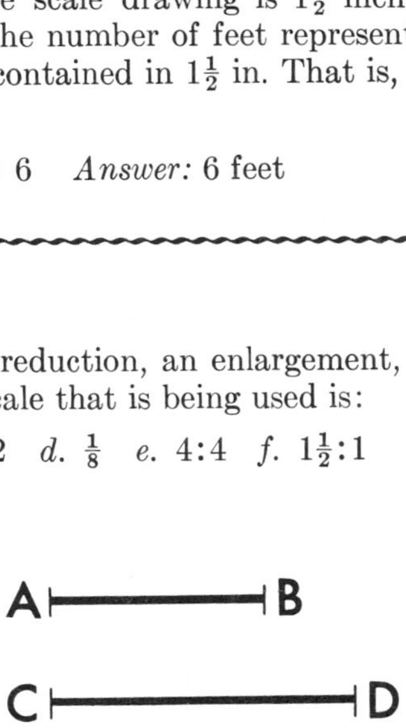

2. If the scale that is being used is $\frac{1}{8}$ in. = 1 in., what is the length of the line represented by AB?

3. If the scale that is being used is $\frac{1}{4}'' = 1'$, what is the length of the line represented by CD?

4. If the scale that is being used in a scale drawing is $\frac{1}{8}$ in. = 1 ft., find the length of the line that should be used to represent:

a. 32 ft. *b.* 12 ft. *c.* 6 ft. *d.* 20 ft. *e.* $\frac{1}{2}$ ft. *f.* $8\frac{1}{2}$ ft.

5. If $\frac{1}{4}$ in. represents 5 yd., how many inches long must a scale length be to represent a football field 100 yd. long?

6. In the scale drawing on the right, the scale used is $\frac{1}{4}$ in. represents 2 yd.

a. Find the length of the side represented by AB.
b. Find the length of the side represented by BC.
c. Find the length of the diagonal represented by AC.

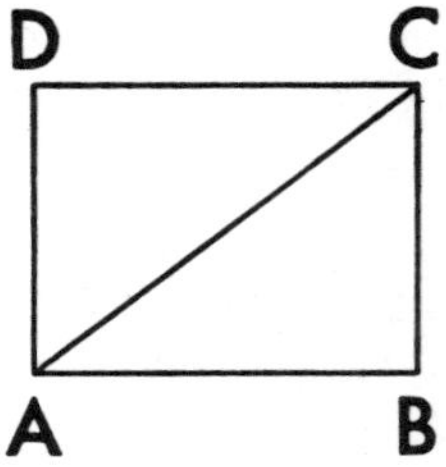

7. Using a scale of $\frac{1}{8}$ in. = 1 ft., make a scale drawing of a rectangle whose length and width are respectively:

a. 16 ft. and 8 ft. *b.* 12 ft. and 9 ft. *c.* 10 ft. and 14 ft.

8. The drawing on the right is part of the floor plan of a house. What was the scale used in drawing the floor plan?

9. A rectangular piece of plywood is 8 ft. long and 6 ft. wide. Make a scale drawing to find the length of the diagonal of the rectangle.

Use the following pictures to answer questions 10–15:

Ostrich $\frac{1}{48}$

Queen Bee 2:1

Postage Stamp 1:1

10. The height of an actual ostrich is how many times the height of the ostrich in the above picture?

11. Find the height of the actual ostrich whose picture appears above.

12. Is the picture of the queen bee a reduction, an enlargement, or the same size as an actual queen bee?

13. Find the length of the body of the actual queen bee pictured above.

14. Is the picture of the postage stamp a reduction, an enlargement, or the same size as an actual postage stamp?

15. Find the length of the longer side of the actual postage stamp.

16. The scale used in a scale drawing is 1 inch = 8 feet. Express this scale as a ratio.

17. A scale given as the ratio 1:4 is used in making a scale drawing for the cover of a book that is 8 inches by 12 inches. What are the dimensions of the cover in the scale drawing?

4. USING SCALES ON MAPS

A ***map*** is a scale drawing of an area that is being pictured. The map below, which shows some of the cities on the main line of the Pennsylvania Railroad, is drawn to scale. Since this map represents a large area, the scale being used is 1 inch = 80 miles.

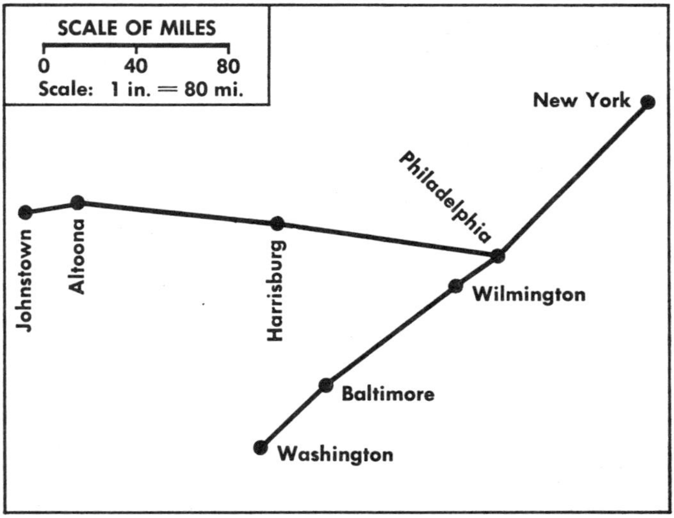

To measure the approximate distance in a straight line between Baltimore and Washington, we use a ruler to measure the length of the line that joins the two points on the map. This line measures $\frac{1}{2}$ inch. Since 1 inch represents 80 miles, $\frac{1}{2}$ inch will represent $\frac{1}{2}$ of 80 miles, or $\frac{1}{2} \times 80$ miles = 40 miles, the distance from Baltimore to Washington. This straight-line distance is also called ***air-line distance.***

EXERCISES

Use the previous map to find the following straight-line distances:

1. New York to Philadelphia
2. Wilmington to Baltimore
3. Altoona to Johnstown
4. New York to Washington

5. The scale at the right appears on a map. How many miles does 1 inch represent on this map?

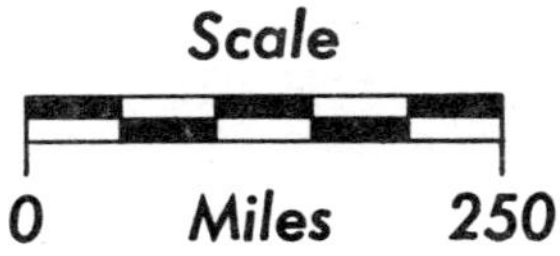

6. A map of a state uses the scale 1 inch = 100 miles. Find the distance represented on the map by each of the following lengths:

a. 2 in. *b.* 3 in. *c.* $2\frac{1}{2}$ in. *d.* $\frac{1}{2}$ in. *e.* $3\frac{3}{4}$ in.

7. The scale of a map is 1 inch = 200 miles. Find the distance in miles between 2 towns $4\frac{1}{4}$ inches apart on the map.

8. A state is mapped on the scale of 1 inch = 80 miles. Find the length of the line on the map that joins two cities 240 miles apart.

9. The scale of a map is 1 inch = 160 miles. How far apart on this map will two cities be if the air-line distance between them is 400 miles?

10. A plane flew over two cities that were $4\frac{1}{2}$ inches apart on a map whose scale was 1 inch = 400 miles. The plane flew over the first city at 8:18 A.M. and over the second city at 4:48 P.M. It made one stop of 30 minutes to refuel at an airport between the two cities.

a. What was the air distance between the two cities?

b. How many hours was the plane in the air while traveling between the two cities?

c. What was the plane's average rate of speed on this flight?

Chapter X

ALGEBRA

1. USING SYMBOLS TO REPRESENT NUMBERS AND OPERATIONS

In our work in arithmetic, we used symbols to represent numbers. In the Roman system, for example, we used the symbol V to express the number five. In the Hindu-Arabic system, we used the symbol 5. In algebra, we use letters in addition to numerical figures to represent numbers. Letters that are used to represent numbers are called ***literal numbers.*** For example, we have seen that the rule, "distance traveled equals the rate of speed multiplied by the time that is spent in traveling," can be expressed more simply by the formula $D = R \times T$.

REPRESENTING OPERATIONS

In algebra as well as in arithmetic, the fundamental operations of addition, subtraction, multiplication, and division are represented by symbols. Addition is represented by $+$, subtraction by $-$, multiplication by $\times$, and division by $\div$ or by a fraction.

REPRESENTING PRODUCTS

If we wish to represent the literal number x multiplied by the arithmetic number 5, we can write $5 \times x$. Because the multiplication symbol $\times$ might be confused with the letter x, we generally avoid using the symbol $\times$ for multiplication. Instead, we sometimes place a center dot, $\cdot$, between the two numbers to be multiplied. Thus, $5 \times x$ may be written $5 \cdot x$, and x times y may be written $x \cdot y$. In fact, when one or both of the numbers to be multiplied are literal numbers, we usually omit the dot and merely place the numbers next to each other. Thus, $5 \cdot x$ may be written as $5x$, and $x \cdot y$ may be written as xy. The expression cd means that the number c is to be multiplied by the number d.

FACTOR; COEFFICIENT

When the two numbers 5 and x are multiplied, the answer, $5x$, is called their ***product.*** The numbers 5 and x, which are being multiplied to give the product $5x$, are called the ***factors*** of $5x$. The 5 is called a ***numerical factor.*** The x is called a ***literal factor.*** Each factor is called the ***coefficient*** of the other factor. In the algebraic expression $5x$, the 5 is also called the ***numerical coefficient*** of x. Similarly, in the expression $\frac{1}{2}b$, $\frac{1}{2}$ is the numerical coefficient of b and b is the coefficient of $\frac{1}{2}$.

In algebra, we usually do not write $1 \cdot x$ as $1x$. Instead, we simply write x. In general, if no coefficient is written before a literal number, the coefficient 1 is understood. Thus, y means $1y$.

REPRESENTING NUMBER EXPRESSIONS

Notice how different number expressions in words can be expressed in algebraic symbols.

Word Expression	*Algebraic Expression*
the sum of c and 5 c increased by 5 5 more than c	$c + 5$
m decreased by 3 m diminished by 3 3 subtracted from m 3 less than m	$m - 3$
twice y the double of y y multiplied by 2 the product of 2 and y	$2y$
y divided by 4 the quotient of y divided by 4 one-fourth of y	$y \div 4$ or $\frac{y}{4}$

MODEL PROBLEMS

1. Mary weighed x pounds. Express algebraically her weight after she gained 10 pounds.

Answer: $x + 10$ (She now weighs 10 more than x pounds.)

2. Sally is y years old now. Represent algebraically her age 5 years ago.

Answer: $y - 5$ (Her age was 5 years less than her present age y.)

3. Perry has n marbles. Represent algebraically the number of marbles that Ted has if he has 4 less than 5 times the number Perry has.

Answer: $5n - 4$ (The number of marbles Ted has is 5 times n decreased by 4.)

4. Express algebraically in terms of x the number of yards in x feet.

Answer: $\frac{x}{3}$ (Since there are 3 feet in a yard, we divide the number of feet, x, by 3 to find the number of yards.)

EXERCISES

1. Represent algebraically:

a. a plus 3
b. s decreased by 10
c. 7 multiplied by p
d. s divided by 3
e. 2 more than x
f. p plus i
g. r more than t
h. y increased by 6
i. 7 less than b
j. c divided by n
k. one-half of h
l. 9 divided by x

2. If n represents a number, express algebraically a number that is:

a. 50 more than n
b. 25 less than n
c. the sum of 20 and n
d. $\frac{2}{3}$ as large as n
e. x times n
f. n increased by 7
g. 5 less than 4 times n
h. $\frac{1}{2}$ of n increased by 5
i. $\frac{2}{3}$ of n decreased by 10
j. 4 more than 7 times n
k. 5 times n decreased by 2
l. twice n increased by 8

3. Read the following and give a meaning for each:

a. bh *b.* $p + i$ *c.* $\frac{d}{t}$ *d.* $s - c$ *e.* $2x + 3$ *f.* $4x - 6$

4. Represent algebraically the number of:

a. inches in f feet *b.* cents in d dimes *c.* weeks in d days

5. If Henry has x dollars and his brother Fred has three times as many, express algebraically how many dollars Fred has.

6. Represent in cents the cost of 5 pencils at c cents apiece.

7. An orange cost c cents. Represent in cents the cost of a dozen oranges.

8. If coal costs x dollars per ton, represent in dollars how much $\frac{1}{4}$ ton will cost.

9. If Tom weighs x pounds, represent his weight after he gains 10 pounds.

10. If Mary weighs m pounds, represent her weight after she loses 5 pounds.

11. Stanley is y years old now. Represent his age 2 years from now.

12. Sheila is x years old now. Represent her age 4 years ago.

13. Jane had 15 cents and lost x cents. How many cents did she have left?

14. If Peter is x years older than his brother, who is 12, express algebraically Peter's age.

15. If a girl's age is x years now, express algebraically her age 10 years ago.

16. If a board is l inches long and a second board is 6 inches longer, what is the length of the second board in terms of l?

17. A car travels m miles per hour. A train travels 40 miles per hour faster than the car. Represent in miles per hour the rate of speed of the train.

18. The width of a rectangle is x feet. Represent its length in feet if the length is 8 times the width.

19. The length of a rectangle is y inches. Represent its width if the width is 5 inches less than the length.

20. The length of a rectangle is l feet. Represent its width if the width is 7 feet more than 3 times its length.

21. The width of a rectangle is represented by w feet. If the length exceeds the width by 4 feet, represent the length. [Hint: "Exceeds" means "is larger than."]

22. Irene is y years old. Represent Sue's age if her age exceeds twice Irene's age by 3 years.

23. If 3 pencils cost x cents, represent in cents the cost of one pencil.

24. If 3 peaches cost 12 cents, represent in cents the cost of p peaches.

25. If 3 toys cost m dollars, represent in dollars the cost of 5 toys.

26. Ben spent d dollars in a store. Represent in dollars the change he received from a $10 bill, assuming that d is less than 10.

27. Eugene bought 3 articles at c cents each. Represent in cents the change he should receive from a $1 bill, assuming that c is not greater than 33.

28. Helen is x years old.
a. If Marion is three times as old as Helen, represent Marion's age.
b. Represent the sum of Helen's age and Marion's age.

29. Robert weighs p pounds.
a. If Paul weighs 20 pounds more than Robert, represent Paul's weight.
b. Represent the total weight of the two boys.

30. Mr. Samuels bought x dollars worth of jewelry.
a. If there is a 10% federal tax on jewelry, represent in dollars the amount of tax Mr. Samuels had to pay.
b. Represent in dollars the total amount that Mr. Samuels had to pay for the jewelry.

2. EVALUATING ALGEBRAIC EXPRESSIONS

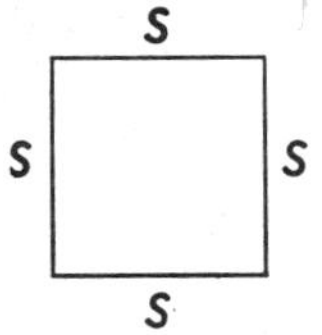

If the side of a square is represented by S, the distance around the square can be represented by $4 \times S$, or $4S$.

As the value of S changes, the value of the expression $4S$ also changes.

When $S = 1$, $4S = 4 \times 1 = 4$.
When $S = 3$, $4S = 4 \times 3 = 12$.
When $S = 5\frac{1}{4}$, $4S = 4 \times 5\frac{1}{4} = 21$.

Whenever we know a value of S, we can find the corresponding value of $4S$, or ***evaluate*** $4S$, by replacing the literal number S by its value. That is, we ***substitute*** an arithmetic value for the literal number.

Procedure. To evaluate an algebraic expression, substitute the given arithmetic values for the literal numbers and perform the indicated operations.

BASES AND EXPONENTS

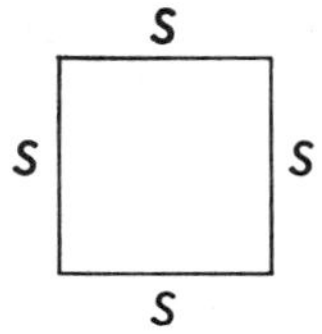

If the side of a square is S linear units, its area is $S \times S$ units of area. In algebra, we write $S \times S$ as S^2. S^2 is read "S squared" or "S to the second power." The small ^{2} above and to the right of S is called an ***exponent.*** It tells us that S is to be used as a factor twice. S, the number used as a factor, is called the ***base.***

x^2 means $x \cdot x$. The base is x; the exponent is 2.
10^2 means 10×10, or 100.
When $d = 5$, $d^2 = d \times d = 5 \times 5 = 25$.
$e \times e \times e$ is written e^3 and is read "e cubed" or "e to the third power."
5^3 means $5 \times 5 \times 5$, or 125. The base is 5; the exponent is 3.
$6e^2$ means $6 \times e \times e$ and is read "six e squared."

MODEL PROBLEM

If $x = 5$, $y = 2$, and $z = 3$, find the numerical value of $4x - 3y + z$.

$$4x - 3y + z$$
$$(4 \times 5) - (3 \times 2) + 3$$
$$20 - 6 + 3$$
$$17$$

$x = 5$, $y = 2$, $z = 3$
Substitute 5 for x, 2 for y, and 3 for z. Then $4 \times 5 = 20$, $3 \times 2 = 6$, and $20 - 6 + 3$ means $(20 - 6) + 3$, or $14 + 3$, giving the answer, 17.

EXERCISES

1. Give the numerical coefficient, base, and exponent in each of the following:

a. $4x^2$ *b.* $7d^3$ *c.* $\frac{3}{4}w^2$ *d.* $3y^2$ *e.* x^3

2. Find the value of each of the following algebraic expressions when $x = 6$:

a. $5x$ *b.* $x + 9$ *c.* $x - 4$ *d.* $2x + 7$ *e.* $3x - 5$ *f.* $\frac{2}{3}x - 1$

3. Tell the meaning of:

a. 4^2 *b.* 6^2 *c.* y^2 *d.* z^2 *e.* 2^3 *f.* s^3 *g.* $4x^2$ *h.* $2y^3$

4. Square each of the following numbers:

a. 8 *b.* 10 *c.* 42 *d.* 125 *e.* $\frac{1}{3}$ *f.* $\frac{5}{4}$ *g.* 8.6 *h.* .36

5. Find the value of x^2 and x^3 when x is:

a. 1 *b.* 7 *c.* 9 *d.* 1.2 *e.* .8 *f.* $\frac{1}{2}$ *g.* $\frac{1}{3}$ *h.* $\frac{3}{4}$

6. If $a = 5$, $b = 4$, and $c = 2$, find the value of:

a. ab	*b.* abc	*c.* $\frac{ab}{2}$	*d.* $\frac{3}{4}bc$
e. $a + 2b$	*f.* $3a + 5c$	*g.* $4a - b$	*h.* $6c - 3b$
i. $2a + 3b + 4c$	*j.* $3a - 2b + 4c$	*k.* $6a - 3b - c$	*l.* $20c - 3a + 4b$
m. a^2	*n.* b^2	*o.* $3c^2$	*p.* $\frac{1}{2}b^2$
q. $a^2 + b^2$	*r.* $2a^2 + c^2$	*s.* $b^2 - c^2$	*t.* $4c^2 - b^2$

7. Find the value of $36x$ when x is: *a.* $\frac{1}{2}$ *b.* 1.5 *c.* .2 *d.* $\frac{2}{3}$

3. UNDERSTANDING AND USING FORMULAS

Mr. Owens was planning to sell cloth at $2 per yard. To find the amount he should charge a customer for any number of yards bought, he used the ***rule:*** "number of dollars charged is equal to $2 multiplied by the number of yards bought." He expressed this rule in algebraic language as a ***formula:***

> Number of dollars charged equals 2 times number of yards bought.
>
> $C = 2 \times n$
>
> or simply: $C = 2n$

$C = 2n$ is called a formula because it is a rule for computation in which letters represent numbers. This formula can be used to find the charge for any number of yards. To find the number of dollars to charge for 2 yards, Mr. Owens said: $C = 2 \times 2 = 4$. To find the number of dollars to charge for 4 yards, he said: $C = 2 \times 4 = 8$. He also found the cost of 6, 8, 10, and 12 yards and arranged his results in the following ***table:***

If $n =$	0	2	4	6	8	10	12
then $C =$	0	4	8	12	16	20	24

Using the previous table, Mr. Owens now made the ***graph*** at the right. On the horizontal scale, he showed the number of yards bought. On the vertical scale, he showed the number of dollars charged. He used each pair of numbers to locate a point on the graph. For example, the pair of numbers, $n = 0$, $C = 0$, was used to locate the first point; the pair, $n = 2$, $C = 4$, was used to locate the second point, etc. All the points are in the same straight line.

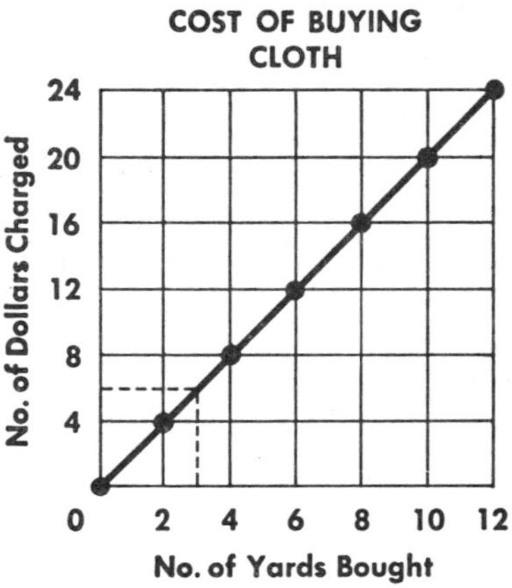

The graph can now be used to find the charge for a purchase of any number of yards. To find the charge for 3 yards, he drew the dotted vertical line at $n = 3$ up to the graph. Then he drew the horizontal line across and discovered that the charge is $6. He checked his result by substituting 3 for n in the formula $C = 2n$ and got the same result, $6.

We see, therefore, that the relationship between the number of dollars charged and the number of yards bought can be represented in any one of four ways: (1) rule (2) formula (3) table (4) graph.

EXERCISES

1. Coffee sells for 75 cents a lb. Write a formula for the cost, C, of n lb.

2. A car travels 40 miles per hour. Write a formula for the distance, d, that it will travel in h hours.

3. In a year, a man earns 6% interest on his investment. Write a formula for the interest, i, his money earns if he invests p dollars.

4. Write as a formula, "The area, A, of a rectangle is equal to the base, b, multiplied by the height, h."

5. Write as a formula, "The area, A, of a triangle is equal to one-half the product of the base, b, and the altitude, h."

6. Write a formula for the number of days, d, in w weeks.

7. Write a formula for the number of minutes, m, in h hours.

8. Write a formula for the average, A, of three numbers, b, c, and d.

9. Write a formula for the perimeter, P, of a rectangle whose length is l and whose width is w.

10. Write as a formula, "The sum of three angles, A, B, and C, of a triangle is 180°."

11. Suits cost $50 each. Which of the following is a correct formula to represent the cost, C, of s suits?

a. $C = s + 50$ *b.* $C = 50s$ *c.* $C = 50 - s$ *d.* $C = \frac{50}{s}$

12. A man makes a trip of 100 miles. Which of the following is a correct formula to represent the time required for the trip, t, if he travels at a rate of r miles per hour?

a. $t = 100r$ *b.* $t = \frac{100}{r}$ *c.* $t = 100 + r$ *d.* $t = 100 - r$

13. If $C = 5p$, find C when p is:

a. \$3 *b.* \$4 *c.* \$1.20 *d.* \$1.98 *e.* $\$1\frac{2}{5}$

14. The following table shows the distance traveled by a car when its rate of speed is 30 miles per hour:

If the number of hours traveled, h, is	0	2	4	6	8
the distance traveled, d, is	0	60	120	180	240

a. To find the value of d, multiply the corresponding value of h by ________.
b. A formula that shows the relationship between d and h is ________.
c. Make a graph from the above table.
d. From the graph, find d when $h = 3, 5, 6\frac{1}{2}$.

15. Study each of the following tables and write a formula to represent the relationship between the two letters in each table:

a.

If r is	20	30	40	50
d is	100	150	200	250

b.

If x is	3	4	5	6
y is	8	9	10	11

16. Make a table for each formula, using the values 6, 8, 10, 12, 14 for x:

a. $y = 2x$ *b.* $y = 1.5x$ *c.* $y = x + 3$ *d.* $y = x - 4$

e. $y = 2x + 3$ *f.* $y = 3x - 1\frac{1}{2}$ *g.* $y = \frac{x}{2}$ *h.* $y = \frac{4x}{5}$

17. The formula for the perimeter of a square, each of whose sides is s, is $P = 4s$. (*a*) Complete the following table. (*b*) Make a graph of this table. (*c*) From the graph, find the perimeter of a square whose sides are 5, 12, 16, 19.5.

If s is	1	2	4	8	10	15	20
$P = 4s$ is	4	8					

4. STUDYING RELATED CHANGES

We know that the formula for the cost, C, of n shirts at \$2 per shirt is $C = 2n$. From this formula, we can prepare the following table:

If n is	1	2	3	4	5
then C is	2	4	6	8	10

This table shows us that if the price per shirt remains \$2, then as the number of shirts sold (n) increases, the cost (C) increases. Also, as the number of shirts sold decreases, the cost decreases.

From the table, we also see that when $n = 1$, $C = 2$ and when $n = 2$, $C = 4$. Notice that when the number of shirts sold is doubled, the cost is also doubled. If the number of shirts sold is tripled, what change takes place in the cost? Although the value of C changes when the value of n changes, the relationship between C and n is always the same, $C = 2n$.

EXERCISES

1. The formula for the perimeter of a hexagon (a six-sided plane figure), each of whose sides is equal to s, is $P = 6s$.

a. Make a table showing the values of P as s changes from 1 to 6.

b. As the value of s increases, what change takes place in the value of P?

c. As the value of s decreases, what change takes place in the value of P?

d. If the value of s is doubled, is the value of P doubled also?

e. If the value of s is tripled, what change takes place in the value of P?

f. If the value of s is multiplied by 5, by what number is the value of P multiplied?

g. Although the value of P changes as the value of s changes, does the relationship between P and s change?

2. The distance, D, that is covered by a car traveling at an average rate of 40 miles per hour for t hours is given by the formula $D = 40t$.

a. Make a table for values of D as t changes from 1 to 10.

b. What change takes place in the value of D when t (1) increases (2) decreases (3) is multiplied by 5 (4) is multiplied by 10?

3. If $V = lwh$, what happens to the value of V when:

a. h is tripled and l and w remain unchanged?

b. l and w are each doubled and h remains unchanged?

5. ADDING AND SUBTRACTING LIKE TERMS

The algebraic expression $5a + 2b - 3c$ has three parts: $5a$, $2b$, and $3c$. These parts, which are separated by + and − signs, are called ***terms.*** An algebraic expression that has only one term, like $5x$, is called a ***monomial.*** An expression that has two terms, like $2x + 5$, is called a ***binomial.***

Like terms, or ***similar terms,*** are terms having a *common literal factor* that appears with the same exponent. $5x$ and $3x$ are like terms; also, $7a^2$ and $9a^2$ are like terms.

Unlike terms, or ***dissimilar terms,*** are terms that do not have a common literal factor or terms in which a common literal factor appears with different exponents. $5x$ and $3y$ are unlike terms; also, $4a^3$ and $6a^2$ are unlike terms.

Let us see how to add and subtract like terms in algebra.

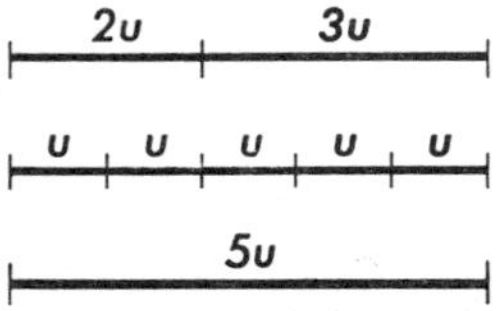

If we wish to add 2 units and 3 units, we can place them next to each other. The total is 5 units. Thus,

$$2u + 3u = 5u$$

Procedure. To add like terms, add their numerical coefficients and multiply the result by their common literal factor.

Similarly, if from 5 units we take off 3 units, we have 2 units left. Thus,

$$5u - 3u = 2u$$

Procedure. To subtract like terms, subtract their numerical coefficients and multiply the result by their common literal factor.

If we wish to add 3 dimes and 4 nickels to 5 dimes and 2 nickels, we can write:

5 dimes and 2 nickels		$5d + 2n$
3 dimes and 4 nickels	*or*	$3d + 4n$
8 dimes and 6 nickels		$8d + 6n$

Procedure. To combine algebraic expressions containing several terms, some of which are alike, combine each set of like terms separately.

MODEL PROBLEMS

1. Add: $\begin{array}{r} 3m \\ \underline{5m} \\ 8m \end{array}$

2. Subtract: $\begin{array}{r} 10\frac{1}{2}C \\ \underline{1\frac{1}{4}C} \\ 9\frac{1}{4}C \end{array}$

3. Subtract: $\begin{array}{r} 12.5t \\ \underline{3.2t} \\ 9.3t \end{array}$

4. Add: $\begin{array}{r} 6x^2 \\ \underline{3x^2} \\ 9x^2 \end{array}$

5. Add: $\begin{array}{r} 6x + 4y \\ \underline{3x + y} \\ 9x + 5y \end{array}$

6. Subtract: $8x - 8x$.

$8x - 8x = 0$

7. Write in simplest form: $9t + 4t - t$.

$9t + 4t - t = 12t$

EXERCISES

In 1–4, state whether the expressions are like terms or unlike terms:

1. $4x$ and $5x$ **2.** d and $7d$ **3.** $7x^2$ and $3x^2$ **4.** s^2 and s^3

5. Add:

a. $\begin{array}{r} 7b \\ \underline{3b} \end{array}$ *b.* $\begin{array}{r} 18d \\ \underline{40d} \end{array}$ *c.* $\begin{array}{r} 9y \\ \underline{y} \end{array}$ *d.* $\begin{array}{r} x^2 \\ \underline{12x^2} \end{array}$ *e.* $\begin{array}{r} 15\frac{1}{3}s \\ \underline{3\frac{5}{6}s} \end{array}$ *f.* $\begin{array}{r} 3.7m \\ \underline{1.4m} \end{array}$

g. $\begin{array}{r} 5x \\ 3x \\ \underline{8x} \end{array}$ *h.* $\begin{array}{r} 9n \\ 4n \\ \underline{n} \end{array}$ *i.* $\begin{array}{r} t^2 \\ 4t^2 \\ \underline{t^2} \end{array}$ *j.* $\begin{array}{r} 4\frac{1}{2}c \\ 5\frac{1}{4}c \\ \underline{6\frac{3}{4}c} \end{array}$ *k.* $\begin{array}{r} 8.3w \\ 9.2w \\ \underline{6.5w} \end{array}$ *l.* $\begin{array}{r} 12.4p \\ 18.7p \\ \underline{20.6p} \end{array}$

6. Subtract:

a. $\begin{array}{r} 40x \\ \underline{25x} \end{array}$ *b.* $\begin{array}{r} 15w \\ \underline{5w} \end{array}$ *c.* $\begin{array}{r} 15y \\ \underline{15y} \end{array}$ *d.* $\begin{array}{r} 8r \\ \underline{r} \end{array}$ *e.* $\begin{array}{r} 12\frac{3}{4}z \\ \underline{9\frac{1}{2}z} \end{array}$ *f.* $\begin{array}{r} 9.4e^2 \\ \underline{1.5e^2} \end{array}$

7. Simplify by combining like terms:

a. $8a + 9a$ *b.* $c + 6c$ *c.* $9d - 8d$ *d.* $12e - 12e$
e. $7z - z$ *f.* $9k^2 + 12k^2$ *g.* $8m - 8m$ *h.* $6f + 5\frac{1}{2}f$
i. $\frac{5}{8}h - \frac{1}{4}h$ *j.* $15i - 9\frac{1}{2}i$ *k.* $r + 5.4r$ *l.* $s - .3s$

8. Simplify by combining like terms:

a. $5x + 13x + 2x$ *b.* $18g - 5g - 13g$ *c.* $12t^2 - 6t^2 + 8t^2$
d. $4\frac{1}{2}h - 3h + 2h$ *e.* $8\frac{2}{3}a - 3\frac{1}{3}a - 2a$ *f.* $8.5d - d + 1.2d$

9. Add:

a. $\begin{array}{r} 6r + 7t \\ \underline{5r + 2t} \end{array}$ *b.* $\begin{array}{r} 8m + 6p \\ \underline{2m + p} \end{array}$ *c.* $\begin{array}{r} 2x + 6 \\ \underline{2x + 6} \end{array}$ *d.* $\begin{array}{r} x + 9 \\ \underline{2x + 2} \end{array}$ *e.* $\begin{array}{r} 5x + y + 6z \\ \underline{x + 3y + 4z} \end{array}$

10. Simplify the following formulas:

a. $P = 5a + 6b + 3b + a$ *b.* $S = 3c + 1\frac{1}{2}d + 1\frac{1}{4}c + 2\frac{3}{4}d$

11. Write the following in simplest form:

a. $8x + 3y + 2z + x + 4y + 6z$ *b.* $6r + s + 4t + 1\frac{1}{2}r + s + \frac{2}{3}t$

6. WRITING EQUATIONS

When we learned how to add numbers in arithmetic, we said that: "when 5 and 3 are added, the result is equal to 8." We wrote this word sentence in symbols as follows: "$5 + 3 = 8$." A statement that two number expressions, or quantities, are equal is called an ***equation.*** Other examples of equations are: $8 - 6 = 2$ and $3 \times 5 = 15$.

Mathematicians frequently use letters like x and y to represent unknown numbers. Questions or statements in words dealing with the equality of two number expressions can be written as equations in algebraic language through the use of letters, known numbers, and other symbols. Study the following examples:

1. *Question:* Three times what number x equals 15?
   Statement: Three times the number x equals 15.
   Equation: $3x = 15$
2. *Question:* What number x increased by 5 equals 9?
   Statement: The number x increased by 5 equals 9.
   Equation: $x + 5 = 9$

In the equation $x + 5 = 9$, the expression $x + 5$, at the left of the equal sign, is called the left ***side,*** or left ***member,*** of the equation. The expression 9, at the right of the equal sign, is called the right side, or right member, of the equation.

MODEL PROBLEMS

1. Write an equation for the following question: What number times 4 is 16?
Solution: Choose any letter to represent the unknown number. Use this letter to translate the statement into algebraic language.

Let n represent the number.
4 times a number is 16.
$$4n = 16$$

2. Write an equation for the following word statement: When 3 times a number is decreased by 2, the result is 30.

Let x = the number.
3 times a number decreased by 2 equals 30.
$$3x - 2 = 30$$

EXERCISES

1. Give a word statement for each of the following equations:

a. $7n = 14$ *b.* $6x = 42$ *c.* $\frac{y}{5} = 4$ *d.* $\frac{1}{2}x = 12$

e. $n + 5 = 18$ *f.* $m - 8 = 14$ *g.* $2n + 1 = 9$ *h.* $3n - 2 = 10$

In 2–5, select the equation that represents the relationship stated in the problem:

2. Five times a boy's weight, w, is 100 pounds.

a. $w + 5 = 100$ *b.* $5w = 100$ *c.* $w - 5 = 100$ *d.* $\frac{1}{5}w = 100$

3. One-quarter of Tom's age, a, is 4 years.

a. $\frac{1}{4}a = 4$ *b.* $4a = 4$ *c.* $a + 4 = 4$ *d.* $a - 4 = 4$

4. A number, x, increased by 5 equals 20.

a. $5x = 20$ *b.* $x - 5 = 20$ *c.* $x + 5 = 20$ *d.* $\frac{1}{5}x = 20$

5. If 10 is subtracted from a number, x, the result is 5.

a. $10 - x = 5$ *b.* $x - 10 = 5$ *c.* $x - 5 = 10$ *d.* $5 - x = 10$

In 6–23, write an equation for each statement. Use the letter n to represent the unknown number.

6. Three times a certain number equals 21.
7. One-third of a number equals 15.
8. A number increased by 8 is 21.
9. A number decreased by 7 is 13.
10. A number divided by 5 equals 8.
11. The product of 5 and a number is 60.
12. Two less than a number equals 40.
13. If 4 is added to a certain number, the sum is 6.
14. If 4 is subtracted from a certain number, the difference is 2.
15. If 4 times a certain number is divided by 3, the quotient is 12.
16. Twice a number increased by 8 gives a result of 18.
17. Six times a number decreased by 1 equals 17.
18. Three times a number decreased by 1 equals 23.
19. The sum of a number and twice the number is 15.
20. One-quarter of a number added to three times the number is 26.
21. One-half a number increased by 5 is 13.
22. Five times a number minus one-third of the number equals 28.
23. 14 increased by two-thirds of a number equals 34.

In 24–27, select the equation that could be used to find the unknown number in the problem:

24. A man had d dollars. After paying a bill of \$15, he had \$6 left. How many dollars did he have at first?

a. $d + 15 = 6$ *b.* $d - 15 = 6$ *c.* $\frac{d}{15} = 6$

25. A caddy had n golf balls. After finding 14 more, he had 37. How many golf balls did he have before he added the ones he found?

a. $n - 14 = 37$ *b.* $\frac{n}{14} = 37$ *c.* $n + 14 = 37$

26. A boy spends \$2 or $\frac{1}{4}$ of his money, m. How much money did he have in the beginning?

a. $\frac{m}{4} = 2$ *b.* $m + \frac{1}{4} = 2$ *c.* $4m = 2$

27. A team played 22 basketball games. It won 4 more games than it lost. If n represents the number of games it lost, which of the following equations could be used to find n?

a. $n - 4 = 22$ *b.* $n + 4 = 22$ *c.* $n + n + 4 = 22$

7. SENTENCES; STATEMENTS; OPEN SENTENCES

A ***sentence*** expresses a complete thought. In a sentence, there must always be a *subject* and a *predicate*. "Our cousin Mary is pretty" is an English sentence. "$2 + 3 = 5$" is a ***mathematical sentence,*** or ***number sentence.*** However, "Our cousin Mary" is not a sentence because there is no predicate, and as a result the words do not express a complete thought. Similarly, "$2 + 3$" is not a sentence because there is no predicate, and therefore the arithmetic symbols do not express a complete thought.

Sometimes a sentence can be judged to be true, or it can be judged to be false. For example,

1. "Albany is the capital of New York State" is a true sentence.
2. "$2 + 3 = 4 + 1$" is a true sentence.
3. "Florida is north of Maine" is a false sentence.
4. "$5 \times 0 = 5$" is a false sentence.

A sentence that can be judged to be true or judged to be false is called a ***statement.***

Sometimes it is impossible to judge whether a sentence is true or false. For example, it is impossible to tell whether the sentence "He discovered America" is true or false until the name of the person to whom "He" refers is known. Similarly, it is impossible to tell whether the sentence "$3 \times ? = 36$" is true or false until the number that is to replace the symbol "?" is known. Sentences such as "He discovered America" and "$3 \times ? = 36$" are called ***open sentences.***

In the open sentence "He discovered America," if "He" is replaced by Columbus, the resulting sentence, "Columbus discovered America," is a true statement. If "He" is replaced by any name other than Columbus—for example, Marconi—the resulting sentence, "Marconi discovered America," is a false statement.

In the open sentence "$3 \times ? = 36$," if "?" is replaced by 12, the resulting sentence, "$3 \times 12 = 36$," is a true statement. If "?" is replaced by any number other than 12—for example, the number 10—the resulting sentence, "$3 \times 10 = 36$," is a false statement.

In an open sentence, when a pronoun such as "he" or a symbol such as "?" is replaced by the name of a particular person or thing, the open sentence becomes a statement.

EXERCISES

1. State whether each of the following is a sentence:

a. The flag is flying
b. Paul, my brother
c. A square has five sides
d. Square $ABCD$
e. It costs two dollars
f. Two dollars
g. $6 + 5 = 11$
h. $9 \times 3 = 25 + 2$
i. $14 - 3$
j. $5 + 4 = 8 - 2$

2. In each of the following, write "true" if the sentence is true or "false" if the sentence is false. When you cannot judge the sentence to be true or false, write "cannot tell."

a. Abraham Lincoln was a president of the United States.
b. Thomas Edison invented the telephone.
c. He is governor of New York State.
d. She is Miss America.
e. A triangle has four sides.
f. A diameter of a circle is longer than its radius.
g. This is a square.
h. $5 \times 7 = 35$
i. $6 + 8 = 2 \times 7$
j. $9 \times 5 = 50 - 10$
k. $10 + ? = 14$
l. $17 - ? = 10$
m. $5 \times 0 = 7 - 2$

3. Is each of the following an open sentence or a statement?

a. Florida is a state of the Union.
b. Marion weighs 110 pounds.
c. He is 6 feet tall.
d. It is a member of the United Nations.
e. William Shakespeare was an American poet.
f. She is the national amateur tennis champion.
g. $7 + 3 = 10$ *h.* $15 \times 2 = 20 - 10$ *i.* $? + 20 = 30$
j. $30 + 4 = 40 - 9$ *k.* $15 \times 3 = 49 - ?$ *l.* $12 + 36 = 12 \times 4$

8. UNDERSTANDING THE MEANING OF SOLVING AN EQUATION

In each of the following equations, we will find the number that can replace the question mark and make the resulting statement true. In each case, there is only one such number. This number is said to ***satisfy*** the equation and is called a ***root*** of the equation.

$7 - 2 = ?$ is a true statement only when the question mark is replaced by 5. $(7 - 2 = 5)$ Therefore, 5 is a root of the equation $7 - 2 = ?$.

$8 + ? = 12$ is a true statement only when the question mark is replaced by 4. $(8 + 4 = 12)$ Therefore, 4 is a root of the equation $8 + ? = 12$.

Similarly, a root of:

$6 \times ? = 18$ is 3, since $6 \times 3 = 18$
$8 \div ? = 4$ is 2, since $8 \div 2 = 4$
$7 + 5 = 4 + ?$ is 8, since $7 + 5 = 4 + 8$

To find a root of the equation $4s = 20$, we must find the value that can replace s and make the resulting statement true. We know that the value of s must be 5 because $4 \times 5 = 20$.

There is no other value for s which will make $4s = 20$. For example, if $s = 3$, does $4s = 20$? No, because $4 \times 3 = 12$.

Since, when 5 is substituted for s, $4s = 20$ becomes a true statement, we say that 5 satisfies the equation $4s = 20$, and 5 is a root of the equation.

The process of finding the numbers that are the roots of an equation is called ***solving the equation.*** The equations we shall solve will have only one root.

Procedure. To determine whether a number is a root of an equation, substitute the number for the unknown in the equation. If the number satisfies the equation, the number is a root of the equation.

MODEL PROBLEM

Is 3 a root of the equation $9x = 27$?

Solution:

$9x = 27$	
$9 \times 3 = 27$	Substitute 3 for the unknown number x.
$27 = 27$	The equation becomes a true statement.

Answer: 3 is a root of the equation.

EXERCISES

In each of the following, find the number that can replace the question mark and make the resulting statement true:

	(a)	(b)	(c)	(d)
1.	$10 - 2 = ?$	$? - 4 = 8$	$15 - ? = 9$	$7 + 8 = ?$
2.	$? + 9 = 17$	$15 + ? = 18\frac{1}{3}$	$8 \times 7 = ?$	$? \times 4 = 36$
3.	$6 \times ? = 48$	$32 \div 4 = ?$	$45 \div ? = 9$	$? \div 2 = 4.5$
4.	$10 + 1 = 8 + ?$	$18 - 6 = 15 - ?$	$16 - ? = 7 + 2$	$2 \times 7 + 1 = ?$
5.	$4 \times ? + 6 = 18$	$\frac{1}{2} \times ? + 5 = 6$	$3 \times 5 - 2 = ?$	$2 \times ? - 4 = 10$
6.	$\frac{1}{2} \times ? - 1 = 2$	$6 \times 6 = 9 \times ?$	$8 + 7 = ? - 5$	$? \times 6 = 36 \times \frac{1}{2}$

In each of the following, find the arithmetic number that can replace the literal number and make the equation a true statement:

7.	$2x = 20$	$72 = 9y$	$\frac{1}{2}x = 8$	$6 = \frac{1}{3}r$
8.	$x + 2 = 8$	$15 = c + 5$	$x - 1 = 9$	$12 = r - 6$
9.	$2x + 1 = 9$	$3y + 2 = 17$	$2x - 1 = 13$	$3r - 2 = 22$

In each of the following, tell whether or not the number in parentheses is a root of the given equation:

10.	$4x = 28$ (7)	$3x = 18$ (3)	$\frac{1}{2}x = 5$ (10)	$\frac{1}{4}y = 6$ (20)
11.	$r + 6 = 12$ (18)	$17 = n + 4$ (13)	$m - 5 = 17$ (12)	$30 = s - 20$ (50)
12.	$2x + 3 = 11$ (4)	$4x + 2 = 26$ (7)	$2x - 3 = 21$ (9)	$3x - 5 = 4$ (3)

9. PREPARING FOR SOLVING EQUATIONS

If we know the product when we *multiply* a number by 2, we *divide* that product by 2 to find the number. For example, if we know twice a number is 8, to find the number, we divide 8 by 2 and get 4. We use the same method with literal numbers. If $2n$ represents twice a number, n, then when $2n$ is divided by 2, the result is the number n itself. Thus,

$$2n \div 2 = n, \quad \text{or} \quad \frac{2n}{2} = n$$

Notice that in the given quantity $2n$, the literal number n and the arithmetic number 2 are related by the operation of multiplication. We used the opposite operation, division, to get n itself as the result. Multiplication and division are called ***inverse operations.***

If we know what the result is when a number has been *divided* by 2, we *multiply* the result by 2 to find the number itself. For example, if we know that a number has been divided by 2 and the result is 8, to find the number, we multiply 8 by 2 and get 16. With literal numbers, if $\frac{n}{2}$ represents the number n divided by 2, then when $\frac{n}{2}$ is multiplied by 2, the result is the number n itself. Thus,

$$2 \times \frac{n}{2} = \frac{2n}{2} = n$$

Notice that in the given quantity $\frac{n}{2}$, the operation of division related the the number n and the number 2. We used the inverse operation, multiplication, to get n itself as the result.

If we know that after 2 has been *added* to a number the result is 8, to find the number, we *subtract* 2 from 8 and get 6. Similarly, if $n + 2$ represents the result after 2 has been added to n, then when 2 is subtracted from $n + 2$, the result will be n itself. Thus,

$$n + 2 - 2 = n$$

Notice that in the given quantity $n + 2$, the operation of addition related the number n and the number 2. We used subtraction, which is the inverse operation of addition, to get n itself as the result.

If we know that after 2 has been *subtracted* from a number the result is 8, to find the number, we *add* 2 to 8 and get 10. Similarly, if $n - 2$ represents the result after 2 has been subtracted from n, then when 2 is added to $n - 2$, the result will be n itself. Thus,

$$n - 2 + 2 = n$$

Notice that in the given quantity $n - 2$, the operation of subtraction related the number n and the number 2. We used addition, which is the inverse operation of subtraction, to get n itself as the result.

We see, therefore, that whenever a literal number, n, is related to an arithmetic number by the operation of addition, subtraction, multiplication, or division, we can use the inverse operation to get the literal number n itself.

EXERCISES

In each of the following: (*a*) State the operation that relates the literal number to the arithmetic number. (*b*) State the operation that you would use to get n itself as the result. (*c*) Give the number that you would use when adding, subtracting, multiplying, or dividing to get n itself as the result. (*d*) Check your answers.

1. $5x$	**2.** $\frac{w}{3}$	**3.** $x + 1$	**4.** $x - 4$	**5.** $10z$
6. $\frac{1}{3}w$	**7.** $\frac{z}{15}$	**8.** $d + 1\frac{1}{2}$	**9.** $p - 1$	**10.** $t - 1.5$

10. EQUIVALENT EQUATIONS

The number 5 is the only number that can replace x in each of the following equations and make the resulting statement true:

$$10x = 50 \qquad 5x = 25 \qquad x = 5$$

Therefore, 5 is the root of each of these equations.

Equations whose roots are the same are called ***equivalent equations.***

DIVISION PRINCIPLE

Notice that if we divide each side of the equation $10x = 50$ by 2 [$10x \div 2 = 50 \div 2$], we get the equivalent equation $5x = 25$. Similarly, if we divide each side of the equation $10x = 50$ by 10 [$10x \div 10 = 50 \div 10$], we get the equivalent equation $x = 5$. These examples illustrate the following principle:

Principle 1. If both sides of an equation are divided by the same number (not zero), an equivalent equation results.

MULTIPLICATION PRINCIPLE

The number 24 is the only root of the following equations, which are therefore equivalent equations:

$$\frac{n}{4} = 6 \qquad \frac{n}{2} = 12 \qquad n = 24$$

Notice that if we multiply each side of the equation $\frac{n}{4} = 6$ by 2 $\left[2 \times \frac{n}{4} = 2 \times 6\right]$, we get the equivalent equation $\frac{n}{2} = 12$. Similarly, if we multiply each side of the equation $\frac{n}{4} = 6$ by 4 $\left[4 \times \frac{n}{4} = 4 \times 6\right]$, we get the equivalent equation $n = 24$. These examples illustrate the following principle:

Principle 2. If both sides of an equation are multiplied by the same number (not zero), an equivalent equation results.

ADDITION PRINCIPLE

The number 8 is the only root of the following equations, which are therefore equivalent equations:

$$y - 5 = 3 \qquad y - 1 = 7 \qquad y = 8$$

Notice that if we add 4 to each side of the equation $y - 5 = 3$ $[y - 5 + 4 = 3 + 4]$, we get the equivalent equation $y - 1 = 7$. Similarly, if we add 5 to each side of the equation $y - 5 = 3$ $[y - 5 + 5 = 3 + 5]$, we get the equivalent equation $y = 8$. These examples illustrate the following principle:

Principle 3. If the same number is added to both sides of an equation, an equivalent equation results.

SUBTRACTION PRINCIPLE

The number 10 is the only root of the following equations, which are therefore equivalent equations:

$$z + 9 = 19 \qquad z + 6 = 16 \qquad z = 10$$

Notice that if we subtract 3 from each side of the equation $z + 9 = 19$ $[z + 9 - 3 = 19 - 3]$, we get the equivalent equation $z + 6 = 16$. Similarly, if we subtract 9 from each side of the equation $z + 9 = 19$ $[z + 9 - 9 = 19 - 9]$, we get the equivalent equation $z = 10$. These examples illustrate the following principle:

Principle 4. If the same number is subtracted from both sides of an equation, an equivalent equation results.

The preceding four principles may be combined to form the ***Rule of the Equation:***

If the same operation with the same number (except division by zero and multiplication by zero) is performed on both sides of an equation, an equivalent equation results.

EXERCISES

In each of the following, are the given equations equivalent equations?

1. $6x = 12$ $\quad x = 6$

2. $\frac{n}{2} = 20$ $\quad n = 40$

3. $x + 8 = 14$ $\quad x = 6$

4. $x - 5 = 12$ $\quad x = 7$

5. $2x + 1 = 9$ $\quad 2x = 8$ $\quad x = 4$

6. $\frac{x}{2} - 4 = 8$ $\quad \frac{x}{2} = 12$ $\quad x = 6$

Write two equations equivalent to each of the following equations:

7. $4x = 32$

8. $12x = 36$

9. $\frac{x}{8} = 6$

10. $\frac{y}{6} = 4$

11. $x + 4 = 17$

12. $y + 9 = 23$

13. $z - 7 = 15$

14. $r - 8 = 12$

15. $2x - 3 = 11$

16. $4x + 5 = 25$

17. $\frac{y}{2} - 4 = 8$

18. $\frac{z}{3} - 2 = 7$

11. SOLVING EQUATIONS BY USING DIVISION

Let us use the Rule of the Equation to solve the following problem: Three times a number equals 15. Find the number.

Let n represent the number.
Then $3n$ represents 3 times the number.

The equation that represents the relationship stated in the problem, "three times a number equals 15," is $3n = 15$. When we solve this equation, we will find the answer to the problem.

In solving the equation, we find a simpler equivalent equation that will reveal the value of n. Since in the expression $3n$, 3 and n are related by the operation of multiplication, we will divide both sides of the equation by 3, the multiplier of n, to obtain the simpler equivalent equation. On the following page, see how we solve the equation.

Method 1	*Method* 2	
$3n = 15$	$3n = 15$	
$\frac{3n}{3} = \frac{15}{3}$	$3n \div 3 = 15 \div 3$	D_3 (Divide both sides of the previous equation by 3.)
$n = 5$	$n = 5$	The answer, 5, is the root of the equation.

Check:

$$3n = 15$$
$$3 \times 5 \stackrel{?}{=} 15$$
$$15 = 15$$

To check, we substitute the answer in place of the unknown number to see whether it will *satisfy* the question.

NOTE

If an equation tells us the value of an unknown quantity multiplied by a number, we can divide both sides of the equation by the multiplier of the unknown quantity, its coefficient, to obtain an equivalent equation that has only the unknown quantity itself on one side of the equation.

MODEL PROBLEMS

1. Solve and check:

$$25 = 6x$$
$$D_6 \quad \frac{25}{6} = \frac{6x}{6}$$
$$4\tfrac{1}{6} = x$$

2. Solve and check:

$$\tfrac{1}{3}x = 6$$
$$D_{\frac{1}{3}} \quad \frac{\frac{1}{3}x}{\frac{1}{3}} = \frac{6}{\frac{1}{3}} \quad [6 \div \tfrac{1}{3} = 6 \times \tfrac{3}{1} = 18]$$
$$x = 18$$

In problems 1 and 2, the check is left to the student.

EXERCISES

Solve and check the following equations:

	(*a*)	(*b*)	(*c*)	(*d*)
1.	$2x = 14$	$10a = 90$	$2x = 0$	$10x = 10$
2.	$180 = 45x$	$2a = 13$	$36 = 8r$	$2x = .8$
3.	$5m = 1.25$	$4t = .6$	$.5x = 10$	$1.2x = 9.6$
4.	$7\tfrac{1}{2}c = 15$	$\tfrac{1}{2}t = 9$	$\tfrac{1}{3}m = 9$	$\tfrac{3}{4}s = 36$

5. Three times a number is 36. Find the number.
6. 4% of a number is 1.2. What is the number?
7. Ten times Elaine's age is 40 years. Find Elaine's age.
8. A man drove his car at an average rate of 40 miles per hour. How long did it take him to drive 180 miles?
9. Baseballs cost \$1.50 each. How many baseballs can be bought for \$12?
10. Find Ted's weight if $\frac{2}{3}$ of his weight is 60 pounds.
11. Pearl bought a coat at a "40% off" sale. If she saved \$24, what was the original price of the coat?

12. SOLVING EQUATIONS BY USING MULTIPLICATION

Let us solve the following problem algebraically: Mr. Jackson divided a sum of money equally among his 4 children. If each child received \$100, what was the sum that Mr. Jackson divided?

Let x represent the number of dollars Mr. Jackson divided.

Then $\frac{x}{4}$ represents the number of dollars each child received.

The equation that represents the relationship stated in the problem, "each child received \$100," is $\frac{x}{4} = 100$. When we solve this equation, we will find the answer to the problem.

In solving the equation, we find a simpler equivalent equation that will reveal the value of x. Since in the expression $\frac{x}{4}$, x and 4 are related by the operation of division, we will multiply both sides of the equation by 4, the divisor of x, to obtain the simpler equivalent equation. See how we solve the equation:

$\frac{x}{4} = 100$	
$4 \times \frac{x}{4} = 4 \times 100$	M_4 (Multiply both sides of the previous equation by 4.)
$x = 400$	The answer is \$400.
Check:	
$\frac{x}{4} = 100$	Substitute the answer to see whether it will satisfy the question.
$\frac{400}{4} \stackrel{?}{=} 100$	
$100 = 100$	

Notice that since the equation $\frac{1}{4}x = 100$ can be written as $\frac{x}{4} = 100$, it can also be solved by the method shown on the preceding page.

NOTE

If an equation tells us the value of an unknown quantity divided by a number, we can multiply both sides of the equation by the divisor to obtain an equivalent equation that has only the unknown quantity itself on one side of the equation.

MODEL PROBLEMS

1. Solve and check:

$$\frac{n}{2} = 12$$

$$M_2 \quad 2 \times \frac{n}{2} = 2 \times 12$$

$$n = 24$$

2. Solve and check:

$$\tfrac{1}{3} = \tfrac{1}{6}x$$

$$M_6 \quad 6 \times \tfrac{1}{3} = 6 \times \tfrac{1}{6}x$$

$$2 = x$$

In problems 1 and 2, the check is left to the student.

EXERCISES

Solve and check the following equations:

	(a)	(b)	(c)	(d)
1.	$\frac{x}{5} = 3$	$9 = \frac{s}{3}$	$\frac{w}{9} = 1$	$\frac{1}{2}x = 8$
2.	$18 = \frac{1}{6}y$	$0 = \frac{1}{5}z$	$\frac{1}{6}x = \frac{1}{2}$	$\frac{t}{3} = \frac{1}{6}$
3.	$\frac{1}{4}r = 37\frac{1}{2}$	$\frac{a}{5} = \frac{1}{6}$	$\frac{x}{.5} = 10$	$\frac{y}{.04} = 200$

4. A number divided by 6 equals 3. What is the number?

5. One-quarter of a number is $5\frac{1}{2}$. What is the number?

6. Harold saved one-eighth of his allowance. If he saved \$.25, how much was his allowance?

7. Mary's age is one-third of Helen's age. If Mary is 6 years old, how old is Helen?

8. Arthur walked one-half of the distance from his home to school. If he walked $\frac{1}{8}$ of a mile, what is the distance from his home to school?

9. If $\frac{1}{5}$ of an angle is 18°, how many degrees are there in the angle?

13. SOLVING EQUATIONS BY USING ADDITION

Let us solve the following problem algebraically: After Blanche lost 4 pounds, she weighed 65 pounds. How much did Blanche weigh originally?

Let w represent Blanche's original weight in pounds.
Then $w - 4$ represents Blanche's weight after she lost 4 pounds.

The equation that represents the relationship stated in the problem, "after Blanche lost 4 pounds, she weighed 65 pounds," is $w - 4 = 65$. When we solve this equation, we will find the answer to the problem.

In solving the equation, we find a simpler equivalent equation that will reveal the value of w. Since in the expression $w - 4$, w and 4 are related by the operation of subtraction, we will add 4 to both sides of the equation to obtain the simpler equivalent equation. See how we solve the equation:

Method 1

$$\begin{array}{rcr} w - 4 &=& 65 \\ +4 &=& +4 \\ \hline w &=& 69 \end{array}$$

Method 2

$$w - 4 = 65$$
$$w - 4 + 4 = 65 + 4$$

A_4 (Add 4 to both sides of the previous equation.)

$$w = 69$$

The answer is 69 pounds.

Check:

$$w - 4 = 65$$
$$69 - 4 \stackrel{?}{=} 65$$
$$65 = 65$$

Substitute the answer to see whether it will satisfy the question.

NOTE

If an equation tells us the result after a number has been subtracted from an unknown quantity, we can add the number to both sides of the equation to obtain an equivalent equation that has only the unknown quantity itself on one side of the equation.

MODEL PROBLEM

Solve and check:

Method 1	*Method* 2
$3.7 = m - 1.4$	$3.7 = m - 1.4$
$A_{1.4}$ $\quad +1.4 = \quad +1.4$	$3.7 + 1.4 = m - 1.4 + 1.4$
$5.1 = m$	$5.1 = m$

The check is left to the student.

EXERCISES

Solve and check the following equations:

1. $t - 3 = 7$
2. $s - 5 = 9$
3. $x - 1 = 5$
4. $8 = s - 4$
5. $15 = a - 3$
6. $x - 15 = 40$
7. $n - 3 = 2\frac{1}{2}$
8. $3\frac{1}{3} = s - 2$
9. $4 = m - 1\frac{1}{3}$
10. $s - \frac{3}{4} = 4\frac{1}{2}$
11. $5\frac{1}{3} = t - \frac{1}{3}$
12. $x - 6 = 1.1$
13. $6.2 = y - 2$
14. $a - 5.5 = 4$
15. $1.4 = y - .6$

16. What number decreased by 10 equals 20?
17. If 5 is subtracted from a number, the result is 13. What is the number?
18. Seven years ago, Elaine was 13 years old. How old is she now?
19. After she lost 15 pounds, Marjorie weighed $110\frac{1}{2}$. What was Marjorie's weight originally?
20. After $3\frac{1}{2}$ feet were cut off a piece of lumber, there were $8\frac{1}{2}$ feet left. What was the original length of the piece of lumber?
21. Mr. Everett's weekly salary is $15 less than Mr. Sherry's. If Mr. Everett's weekly salary is $75, what is Mr. Sherry's weekly salary?

14. SOLVING EQUATIONS BY USING SUBTRACTION

Let us solve the following problem algebraically: Five years from now Tom will be 15 years of age. How old is Tom now?

Let x represent Tom's age now in years.
Then $x + 5$ represents Tom's age 5 years from now.

The equation that represents the relationship stated in the problem, "five years from now Tom will be 15 years of age," is $x + 5 = 15$. When we solve this equation, we will find the answer to the problem.

In solving the equation, we find a simpler equivalent equation that will reveal the value of x. Since in the expression $x + 5$, x and 5 are related by the operation of addition, we will subtract 5 from both sides of the equation to obtain the simpler equivalent equation. See how we solve the equation:

Method 1	*Method* 2	
$x + 5 = 15$	$x + 5 = 15$	
$-5 = -5$	$x + 5 - 5 = 15 - 5$	S_5 (Subtract 5 from both sides of the previous equation.)
$x = 10$	$x = 10$	The answer is 10 years.

Check:

$$x + 5 = 15$$
$$10 + 5 \stackrel{?}{=} 15$$
$$15 = 15$$

Substitute the answer to see whether it will satisfy the question.

NOTE

If an equation tells us the result after a number has been added to an unknown quantity, we can subtract the number from both sides of the equation to obtain an equivalent equation that has only the unknown quantity itself on one side of the equation.

MODEL PROBLEM

Solve and check:

Method 1	*Method* 2	*Check:*
$y + 2\frac{3}{4} = 6$	$y + 2\frac{3}{4} = 6$	$y + 2\frac{3}{4} = 6$
$S_{2\frac{3}{4}}$ $-2\frac{3}{4} = -2\frac{3}{4}$	$y + 2\frac{3}{4} - 2\frac{3}{4} = 6 - 2\frac{3}{4}$	$3\frac{1}{4} + 2\frac{3}{4} \stackrel{?}{=} 6$
$y = 3\frac{1}{4}$	$y = 3\frac{1}{4}$	$6 = 6$

EXERCISES

Solve and check the following equations:

1. $a + 5 = 7$ **2.** $b + 8 = 18$ **3.** $19 = r + 10$

4. $x + 8 = 8$ **5.** $12 + y = 12$ **6.** $49 = 13 + w$

7. $y + \frac{1}{3} = 9\frac{2}{3}$ **8.** $20\frac{5}{6} = v + 1\frac{1}{6}$ **9.** $r + 3 = 6\frac{1}{2}$

10. $5\frac{7}{8} = x + 1\frac{1}{2}$ **11.** $y + 1\frac{3}{4} = 8\frac{1}{2}$ **12.** $y + .3 = 9.1$

13. $9 = t + .7$ **14.** $9.6 = x + 9.6$ **15.** $y + 6.4 = 15$

16. What number increased by 8 is 12?

17. If 12 is added to a number, the result is 28. What is the number?

18. Ten years from now, Marie will be 25 years old. How old is Marie now?

19. After the price of steak rose 15 cents a pound, it was sold for 70 cents a pound. What was the original price?

20. After a train increased its rate of speed by 20 miles per hour, it was traveling 70 miles per hour. What was its original rate of speed?

21. During a Red Cross drive, the boys in a class contributed $1.25 more than the girls. If the boys contributed $5.50, how much did the girls contribute?

15. SOLVING MORE DIFFICULT EQUATIONS

In solving more difficult equations, it is sometimes necessary to use two or more operations to obtain an equivalent equation that has only the unknown quantity itself on one side of the equation. Study the following illustrative examples to see how this is done.

MODEL PROBLEMS

1. Solve and check:

$$3x + 2x = 35$$
$$5x = 35 \quad \text{Collect like terms.}$$
$$\frac{5x}{5} = \frac{35}{5} \quad D_5$$
$$x = 7$$

2. Solve and check:

$$\frac{5y}{4} = 20$$
$$4 \times \frac{5y}{4} = 4 \times 20 \quad M_4$$
$$5y = 80$$
$$\frac{5y}{5} = \frac{80}{5} \quad D_5$$
$$y = 16$$

3. Solve and check:

$$3 = 4x - 5$$
$$+5 = \quad +5 \quad A_5$$
$$8 = 4x$$
$$\frac{8}{4} = \frac{4x}{4} \quad D_4$$
$$2 = x$$

4. Solve and check:

$$9x = 3x + 30$$
$$9x - 3x = 3x + 30 - 3x \quad S_{3x}$$
$$6x = 30$$
$$\frac{6x}{6} = \frac{30}{6} \quad D_6$$
$$x = 5$$

In problems 1–4, the check is left to the student.

Procedure. **To solve equations when more than one operation is required:**

1. If there are any like terms on the same side of the equation, collect them.

2. If the operation of addition or subtraction appears in the equation, use the inverse operation to obtain a simpler equivalent equation.

3. If the operation of division or multiplication appears in the equation, use the inverse operation to obtain a simpler equivalent equation.

NOTE

In the process of obtaining simpler equivalent equations, we try to have all the terms containing the unknown on one side of the equation and all other terms on the other side of the equation.

EXERCISES

Solve and check the following equations:

Group A

1. $3x + 2x = 50$
2. $42 = 6x + x$
3. $10x - 3x = 21$
4. $5x - x = 64$
5. $10 = 6x - x$
6. $90 = 8c + 7c - 5c$
7. $x + 3x + x + 3x = 48$
8. $2x + 3x + 2x + 3x = 60$

Group B

9. $\frac{3x}{4} = 6$
10. $\frac{2y}{3} = 14$
11. $\frac{4t}{5} = 16$
12. $\frac{2}{3}a = 20$
13. $\frac{5}{9}b = 35$
14. $\frac{7}{10}r = 28$
15. $\frac{15x}{4} = 60$
16. $\frac{5w}{4} = \frac{15}{2}$
17. $\frac{3t}{4} = \frac{9}{4}$

Group C

18. $2a + 9 = 21$
19. $3b + 7 = 37$
20. $33 = 2x + 3$
21. $13 = 10x + 8$
22. $4x = x + 15$
23. $2y + 18 = 5y$
24. $\frac{c}{5} + 4 = 9$
25. $6 = \frac{y}{7} + 5$
26. $5x + \frac{1}{2} = 10\frac{1}{2}$
27. $3m + .8 = 2$
28. $\frac{1}{5}x + 2 = 10$
29. $\frac{2n}{3} + 5 = 7$
30. $\frac{3x}{4} + 6 = 15$
31. $\frac{4x}{5} + 3 = 31$
32. $\frac{x}{3} + 1\frac{1}{2} = 4\frac{1}{2}$
33. $3x + 2x + 1 = 41$
34. $9x + 3 + 2 = 32$
35. $4x - x + 4 = 31$
36. $6x + 1 + 3x = 19$
37. $7x = 5x + 12$
38. $12 + 3x = 9x$

Group D

39. $2x - 5 = 13$
40. $4a - 2 = 22$
41. $5b - 1 = 24$
42. $35 = 6c - 7$
43. $17 = 9t - 10$
44. $4a = 60 - 2a$
45. $18 - a = 2a$
46. $\frac{x}{2} - 3 = 5$
47. $2y - \frac{1}{2} = 9\frac{1}{2}$
48. $5t - .5 = 6$
49. $\frac{1}{3}x - 4 = 12$
50. $7 = \frac{1}{2}y - 5$
51. $\frac{2x}{3} - 4 = 16$
52. $\frac{3m}{4} - 3 = 21$
53. $4x + x - 1 = 24$
54. $7x + 1 + 2 = 17$
55. $8x - 5x - 3 = 24$
56. $8x = 60 - 2x$
57. $5x + 4 = 3x + 20$
58. $6y - 2 = 2y + 26$
59. $x + 4 = 13 - 2x$

More Practice in Solving Equations

Solve and check the following equations:

1. $5a = 40$
2. $.7b = 84$
3. $6c + 3 = 27$
4. $36 = 3m - 3$
5. $5b = 3b + 40$
6. $6x - 4 = 38$
7. $9x - 7x = 32$
8. $d - 5\frac{1}{4} = 7\frac{1}{2}$
9. $.25x = 6.50$
10. $6b = 56 - 2b$
11. $12 = 3a + 6$
12. $t + \frac{3}{8} = \frac{7}{8}$
13. $\frac{1}{2}a - 1 = 7$
14. $\frac{3}{5}b = 21$
15. $.2b + .3b = 4$
16. $5x + 7 = 7$
17. $14d = 7$
18. $17 = 9x - 1$
19. $2d - 4\frac{1}{2} = 5\frac{1}{2}$
20. $3x + 5x = 64$
21. $.2y - .4 = 2$
22. $9 = \frac{1}{2}y + 3$
23. $\frac{9}{4} = \frac{b}{16}$
24. $3x + 16 = 67$
25. $w + 3 = 3$
26. $10 = 20a$
27. $y + 12 = 4y$
28. $11y + 12 = 39$
29. $19 = x - 3.2$
30. $5s - s = 36$
31. $2m - \frac{1}{2} = 3\frac{1}{2}$
32. $.8z - 5 = .6$
33. $42 = 5m - 3$
34. $y + 5 = 6$
35. $3y + 16 = 40$
36. $90 = 4x - 10$
37. $3y - 8 = 25$
38. $\frac{2d}{3} = 18$
39. $6x + x = 49$
40. $8b - 6b - 2 = 18$
41. $\frac{1}{2}c + \frac{1}{4}c = 6$
42. $\frac{3}{10}x = 36$
43. $24 = 3.6a - 2.4a$
44. $\frac{3x}{4} = 27$
45. $\frac{y}{4} - 3 = 5$
46. $3n + 4n - 5 = 30$
47. $\frac{3x}{5} + 4 = 34$
48. $\frac{m}{5} + 3\frac{1}{2} = 6\frac{1}{2}$
49. $5x - 3x + 5 = 21$
50. $.4c - 4 = 3.6$
51. $\frac{b}{4} = \frac{5}{2}$
52. $x + 2x + 3 = 27$
53. $\frac{x}{10} = \frac{3}{5}$
54. $\frac{3y}{2} = 16\frac{1}{2}$
55. $m + 2m - 7 = 11$
56. $\frac{3a}{4} - 3 = 9$
57. $5b + 2 - b = 22$
58. $7x - 3 = 5x + 7$
59. $4y + 5 = y + 35$
60. $3b - 3 = 37 - 2b$

16. USING FORMULAS AS EQUATIONS

Since a formula states that two quantities are equal, it can be used as an equation.

MODEL PROBLEMS

1. Find the height of a rectangle if its area is 36 sq. in. and its base is 9 in.

Solution:

	$A = bh$	Write the proper formula.
	$A = 36,\ b = 9$	Write the given values of A and b.
	$36 = 9h$	Substitute these values in the formula.
D_9	$4 = h$	Solve the equation.

Answer: 4 inches

2. Given the formula $d = rt$, solve for t.

	$d = rt$	A formula is an equation.
D_r	$\frac{d}{r} = \frac{rt}{r}$	Divide both sides of the equation by r, the multiplier of t.
	$\frac{d}{r} = t$	

EXERCISES

1. Using the formula $C = 5n$, find n when:

a. $C = 50$ *b.* $C = 35$ *c.* $C = 150$ *d.* $C = 200$

2. Using the formula $A = bh$, find:

a. h when $A = 100,\ b = 5$ *b.* h when $A = 56,\ b = 8$

c. b when $A = 80,\ h = 10$ *d.* b when $A = 30,\ h = 4$

3. Using the formula $P = br$, find:

a. r when $P = 10,\ b = 25$ *b.* r when $P = 3,\ b = 150$

c. b when $P = 12,\ r = .06$ *d.* b when $P = 20,\ r = .05$

4. Using the formula $A = \frac{1}{2}bh$, find:

a. h when $A = 36,\ b = 4$ *b.* b when $A = 28,\ h = 14$

5. Using the formula $V = lwh$, find h when $V = 144$, $l = 3$, $w = 8$.

6. Using the formula $I = prt$, find t when $I = 36$, $p = 200$, $r = .06$.

7. Using the formula $P = a + b + c$, find a when $P = 50$, $b = 20$, and $c = 18$.

8. Using the formula $P = 2a + b$, find a when $P = 36$ and $b = 10$

9. Using the formula $R = \frac{D}{T}$, find D when $R = 30$ and $T = 5$.

10. Using the formula $P = M - C$, find M when $P = 20$ and $C = 80$.

11. In the formula $C = np$ (*a*) solve for p (*b*) solve for n.

12. In the formula $P = br$ (*a*) solve for b (*b*) solve for r.

13. In the formula $A = bh$ (*a*) solve for h (*b*) solve for b.

17. PROPORTIONS

Since each of the ratios $\frac{4}{2}$ and $\frac{6}{3}$ is equal to $\frac{2}{1}$, the ratio $\frac{4}{2}$ is equal to the ratio $\frac{6}{3}$. This equality, which may be written $\frac{4}{2} = \frac{6}{3}$, is read "4 is to 2 as 6 is to 3" and is called a ***proportion.***

A proportion is a sentence which states that two ratios are equal.

A proportion in which the two ratios *are* equal is called a ***true proportion.***

Another way of writing the proportion $\frac{4}{2} = \frac{6}{3}$ is $4:2 = 6:3$, which is also read "4 is to 2 as 6 is to 3."

Notice that in the true proportion $\frac{4}{2} = \frac{6}{3}$, the first term and second term of the first ratio are related in the same way as the first term and second term of the second ratio. In each of the ratios $\frac{4}{2}$ and $\frac{6}{3}$, the first term is twice the second term, that is, 4 is twice 2, and 6 is twice 3. Also, the second term is one-half of the first term, that is, 2 is one-half of 4, and 3 is one-half of 6. Observe, also, that the two first terms of the two ratios are related in the same way as the two second terms. The first term 4 is two-thirds of the first term 6, and the second term 2 is two-thirds of the second term 3. These relationships are helpful in solving problems that deal with proportions.

MODEL PROBLEMS

1. Arrange the numbers 1, 3, 15, 5 to form a true proportion.

$\frac{1}{3} = \frac{5}{15}$ *or* $1:3 = 5:15$	Arrange the numbers so that they will form two equal ratios.

2. Solve the proportion $x:8 = 3:4$ for x.
Solution: To solve the proportion $x:8 = 3:4$ means to find the value that can replace x and make the result a true proportion.

Method 1	*Method* 2
$x:8 = 3:4$ *or* $\frac{x}{8} = \frac{3}{4}$ Since 8, the second term of the first ratio, is twice 4, the second term of the second ratio, then x, the first term of the first ratio, must be twice 3, the first term of the second ratio: $x = 2 \times 3 = 6$. *Answer:* $x = 6$	$x:8 = 3:4$ *or* $\frac{x}{8} = \frac{3}{4}$ M_8 $8 \times \frac{x}{8} = 8 \times \frac{3}{4}$ $x = 6$ *Check:* $\frac{6}{8} \stackrel{?}{=} \frac{3}{4}$ $\frac{3}{4} = \frac{3}{4}$

3. On a map, two cities are 4 inches apart. On this map, 1 inch represents a distance of 50 miles. How many miles apart are the two cities?

Solution: $\dfrac{\text{first measurement on map}}{\text{second measurement on map}} = \dfrac{\text{first distance on ground}}{\text{second distance on ground}}$

Let x = the unknown distance on the ground. Then $\frac{4}{1} = \frac{x}{50}$

Method 1	*Method* 2
Since 4, the first term of the first ratio, is 4 times 1, the second term of the first ratio, then x, the first term of the second ratio, must be 4 times 50, the second term of the second ratio: $x = 4 \times 50 = 200$. *Answer:* 200 miles	$\frac{4}{1} = \frac{x}{50}$ M_{50} $50 \times \frac{4}{1} = 50 \times \frac{x}{50}$ $200 = x$

A PRINCIPLE OF PROPORTIONS

In the proportion $4:2 = 6:3$, also written $\frac{4}{2} = \frac{6}{3}$, the first and fourth terms, 4 and 3, are called the ***extremes;*** the second and third terms, 2 and 6, are called the ***means.***

Notice that in the true proportion $4:2 = 6:3$, the product of the two means, 2×6, is 12; also, the product of the two extremes, 4×3, is 12. Observe that the product of the means is equal to the product of the extremes. This example illustrates the following principle:

Principle. In a true proportion, the product of the means is equal to the product of the extremes.

A proportion in which the product of the means does not equal the product of the extremes is not a true proportion.

MODEL PROBLEMS

1. Is $3:4 = 15:20$ a true proportion?

Solution:

In the proportion $3:4 = 15:20$, the product of the means (4 and 15) is 4×15, or 60, and the product of the extremes (3 and 20) is 3×20, or 60.

Since the product of the means is equal to the product of the extremes, $3:4 = 15:20$ is a true proportion. *Answer:* Yes

2. Solve the proportion $4:5 = 16:x$ for x.

Solution:

If $4:5 = 16:x$ is to be a true proportion, then $5 \times 16 = 4x$. (The product of the means must equal the product of the extremes.)

	Check:
$80 = 4x$	$4:5 \stackrel{?}{=} 16:20$
$20 = x$	$4:5 = 4:5$

Answer: $x = 20$

EXERCISES

1. In each of the following, state whether or not the ratios may form a true proportion:

a. $\frac{1}{4}, \frac{4}{16}$ *b.* $\frac{3}{5}, \frac{9}{15}$ *c.* $\frac{2}{3}, \frac{4}{9}$ *d.* $\frac{10}{15}, \frac{4}{6}$ *e.* $\frac{3}{4}, \frac{6}{10}$

f. $\frac{4}{1}, \frac{28}{7}$ *g.* $\frac{2}{3}, \frac{3}{2}$ *h.* $\frac{4}{12}, \frac{10}{15}$ *i.* $\frac{14}{18}, \frac{28}{36}$ *j.* $\frac{15}{12}, \frac{30}{36}$

2. Use the following sets of numbers to form true proportions:

a. 1, 4, 20, 5 *b.* 1, 2, 12, 6 *c.* 2, 3, 9, 6
d. 5, 4, 20, 25 *e.* 3, 1, 5, 15 *f.* 6, 7, 28, 24
g. 2, 12, 18, 3 *h.* 4, 3, 6, 2 *i.* 8, 5, 4, 10

3. In each of the following, find the number that can replace the question mark and make the result a true proportion:

a. $\frac{1}{2} = \frac{?}{10}$ *b.* $\frac{3}{4} = \frac{18}{?}$ *c.* $\frac{5}{?} = \frac{15}{60}$ *d.* $\frac{?}{9} = \frac{35}{63}$

e. 1:5 = 6:? *f.* 4:6 = ?:42 *g.* ?:30 = 6:10 *h.* 8:? = 12:9

4. Find and check the value of x in each of the following proportions:

a. $\frac{x}{4} = \frac{9}{2}$ *b.* $\frac{x}{50} = \frac{7}{10}$ *c.* $\frac{5}{3} = \frac{x}{9}$ *d.* $\frac{24}{10} = \frac{x}{30}$

e. $\frac{x}{8} = \frac{14}{56}$ *f.* $\frac{x}{36} = \frac{12}{27}$ *g.* $\frac{24}{40} = \frac{x}{20}$ *h.* $\frac{21}{35} = \frac{x}{25}$

i. $x:5 = 4:20$ *j.* $x:6 = 12:18$ *k.* $x:32 = 5:8$
l. $2:3 = x:18$ *m.* $3:7 = x:28$ *n.* $5:1 = x:15$
o. $x:69 = 6:18$ *p.* $x:14 = 8:28$ *q.* $10:15 = x:12$

5. If 3 apples cost 20 cents, find the cost of 15 apples at the same rate.
6. If 4 oranges cost 23 cents, find the cost of 2 dozen oranges at the same rate.
7. If 2 tickets cost $8.80, find the cost of 9 tickets.
8. Choice apples are being sold in a grocery store at 3 for 25 cents. How many apples can you buy for 75 cents?
9. How much would you pay for 5 oranges at the rate of 60 cents a dozen?
10. If a machine can print 3000 copies of a certain poster in 1 hour, how many minutes will it take, at the same rate, for the machine to print 1000 copies?
11. A 20-acre field yields 300 bushels of wheat. At the same rate, how many bushels will a 50-acre field yield?
12. If a boy walks 6 miles in 2 hours, how far will he walk at the same rate in 5 hours?
13. A plane traveled 600 miles in 3 hours. At the same rate of speed, how many hours will it take to travel 1000 miles?
14. The scale on a map is 1 inch = 200 miles. What is the distance in miles between two cities $4\frac{1}{2}$ inches apart on the map?
15. The scale on a map is $\frac{1}{2}$ inch = 50 miles. Two towns are 3 inches apart on the map. What is the distance in miles between the two towns?
16. In a certain concrete mixture, the ratio of cement to sand is 1 to 4. At this ratio, how many shovelfuls of cement would be needed to be mixed with 80 shovelfuls of sand?
17. In a school, the ratio of the number of girls to the number of boys is 2 to 3. If 630 boys attend the school, what is the number of girls attending the school?

18. SOLVING PROBLEMS BY USING ALGEBRAIC EQUATIONS

The following suggestions will be found helpful in learning how to use algebraic equations to solve problems.

1. Read the problem carefully once or twice.
2. Draw a diagram if it helps you better to understand the problem.
3. Discover what is given and what is to be found.
4. Represent one of the unknown numbers by a letter.
5. Use relationships among the unknowns to represent the remaining unknowns in terms of this letter.
6. Write the equation that expresses the relationship stated in the problem.
7. Solve the equation.
8. Write your answer to the problem.
9. Check to see whether your answer satisfies the relationships given in the original problem.

MODEL PROBLEMS

1. Four times a number decreased by 3 equals 25. Find the number.

Solution: Let n represent the unknown number.

Four times a number decreased by 3 equals 25.

			Check:
1.	$4n - 3 = 25$		$(4 \times 7) - 3 \stackrel{?}{=} 25$
2.	$4n - 3 + 3 = 25 + 3$	A_3	$28 - 3 \stackrel{?}{=} 25$
3.	$4n = 28$		$25 = 25$
4.	$n = 7$	D_4	

Answer: 7

2. In a class, there are 36 pupils. There are three times as many boys as there are girls. Find the number of girls and the number of boys in the class.

Solution: Let x represent the number of girls.

Then $3x$ represents the number of boys.

The total number of pupils in the class is 36.

			Check:
1.	$x + 3x = 36$		
2.	$4x = 36$	Collect like terms.	$9 + 27 \stackrel{?}{=} 36$
3.	$x = 9$	D_4	$36 = 36$
4.	$3x = 27$		

Answer: There are 9 girls and 27 boys in the class.

EXERCISES

1. Three times a certain number is 36. Find the number.
2. A number divided by 5 is 17. Find the number.
3. A number increased by 13 is 30. Find the number.
4. A number decreased by 7 is 14. Find the number.
5. Three-sevenths of a number is 21. Find the number.
6. Two times a number increased by 12 is 52. Find the number.
7. Four times a number decreased by 7 is 41. Find the number.
8. If two times a certain number is diminished by 8, the result is 4. Find the number.
9. If three times a certain number increased by 4 is equal to 19, what is the number?
10. If 9 is subtracted from two-thirds of a number, the result is 7. Find the number.
11. The larger of two numbers is 5 times the smaller. The sum of the two numbers is 48. Find the numbers.
12. If one-quarter of a number is added to the number, the result is 50. Find the number.
13. What number increased by $\frac{2}{3}$ of itself is equal to 40?
14. Six times Peter's savings is $72. How much has Peter saved?
15. A basketball player scored $\frac{2}{3}$ of his team's points. If he scored 24 points, how many points did the team score?
16. In 10 days, a man earned $95. How much did he earn each day?
17. One number is 8 times another number. Their difference is 70. Find the numbers.
18. Tom spent 4 times as much as his sister. Together they spent $15.50. How much did each spend?
19. After Larry had worked 30 hours, he had finished $\frac{3}{4}$ of a job. How many hours did he require to do the entire job?
20. After spending $15, Sally had $4 left. How much money did she have originally?
21. Mr. James traveled 150 miles in 5 hours. How far did he travel in one hour?
22. How much money must be invested at 6% to yield $42 interest in one year?
23. A car traveled 35 miles an hour. How many hours did it require to travel 280 miles?
24. Harvey is 3 times as old as Jane. The sum of their ages is 48 years. Find the age of each.
25. In a track meet, Sam scored 9 points. This represented $\frac{1}{3}$ of his team's final score. How many points did the team score?
26. The Bayview Junior High School baseball team won 3 games more than it lost. If it played 15 games, how many games did it lose?

27. John and Frank worked a total of 36 hours. If Frank worked 10 hours longer than John, how many hours did each boy work?
28. A carpenter sawed off $\frac{5}{8}$ of a piece of lumber. If he sawed off 10 feet, what was the length of the original piece of lumber?
29. Eight years ago, Alice was 15 years old. How old is she now?
30. Ronald is 7 years older than Marie. If Ronald is 21, how old is Marie?
31. A man left \$8000 to his son and daughter. He left three times as much to his son as he left to his daughter. How much did he leave to each?
32. Frank is 5 years younger than Charles. If Frank is 24, how old is Charles?
33. Thomas had \$15 in the bank. He worked for 20 hours and put his earnings in the bank. He then had \$40. How much did he earn each hour he worked?
34. Edgar and Arthur worked a total of 48 hours. If Arthur worked 10 hours more than Edgar, how many hours did each work?
35. Peter drew half of his money out of the bank to go shopping. After spending \$9, he had \$3 left. How much did he have in the bank originally?
36. If Carl adds \$15 to twice the amount he has, he will have \$85. How much money does Carl have?
37. Separate 81 into two parts so that one part will be 8 times the other.
38. Separate 52 into two parts so that one part will be 3 times the other.
39. A boy earned \$10 more than twice as much as his sister. Together, they earned \$40. How much did each one earn?
40. Ted weighs twice as much as Bob. Martin weighs 3 times as much as Bob. The three boys together weigh 210 pounds. What is the weight of each boy?
41. A man divided \$9,000 among his wife, son, and daughter. The wife received twice as much as the daughter, and the son received \$1,000 more than the daughter. How much did each receive?

Geometric Problems

1. Find the side of a square whose perimeter is 40 feet.
2. Find the side of an equilateral triangle whose perimeter is 45 inches.
3. The length of a rectangle is 14 feet. This represents $\frac{1}{3}$ of the perimeter. Find the perimeter.
4. The sides of a triangle are represented by the numbers $3x$, $4x$, and $5x$. The perimeter is 60 feet. Find each side.
5. The sides of a triangle are represented by the numbers x, $2x$, and $2x + 2$. The perimeter is 22 inches. Find each side.
6. The sides of a rectangle are represented by the numbers x, $x + 5$, x, and $x + 5$. The perimeter is 46 yards. Find each side.
7. The length of a rectangle is twice its width. The perimeter is 60 feet. Find the width and the length.
8. The perimeter of a rectangle is 54 inches. The length is $\frac{1}{2}$ of the width. Find the length and the width.

9. The perimeter of a rectangle is 34 feet. The width is 7 feet more than the length. Find the length and the width.
10. The width of a rectangle is 3 feet less than its length. If the perimeter of the rectangle is 54 feet, find the length and the width.
11. The area of a rectangle is 48 square inches. If the length is 12 inches, find the width.
12. The sum of the angles of a triangle is 180°. If the three angles of a triangle are equal to each other, find the number of degrees in each angle.
13. If the first angle of a triangle is equal to twice the second angle, and the third angle is equal to three times the first angle, find each angle of the triangle.
14. If each of the equal angles of an isosceles triangle is 30°, find the vertex angle.
15. If the vertex angle of an isosceles triangle is 80°, find each of the equal base angles.
16. If each of the equal angles of an isosceles triangle is 4 times the vertex angle, find the three angles of the triangle.
17. If the sum of two angles is 90°, they are *complementary*. If one angle is 4 times as large as its complementary angle, find both angles.
18. Two angles are complementary. One contains 30° more than the other. Find both angles.
19. If the sum of two angles is 180°, they are *supplementary*. If one angle is $\frac{1}{2}$ as large as its supplementary angle, find both angles.
20. Two angles are supplementary. One angle contains 20° more than 4 times the other. Find both angles.

19. UNDERSTANDING THE MEANING AND USE OF SIGNED NUMBERS

A newspaper reported that the temperature at noon the previous day was +3°, or 3° above zero. The temperature at 1 P.M. was reported as −3°, or 3° below zero. Notice that these two temperatures extend in opposite directions from the zero reading on a thermometer. In our daily life, we frequently talk about things that are opposites. For example, we talk about profit and loss, riding up and down elevators, depositing money in and withdrawing money from a bank, traveling north and south, etc.

In mathematics, when we talk about quantities that are in opposite directions from a reference point, the ***zero point,*** or ***origin,*** we consider one of the quantities as ***positive*** (+) and the other as ***negative*** (−).

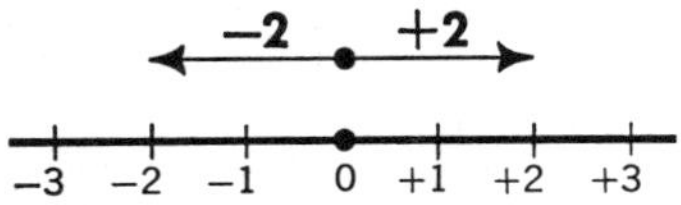

In the drawing at the left, if we start at the zero point, the origin, and move 2 units to the right, we say that we have traveled +2 units. However, if we start at the zero point and move 2 units to the left, we say that we have traveled −2 units.

If winning \$5 is indicated by +\$5 (positive \$5), losing \$5 is indicated by −\$5 (negative \$5).

The positive and negative numbers are called ***signed numbers,*** or ***directed numbers.*** If no sign appears before a number, it is understood to be positive. Thus, 8 means +8 or positive 8.

In an arithmetic signed number, the + or − sign shows in which *direction* from the origin the number is located. When the + or − sign is removed in such a number, the result is called the ***absolute value*** of the signed number. Thus, in the temperature reading +3°, the + indicates the direction (above zero), and the 3 is the absolute value of the number. Likewise, in the temperature reading −3°, the − indicates the direction (below zero), and the 3 is the absolute value of the number. The numbers +3 and −3 have the same absolute value, 3. However, they do not represent the same number because they represent quantities that are measured in opposite directions from the zero point.

We must be very careful not to confuse the + sign of a directed number with the symbol + that we have always used to represent the operation of addition. We must also be very careful not to confuse the − sign of a directed number with the symbol − that we have always used to represent the operation of subtraction.

MODEL PROBLEMS

1. If the temperature rose from 3° below zero to 5° above zero, how many degrees did it rise?

Solution: When the temperature rose from 3° below zero to 0°, it rose 3°. When it rose from 0° to 5° above zero, it rose 5° more. Therefore, in all it rose 3° + 5°, or 8°. *Answer:* 8°

2. If the temperature fell from +20° to −5°, how many degrees did it fall?

Solution: When the temperature fell from +20° to 0°, it fell 20°. When it fell from 0° to −5°, it fell 5° more. Therefore, in all it fell 20° + 5°, or 25°. *Answer:* 25°

EXERCISES

1. Give the opposite of:

a. rising in the air *b.* above sea level *c.* north of the equator
d. making a profit *e.* rise in price *f.* traveling east
g. gain in weight *h.* west longitude *i.* south latitude

2. If +7 means gaining 7 pounds, what does −7 mean?
3. If +$100 means depositing $100 in a bank, what does −$100 mean?
4. If +$50 means a profit of $50, what does −$50 mean?
5. If +15° means 5° north of the equator, what does −15° mean?
6. If +200 means 200 miles east, what does −200 mean?
7. Express as a signed number:

   a. 15° above zero *b.* a profit of $10
   c. a $50 withdrawal from a bank *d.* 10° below zero
   e. a 7-yard loss in football *f.* 35 feet below sea level

8. What is meant by the statement that the altitude of the Dead Sea is −1290 feet?
9. What is meant by the statement that the change in value of a stock is:

   a. +$2 *b.* −$1 *c.* $+\$\frac{1}{2}$ *d.* $-\$\frac{3}{4}$ *e.* $+\$1\frac{7}{8}$ *f.* $-\$3\frac{5}{8}$

10. If the temperature changes from 5° above zero to 8° above zero, is the change positive or negative?
11. If the temperature changes from 7° below zero to 15° below zero, is the change positive or negative?
12. Find the number of degrees the temperature rises when it changes from:

    a. 50° above zero to 68° above zero *b.* 12° below zero to 6° below zero
    c. 7° below zero to 0° *d.* 5° below zero to 8° above zero
    e. +20° to +35° *f.* −6° to −2° *g.* −9° to 0° *h.* −15° to +8°

13. Find the number of degrees the temperature drops when it changes from:

    a. 72° above zero to 54° above zero *b.* 3° below zero to 10° below zero
    c. 13° above zero to 0° *d.* 4° above zero to 8° below zero
    e. +45° to +10° *f.* −10° to −15° *g.* +18° to 0° *h.* +12° to −7°

14. In each of the following exercises, use the number line below to discover the point at which you have arrived:

    a. Start at the number zero and move a distance represented by +2. From that point, move a distance represented by +6.
    b. Start at the number zero and move a distance represented by −3. From that point, move a distance represented by −7.
    c. Start at the number zero and move a distance represented by +9. From that point, move a distance represented by −5.
    d. Start at the number zero and move a distance represented by +5. From that point, move a distance represented by −7.
    e. Start at the number zero and move a distance represented by −8. From that point, move a distance represented by +5.
    f. Start at the number zero and move a distance represented by −4. From that point, move a distance represented by +9.

20. ADDING SIGNED NUMBERS

If the temperature rose 5° (represented by +5) in one hour and rose 3° (represented by +3) more in the next hour, to find the combined effect during the two-hour period, we can combine, or add, the two signed numbers +5 and +3. We know that the total rise was 8° (represented by +8). The addition may be written in either of the following ways:

$$\text{Add: } \begin{array}{r} +5 \\ \underline{+3} \\ +8 \end{array} \quad \textit{or} \quad (+5) + (+3) = +8$$

Observe that the sum of the positive numbers is a positive number and that 8, the absolute value of the sum, is the sum of the absolute values 5 and 3.

If the temperature dropped 2° (represented by −2) in one hour and dropped 4° (represented by −4) in the next hour, to find the combined effect during the two-hour period, we can add the two signed numbers −2 and −4. We know that the total drop was 6° (represented by −6). The addition may be written in either of the following ways:

$$\text{Add: } \begin{array}{r} -2 \\ \underline{-4} \\ -6 \end{array} \quad \textit{or} \quad (-2) + (-4) = -6$$

Observe that the sum of the negative numbers is a negative number and that 6, the absolute value of the sum, is the sum of the absolute values 4 and 2.

Procedure. To add signed numbers that have like signs, add the absolute values of the numbers and place the common sign before the sum.

If the temperature rose 6° (represented by +6) in one hour and dropped 1° (represented by −1) in the next hour, to find the combined effect during the two-hour period, we can add the signed numbers +6 and −1. We know that the change during this period would be a rise of 5° (represented by +5). The addition may be written as follows:

$$\text{Add: } \begin{array}{r} +6 \\ \underline{-1} \\ +5 \end{array} \quad \textit{or} \quad (+6) + (-1) = +5$$

Observe that the sum +5 is positive and has the same sign as +6, the number with the larger absolute value. Also observe that 5, the absolute value of the sum, is the difference of the absolute values 6 and 1.

If the temperature dropped 8° (represented by -8) in one hour and rose 5° (represented by $+5$) in the next hour, to find the combined effect during the two-hour period, we can add the signed numbers -8 and $+5$. We know that the change during this period would be a drop of 3° (represented by -3). The addition may be written as follows:

$$\text{Add: } \begin{array}{r} -8 \\ \underline{+5} \\ -3 \end{array} \quad \textit{or} \quad (-8) + (+5) = -3$$

Observe that the sum -3 is negative and has the same sign as -8, the number with the larger absolute value. Also observe that 3, the absolute value of the sum, is the difference of the absolute values 8 and 5.

Procedure. To add signed numbers that have unlike signs, find the difference of the absolute values and place the sign of the number that has the greater absolute value before the difference.

If the temperature rises 3° (represented by $+3$) in one hour and drops 3° (represented by -3) in the next hour, the combined effect over the two-hour period is neither a rise nor a fall (represented by 0). Thus, $(+3) + (-3) = 0$.

Principle. The sum of two signed numbers that have the same absolute value but opposite signs is 0.

The sum of two signed numbers is called their ***algebraic sum.***

We have now seen that the symbol + is used for two purposes. It indicates the operation of addition when it is placed between two signed numbers, and it expresses a positive number when it is placed in front of a number. Thus, $(+6) + (-1)$ means that positive 6 and negative 1 are to be added.

It is also possible to write $(+6) + (-1)$ as $+6 - 1$. Notice that the parentheses and the symbol of addition have been omitted. In algebra, when signed numbers are written next to each other, it is understood that they are to be added algebraically. Thus, $+8 - 4$ means that $+8$ and -4 are to be added; and $+6 - 2 + 5$ means that $+6$, -2, and $+5$ are to be added.

MODEL PROBLEMS

Add:

	1.	2.	3.	4.	5.
	$+15$	-12	$+15$	-22	$+6$
	$+12$	$-\ 9$	$-\ 9$	$+13$	-6
	$+27$	-21	$+\ 6$	$-\ 9$	0

THE COMMUTATIVE PRINCIPLE FOR ADDITION OF SIGNED NUMBERS

Since $(-2) + (-4) = (-6)$ and $(-4) + (-2) = (-6)$,
then $\mathbf{(-2) + (-4) = (-4) + (-2)}$.

Since $(+6) + (-1) = (+5)$ and $(-1) + (+6) = (+5)$,
then $\mathbf{(+6) + (-1) = (-1) + (+6)}$.

The above examples illustrate the ***commutative principle for addition of signed numbers.*** This means that when two signed numbers are added, changing the order in which they are added does not change their sum.

In general, if a and b are signed numbers, then:

$$(a) + (b) = (b) + (a)$$

THE ASSOCIATIVE PRINCIPLE FOR ADDITION OF SIGNED NUMBERS

Since $[(+8) + (-3)] + (+2) = (+5) + (+2) = (+7)$
and $(+8) + [(-3) + (+2)] = (+8) + (-1) = (+7)$,
then $\mathbf{[(+8) + (-3)] + (+2) = (+8) + [(-3) + (+2)]}$.

The above example illustrates the ***associative principle for addition of signed numbers.*** This means that when we are finding the sum of three or more signed numbers, changing the way in which we group the numbers does not change their sum.

In general, if a, b, and c are signed numbers, then:

$$[(a) + (b)] + (c) = (a) + [(b) + (c)]$$

EXERCISES

Find the algebraic sum:

	(a)	(b)	(c)	(d)	(e)	(f)
1.	$+10$ $+7$	$+4$ $+12$	$+18$ $+9$	$+29$ $+13$	$+8.6$ $+5.4$	$+6\frac{5}{8}$ $+3\frac{2}{8}$
2.	-9 -15	-25 -47	-40 -34	-5.8 -6.7	-9.9 -8.1	$-8\frac{3}{4}$ $-5\frac{1}{2}$
3.	$+25$ -10	-40 $+22$	$+9$ -9	-6.9 $+5.3$	-4.7 $+8.4$	$+12\frac{1}{3}$ $-8\frac{2}{3}$
4.	-18 $+24$	$+36$ -48	$+40$ -60	$+7.7$ -7.7	$+6.3$ -9.8	$-18\frac{1}{6}$ $+7\frac{5}{6}$

5. $(+10) + (+7)$ **6.** $(+15) + (-20)$ **7.** $(-9) + (-13)$

8. $(+4) + (-10)$ **9.** $(-18) + (+25)$ **10.** $(+50) + (-9)$

11. $(-3) + (-12)$ **12.** $(+8) + (-8)$ **13.** $(-22) + (+18)$

14. Add $+7$ to each of the following numbers:

a. $+3$ *b.* $+8$ *c.* $+10$ *d.* -2 *e.* -15 *f.* -20 *g.* -7

15. Add -10 to each of the following numbers:

a. -5 *b.* -2 *c.* -7 *d.* $+8$ *e.* $+6$ *f.* $+12$ *g.* $+10$

Add:

	(*a*)	(*b*)	(*c*)	(*d*)	(*e*)	(*f*)
16.	$+20$	$+25$	-18	$+13$	$-.6$	$+7\frac{1}{2}$
	$+8$	-12	$+6$	-8	$+1.2$	$-1\frac{1}{4}$
	-4	-10	$+5$	-5	-1.5	$+9\frac{3}{4}$
17.	$+8$	-14	-18	$+12$	$-.5$	$+5\frac{1}{3}$
	$+7$	-12	$+8$	-8	$+1.1$	$-15\frac{5}{6}$
	-3	$+15$	$+22$	$+6$	-1.7	$+3\frac{1}{6}$
	-6	$+9$	-14	-10	$+.2$	$-6\frac{1}{6}$

18. $(+10) + (-12) + (+8)$ **19.** $(+15) + (-8) + (-7)$

20. $(-15) + (+7) + (-13)$ **21.** $(-9) + (-12) + (+30)$

22. $+10 + 8 + 6 - 4$ **23.** $+12 - 6 - 9 - 1$

24. $-20 + 8 + 22 - 10$ **25.** $-12 - 18 + 7 - 8$

26. In each of the following, represent the quantities by signed numbers and then find the algebraic sum:

a. The temperature rose 5° and then fell 2°.

b. A man deposited \$400 in a bank and then withdrew \$150.

c. A man earned \$50 and then earned \$200.

d. A plane flew 900 miles south and then flew 475 miles north.

e. A boy gained 12 pounds and then lost 9 pounds.

21. ADDING ALGEBRAIC EXPRESSIONS

We are now ready to learn how to add algebraic expressions.

$(+5x) + (+7x) = +12x$ $(-2a) + (-4a) = -6a$ $-8b + 2b = -6b$

Procedure. To add like terms that involve signed numbers, add their numerical coefficients and multiply the result by their common literal factor.

If we wish to represent the perimeter of a rectangle whose length is represented by x and whose width is represented by $x - 4$, we must add all the sides. The perimeter, p, can be represented as:

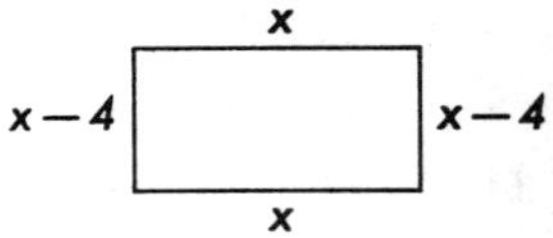

$$p = (x) + (x - 4) + (x) + (x - 4)$$
$$\textit{or, } p = x + x - 4 + x + x - 4$$
$$\textit{or, } p = 4x - 8 \quad \text{(Combine like terms.)}$$

Procedure. To add algebraic expressions that contain several terms, some of which are alike, combine each set of like terms separately.

MODEL PROBLEMS

Add:

1.	2.	3.	4.	5.	6.
$+4x$	$-8a$	$+2b$	$+7d$	$+2x + 5$	$10a^2 - 5b^2$
$+2x$	$-\ a$	$-5b$	$-7d$	$+6x - 2$	$a^2 + 3b^2$
$+6x$	$-9a$	$-3b$	0	$+8x + 3$	$11a^2 - 2b^2$

7. Combine like terms: $4x + 8y - 6z + 4y - x - 6z$

Answer: $3x + 12y - 12z$

EXERCISES

Add:

	(a)	(b)	(c)	(d)	(e)
1.	$+3a$ $+4a$	$-6b$ $-2b$	$-\ 8m$ $+10m$	$+7c^2$ $-9c^2$	$+4d$ $-4d$
2.	$-12t$ $-10t$	$+13r$ $-15r$	$+7w$ $+\ w$	$-\ m$ $+3m$	$-10y^2$ $+\ y^2$
3.	$+9.5a$ $+2.7a$	$+8.3d$ $-9.5d$	$-2.4r$ $-3.6r$	$-5.6t$ $+7.8t$	$+6.4b^2$ $-1.2b^2$
4.	$+9\frac{1}{2}x$ $-7\frac{1}{2}x$	$-1\frac{1}{4}y$ $-2\frac{1}{4}y$	$+8\frac{3}{8}a$ $-2\frac{1}{8}a$	$-6\frac{1}{3}b$ $+4\frac{2}{3}b$	$+7\frac{2}{3}t^2$ $-4\frac{5}{6}t^2$
5.	$+5x$ $-4x$ $+2x$	$-8c$ $-9c$ $-3c$	$+12y$ $-\ 4y$ $-\ 9y$	$-10d$ $+\ 8d$ $-14d$	$+\frac{1}{3}a^2$ $-\frac{5}{6}a^2$ $-\frac{2}{3}a^2$

6.	7.	8.	9.
$3x + 4$ $2x + 5$	$9x - 6$ $x - 8$	$7x - 9$ $-2x + 5$	$-6x^2 + 10$ $9x^2 - 10$

10. $\begin{array}{r} +5x - 6y \\ \underline{-3x - 4y} \end{array}$

11. $\begin{array}{r} -4a + 8b \\ \underline{+a - 10b} \end{array}$

12. $\begin{array}{r} 13c - 4d \\ \underline{-c + 9d} \end{array}$

13. $\begin{array}{r} -8m^2 + 6n^2 \\ \underline{10m^2 - 6n^2} \end{array}$

14. $\begin{array}{r} +8r - 4s - 5t \\ \underline{+2r - 5s + 6t} \end{array}$

15. $\begin{array}{r} +x + 6y - 7z \\ \underline{-3x - 6y + 5z} \end{array}$

16. $\begin{array}{r} 6a^2 - b^2 + 8c^2 \\ \underline{-9a^2 + b^2 - c^2} \end{array}$

By combining like terms, write the following in simplest form:

17. $+6a - 5b + 3a - 2b$

18. $-7c - 5d - c + 4d$

19. $-10r + s + 10r - 5s$

20. $+4m + 2n - 2m + n$

21. $-3a + 8d + 3a - 7d$

22. $+4a - 4b + 3a - 2b - 5a + b$

23. $-3c + d - c - 4d + 4c + 8d$

24. $-r + 8s + 4r - 4s - 5r - 2s$

25. $+7c - 4d + e - 8c + 4d - 5e$

26. $-9r + s - 4t + r - 6s + 4t$

22. UNDERSTANDING THE MEANING OF SQUARE ROOT

When we multiply a number by itself, that is, when we use a number as a factor twice, the product is called the *square* of the number. For example:

If $3 \times 3 = 9$, then 9 is the square of 3.
If $8 \times 8 = 64$, then 64 is the square of 8.

In each of these examples, two equal numbers were multiplied to give a product. Either of these equal numbers is called the ***square root*** of the product.

Finding the square root of a product means finding one of its two equal factors.

The square root of 9 is 3 because $3 \times 3 = 9$.
The square root of 64 is 8 because $8 \times 8 = 64$.

The symbol $\sqrt{}$ is used to indicate the square root of a number. Thus, $\sqrt{9}$ is read "the square root of 9."

If we have to find the value of $\sqrt{484}$, we must realize that the answer will be more than 20 because $20 \times 20 = 400$. Since 484 is close to but greater than 400, let us try 21. However, $21 \times 21 = 441$. Then let us try 22. We find that $22 \times 22 = 484$. Now we know that $\sqrt{484} = 22$.

MODEL PROBLEMS

Find the square root and check the answer:

1. $\sqrt{49} = 7$

   Check:

   $7 \times 7 = 49$

2. $\sqrt{121} = 11$

   Check:

   $11 \times 11 = 121$

3. $\sqrt{\frac{4}{9}} = \frac{2}{3}$

   Check:

   $\frac{2}{3} \times \frac{2}{3} = \frac{4}{9}$

4. $\sqrt{361} = 19$

   Check:

   $19 \times 19 = 361$

EXERCISES

In 1–7, find the square root of the number and check the answer:

1. 25 **2.** 1 **3.** 81 **4.** 49 **5.** 1600 **6.** $\frac{16}{25}$ **7.** 324

Find the value of each of the following and check the answer:

8. $\sqrt{9}$ **9.** $\sqrt{36}$ **10.** $\sqrt{100}$ **11.** $\sqrt{144}$ **12.** $\sqrt{\frac{9}{16}}$ **13.** $\sqrt{784}$

Find the length of each side of a square plot of land whose area is:

14. 169 sq. ft. **15.** 1600 sq. ft. **16.** 961 sq. ft. **17.** 2304 sq. ft.

23. ESTIMATING APPROXIMATE SQUARE ROOTS

If we wish to find the side of a square whose area is 38 square inches, we must find the value of $\sqrt{38}$.

$\sqrt{38}$ is not a whole number because $6^2 = 36$ and $7^2 = 49$. Since 38 is between 36 and 49, $\sqrt{38}$ is between 6 and 7. Since $(6.5)^2 = 42.25$, and 38 is less than 42.25, then $\sqrt{38}$ is less than 6.5 and must be closer to 6 than to 7. Therefore, $\sqrt{38}$, correct to the nearest whole number, is 6. The side of the square, correct to the nearest inch, is 6 inches.

If we wish, we can also find the value of $\sqrt{38}$ correct to the nearest tenth. Since we know that $\sqrt{38}$ is a little larger than 6, let us test 6.1. When we square 6.1, we get $6.1 \times 6.1 = 37.21$, which is less than 38. Then let us test 6.2. When we square 6.2, we get $6.2 \times 6.2 = 38.44$. Since $(6.15)^2 = 37.8225$, and 38 is greater than 37.8225, then $\sqrt{38}$ is greater than 6.15 and must be closer to 6.2 than to 6.1. Therefore, the value of $\sqrt{38}$, correct to the nearest tenth, is 6.2. The side of the square, correct to the nearest tenth of an inch, is 6.2 inches.

EXERCISES

Between which two consecutive whole numbers is the square root of each of the following numbers?

1. 5 **2.** 21 **3.** 43 **4.** 76 **5.** 115 **6.** 157 **7.** 240

Estimate the square root of each of the following numbers correct to the nearest whole number:

8. 7 **9.** 18 **10.** 32 **11.** 56 **12.** 95 **13.** 116 **14.** 208

Find the value of each of the following correct to the nearest tenth:

15. $\sqrt{2}$ **16.** $\sqrt{13}$ **17.** $\sqrt{46}$ **18.** $\sqrt{86}$ **19.** $\sqrt{108}$ **20.** $\sqrt{140}$

Find, to the nearest whole number, the side of a square whose area is:

21. 8 sq. in. **22.** 55 sq. ft. **23.** 93 sq. ft. **24.** 131 sq. yd.

Find, correct to the nearest tenth, the side of a square whose area is:

25. 14 sq. in. **26.** 29 sq. ft. **27.** 61 sq. ft. **28.** 113 sq. yd.

24. USING DIVISION TO FIND SQUARE ROOTS

Division may be used to find the square root of a number correct to the nearest tenth.

Let us find $\sqrt{14}$ correct to the nearest tenth.

$\sqrt{14}$ is not a whole number because $3^2 = 9$ and $4^2 = 16$. Since 14 is between 9 and 16, $\sqrt{14}$ is between 3 and 4. Since 14 is much closer to 16 than it is to 9, let us estimate that $\sqrt{14}$ is 3.8.

```
          3.68
3.8^)14.0^00
     11 4
      2 60
      2 28
        320
        304
         16
```

If this estimate is correct, when 14 is divided by 3.8, the quotient should also be 3.8. At the right, we see the division in which the quotient found to two decimal places is 3.68.

Therefore, 3.8 is too large for $\sqrt{14}$ and 3.68 is too small.

```
            3.74
3.74^)14.00^00
      11 22
       2 780
       2 618
         1620
         1496
          124
```

Let us take the average of 3.8 and 3.68 as our next estimate for $\sqrt{14}$. The average of 3.8 and 3.68 is $\frac{3.8 + 3.68}{2} = \frac{7.48}{2} = 3.74$.

When we divide 14 by 3.74, the quotient is 3.74. Since both the quotient and the divisor are 3.7, correct to the nearest tenth, $\sqrt{14}$, correct to the nearest tenth, is 3.7.

EXERCISES

Use the division method to find the square root of each of the following numbers correct to the nearest tenth:

1. 6	**2.** 11	**3.** 18	**4.** 21	**5.** 34
6. 53	**7.** 90	**8.** 108	**9.** 19.5	**10.** 41.7

Use the division method to find the value of each of the following correct to the nearest tenth:

11. $\sqrt{7}$	**12.** $\sqrt{19}$	**13.** $\sqrt{28}$	**14.** $\sqrt{61}$	**15.** $\sqrt{74}$
16. $\sqrt{106}$	**17.** $\sqrt{111}$	**18.** $\sqrt{127}$	**19.** $\sqrt{23.5}$	**20.** $\sqrt{88.2}$

25. FINDING SQUARES AND SQUARE ROOTS BY USING A TABLE

People who often have to square numbers or find the square roots of numbers can save much time by using a table of squares and square roots such as the one that appears on page 233.

In this table, next to each number from 1 to 150 we find the square and the square root of the number. If we wish to find the square of 32, we look to the right of 32 in the column headed "Square" and find 1,024. If we wish to find the square root of 32, we look to the right of 32 in the column headed "Square Root" and find 5.657, which is given correct to thousandths. We may round off 5.657 to the nearest hundredth and get 5.66, or to the nearest tenth and get 5.7.

We can also use the table to find the square root of a number greater than 150. For example, let us find the square root of 1,369. We look down the column headed "Square" until we come to 1,369. To the left of 1,369, we find 37 in the column headed "Number." Since 37 is the number whose square is 1,369, the square root of 1,369 is 37.

No.	*Square*
55	3,025
?	3,000
54	2,916

However, if we try to find the square root of 3,000 in the same way, we find that 3,000 is not in the column headed "Square." In that column, the number just smaller than 3,000 is 2,916, which is 54^2; and the number just larger than 3,000 is 3,025, which is 55^2. Since $(54.5)^2 = 2970.25$ and 3,000 is greater than 2970.25, the square root of 3,000 is greater than 54.5 and must be closer to 55 than to 54. Therefore, the square root of 3,000, correct to the nearest whole number, is 55.

EXERCISES

(Use the Table of Squares and Square Roots on page 233.)

Find the square of each of the following numbers:

1. 19 **2.** 34 **3.** 65 **4.** 87 **5.** 106 **6.** 129 **7.** 138

Find the square root of each number correct to the nearest hundredth:

8. 7 **9.** 33 **10.** 67 **11.** 89 **12.** 97 **13.** 111 **14.** 136

Find the value of each of the following correct to the nearest tenth:

15. $\sqrt{5}$ **16.** $\sqrt{29}$ **17.** $\sqrt{58}$ **18.** $\sqrt{74}$ **19.** $\sqrt{108}$ **20.** $\sqrt{143}$

Find the square root of each of the following numbers:

21. 784 **22.** 1,681 **23.** 5,476 **24.** 9,409 **25.** 19,881

Find the square root of each number correct to the nearest whole number:

26. 534 **27.** 910 **28.** 2,125 **29.** 8,326 **30.** 15,225

TABLE OF SQUARES AND SQUARE ROOTS

No.	Square	Square Root	No.	Square	Square Root	No.	Square	Square Root
1	1	1.000	**51**	2,601	7.141	**101**	10,201	10.050
2	4	1.414	**52**	2,704	7.211	**102**	10,404	10.100
3	9	1.732	**53**	2,809	7.280	**103**	10,609	10.149
4	16	2.000	**54**	2,916	7.348	**104**	10,816	10.198
5	25	2.236	**55**	3,025	7.416	**105**	11,025	10.247
6	36	2.449	**56**	3,136	7.483	**106**	11,236	10.296
7	49	2.646	**57**	3,249	7.550	**107**	11,449	10.344
8	64	2.828	**58**	3,364	7.616	**108**	11,664	10.392
9	81	3.000	**59**	3,481	7.681	**109**	11,881	10.440
10	100	3.162	**60**	3,600	7.746	**110**	12,100	10.488
11	121	3.317	**61**	3,721	7.810	**111**	12,321	10.536
12	144	3.464	**62**	3,844	7.874	**112**	12,544	10.583
13	169	3.606	**63**	3,969	7.937	**113**	12,769	10.630
14	196	3.742	**64**	4,096	8.000	**114**	12,996	10.677
15	225	3.873	**65**	4,225	8.062	**115**	13,225	10.724
16	256	4.000	**66**	4,356	8.124	**116**	13,456	10.770
17	289	4.123	**67**	4,489	8.185	**117**	13,689	10.817
18	324	4.243	**68**	4,624	8.246	**118**	13,924	10.863
19	361	4.359	**69**	4,761	8.307	**119**	14,161	10.909
20	400	4.472	**70**	4,900	8.367	**120**	14,400	10.954
21	441	4.583	**71**	5,041	8.426	**121**	14,641	11.000
22	484	4.690	**72**	5,184	8.485	**122**	14,884	11.045
23	529	4.796	**73**	5,329	8.544	**123**	15,129	11.091
24	576	4.899	**74**	5,476	8.602	**124**	15,376	11.136
25	625	5.000	**75**	5,625	8.660	**125**	15,625	11.180
26	676	5.099	**76**	5,776	8.718	**126**	15,876	11.225
27	729	5.196	**77**	5,929	8.775	**127**	16,129	11.269
28	784	5.291	**78**	6,084	8.832	**128**	16,384	11.314
29	841	5.385	**79**	6,241	8.888	**129**	16,641	11.358
30	900	5.477	**80**	6,400	8.944	**130**	16,900	11.402
31	961	5.568	**81**	6,561	9.000	**131**	17,161	11.446
32	1,024	5.657	**82**	6,724	9.055	**132**	17,424	11.489
33	1,089	5.745	**83**	6,889	9.110	**133**	17,689	11.533
34	1,156	5.831	**84**	7,056	9.165	**134**	17,956	11.576
35	1,225	5.916	**85**	7,225	9.220	**135**	18,225	11.619
36	1,296	6.000	**86**	7,396	9.274	**136**	18,496	11.662
37	1,369	6.083	**87**	7,569	9.327	**137**	18,769	11.705
38	1,444	6.164	**88**	7,744	9.381	**138**	19,044	11.747
39	1,521	6.245	**89**	7,921	9.434	**139**	19,321	11.790
40	1,600	6.325	**90**	8,100	9.487	**140**	19,600	11.832
41	1,681	6.403	**91**	8,281	9.539	**141**	19,881	11.874
42	1,764	6.481	**92**	8,464	9.592	**142**	20,164	11.916
43	1,849	6.557	**93**	8,649	9.644	**143**	20,449	11.958
44	1,936	6.633	**94**	8,836	9.695	**144**	20,736	12.000
45	2,025	6.708	**95**	9,025	9.747	**145**	21,025	12.042
46	2,116	6.782	**96**	9,216	9.798	**146**	21,316	12.083
47	2,209	6.856	**97**	9,409	9.849	**147**	21,609	12.124
48	2,304	6.928	**98**	9,604	9.899	**148**	21,904	12.166
49	2,401	7.000	**99**	9,801	9.950	**149**	22,201	12.207
50	2,500	7.071	**100**	10,000	10.000	**150**	22,500	12.247

Chapter XI

GEOMETRY

I. GEOMETRIC FIGURES

COMMON GEOMETRIC SOLID FIGURES

Geometric solid figures occupy space. Below are some of the geometric solid figures that we often see in nature and in objects made by man. Study their shapes and learn their names.

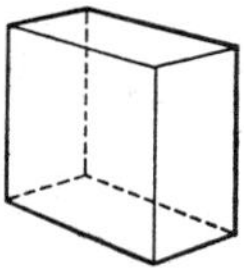

Rectangular Solid

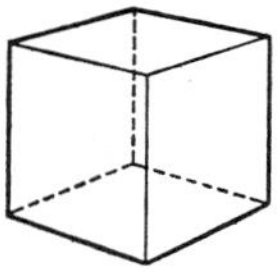

Cube

Triangular Prism

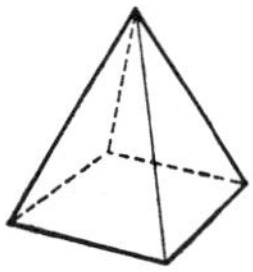

Pyramid

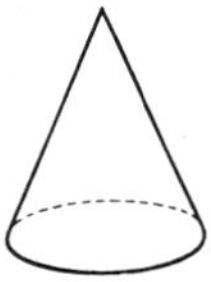

Cone

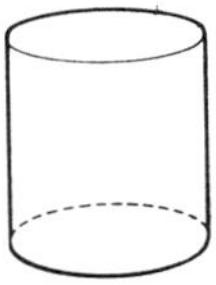

Cylinder

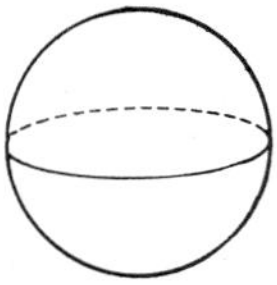

Sphere

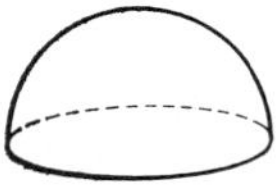

Hemisphere

EXERCISES

1. Name two familiar objects that are shaped like a:

a. cube *b.* sphere *c.* cylinder
d. cone *e.* hemisphere *f.* rectangular solid

2. Name one familiar object that is shaped like a pyramid.
3. Name one familiar object that is shaped like a triangular prism.
4. Each of the following is shaped like a familiar geometric solid. Name it.

a. a baseball *b.* a box of chalk *c.* a child's wood block
d. an orange *e.* a crate of apples *f.* an ice cream cone
g. a tepee *h.* a water pipe *i.* half of a grapefruit
j. a funnel *k.* a tree trunk *l.* a lump of sugar

5. From a newspaper or magazine, cut out a picture containing some geometric solids. Name them.

COMMON GEOMETRIC PLANE FIGURES

Each surface of a cube or a rectangular solid is a flat surface, or a *plane*. The plane has no thickness. A geometric figure that can be drawn on a plane is called a *plane figure*. ***Geometric plane figures*** lie on a flat surface. Below are some of the common geometric plane figures. Study their shapes and learn their names.

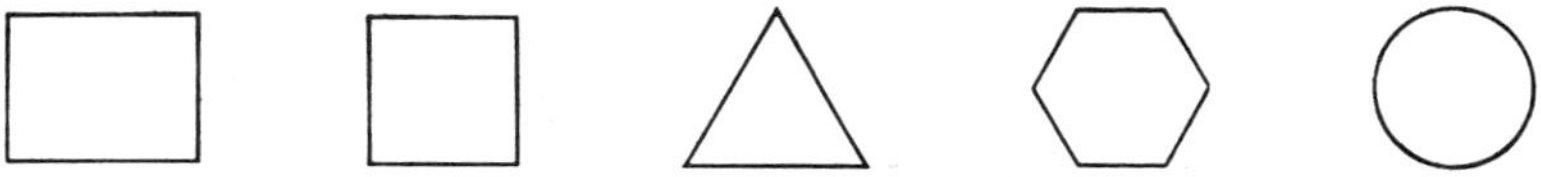

Rectangle   Square   Triangle   Hexagon   Circle

EXERCISES

1. Name the geometric solids on page 234 that have surfaces shaped like:

a. circles *b.* triangles *c.* squares *d.* rectangles

2. What is the number of equal squares that form the surface of a cube?

3. What is the number of circles that form the surface of (*a*) a cylinder (*b*) a cone (*c*) a hemisphere?

4. Name two familiar objects that are shaped like a:

a. square *b.* triangle *c.* circle *d.* rectangle *e.* hexagon

5. Each of the following is shaped like a geometric plane figure. Name the figure.

a. a wheel	*b.* a dollar bill	*c.* the cells of a honeycomb
d. a door	*e.* a football field	*f.* a dime
g. a snowflake	*h.* a baseball diamond	*i.* the sun

2. THE STRAIGHT LINE

The word "point" is often used to mean "place" or "position." In geometry, a ***point*** is merely an idea. It has no length, no width, no thickness—only position. A point is represented by a small dot and named by a capital letter. The point at the right is called "point *P*."

A straight line is suggested by the edge of a ruler. It is also suggested by the top of this page or by the boundaries of a tennis court. In geometry, a ***straight line*** is a set of points that extends endlessly in both directions. We will use the term ***line*** to mean straight line.

A line can be named by naming two points on it. At the right, see the two ways that are commonly used to represent a line, which may be called "line AB."

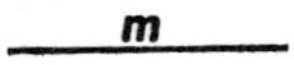

A line may also be named by placing a single small letter above it. The line at the left is called "line m."

A line is indefinite in length. If two points are selected on a line, these two points and the part of the line included between these points form a ***line segment***, or ***segment***. Therefore, a segment has a beginning point, an end point, and a definite length. At the right, see the two ways that are commonly used to represent a line segment. A segment, such as RS, may also be considered as the set of points consisting of point R, point S, and all the points between R and S.

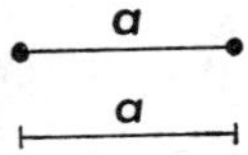

A line segment may also be named by placing a single small letter above it. Each segment at the left is called "segment a."

A line segment has not only a definite length but a fixed direction. We use the words *vertical*, *horizontal*, and *oblique* to describe the direction of a line or a line segment.

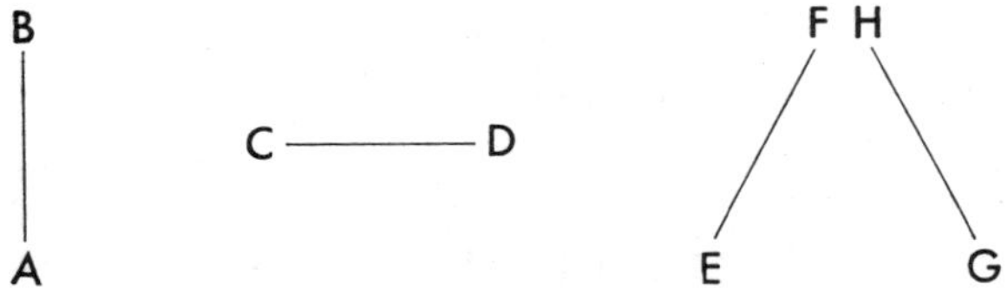

Vertical Horizontal Oblique

EXERCISES

1. Name the line segments in each of the following diagrams:

(a) (b) (c) (d)

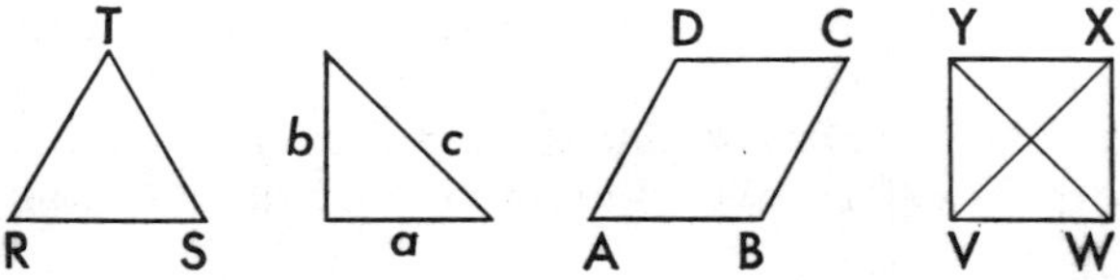

2. Use the word *horizontal*, *oblique*, or *vertical* to complete each of the following statements:

a. In 1*a*, segment RS is ______; segment RT is ______; segment ST is ______.

b. In 1*b*, segment a is ______; segment b is ______; segment c is ______.

c. In 1*c*, segments AB and DC are ______; segments AD and BC are ______.

d. In 1*d*, segments VY and WX are ______; segments VW and YX are ______; segments VX and WY are ______.

e. A telephone pole is usually in a (an) ______ position.

f. The ruled lines on the page of a notebook are usually in a (an) ______ position.

PROPERTIES OF STRAIGHT LINES

1. Place a point on your paper. Draw a vertical line through the point. Draw a horizontal line through the point. Draw several oblique lines through it. How many straight lines can be drawn through the point?

2. Place two points on your paper. Draw a straight line between them with a ruler. Can you draw a different straight line between them? How many different straight lines can be drawn between any two points?

3. Draw any two straight lines that cross each other. We call these ***intersecting lines.*** At how many points do they intersect? Can they intersect at any other point? At how many points can two straight lines intersect?

4. At the right we see two curved lines that enclose a space. We also see three straight lines that enclose a space. They enclose a triangle. Can two straight lines enclose a space?

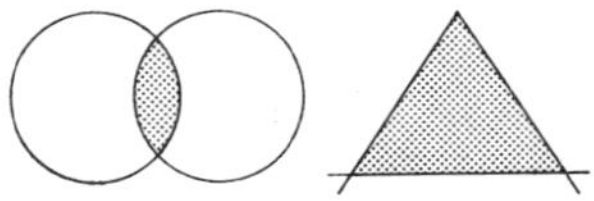

FACTS ABOUT STRAIGHT LINES

1. Any number of straight lines can be drawn through a given point.
2. Only one straight line can be drawn through two given points.
3. Two straight lines can intersect at only one point.
4. Two straight lines cannot enclose a space.

MEASURING THE LENGTH OF A LINE SEGMENT

If we wish to find the length of a line segment, we must first select a *standard unit of measure.* A standard unit of measure is a definite, fixed measure that means the same to all people who use it in all places. When we measure the length of a line segment, we find the number of times the unit is contained in the line segment. The *yard,* the *foot,* the *inch* are standard units for measuring lengths. We call them units of ***linear measure.*** We use a *linear scale,* or a *ruler,* to measure lengths of straight lines.

Let us measure the length of line segment GH.

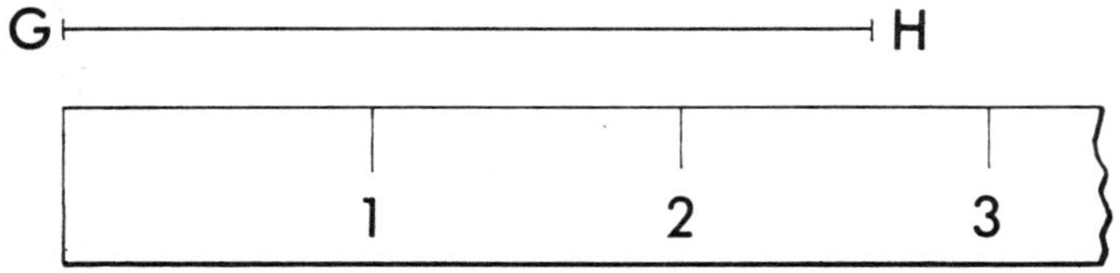

If we use a scale marked off in inches, we see that line segment GH is between 2 and 3 inches in length. Since its end is nearer to 3 inches than it is to 2 inches, we say that the length of GH is 3 inches, *to the nearest inch.*

Let us again measure the length of line segment GH.

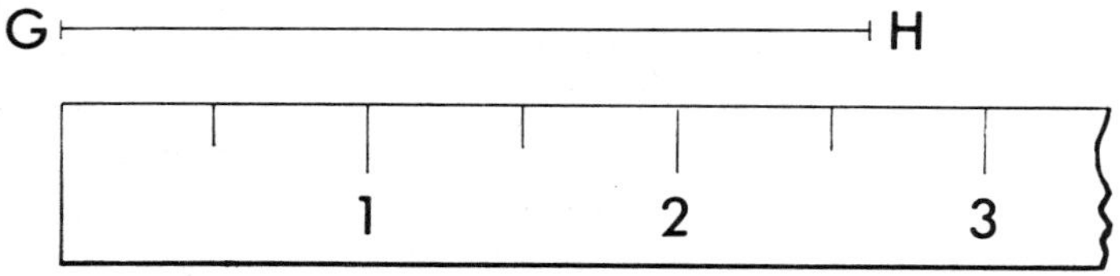

If we use a scale marked off in half inches, we see that GH is between $2\frac{1}{2}$ and 3 inches long. Since its end is nearer to $2\frac{1}{2}$ inches, we say that the length of GH is $2\frac{1}{2}$ inches, *to the nearest half inch.*

Therefore, we see that when we measure the length of a line segment with a scale, we can find only its approximate length. We can increase the *precision* of our measurements by using a scale in which each inch is divided into a greater number of equal parts.

EXERCISES

1. Measure each of the following line segments to the nearest inch, $\frac{1}{2}$ inch, $\frac{1}{4}$ inch, and $\frac{1}{8}$ inch.

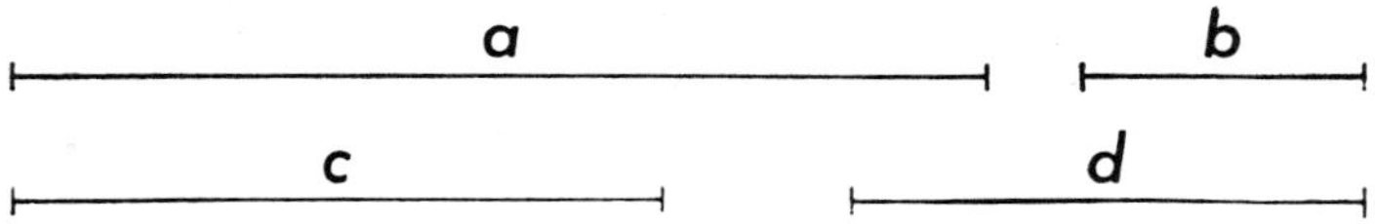

2. Find, to the nearest quarter of an inch, the length of the piece of cardboard shown at the right.

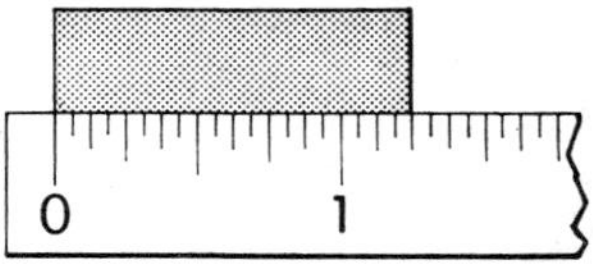

3. What measurement is indicated by the arrow on the part of a ruler pictured at the right?

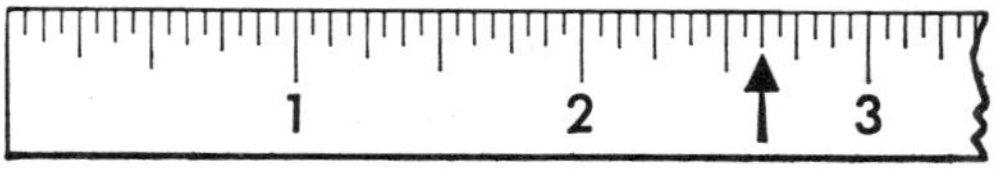

4. A ruler is placed against a metal plate as shown in the diagram. What is the length of the plate?

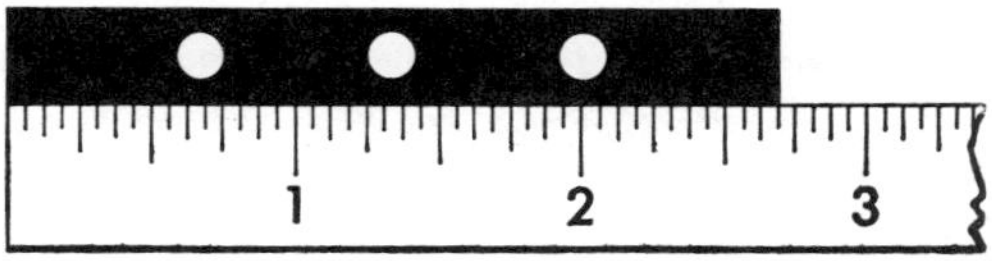

5. Estimate the length of each of the following line segments to the nearest $\frac{1}{4}$ inch. Measure each segment with a ruler. Find your error in each estimate.

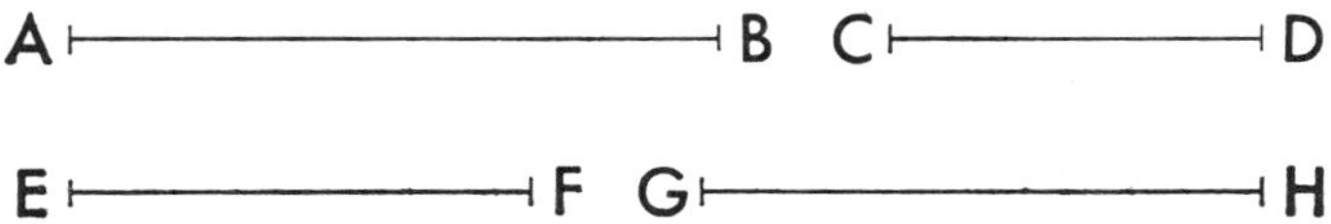

	AB	*CD*	*EF*	*GH*
Estimate				
Measurement				
Error				

6. Use a ruler to draw line segments of the following lengths: (*a*) 4 in. (*b*) 5 in. (*c*) $2\frac{3}{4}$ in. (*d*) $3\frac{5}{8}$ in. Estimate the center of each segment. Check with a ruler to see how close the estimated midpoint is to the actual midpoint.

USING THE STRAIGHTEDGE AND COMPASSES

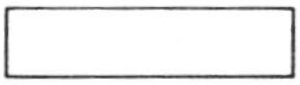

Straightedge

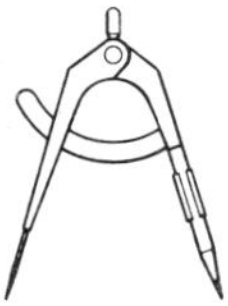

Compasses

Two instruments used in geometry are the straightedge and compasses. The ***straightedge*** is an instrument that is used to draw straight lines. It looks like a ruler but does not have units of measure marked off on it. The ***compasses*** are used to draw circles or parts of circles and to copy a line segment.

If on line *CD* we wish to copy a line segment equal in length to line segment *AB*, we begin by placing the point of the compasses on *A* and opening the compasses so that the point of the pencil is on *B*. Then we place the point of the compasses on *C* and swing the pencil point so that it cuts across line *CD* at *E*, being careful not to change the distance between the points of the compasses. The line segment *CE* is equal to the line segment *AB*.

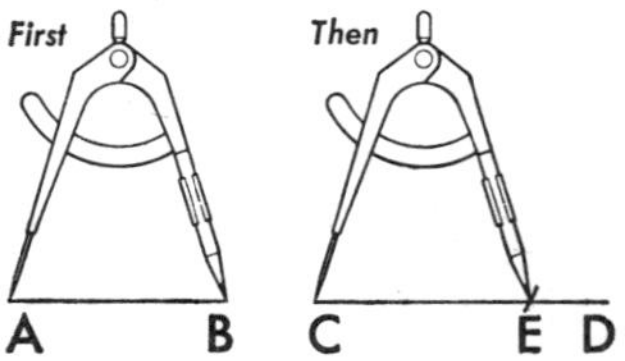

EXERCISES

1. Draw a line. On this line, with the use of compasses, make a line segment whose length is:

a. $\frac{1}{2}$ inch *b.* $\frac{3}{4}$ inch *c.* 1 inch *d.* $1\frac{1}{2}$ inches *e.* $1\frac{5}{8}$ inches *f.* 2 inches

2. Draw a line. On this line, copy line segments equal in length to each of the following line segments:

3. THE CIRCLE

In the diagram at the right, if point *P* moves in such a way that it is always $\frac{3}{8}$ of an inch from point *O*, it will make a circle. Point *O* is called the center of the circle.

A ***circle*** is a curved line that is a set of points each of which is the same distance from a fixed point inside, called the ***center.*** The circle is drawn with compasses.

DEFINITIONS OF IMPORTANT TERMS

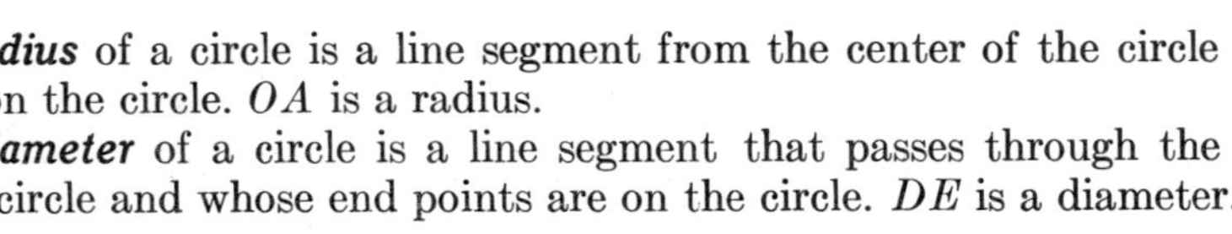

A ***radius*** of a circle is a line segment from the center of the circle to any point on the circle. *OA* is a radius.

A ***diameter*** of a circle is a line segment that passes through the center of the circle and whose end points are on the circle. *DE* is a diameter.

The ***circumference*** of a circle is the distance around the circle.

An ***arc*** of a circle is any part of the circle. The curved line FG is an arc of the circle.

A ***semicircle*** is one-half of a circle.

A ***quadrant*** is one-quarter of a circle.

FACTS ABOUT CIRCLES

1. A circle has any number of radii, all of which are equal to each other.
2. A circle has any number of diameters, all of which are equal to each other.
3. A diameter of a circle is twice as large as a radius.
4. A diameter divides the circle into two equal parts.

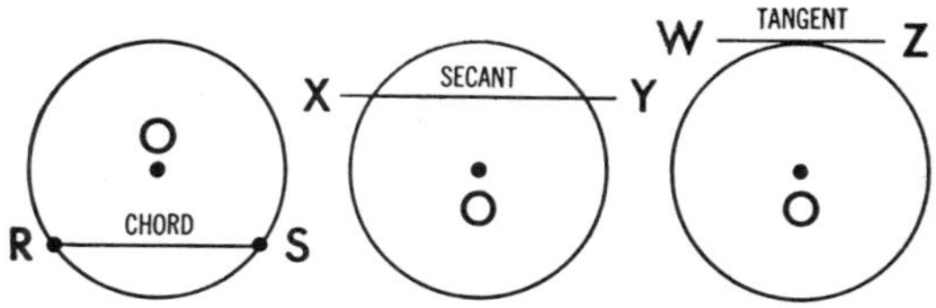

A ***chord*** is a line segment that joins two points on a circle. RS is a chord in circle O.

A ***secant*** is a line that cuts across a circle in two points. XY is a secant to circle O.

A ***tangent*** is a line that touches a circle at only one point no matter how far the line is extended. WZ is a tangent to circle O.

EXERCISES

1. Use compasses to draw a circle whose radius is:

a. $\frac{1}{2}$ inch *b.* $\frac{3}{4}$ inch *c.* 1 inch *d.* $1\frac{3}{8}$ inches *e.* $1\frac{5}{8}$ inches

2. Find the diameter of a circle whose radius is:

a. 2 inches *b.* $1\frac{1}{2}$ inches *c.* $2\frac{5}{8}$ inches *d.* 2.5 inches *e.* 3.75 inches

3. Find the radius of a circle whose diameter is:

a. 4 inches *b.* 5 inches *c.* $3\frac{1}{2}$ inches *d.* $3\frac{7}{8}$ inches *e.* 1.4 inches

4. What is the ratio of the radius of a circle to its diameter?

5. A quadrant is what per cent of the whole circle?

6. A semicircle is what per cent of the whole circle?

4. THE ANGLE

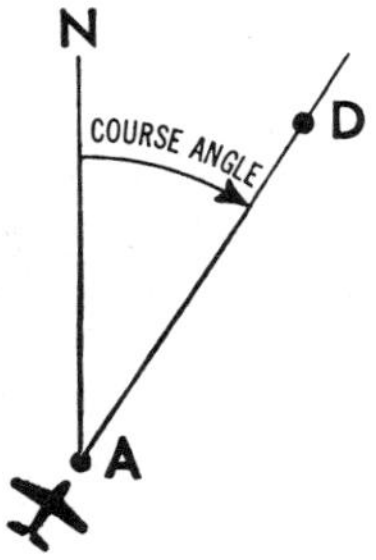

When a plane leaves an airport, A, for a given destination, D, the navigator tells the pilot the direction in which he is to fly the plane by giving him the proper course ***angle.***

AN is a north-south line through the airport, A. AD is a straight line passing through the point of departure, A, and the destination, D. If line AN is rotated about point A in the direction of the arrow until line AN takes the position of line AD, an angle (angle NAD) is formed. The measure of angle NAD is the ***amount of rotation.*** Point A is called the ***vertex*** of the angle. Lines AN and AD are called the ***sides*** of the angle.

TYPES OF ANGLES

Right Angle | Straight Angle | Acute Angle | Obtuse Angle | Reflex Angle

A ***right angle*** is formed by one-quarter of a complete rotation.

A ***straight angle*** is formed by one-half of a complete rotation.

An ***acute angle*** is smaller than a right angle.

An ***obtuse angle*** is greater than a right angle but less than a straight angle.

A ***reflex angle*** is formed by more than one-half of a complete rotation but less than a complete rotation.

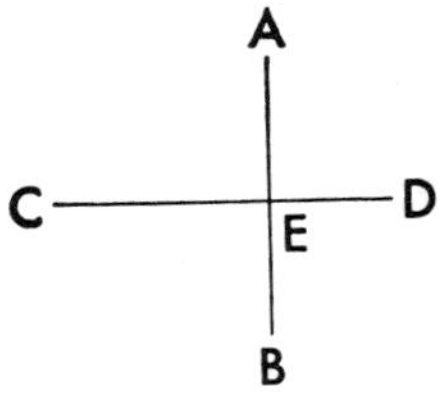

If two straight lines intersect at right angles, they are ***perpendicular*** to each other. The symbol for perpendicular is $\perp$. If $AB \perp CD$ at E, then the four angles at E are right angles.

NAMING ANGLES

There are several ways of naming angles.

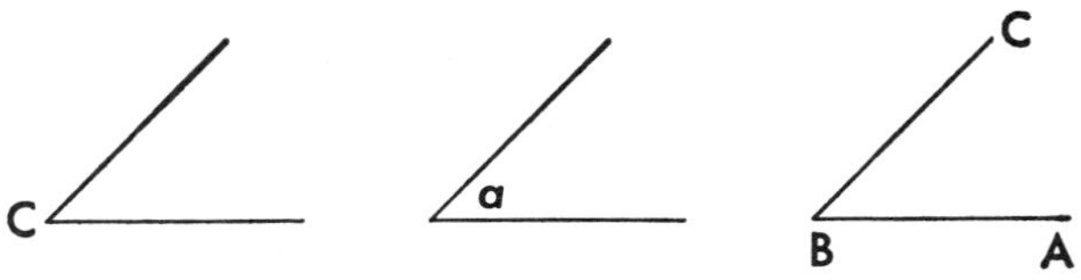

Fig. 1 Fig. 2 Fig. 3

The angle in figure 1 is named by the capital letter *C* at its vertex. It is called "angle *C*." The symbol for angle is ∠. Angle *C* may be written ∠*C*.

The angle in figure 2 is named by the small letter *a* inside the angle. It is called "angle *a*," or ∠*a*.

The angle in figure 3 is named by three capital letters. In reading the angle, we always use the letter at its vertex as the middle letter. The angle is called "angle *ABC*" (∠*ABC*), or "angle *CBA*" (∠*CBA*).

In diagrams where several angles have the same vertex, it is best to name these angles by using three capital letters.

EXERCISES

1. If one road runs north and south and another road runs east and west, what kind of angle is formed at the intersection of the two roads?

2. Name the kind of angle that is formed between the hands of a clock when they show (*a*) 3 o'clock (*b*) 6 o'clock (*c*) 1 o'clock (*d*) 5 o'clock.

3. Which of the following is the smallest angle: a right angle, an obtuse angle, or an acute angle?

4. Which of the following is the largest angle: a right angle, a straight angle, an obtuse angle, or a reflex angle?

5. What kind of angle is each angle at the right?

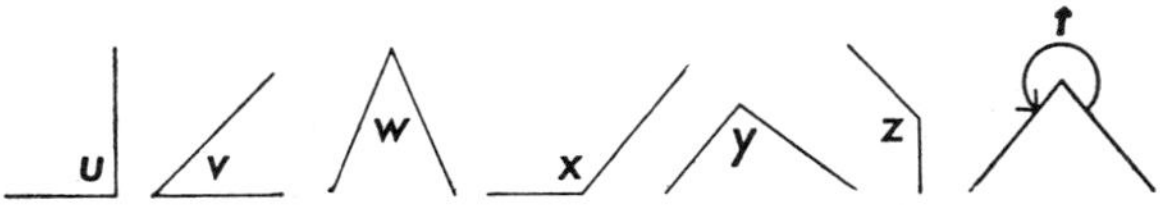

6. Draw an acute angle. Name it, using one capital letter.

7. Draw an obtuse angle. Name it, using one small letter.

8. Draw a right angle. Name it, using three capital letters.

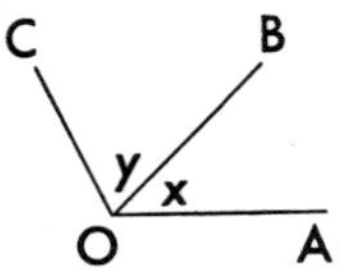

9. Using the figure at the left:
   a. Name angle x, using three capital letters.
   b. Give the shorter name for angle COB.
   c. Name one acute angle.
   d. Name one obtuse angle.

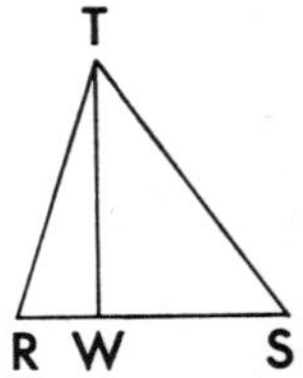

10. In triangle RST, line TW is perpendicular to RS. Angle TWR is (*a*) an acute angle (*b*) an obtuse angle (*c*) a right angle.

MEASURING ANGLES

Since we know that the size of an angle depends upon the amount of rotation, we can now learn how to measure angles.

If a complete rotation is divided into 360 equal parts, we have 360 small angles, each of which is $\frac{1}{360}$ of a complete rotation. Each of these angles is called ***one angle degree,*** or ***one degree.*** The symbol for degree or degrees is °.

If the circumference of a circle is divided into 360 equal parts, we have 360 small arcs, each of which is $\frac{1}{360}$ of the circumference. Each of these arcs is called ***one arc degree,*** or ***one degree.***

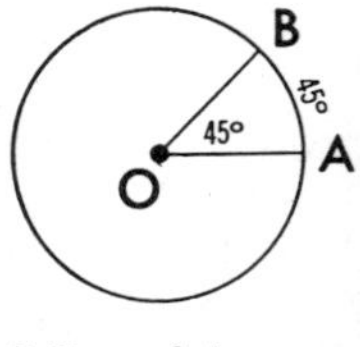

Angle AOB has its vertex at O, the center of the circle. The two sides of the angle are radii of the circle. Such an angle is called a ***central angle.*** Arc AB is the arc that the angle cuts off on the circle. If $\angle AOB$ is $\frac{45}{360}$ of a complete rotation, which means that it contains 45 angle degrees, then arc AB is $\frac{45}{360}$ of the whole circumference, which means that it contains 45 arc degrees. Notice that central angle AOB and its arc on the circle each contain 45°. We can now understand the following principle:

Principle. In a circle, a central angle contains the same number of degrees as the arc it cuts off on the circle.

When we measure an angle, we discover the number of degrees it contains. The ***protractor*** is an instrument used to measure angles. Study the model protractor below.

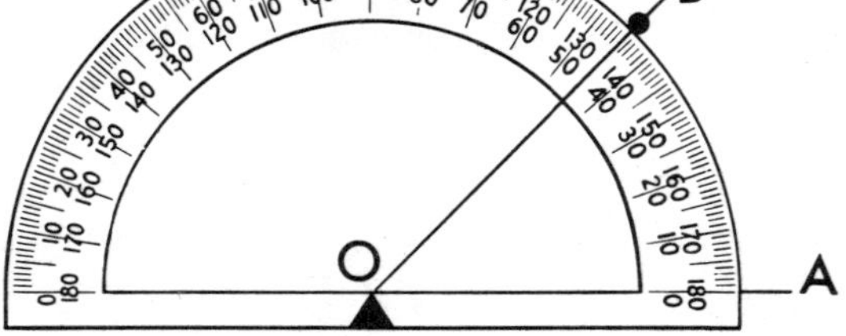

The two scales on the protractor show divisions starting at 0° and going up to 180°. One scale starts at 0° at the right and continues around to 180° at the left. The other scale starts at 0° at the left and continues around to 180° at the right. These two scales make it convenient to measure any angle easily, regardless of its position.

To measure an angle, place the center of the protractor on the vertex of the angle. Place the upper part of the base of the protractor on one side of the angle. This side of the angle cuts across 0° on one scale. Notice that the other side of the angle also cuts across that scale. At the point that the other side of the angle cuts across the scale, there is a reading that gives the measure of the angle.

In measuring angle AOB in the previous illustration, place the protractor so that side OA passes through the 0° reading on the inner scale. Side OB cuts across this scale at a point marked 45°. Thus, angle AOB contains 45°.

EXERCISES

1. Use a protractor to measure each of the following angles:

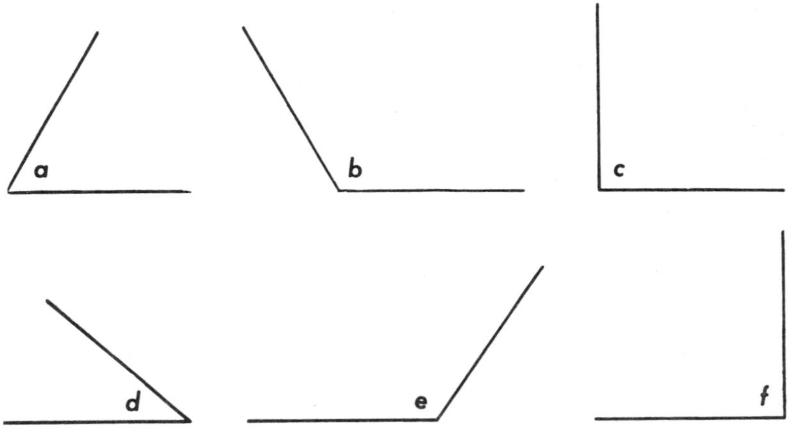

2. How many degrees are there in a right angle?

3. How many degrees does a straight angle contain?

4. Complete: An acute angle contains less than ______ degrees.

5. Find the number of degrees in the angle formed by the hands of a clock at (*a*) 3 o'clock (*b*) 10 o'clock (*c*) 5 o'clock.

6. What kind of angle is formed when a vertical line intersects a horizontal line?

7. How many degrees are there in $\frac{1}{3}$ of a right angle?

8. How many degrees are there in $\frac{3}{5}$ of a right angle?

9. How many degrees are there in $\frac{3}{4}$ of a straight angle?

10. Find the number of degrees in the arc cut off on a circle by a central angle that contains:

a. 40° *b.* 90° *c.* 120° *d.* 150° *e.* 165° *f.* 180°

11. If two straight lines intersect and are perpendicular to each other, how many degrees are there in each of the four resulting angles?

STUDYING PAIRS OF ANGLES

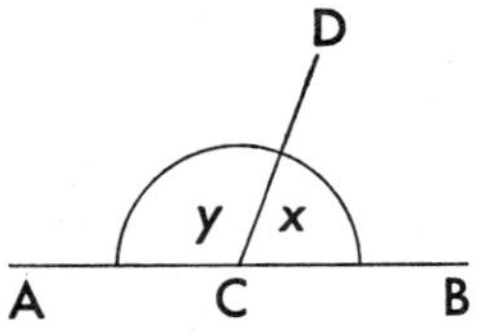

If the sum of two angles is 180°, the angles are called ***supplementary angles.*** If $\angle ACB$ is a straight angle, $\angle x + \angle y = 180°$. Therefore, $\angle x$ and $\angle y$ are supplementary. We say that $\angle x$ is the supplement of $\angle y$, and $\angle y$ is the supplement of $\angle x$. Thus, if $\angle x = 70°$, its supplement $\angle y$ is $180° - 70°$, or 110°.

If the sum of two angles is 90°, the angles are called ***complementary angles.*** If $\angle EFG$ is a right angle, $\angle a + \angle b = 90°$. Therefore, $\angle a$ and $\angle b$ are complementary. We say that $\angle a$ is the complement of $\angle b$, and $\angle b$ is the complement of $\angle a$. Thus, if $\angle a = 40°$, its complement $\angle b$ is $90° - 40°$, or 50°.

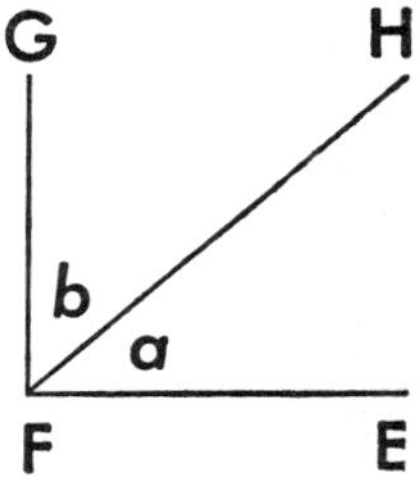

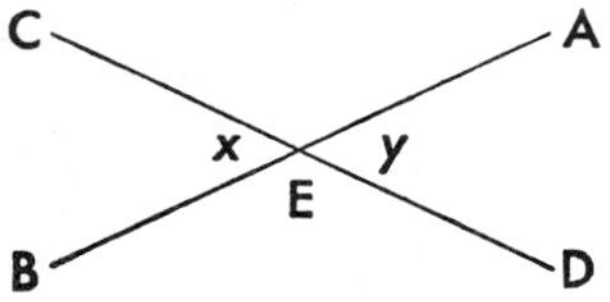

If two straight lines AB and CD intersect at E, $\angle x$ and $\angle y$ are opposite each other. They are called a pair of ***vertical angles.*** When we measure $\angle x$ and $\angle y$, we find that each angle is 50°. Therefore, $\angle x = \angle y$. Also, $\angle CEA$ and $\angle BED$ are a pair of vertical angles, each of which measures 130°. Therefore, $\angle CEA = \angle BED$. We can now understand the following principle:

Principle. If two straight lines intersect, the vertical angles formed are equal.

EXERCISES

1. Find the supplement of an angle of:

a. 50° *b.* 75° *c.* 90° *d.* 110° *e.* 150° *f.* 162°

2. Find the complement of an angle of:

a. 10° *b.* 40° *c.* 65° *d.* 75° *e.* 80° *f.* 87°

3. If AB and CD intersect at E, find $\angle AED$ when $\angle CEB$ contains:

a. 20° *b.* 70° *c.* 90° *d.* 140° *e.* 170° *f.* 175°

4. Represent the supplement of an angle of $x°$.
5. Represent the complement of an angle of $a°$.
6. Find the number of degrees contained in an angle that is (*a*) equal to its supplementary angle (*b*) equal to its complementary angle.
7. By how many degrees is the supplement of an acute angle larger than the complement of the acute angle?

DRAWING ANGLES

The protractor is used not only for measuring angles, but also for drawing them. Let us draw an angle containing 65°.

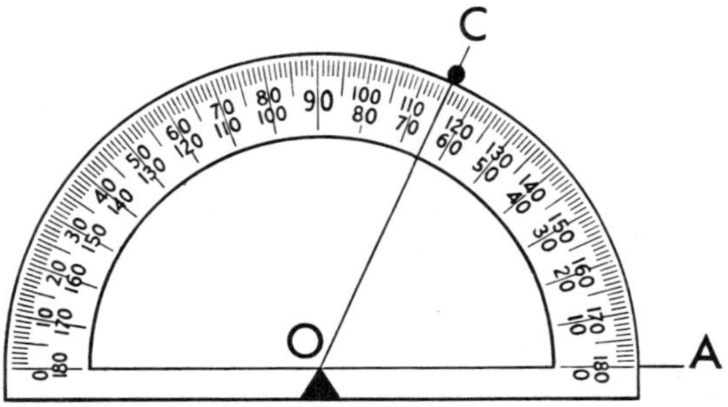

Draw a line and name it OA. Place the protractor so that its center is at O and the zero on one of its scales lies on OA. Place a point next to the number 65 on that scale. Remove the protractor and draw line OC from O through the point you made. Angle AOC at the center of the protractor contains the same number of degrees as the arc that it cuts off on the protractor. Therefore, angle AOC contains 65°.

EXERCISES

Use your protractor to draw an angle of:

1. 30°	**2.** 70°	**3.** 90°	**4.** 75°	**5.** 47°	**6.** 23°
7. 120°	**8.** 150°	**9.** 175°	**10.** 115°	**11.** 98°	**12.** 136°

PARALLEL LINES

Two lines such as AB and CD, which are in the same plane, which run in the same direction, and which will not meet no matter how far they are extended, are called ***parallel lines***. The symbol for parallel is $\parallel$. $AB \parallel CD$ is read "AB is parallel to CD."

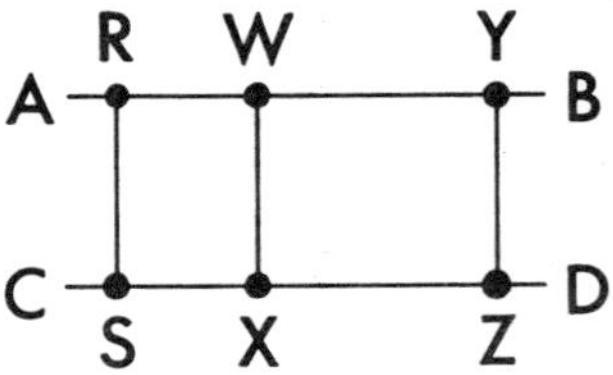

If we wish to measure the distance between two parallel lines AB and CD, we draw a line segment SR perpendicular to line CD. We then measure line segment SR and find it to be $\frac{1}{2}$ inch. Therefore, the distance between the two parallel lines AB and CD is $\frac{1}{2}$ inch. If in a similar manner we measure the distance between AB and CD at point X or point Z, we would also find the result to be $\frac{1}{2}$ inch. Now we are ready to understand the following principle:

Principle. Parallel lines are everywhere the same distance apart.

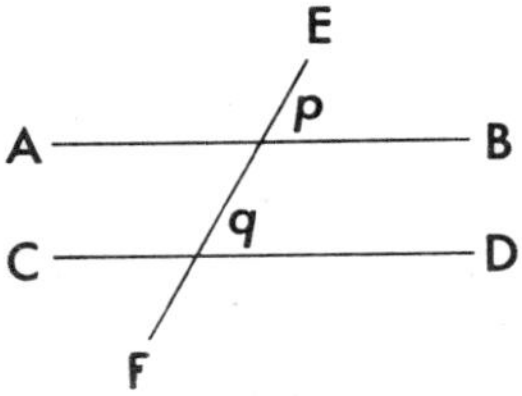

Line EF, which cuts across the two parallel lines AB and CD, is called a ***transversal.*** Notice that $\angle p$ and $\angle q$ are similarly placed in the diagram. Each angle is above one of the parallel lines and to the right of the transversal EF. We call such angles a pair of ***corresponding angles.*** If we measure $\angle p$ and $\angle q$, we find that each angle contains 60°. Therefore, $\angle p = \angle q$. Now we are ready to understand:

Principle. If two parallel lines are cut by a transversal, the corresponding angles are equal.

If Pine Street and Spruce Street each make an angle of 40° with Broadway, they run in the same direction and are parallel.

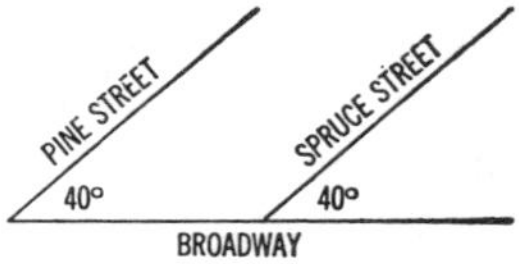

Principle. If a transversal to two lines makes a pair of corresponding angles equal, the lines are parallel.

EXERCISES

1. Which of the following pairs of lines appear parallel?

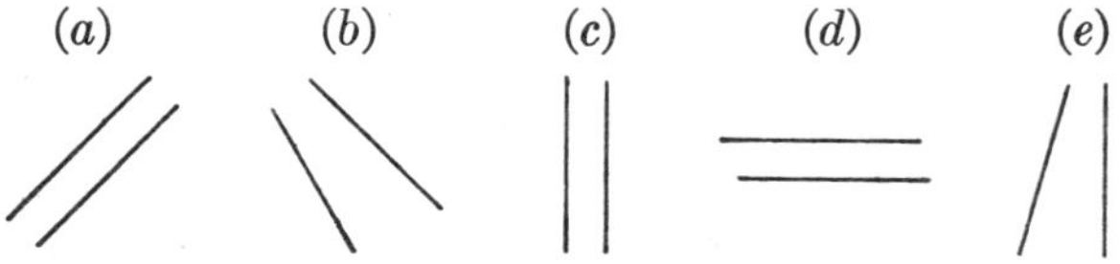

2. Broadway runs north and south. Main Street runs east and west. Are they parallel?

3. If line AB is parallel to line CD, find the number of degrees $\angle y$ contains when $\angle x$ contains: *a.* 20° *b.* 30° *c.* 60° *d.* 75° *e.* 80° *f.* 90°

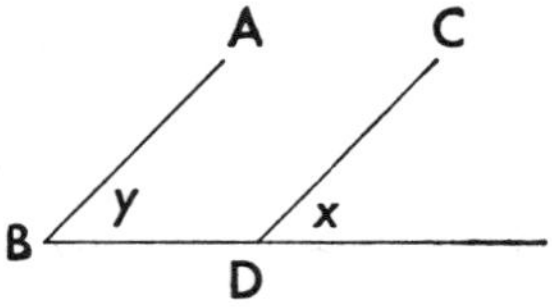

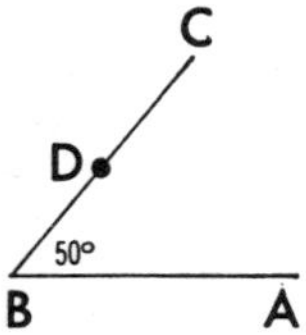

4. Copy the diagram at the left. Then use your protractor to draw a line DE parallel to line BA.

In 5–7, refer to the diagram at the right:

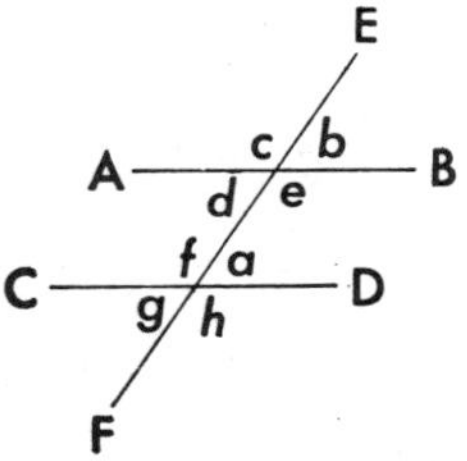

5. If $AB \parallel CD$ and $\angle a = 65°$, find $\angle b$, $\angle c$, $\angle d$, $\angle e$, $\angle f$, $\angle g$, and $\angle h$.
6. $\angle a = 55°$ and $\angle c = 125°$. (*a*) Find $\angle b$. (*b*) Is $AB \parallel CD$? (*c*) Why?
7. $\angle d = 70°$ and $\angle a = 70°$. (*a*) Find $\angle b$. (*b*) Is $AB \parallel CD$? (*c*) Why?

5. THE TRIANGLE

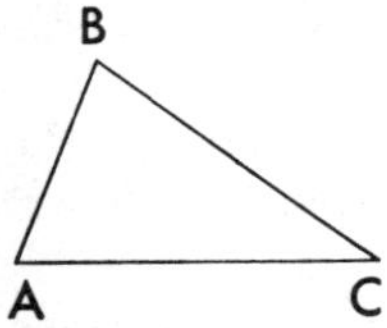

A ***triangle*** is a polygon (a closed plane figure formed by straight lines) having three ***sides.*** These three sides meet in three points, each of which is called a ***vertex*** of the triangle. At each vertex of the triangle, two sides meet to form an ***angle.*** The triangle, therefore, has three angles.

A triangle is named by placing a capital letter at each vertex. The triangle at the left would be called "triangle ABC." The symbol for a triangle is $\triangle$. Triangle ABC may be written as $\triangle ABC$.

CLASSIFYING TRIANGLES ACCORDING TO ANGLES

An ***acute triangle*** has three acute angles.

An ***obtuse triangle*** has one obtuse angle.

A ***right triangle*** has one right angle.

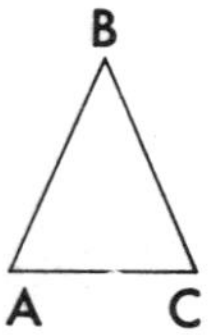

Acute Triangle

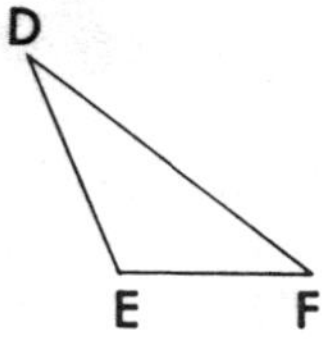

Obtuse Triangle

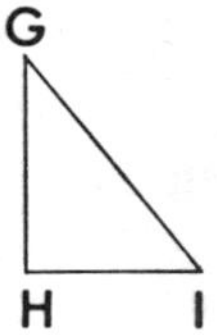

Right Triangle

In right triangle GHI above, the two sides of the triangle that form the right angle, GH and HI, are called the ***arms,*** or ***legs,*** of the right triangle. GI, the side opposite the right angle, is called the ***hypotenuse.***

CLASSIFYING TRIANGLES ACCORDING TO SIDES

An ***equilateral triangle*** has three equal sides.

An ***isosceles triangle*** has two equal sides.

A ***scalene triangle*** has no sides equal.

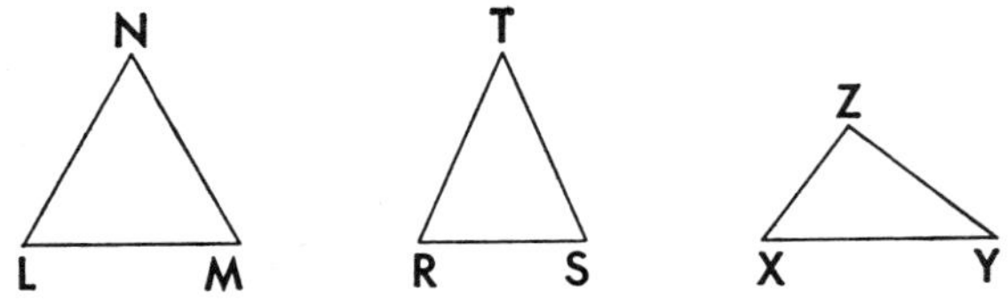

Equilateral Triangle   Isosceles Triangle   Scalene Triangle

THE ISOSCELES TRIANGLE

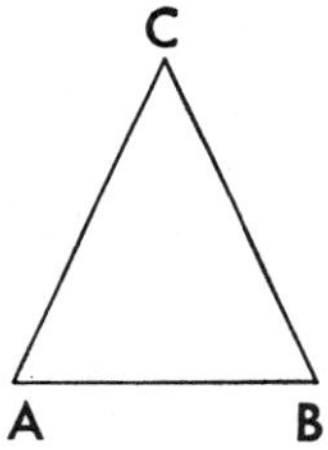

In isosceles triangle ABC, the two equal sides, AC and BC, are called the ***legs.*** The remaining side, AB, is called the ***base.*** The angle formed by the two equal sides, $\angle C$, is called the ***vertex angle.*** The two angles at the ends of the base, $\angle A$ and $\angle B$, are called the ***base angles.***

In isosceles triangle ABC, if we measure the base angles, $\angle A$ and $\angle B$, we find that each angle contains 65°. Therefore, base $\angle A$ = base $\angle B$. We are now ready to understand the following principle:

Principle. The base angles of an isosceles triangle are equal.

Since the base angles A and B in isosceles triangle ABC are opposite the two equal sides AC and BC, this principle may also be stated as follows:

Principle. If two sides of a triangle are equal, the angles opposite these sides are equal.

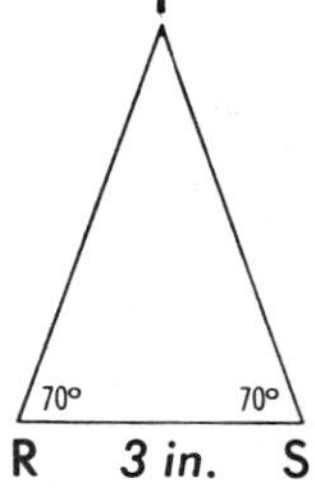

Draw a triangle in which side RS = 3 in., $\angle R$ = 70°, and $\angle S$ = 70°, as shown at the right. Use a ruler to measure the lengths of sides RT and ST. Is the length of RT equal to the length of ST? It should be. Now we are ready to understand the following principle:

Principle. If two angles of a triangle are equal, the sides opposite these angles are equal and the triangle is an isosceles triangle.

EXERCISES

1. Classify the following triangles according to *angles:*

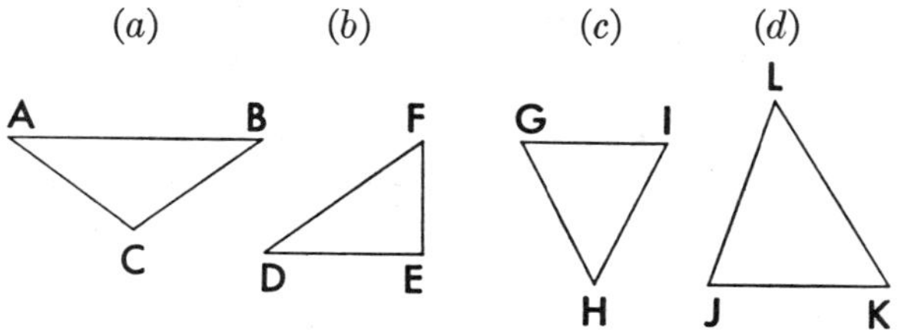

2. Classify the following triangles according to *sides:*

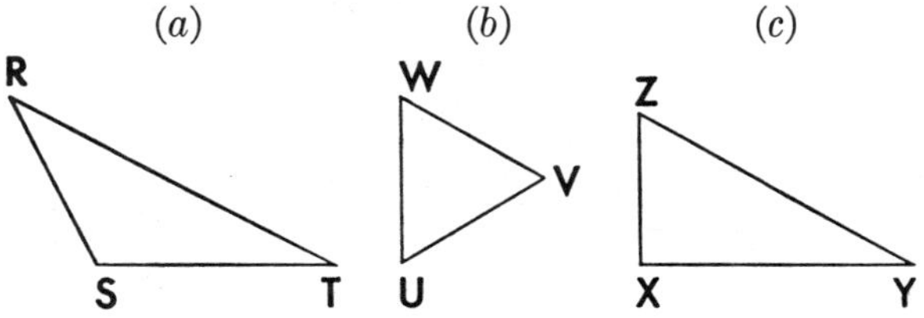

3. Name the legs and the hypotenuse in each of the following right triangles:

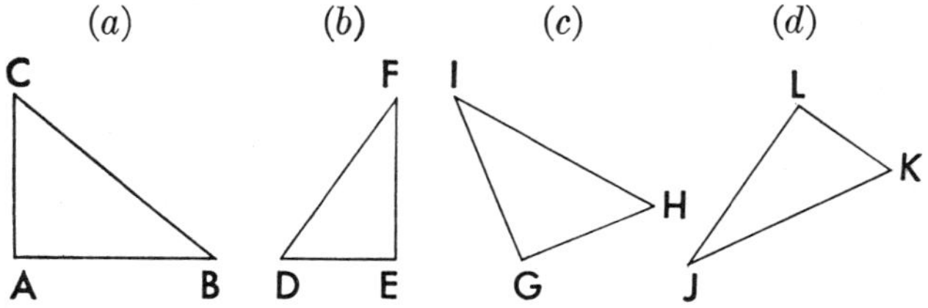

4. Name the legs, base, vertex angle, and base angles in each of the following isosceles triangles:

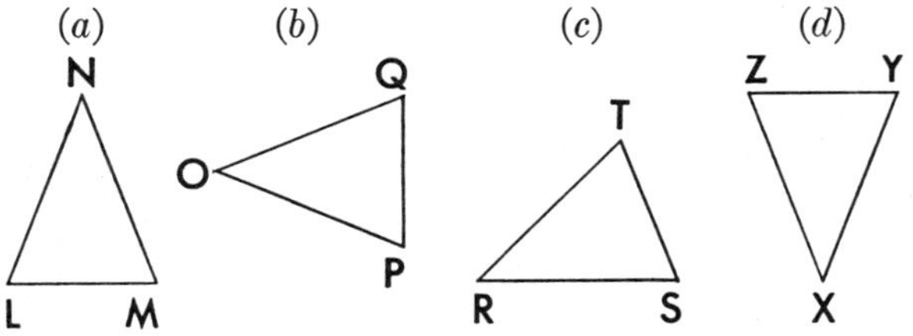

5. Draw an isosceles triangle that is (*a*) an acute triangle (*b*) an obtuse triangle (*c*) a right triangle.

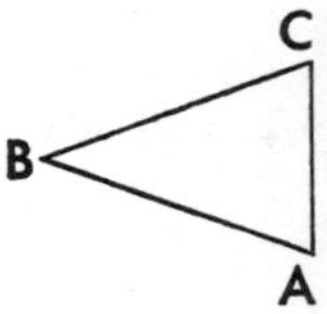

6. In $\triangle ABC$, $AC = 4$ in., $CB = 6$ in., and $AB = 6$ in.
a. What type of triangle is $\triangle ABC$?
b. Name two equal angles in $\triangle ABC$.
c. Why are they equal?

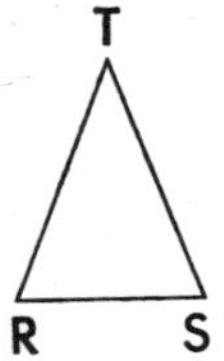

7. In $\triangle RST$, $\angle R = 70°$, $\angle S = 70°$, and $\angle T = 40°$.
a. Name two equal sides in $\triangle RST$.
b. Why are they equal?
c. What type of triangle is $\triangle RST$?

THE RELATIONSHIP AMONG THE SIDES OF A TRIANGLE

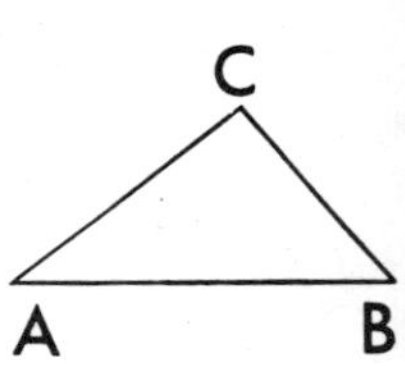

Draw any triangle ABC; measure the three sides AB, BC, CA; and compare the sum of any two sides with the third side. Is the sum of any two sides greater than the third side? It should be. This is in agreement with our common experience that if we wish to travel from one point to another in the shortest possible way, we travel in a straight line connecting these two points. Now we are ready to understand:

Principle. In a triangle, the sum of the lengths of any two sides is greater than the length of the third side.

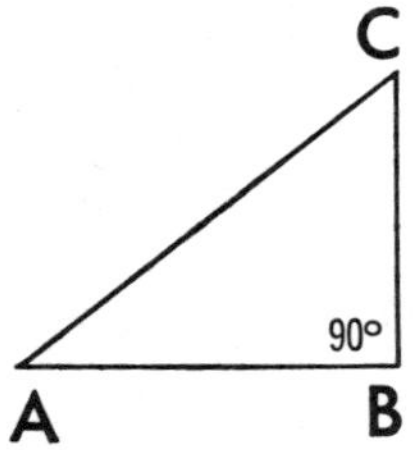

Draw any right triangle ABC and measure the hypotenuse AC and sides AB and BC. How do they compare? Is the hypotenuse the greatest side? It should be. Now we are ready to understand:

Principle. In a right triangle, the hypotenuse is the greatest side.

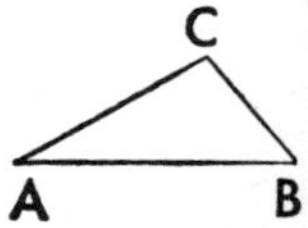

Draw a scalene triangle ABC. Measure side AB and its opposite angle C, side BC and its opposite angle A, side AC and its opposite angle B. Is the largest angle opposite the largest side? It should be. Is the largest side opposite the largest angle? It should be. Now we are ready to understand:

Principle 1. In a triangle, the greatest angle is opposite the greatest side.

Principle 2. In a triangle, the greatest side is opposite the greatest angle.

EXERCISES

1. Tell whether it is possible for the lengths of the sides of a triangle to be:
a. 7″, 8″, 9″ *b.* 3′, 6′, 10′ *c.* 9″, 5″, 8″ *d.* 4″, 3″, 7″ *e.* 6′, 9′, 7′

2. In right triangle XYZ, $\angle Y$ is the right angle. Name the greatest side of the triangle.

3. Tell which length would be the hypotenuse of a right triangle, the lengths of whose sides are (*a*) 3′, 4′, 5′ (*b*) 8″, 10″, 6″ (*c*) 5′, 13′, 12′

4. In triangle XYZ, $XY = 5$ in., $YZ = 6$ in., and $XZ = 4$ in. Name the greatest angle in the triangle.

5. In triangle ABC, $\angle A = 45°$, $\angle B = 60°$, and $\angle C = 75°$. Name the greatest side of the triangle.

THE SUM OF THE ANGLES OF A TRIANGLE

1. Measure $\angle A$, $\angle B$, and $\angle C$ in triangle ABC.

2. What is the total, or sum, of the degrees in the three angles?

3. Draw a triangle DEF different in shape from triangle ABC. Measure $\angle D$, $\angle E$, and $\angle F$ and find their sum.

4. How do your answers to questions 2 and 3 compare?

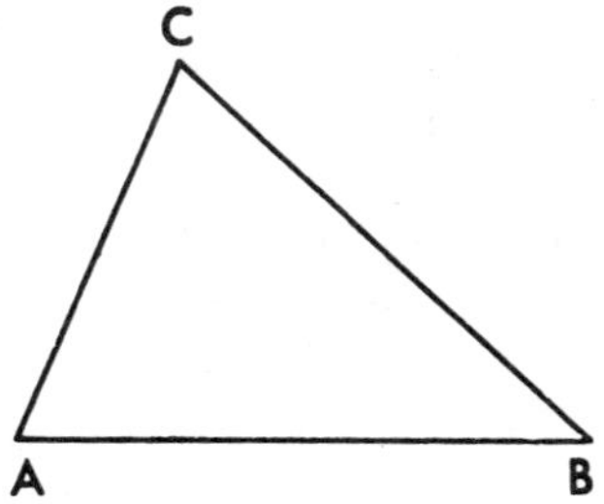

If you measured accurately, you should have found that in each triangle the sum of the angles was very close to 180°. Now we are ready to understand:

Principle. The sum of the angles of a triangle is 180°.

MODEL PROBLEM

Two angles of a triangle measure 25° and 85°. How many degrees are there in the third angle?

Solution:

1. The sum of the three angles of a triangle is 180°.
2. The sum of the two given angles is $25° + 85°$, or 110°.
3. Therefore, the third angle equals $180° - 110°$, or 70°. *Answer:* 70°

EXERCISES

1. Discover whether each of the following sets of angles can be the three angles of a triangle:
   a. 80°, 60°, 40° *b.* 100°, 20°, 60° *c.* 90°, 110°, 20° *d.* 90°, 10°, 90°
2. Find the third angle of a triangle if the first two angles contain:
   a. 70°, 20° *b.* 130°, 40° *c.* 60°, 95° *d.* 60°, 60° *e.* 45°, 45°
3. Find the third angle of a triangle if the sum of the first two angles is:
   a. 50° *b.* 90° *c.* 120° *d.* 135° *e.* 95°
4. Two angles of a triangle are 38° and 52°. What kind of triangle is it?
5. Find the vertex angle of an isosceles triangle if each base angle is:
   a. 80° *b.* 55° *c.* 42° *d.* $22\frac{1}{2}$° *e.* 51.5°
6. Find each base angle of an isosceles triangle if the vertex angle is:
   a. 40° *b.* 50° *c.* 76° *d.* 100° *e.* 65°
7. What is the number of degrees in each angle of an equiangular triangle?
8. What is the number of degrees in each acute angle of an isosceles right triangle?
9. Can a triangle have two right angles? Why?
10. Can a triangle have two obtuse angles? Why?
11. Can a base angle of an isosceles triangle be a right angle? Why?
12. What is the number of degrees in each angle of an equilateral triangle?
13. In triangle RST, $\angle R = 50°$ and $\angle S = 62°$. Name the greatest side of the triangle.

THE RULE OF PYTHAGORAS

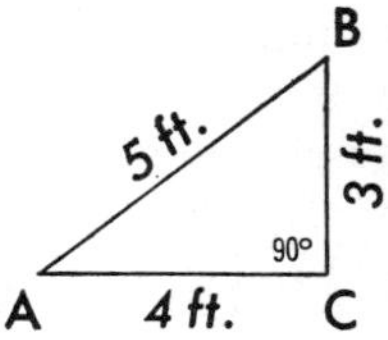

The figure at the right represents a right triangle in which the length of side BC is 3 ft., the length of side AC is 4 ft., and the length of hypotenuse AB is 5 ft.

If we square the length of side BC, we get $3^2 = 9$.
If we square the length of side AC, we get $4^2 = 16$.
If we square the length of the hypotenuse AB, we get $5^2 = 25$.
Do you notice that $25 = 9 + 16$, or that $5^2 = 3^2 + 4^2$?

When we square the length of the hypotenuse of right triangle ABC, the result is equal to the total we get when we add the squares of the lengths of the other two sides. This relationship is true in all right triangles, a fact shown more than 2000 years ago by the Greek mathematician Pythagoras. This *rule of the right triangle*, also called the *Rule of Pythagoras*, may be stated:

Principle. In a right triangle, the square of the length of the hypotenuse is equal to the sum of the squares of the lengths of the other two sides.

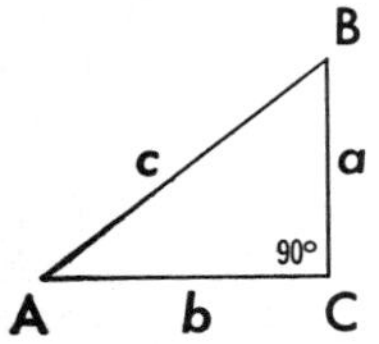

If we represent the length of the hypotenuse of right triangle ABC by c and the lengths of the other two sides by a and b, the Rule of Pythagoras may be written as the following formula:

$$c^2 = a^2 + b^2$$

MODEL PROBLEM

Find the hypotenuse of a right triangle whose other two sides are 5 inches and 12 inches.

Solution: hypotenuse = c, side a = 5, side b = 12

$c^2 = a^2 + b^2$ Rule of Pythagoras
$c^2 = 5^2 + 12^2$
$c^2 = 25 + 144$
$c^2 = 169$
$c = \sqrt{169} = 13$ If c is the number whose square is 169, then c is the square root of 169.

Answer: 13 inches

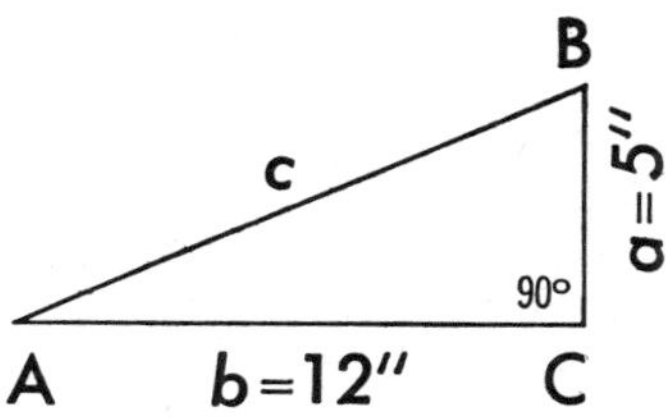

EXERCISES

1. Find the hypotenuse of a right triangle whose sides are:
   a. 6 in., 8 in. *b.* 9 ft., 12 ft. *c.* 30 ft., 40 ft. *d.* 12 in., 16 in.

2. If diagonal AC is drawn in rectangle $ABCD$, two right triangles are formed. In rectangle $ABCD$, find diagonal AC when:
   a. AB = 4 ft., BC = 3 ft. *b.* AB = 6 in., BC = 8 in.
   c. AB = 5 in., BC = 12 in.

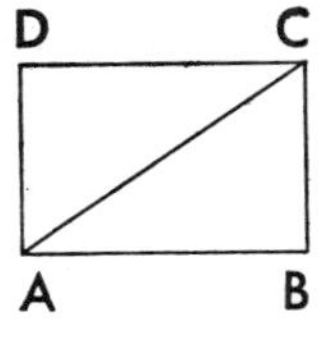

3. The hypotenuse of a right triangle is 5 feet in length. One of the other sides of the triangle is 4 feet in length. What is the length of the third side?

4. The diagonal of a rectangle is 25 ft. If the length of the rectangle is 15 ft., what is the width of the rectangle?

5. Find, correct to the nearest inch, the diagonal of a square whose side is:
   a. 4 inches *b.* 5 inches *c.* 8 inches *d.* 10 inches *e.* 12 inches

6. A baseball diamond is a square 90 ft. on each side. Find, correct to the nearest foot, the distance from home plate to second base. [Hint: Use the square root table.]

7. Find, correct to the nearest tenth of a foot, the length of each diagonal piece of lumber used to strengthen a rectangular gate that is 8 feet wide and 4 feet high.

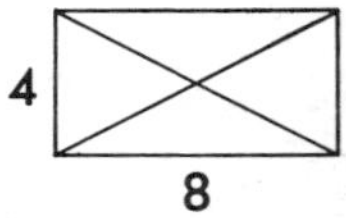

CONGRUENT TRIANGLES

In modern industry, machines can stamp out many triangular pieces of metal, all of which have exactly the same size and shape. Such triangles are called ***congruent triangles.*** The symbol for congruent is $\cong$.

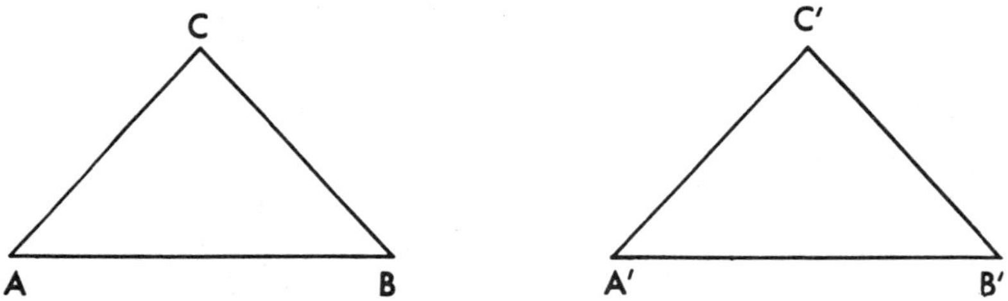

Triangle ABC and triangle $A'B'C'$ above are congruent triangles, which may be written $\triangle ABC \cong \triangle A'B'C'$. This implies that the corresponding lines and the corresponding angles of these triangles have equal measures, that is, $AB = A'B'$, $BC = B'C'$, $CA = C'A'$, $\angle A = \angle A'$, $\angle B = \angle B'$, and $\angle C = \angle C'$. Check this statement with the use of a ruler and a protractor. We can now understand the following principle:

Principle. In two congruent triangles, all corresponding sides and all corresponding angles are equal.

It is easy to see that if we were to cut out $\triangle ABC$ and $\triangle A'B'C'$ and place one on top of the other, they would fit exactly.

EXERCISES

1. Triangle ABC is congruent to triangle RST. If $AB = 8$ inches, $BC = 7$ inches, $CA = 6$ inches, $RS = 8$ inches, $ST = 7$ inches, and $TR = 6$ inches, name three pairs of equal angles.

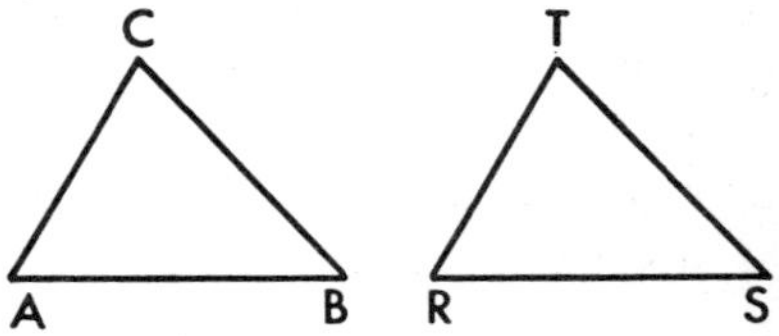

2. $\triangle ABC \cong \triangle RST$. If $\angle A = \angle R$, $\angle B = \angle S$, and $\angle C = \angle T$, name three pairs of equal sides.

6. THE QUADRILATERAL

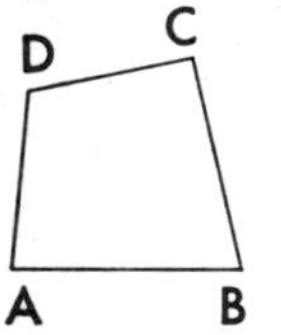

A ***quadrilateral*** is a polygon that has four sides. A point at which any two sides of the quadrilateral meet is called a ***vertex*** of the quadrilateral. At each vertex, the two sides that meet form an angle of the quadrilateral. Thus, $ABCD$ is a quadrilateral whose sides are AB, BC, CD, and DA. Its vertices are A, B, C, and D. Its angles are $\angle ABC$, $\angle BCD$, $\angle CDA$, and $\angle DAB$.

When we vary the shape of the quadrilateral by making some of its sides parallel, by making some of its sides equal, or by making its angles become right angles, we get different members of the family of quadrilaterals, as shown below:

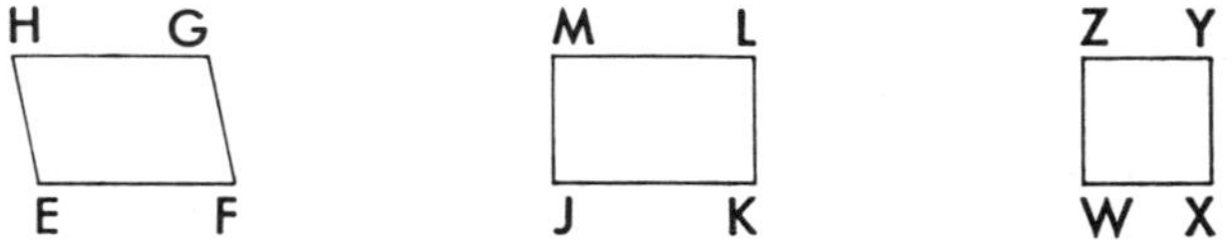

Parallelogram Rectangle Square

A ***parallelogram*** is a quadrilateral in which both pairs of opposite sides are parallel. In parallelogram $EFGH$, $EF \parallel GH$ and $EH \parallel FG$. The symbol for parallelogram is $\square$.

A ***rectangle*** is a parallelogram in which all four angles are right angles. Rectangle $JKLM$ is a parallelogram in which $\angle J$, $\angle K$, $\angle L$, and $\angle M$ are right angles. The symbol for rectangle is $\square$.

A ***square*** is a rectangle all of whose sides are equal. Therefore, square $WXYZ$ is also a parallelogram in which $\angle W$, $\angle X$, $\angle Y$, and $\angle Z$ are right angles, and side WX = side XY = side YZ = side ZW.

FAMILY OF PARALLELOGRAMS

Let us discover some of the relationships that hold true in parallelograms, rectangles, and squares.

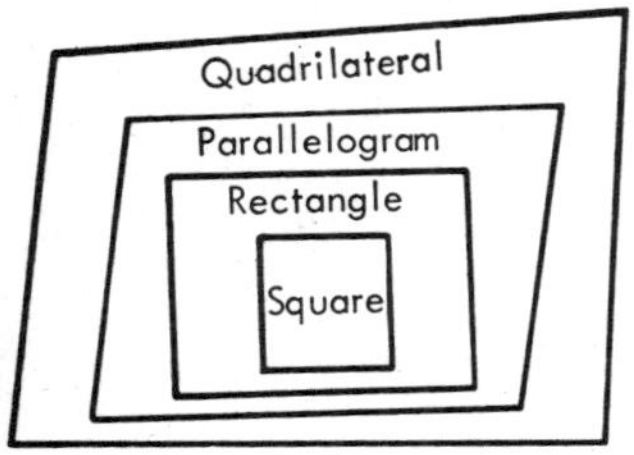

1. Rectangles and squares are members of the family of parallelograms. Therefore, any property of the family of parallelograms must also be a property of rectangles and squares.

2. A square is a member of the family of rectangles. Therefore, any property of the family of rectangles must also be a property of squares.

Draw a large parallelogram $ABCD$ like the one shown at the right. Measure the lengths of side AD and its opposite side BC. Are the lengths equal? They should be. Measure the lengths of side AB and its opposite side DC. Are these lengths equal? They should be. Now we are ready to understand the following principle:

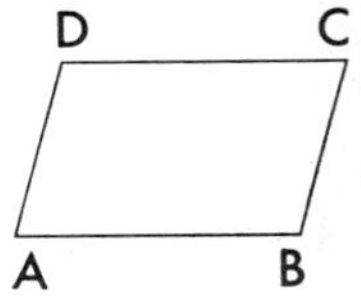

Principle. The lengths of the opposite sides of a parallelogram are equal.

Since rectangles and squares are members of the family of parallelograms, the above principle is also true for rectangles and squares.

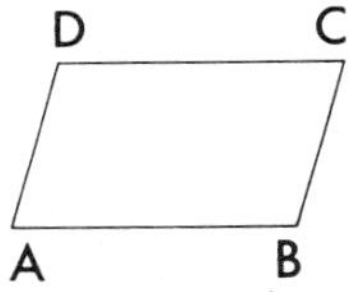

Draw a large parallelogram $ABCD$ like the one shown at the left. Measure the opposite acute angles, $\angle A$ and $\angle C$. Are their measures equal? They should be. Measure the opposite obtuse angles, $\angle B$ and $\angle D$. Are their measures equal? They should be. Now we are ready to understand:

Principle. The measures of the opposite angles of a parallelogram are equal.

Since rectangles and squares are members of the family of parallelograms, the above principle is also true for rectangles and squares.

Draw a large parallelogram $ABCD$ like the one shown at the right. Draw diagonal AC. Cut the parallelogram into two triangles by cutting along the diagonal AC. Fit one triangle upon the other. Do they fit exactly? They should. Now we are ready to understand the following principle:

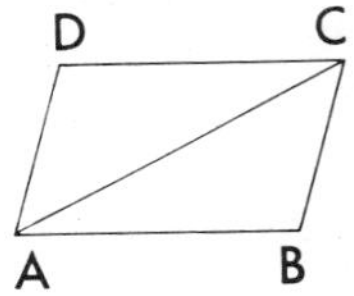

Principle. A diagonal of a parallelogram divides the parallelogram into two congruent triangles.

Since rectangles and squares are members of the family of parallelograms, the above principle is also true for rectangles and squares.

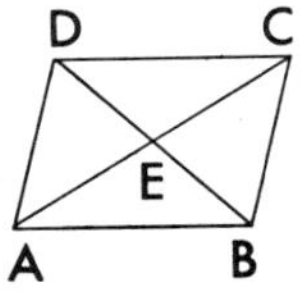

Draw a large parallelogram $ABCD$ like the one shown at the left. Draw diagonals AC and BD, which intersect at E. Measure the lengths of segments AE and EC, the two parts of diagonal AC. Are their lengths equal? They should be. Since diagonal BD divides diagonal AC into two equal parts, we say that BD bisects AC. To ***bisect*** a line or an angle means to divide the line or the angle into two equal parts. Measure the lengths of BE and ED, the two parts of diagonal BD. Are their lengths equal? They should be. Therefore, AC bisects BD. Now we are ready to understand the following principle:

Principle. The diagonals of a parallelogram bisect each other.

Since rectangles and squares are members of the family of parallelograms, the above principle is also true for rectangles and squares.

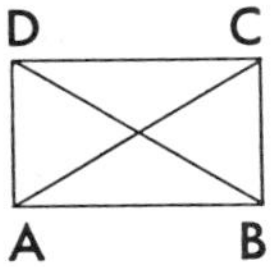

Draw a large rectangle $ABCD$, as shown at the left. Draw diagonals AC and BD. Measure the lengths of diagonals AC and BD. Are their lengths equal? They should be. Now we are ready to understand the following principle:

Principle. The diagonals of a rectangle are equal.

Since squares are members of the family of rectangles, the previous principle is also true for squares.

The chart below lists some of the properties of parallelograms, rectangles, and squares:

PROPERTIES OF PARALLELOGRAMS

Property	*Parallelogram*	*Rectangle*	*Square*
Opposite sides are parallel.	Yes	Yes	Yes
Opposite sides are equal.	Yes	Yes	Yes
Opposite angles are equal.	Yes	Yes	Yes
Diagonals bisect each other.	Yes	Yes	Yes
A diagonal forms two congruent triangles.	Yes	Yes	Yes
All angles are equal.	No	Yes	Yes
All sides are equal.	No	No	Yes
Diagonals are equal.	No	Yes	Yes

EXERCISES

In 1–12, tell whether each of the statements is true or false:

1. If a figure is a rectangle, it must be a parallelogram.
2. If a figure is a parallelogram, it must be a rectangle.
3. If a figure is a square, it must be a rectangle.

4. If a figure is a rectangle, it must be a square.
5. If a figure is a square, it must be a parallelogram.
6. If a figure is a parallelogram, it must be a square.
7. The adjacent sides of a parallelogram must be perpendicular to each other.
8. The adjacent sides of a rectangle must be perpendicular to each other.
9. The angles of a square must be right angles.
10. The diagonals of a parallelogram must be equal.
11. The diagonals of a square must be equal.
12. The opposite angles of a parallelogram must be equal.

In 13–16, refer to the parallelogram at the right:

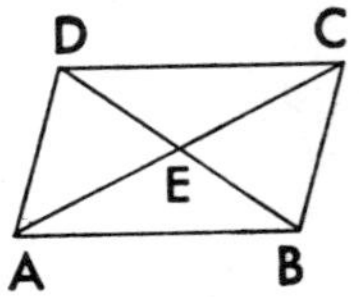

13. If $AD = 7$ in. and $AB = 10$ in., find BC and DC.
14. If $\angle DAB = 60°$ and $\angle ABC = 120°$, find $\angle BCD$ and $\angle CDA$.
15. If $AE = 12$ in. and $BE = 8$ in., find EC, ED, AC, and BD.
16. If $AB = 8$ in. and $AD = 6$ in., is $AC \perp BD$?

In 17–23, refer to the rectangle at the left:

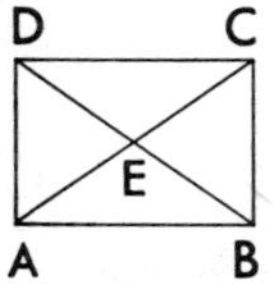

17. If $AD = 6$ in. and $AB = 8$ in., find BC and DC.
18. If $AC = 10$ in., find BD.
19. If $AC = 10$ in., find AE, EC, EB, and ED.
20. To which lines is DA perpendicular?
21. What type of triangle is triangle ABC?
22. If $AB = 40$ in. and $BC = 30$ in., find the length of AC.
23. Triangle AEB must be (*a*) a right triangle (*b*) an isosceles triangle (*c*) an equilateral triangle.

In 24–28, refer to the square at the right:

24. If $AB = 8$ ft., find BC, CD, and DA.
25. If $AE = 14$ in., find EC, AC, BD, BE, and ED.
26. To which lines is DC perpendicular?
27. Triangle DAB is (*a*) a right triangle (*b*) an obtuse triangle (*c*) an acute triangle.
28. Triangle AEB is (*a*) a scalene triangle (*b*) an isosceles triangle.

7. SIMILAR FIGURES AND SCALE DRAWINGS

When we studied maps and scale drawings, we saw that two figures may have the same shape although they do not have the same size. We call such figures ***similar figures***. The symbol for similar is $\sim$.

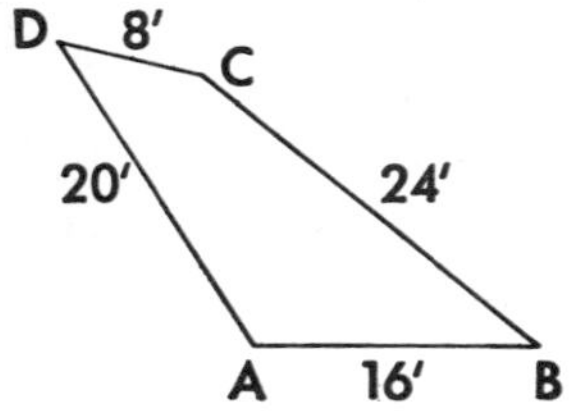

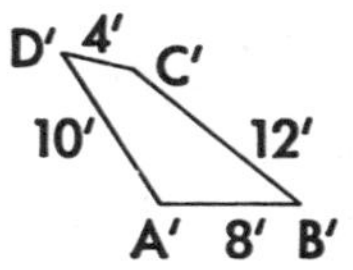

The smaller figure on the left is a scale drawing of the larger figure. They are similar figures. Notice that they have the same shape. Let us study some relationships that are true in similar figures.

Measure $\angle A$ and the corresponding $\angle A'$. How do they compare? Also measure $\angle B$ and $\angle B'$, $\angle C$ and $\angle C'$, $\angle D$ and $\angle D'$. What do you notice about all pairs of corresponding angles? They should be equal.

The ratio of side AB to side $A'B'$ is $\frac{16}{8} = \frac{2}{1}$.
The ratio of side BC to side $B'C'$ is $\frac{24}{12} = \frac{2}{1}$.
The ratio of side CD to side $C'D'$ is _______.
The ratio of side DA to side $D'A'$ is _______.

What do you notice about the ratios of all pairs of corresponding sides? They should all be equal. We are now ready to understand the following principle:

Principle. If two figures are similar, their corresponding angles are equal and the ratio of any pair of corresponding sides is equal to the ratio of any other pair of corresponding sides.

Note. We can also say that the corresponding sides are in proportion.

MODEL PROBLEM

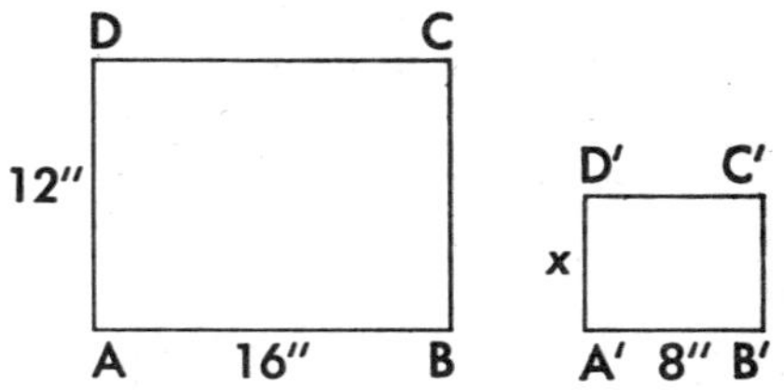

If rectangle $ABCD$ is similar to rectangle $A'B'C'D'$, find the length of $A'D'$.

Solution:

Let x represent the length of $A'D'$.
The ratio of $A'D'$ to $AD = \frac{x}{12}$.
The ratio of $A'B'$ to $AB = \frac{8}{16}$.
Since the figures are similar,

$\frac{x}{12} = \frac{8}{16}$ Why?

$12 \times \frac{x}{12} = 12 \times \frac{8}{16}$ M_{12} $[12 \times \frac{8}{16} = \frac{96}{16} = 6]$

$x = 6$ *Answer:* $A'D' = 6''$

EXERCISES

1. An architect's floor plan of a room and the floor of the room itself are an example of (*a*) congruent figures (*b*) similar figures.

2. Are all squares similar? Why?

3. Are all equilateral triangles similar? Why?

4. Are all rectangles similar? Why?

5. Are all congruent triangles similar? Why?

6. If the two triangles at the right are similar, find side x.

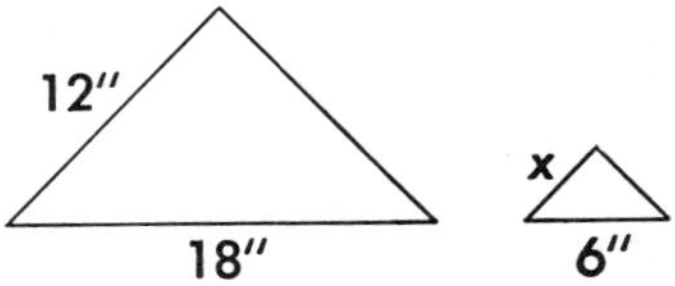

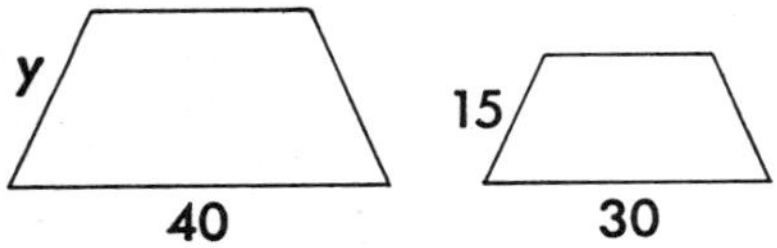

7. If the two figures at the left are similar, find side y.

8. John wishes to enlarge a rectangular photograph that is $4\frac{1}{2}$ inches long and $2\frac{3}{4}$ inches wide so that it will be 9 inches long. How wide will the enlargement be?

9. Thomas wishes to reduce the size of a picture that is 10 inches long and 8 inches wide so that it will be 4 inches wide. What will be the length of the reduced picture?

10. On a certain map, a distance of $\frac{3}{8}$ of an inch represents 300 miles. How many miles are represented by $1\frac{1}{4}$ inches?

11. The diagram at the right represents a rectangular garden. It is drawn to a scale of 1″ = 80′.

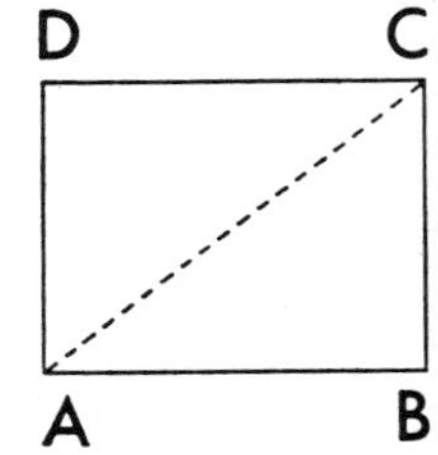

a. Find the number of feet in the length of the garden represented by AB.

b. Find the number of feet in the width of the garden represented by BC.

c. Find the number of feet in the diagonal of the garden represented by AC.

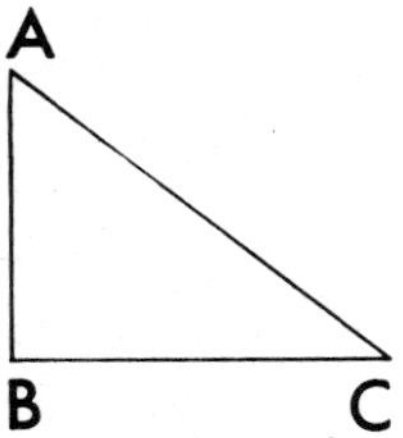

12. In the drawing at the left, A is Arthur's house, B is Burton's house, and C is Carl's house. The scale is 1 inch equals 1000 feet.

a. What is the distance from Arthur's house to Burton's house?

b. What is the distance from Burton's house to Carl's house?

c. What is the distance from Arthur's house to Carl's house?

d. If Arthur visits Carl, how much farther does he walk if he goes by way of Burton's house than if he goes directly from his house to Carl's house?

8. MEASURING PERIMETERS

The ***perimeter*** of a polygon is the sum of the lengths of all its sides.

THE TRIANGLE

When the sides of a triangular piece of land are 100 ft., 80 ft., and 60 ft., we find its perimeter by adding the lengths of the three sides. Thus, 100 ft. + 80 ft. + 60 ft., or 240 ft., is the perimeter of the triangular piece of land.

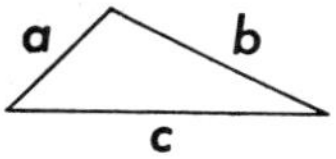

When the sides of a triangle are represented by a, b, and c, we find its perimeter in the same way. The formula for the perimeter of the triangle whose sides are represented by a, b, and c is:

$$\mathbf{P = a + b + c}$$

MODEL PROBLEM

Find the perimeter of a triangle whose sides are 5 in., 7 in., and 9 in.

$P = a + b + c \qquad a = 5, b = 7, c = 9$

$P = 5 + 7 + 9 = 21 \qquad$ *Answer:* Perimeter is 21 in.

EXERCISES

1. Find the perimeter of a triangle whose sides are:

a. 3″, 8″, 7″ *b.* 6″, 8″, 10″ *c.* 8.1″, 9.5″, 7.4″

d. 6.4″, 9.7″, 8.1″ *e.* $2\frac{3}{4}'$, $5\frac{1}{4}'$, $6\frac{1}{4}'$ *f.* $7\frac{1}{3}'$, $4\frac{1}{6}'$, $5\frac{1}{2}'$

2. How much fencing is required to enclose a triangular park whose sides are 153 ft., 256 ft., and 210 ft.?

3. The perimeter of a triangle is 120 ft. If two of the sides are 30 ft. and 40 ft., find the third side of the triangle.

THE ISOSCELES TRIANGLE

The formula for the perimeter of an isosceles triangle each of whose equal sides is represented by a and whose base is represented by b is:

$$P = a + a + b, \text{ or } \mathbf{P = 2a + b}$$

MODEL PROBLEM

Find the perimeter of an isosceles triangle each of whose equal sides is 9.5 in. and whose base is 15 in.

$P = 2a + b \qquad a = 9.5,\ b = 15$
$P = (2 \times 9.5) + 15$
$P = 19 + 15 = 34$ *Answer:* Perimeter is 34 in.

EXERCISES

1. Find the perimeter of each of the following isosceles triangles in which a represents each of the equal legs and b represents the base:

a. $a = 8',\ b = 6'$ *b.* $a = 9'',\ b = 7''$
c. $a = 5.5'',\ b = 7.3''$ *d.* $a = 6.2'',\ b = 8.6''$
e. $a = 7\frac{1}{2}'',\ b = 4''$ *f.* $a = 5\frac{2}{3}',\ b = 9\frac{2}{3}'$

2. The perimeter of an isosceles triangle is 25 ft. If each of the equal sides is 8 ft., find the base of the triangle.

3. The perimeter of an isosceles triangle is 40 ft. If the base of the triangle is 10 ft., find each of the equal sides.

THE EQUILATERAL TRIANGLE

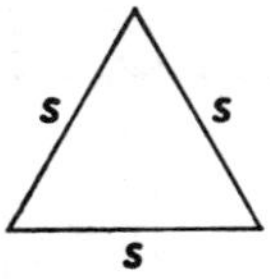

The formula for the perimeter of an equilateral triangle each of whose sides is represented by s is:

$$P = s + s + s, \text{ or } \mathbf{P = 3s}$$

MODEL PROBLEM

Find the perimeter of an equilateral triangle each of whose sides is 6 ft.

$P = 3s \qquad s = 6$
$P = 3 \times 6 = 18$ *Answer:* Perimeter is 18 ft.

EXERCISES

1. Find the perimeter of the equilateral triangle each of whose sides is:
   a. $4''$ *b.* $13'$ *c.* $125'$ *d.* $8.5''$ *e.* $5\frac{2}{3}'$
2. What is the ratio of one side of an equilateral triangle to the perimeter of the triangle?
3. Find the length of a side of an equilateral triangle whose perimeter is:
   a. $75'$ *b.* $36''$ *c.* $16.5''$ *d.* $22.2''$ *e.* $19\frac{1}{3}'$
4. If each side of an equilateral triangle is made three times its original size, what change takes place in the perimeter?
5. What per cent of the perimeter of an equilateral triangle is each side of the triangle?

THE RECTANGLE

The formula for the perimeter of a rectangle whose length is represented by l and whose width is represented by w is:

$$P = l + l + w + w, \text{ or } \mathbf{P = 2l + 2w}$$

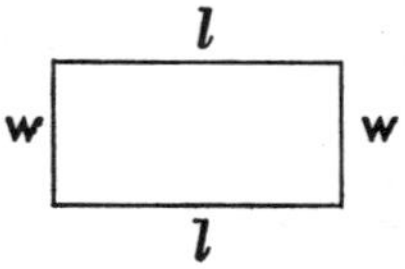

MODEL PROBLEM

Find the perimeter of a rectangle whose length is $4\frac{3}{4}$ in. and whose width is $2\frac{1}{4}$ in.

$P = 2l + 2w \qquad l = 4\frac{3}{4},\ w = 2\frac{1}{4}$

$P = (2 \times 4\frac{3}{4}) + (2 \times 2\frac{1}{4})$

$P = 9\frac{1}{2} + 4\frac{1}{2} = 14$ *Answer:* Perimeter is 14 in.

EXERCISES

1. Find the perimeter of a rectangle whose measurements are:
   a. $l = 10'', w = 7''$ *b.* $l = 86', w = 23'$ *c.* $l = 9.3'', w = 8.2'$
   d. $l = 8.9', w = 6.7'$ *e.* $l = 5\frac{3}{4}'', w = 3\frac{1}{4}''$ *f.* $l = 6\frac{2}{3}', w = 4\frac{1}{6}'$
2. How many feet of fencing are necessary to enclose a rectangular garden 25 feet by 32 feet?

3. The perimeter of a rectangle is 64 feet. If its length is 21 feet, find its width.

4. The perimeter of a rectangle is 35 feet. If its width is 5 feet, find its length.

THE SQUARE

The formula for the perimeter of a square each of whose sides is represented by s is:

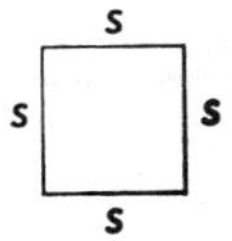

$$P = s + s + s + s, \text{ or } \boldsymbol{P = 4s}$$

MODEL PROBLEM

Find the perimeter of a square each of whose sides is 8.25 inches.

$P = 4s$ $\quad s = 8.25$

$P = 4 \times 8.25 = 33$ $\quad$ *Answer:* Perimeter is 33 in.

EXERCISES

1. Find the perimeter of a square each of whose sides is:

a. 7 in. *b.* 15 ft. *c.* $\frac{7}{8}$ in. *d.* 5.4 in. *e.* $10\frac{1}{2}$ in.

2. Find the side of a square whose perimeter is:

a. 80 in. *b.* 64 ft. *c.* 132 ft. *d.* 12.2 in. *e.* 366 ft.

3. What is the ratio of the perimeter of a square to a side of the square?

4. If each side of a square is made twice its original size, what change takes place in its perimeter?

5. What per cent of the perimeter of a square is each side of a square?

6. Which is longer, the perimeter or the diagonal of a square? Why?

THE CIRCUMFERENCE OF A CIRCLE

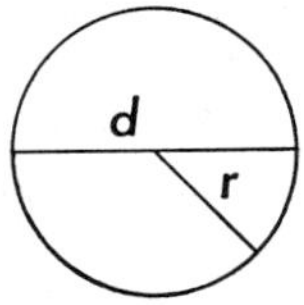

The distance around a circle, or the length of a circle, is called its ***circumference.*** Mathematicians have found that the circumference of a circle is a little more than 3 times the diameter of the circle.

The formula for the circumference of a circle whose diameter is represented by d is:

$$\boldsymbol{C = \pi d}$$

The symbol π is a Greek letter that represents a number whose approximate value is $3\frac{1}{7}$, $\frac{22}{7}$, or 3.14.

Since in a circle the length of a diameter is equal to the length of 2 radii, the formula for the circumference of a circle whose radius is represented by r is:

$$C = \pi \cdot 2r, \text{ or } \mathbf{C = 2\pi r}$$

MODEL PROBLEM

Find the circumference of a circle whose radius is 14 ft. [Use $\pi = \frac{22}{7}$.]

Method 1	*Method* 2
$C = 2\pi r \qquad r = 14$	$C = \pi d \qquad \text{If } r = 14, d = 28.$
$C = 2 \times \frac{22}{\cancel{7}_1} \times \cancel{14}^{2} = 88$	$C = \frac{22}{\cancel{7}_1} \times \cancel{28}^{4} = 88$

Answer: Circumference is 88 ft.

EXERCISES

1. Find the circumference of each of the following circles whose radius or diameter is given: [Use $\pi = \frac{22}{7}$.]

a. $r = 7$ ft. *b.* $d = 42$ in. *c.* $d = 35$ in. *d.* $r = 42$ in.

e. $r = 10.5$ in. *f.* $d = 15.4$ ft. *g.* $d = 17\frac{1}{2}$ in. *h.* $r = 3\frac{1}{2}$ in.

2. Find, correct to the nearest tenth, the circumference of each of the following circles whose radius or diameter is given: [Use $\pi = 3.14$.]

a. $r = 5.4$ ft. *b.* $d = 8.2$ in. *c.* $d = 9.7$ in. *d.* $r = 13.5$ ft.

e. $r = 8.75$ ft. *f.* $d = 6.25$ in. *g.* $d = 12.50$ ft. *h.* $r = 7.15$ ft.

3. The diameter of a circle is 35 inches. What is its circumference?

4. How many inches does Joan's bicycle go in one turn of the wheels if the diameter of the wheels is 28 inches?

5. A circular flower bed has a diameter of 7 ft. How many feet of fencing will be required to enclose this garden?

6. The diameter of a wheel is $3\frac{1}{2}$ feet. How many feet will the wheel travel in a complete turn? How many times must the wheel turn to travel 880 ft.? [Use $\pi = \frac{22}{7}$.]

9. MEASURING AREAS

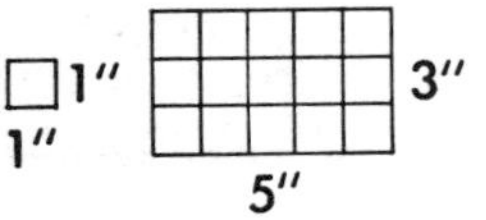

The number of units of square measure that are contained in a given figure is called the ***area*** of the figure. In the diagram at the left, the smaller figure is a unit of measure that represents the *square inch*. A square inch is a square whose length is 1 inch and whose width is 1 inch. Other units of square measure are the *square foot*, *square yard*, *square mile*, *square centimeter*, and *square meter*.

The larger of the above figures represents a rectangle whose base is 5 inches and whose altitude is 3 inches. By counting the number of square inches in the rectangle, we find that the area enclosed by the rectangle is 15 square inches. More often, we simply say that the area of the rectangle is 15 square inches.

It is often difficult or impossible to find the area of a figure by counting the number of units of square measure it encloses. Therefore, mathematicians have developed formulas that help us find the areas of familiar geometric figures.

UNITS OF SQUARE MEASURE

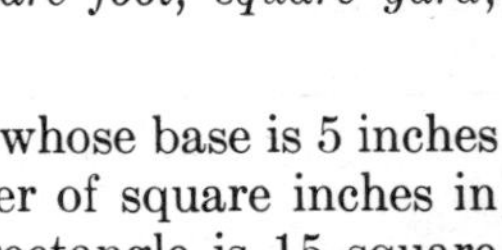

At the right, the smaller of the two diagrams represents a square whose length is 1 foot and whose width is 1 foot. This square is a unit of square measure called 1 square foot. The larger of the two diagrams represents a square whose length is 1 yard and whose width is 1 yard. This square is a unit of square measure called 1 square yard. Since the square foot is contained in the square yard 9 times, we see that 1 sq. yd. = 9 sq. ft.

Below we see a table of the most frequently used units of square measure:

Square Measure

144 square inches (sq. in.)	=	1 square foot (sq. ft.)
9 sq. ft.	=	1 square yard (sq. yd.)
30.25 sq. yd.	=	1 square rod (sq. rd.)
160 sq. rd.	=	1 acre (A.)
43,560 sq. ft.	=	1 acre (A.)
640 acres	=	1 square mile (sq. mi.)

MODEL PROBLEMS

1. Find the number of square inches in 5 square feet.

Solution: Since each square foot contains 144 square inches, we multiply 144 by 5.

$5 \times 144 = 720$ *Answer:* 720 sq. in.

2. A man's farm contains 32,000 square rods. How many acres are there on his farm?

Solution: Since it requires 160 square rods to make 1 acre, we find how many times 160 is contained in 32,000.

$$160\overline{)32{,}000}\quad 200$$

Answer: 200 acres

EXERCISES

1. To convert a given number of square inches to square feet, would you *multiply* by 144 or *divide* by 144?

2. To convert a given number of square yards to square feet, would you *multiply* by 9 or *divide* by 9?

3. To convert a given number of acres to square rods, would you *multiply* by 160 or *divide* by 160?

4. To convert a given number of acres to square miles, would you *multiply* by 640 or *divide* by 640?

In 5–22, supply the missing numbers:

5. 3 sq. ft. = _____ sq. in.

6. $6\frac{1}{2}$ sq. ft. = _____ sq. in.

7. 288 sq. in. = _____ sq. ft.

8. 360 sq. in. = _____ sq. ft.

9. 72 sq. in. = _____ sq. ft.

10. 48 sq. in. = _____ sq. ft.

11. 4 sq. yd. = _____ sq. ft.

12. 54 sq. ft. = _____ sq. yd.

13. $5\frac{1}{3}$ sq. yd. = _____ sq. ft.

14. 6 sq. ft. = _____ sq. yd.

15. 10 A. = _____ sq. rd.

16. 800 sq. rd. = _____ A.

17. 121 sq. yd. = _____ sq. rd.

18. $\frac{1}{4}$ A. = _____ sq. rd.

19. 4 sq. mi. = _____ A.

20. 960 A. = _____ sq. rd.

21. 5 A. = _____ sq. ft.

22. 87,120 sq. ft. = _____ A.

23. What fractional part of a square yard is a square foot?

THE RECTANGLE

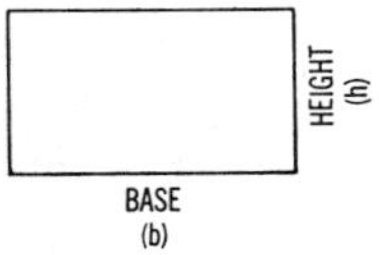

A rule for finding the area of a rectangle is: "The area of a rectangle is equal to the length of its base multiplied by the length of its height." Thus, the formula for the area, A, of a rectangle of base b and height h is:

$$A = bh$$

The height of a rectangle is also called its ***altitude.***
The length and width of a rectangle are also called its ***dimensions.***

MODEL PROBLEMS

1. Find the area of a rectangle whose base is 15 inches and whose height is 4 inches.

$$A = bh \qquad b = 15,\ h = 4$$
$$A = 15 \times 4 = 60$$

Answer: Area is 60 square inches.

2. Find the area of a rectangle whose length is 2 feet and whose width is 4 inches.

$A = lw$ $\qquad l = 2$ ft. $= 24$ in., $w = 4$ in.

$A = 24 \times 4 = 96$ $\qquad$ (Convert 2 ft. to 24 in. because all measurements must be in the same unit when we find the area.)

Answer: Area is 96 square inches.

EXERCISES

Find the area of a rectangle whose dimensions are:

	(*a*)	(*b*)	(*c*)	(*d*)
1.	35′ by 10′	24″ by 7″	80″ by 45″	41″ by 25″
2.	24′ by 36′	70′ by 25′	8′ by $4\frac{1}{2}$′	33′ by $10\frac{2}{3}$′
3.	$12\frac{5}{8}$″ by 48″	1.5 yd. by 8.3 yd.	.75 mi. by .5 mi.	1.2 ft. by 8.5 yd.

4. Find the area of a rectangular tabletop that is 41 inches long and 25 inches wide.

5. How many square feet of linoleum will be required to cover the floor of a room 15 feet by 24 feet?

6. If the length of one side of a rectangle is 8 inches and the width is $2\frac{1}{2}$ inches, how many square inches are there in the area of the rectangle?
7. Find the area of a rectangle whose length is 75 centimeters and whose width is 50 centimeters.
8. Find the area of a rectangle whose length is 8 meters and whose width is $4\frac{1}{2}$ meters.
9. How many square yards of carpeting are required to cover a floor 18 ft. long by 12 ft. wide?
10. The dimensions of a rectangular living room are 24 feet by 12 feet. How many square yards of carpet are needed to cover the whole floor?
11. How many tiles, each of which is 1 sq. ft., are needed to cover the floor of a pool 50 ft. wide and 125 ft. long?
12. How many tiles, each of which is 1 sq. ft., are needed to cover the floor of a hall 3 yd. wide and 20 yd. long?
13. A rectangular garden plot is 12 feet wide. If it contains 288 square feet, how long is it?
14. A room is to have an area of 500 sq. ft. If its length is 25 ft., what must be its width?
15. Which rectangle has the larger area, one that is 12′ by 20′ or one that is 10′ by 25′?
16. *a.* What is the area of a lawn 50 feet by 40 feet?
    b. Allowing 1 pound of grass seed to 400 square feet, find how many pounds of seed will be required to seed the lawn.
17. At 15 cents a square foot, what is the cost of cleaning a rug 9 feet long and 6 feet wide?
18. At $5.40 per square yard, find the cost of constructing a concrete walk 144 feet long and 5 feet wide.
19. What change takes place in the area of a rectangle if its height is doubled and its width is unchanged?
20. What change takes place in the area of a rectangle if its base is tripled and its height remains unchanged?
21. Tell the change that takes place in the area of a rectangle if both the base and height are:

    a. doubled *b.* tripled *c.* multiplied by 10 *d.* halved

THE SQUARE

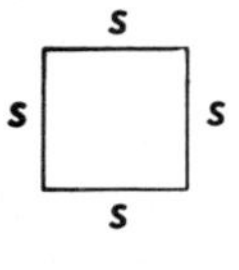

A rectangle whose base is equal to its height is a square. To find its area, we use the rule: "The area of a square is equal to its base multiplied by its height."

If the length of a side of the square is represented by s, this rule gives us the following formula:

$$A = s \times s, \text{ or } A = s^2$$

MODEL PROBLEMS

1. Find the area of a square each of whose sides is 5 feet.

$A = s^2 \qquad s = 5$

$A = 5 \times 5 = 25$ *Answer:* Area is 25 square feet.

2. Find the side of a square whose area is 64 square inches.

$A = s^2 \qquad A = 64$

$64 = s^2$ A number multiplied by itself equals 64.

$8 = s$ 8 multiplied by itself equals 64.

Answer: Side is 8 inches.

EXERCISES

1. Find the area of a square whose side is:

a. 4 in. *b.* 15 in. *c.* 27 ft. *d.* 7 yd. *e.* $\frac{2}{3}$ ft. *f.* 6.5 ft.

2. Find the side of a square whose area is:

a. 100 sq. in. *b.* 36 sq. ft. *c.* 81 sq. yd. *d.* 49 sq. in. *e.* 400 sq. ft.

3. Find the area of a square whose perimeter is:

a. 40 in. *b.* 16 ft. *c.* 28 yd. *d.* 14 in. *e.* 5 ft.

4. Find the perimeter of a square whose area is:

a. 4 sq. in. *b.* 25 sq. ft. *c.* 64 sq. yd. *d.* 1 sq. ft.

5. A baseball diamond is a square 90 ft. on a side. What is the area enclosed by the baselines?

6. A square garden has an area of 144 sq. ft. How many feet of fencing are required to enclose it?

7. Tom used 40 feet of fencing to enclose his square vegetable patch. What was the area of his patch?

8. Tell the change that takes place in the area of a square if each side is:

a. doubled *b.* tripled *c.* multiplied by 10 *d.* halved

THE PARALLELOGRAM

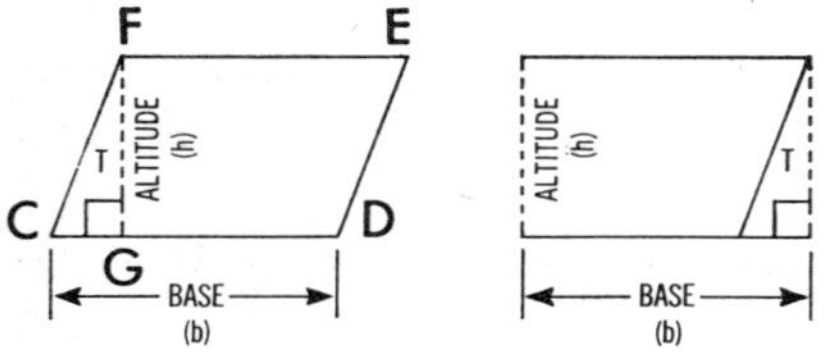

The side on which a parallelogram rests is called the ***base.*** In ▱*CDEF*, the base is line *CD*. The distance between the two parallel lines *CD* and *FE* is called the ***height,*** or ***altitude,*** of the parallelogram. The distance between two parallel lines is measured on a line that is perpendicular to both of them. Thus, *FG* is the height, or altitude, of ▱*CDEF*.

Notice that if we cut off triangle CFG from the left side of the parallelogram and fit it onto the right side of the remaining figure, with FC fitting on ED, we obtain the rectangle pictured to the right of the parallelogram. The rectangle and the parallelogram have exactly the same area, exactly the same base, and exactly the same altitude.

Since the formula for the area of a rectangle is $A = bh$, the formula for the area of a parallelogram also is:

$$A = bh$$

MODEL PROBLEM

Find the area of parallelogram $ABCD$.

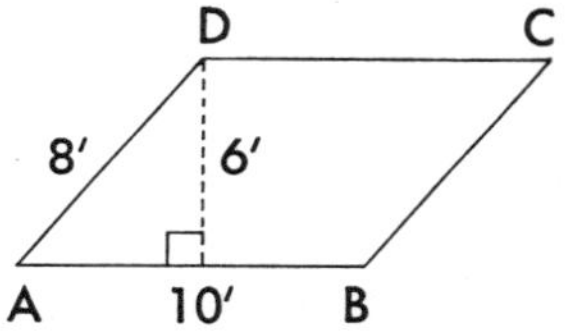

$A = bh$ $\qquad b = 10,\ h = 6$
$A = 10 \times 6 = 60$

Answer: Area is 60 square feet.

EXERCISES

1. Find the area of each of the following parallelograms:

a. $b = 7$ in., $h = 15$ in. *b.* $b = 32$ ft., $h = 14$ ft.
c. $b = 8$ yd., $h = 5$ yd. *d.* $b = 24$ ft., $h = 12\frac{1}{3}$ ft.
e. $b = 10.2$ ft., $h = 5.5$ ft. *f.* $b = 28$ in., $h = 2$ ft.

2. The area of a parallelogram is 360 square inches. Find its altitude if the base is 24 inches.

3. The area of a parallelogram is equal to the area of a rectangle. The base of the parallelogram is 24 in. and the altitude drawn to the base is 10 in. If one side of the rectangle is 20 in., find an adjacent side of the rectangle.

THE TRIANGLE

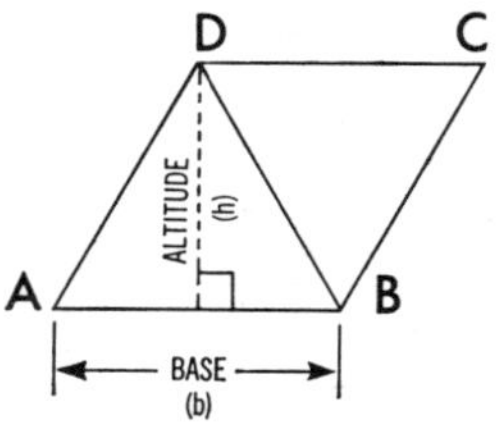

If in $\square ABCD$ we draw diagonal BD, two triangles are formed, $\triangle ABD$ and $\triangle BCD$, which are equal in area. Therefore, the area of $\triangle ABD$ is equal to one-half the area of $\square ABCD$. Since the area of a parallelogram is equal to its base multiplied by its altitude, we get the following rule for finding the area of a triangle: "The area, A, of a triangle is equal to one-half its base, b, multiplied by its altitude, h." This rule gives us the following formula:

$$A = \frac{1}{2}bh, \quad \text{or} \quad A = \frac{bh}{2}$$

MODEL PROBLEM

Find the area of a triangle whose altitude is 20 in. and whose base is 3 ft.

$A = \frac{1}{2}bh$ $\qquad b = 3$ ft. $= 36$ in., $h = 20$ in.

$$A = \frac{1}{\cancel{2}_{1}} \times \overset{18}{\cancel{36}} \times 20 = 360$$

Answer: Area is 360 square inches.

EXERCISES

1. Find the area of each of the following triangles:

a. $b = 18$ in., $h = 8$ in. *b.* $b = 14$ ft., $h = 5$ ft.
c. $b = 9$ yd., $h = 15$ yd. *d.* $b = 16$ ft., $h = 12$ ft.
e. $b = 24$ ft., $h = 16$ ft. *f.* $b = 19$ in., $h = 25$ in.
g. $b = 6.8$ in., $h = 12.5$ in. *h.* $b = 16$ ft., $h = 5\frac{3}{4}$ ft.

2. Find the area of a triangular gas station whose base is 80 ft. and whose altitude is 65 ft.

3. Find the area of a right triangle if the sides that form the right angle are:
a. 6 in., 8 in. *b.* 30 in., 40 in. *c.* 5 in., 12 in. *d.* 3 ft., 7 ft. *e.* 9 in., 5 in.

4. *ABCD* is a rectangle. Find the area of the shaded triangle:

(*a*) (*b*) (*c*) (*d*)

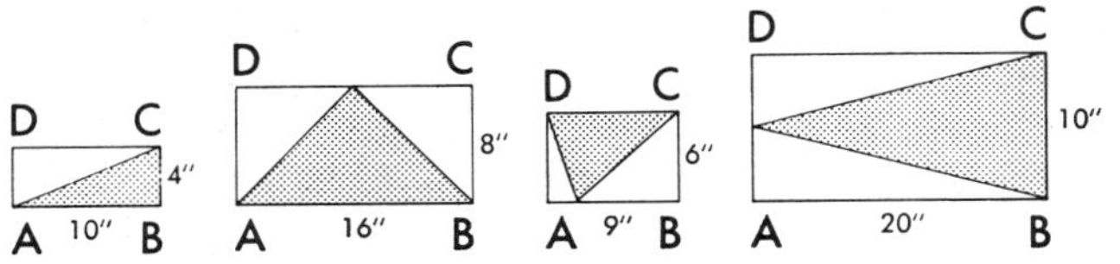

THE CIRCLE

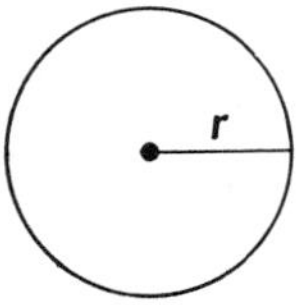

Mathematicians have discovered the following rule for finding the area of a circle: "The area, A, of a circle is equal to π multiplied by the square of the radius, r."

This rule gives us the following formula:

$$A = \pi r^2$$

MODEL PROBLEM

Find the area of a circle whose radius is 14 inches. [Use $\pi = \frac{22}{7}$.]

$A = \pi r^2$ $\qquad\qquad r = 14$

$$A = \frac{22}{\cancel{7}_1} \times \overset{2}{\cancel{14}} \times 14$$

$A = 616$ $\qquad\qquad$ *Answer:* Area is 616 square inches.

EXERCISES

1. Find the area of a circle whose radius is: [Use $\pi = \frac{22}{7}$.]

a. 7 ft. *b.* 14 in. *c.* 21 yd. *d.* 28 in. *e.* 5.6 ft.
f. 4.9 ft. *g.* $3\frac{1}{2}$ yd. *h.* $4\frac{1}{5}$ ft. *i.* 10 in. *j.* 6 ft.

2. Find the area of a circular flower bed if the radius of the circle is 3.5 ft. [Use $\pi = \frac{22}{7}$.]

3. A circular mirror has a diameter of 14 inches.
a. Find the length of the radius.
b. Find the area of the mirror. [Use $\pi = \frac{22}{7}$.]

4. A circular swimming pool is 70 feet in diameter. Find the area of the pool. [Use $\pi = \frac{22}{7}$.]

5. The bottom of a water glass is a circle whose radius is 1.25 inches. Find its area to the nearest tenth of a square inch. [Use $\pi = 3.14$.]

6. The radius of circle O is 14 in., and the radius of circle O' is 7 in.
a. Find the circumference of circle O. [Use $\pi = \frac{22}{7}$.]
b. Find the circumference of circle O'. [Use $\pi = \frac{22}{7}$.]
c. Find the ratio of the circumference of circle O to the circumference of circle O'.
d. Find the ratio of the radius of circle O to the radius of circle O'.
e. In circles O and O' when the ratio of the radii is 2:1, what is the ratio of the circumferences?
f. If the radius of a circle is multiplied by 2, by what number is the circumference multiplied?

g. Find the area of circle O. [Use $\pi = \frac{22}{7}$.]
h. Find the area of circle O'. [Use $\pi = \frac{22}{7}$.]
i. Find the ratio of the area of circle O to the area of circle O'.
j. In circles O and O' when the ratio of the radii is $2:1$, what is the ratio of the areas?
k. If the radius of a circle is multiplied by 2, by what number is the area multiplied?

7. *a.* What fractional part of each circle has been shaded?
b. What per cent of each circle has been shaded?

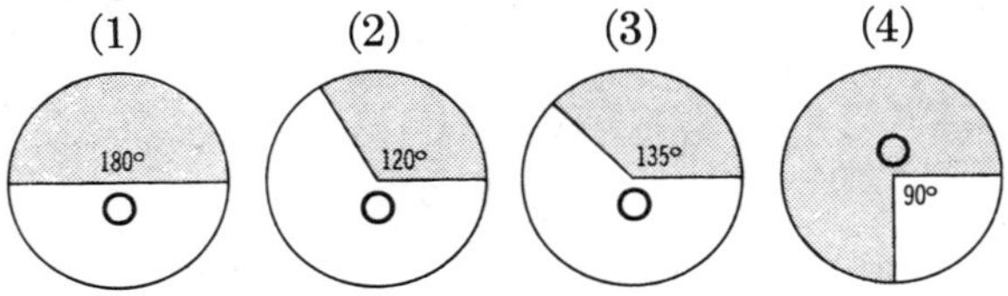

MORE DIFFICULT AREA PROBLEMS

EXERCISES

1. The accompanying sketch represents the side of a house.
a. What two different geometric figures are represented in the sketch?
b. What is the area of the side of the house?
c. How much paint will be required to cover the side with one coat if one gallon of paint will cover 336 square feet of surface?

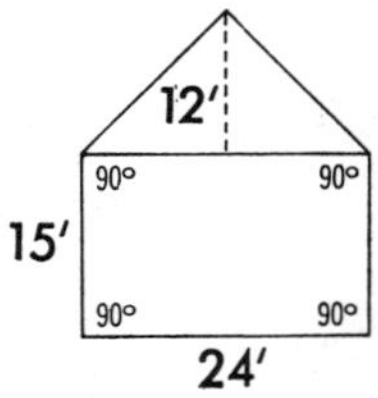

2. The accompanying sketch represents Mr. Brown's house and lot. Grass seed costs 80 cents a pound, and one pound will cover 100 square feet of ground. Making no allowance for paths or driveways, find the following:
a. How many square feet are there in Mr. Brown's lot?

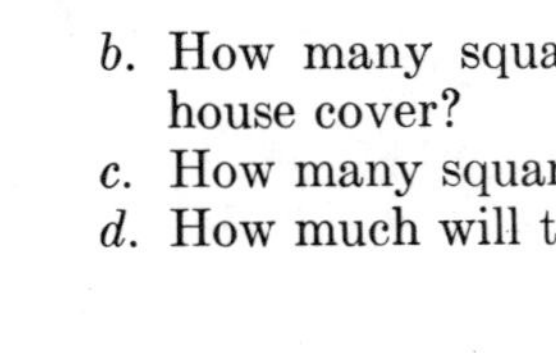

b. How many square feet does Mr. Brown's house cover?
c. How many square feet of ground will Mr. Brown have to seed?
d. How much will the grass seed cost?

3. The figure at the right represents the outline of a floor.

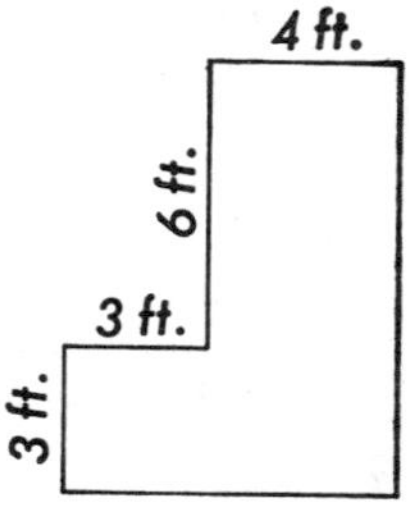

a. What is the length of the longest side?
b. What is the total number of square feet in the floor?
c. What would it cost to refinish the floor at $1.25 per sq. ft.?

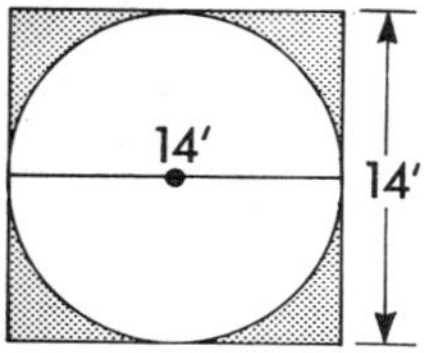

4. In the figure at the left:
a. What is the area of the square?
b. What is the area of the circle?
c. What is the area of the shaded portion?

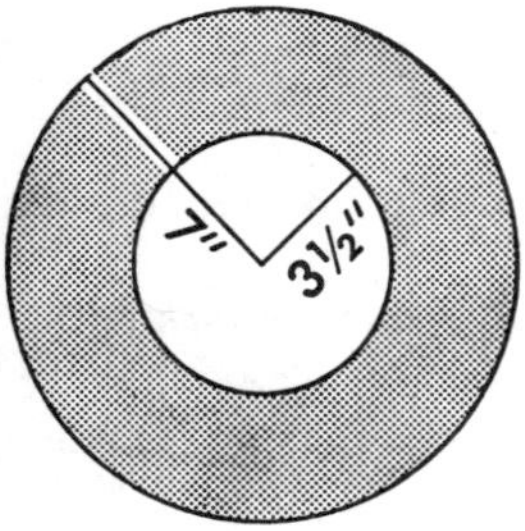

5. In the figure at the left:
a. What is the area of the large circle?
b. What is the area of the small circle?
c. What is the area of the shaded portion?

6. Mr. Sullivan has a piece of ground that is 50 ft. by 100 ft. He plants a lawn that is 40 ft. by 90 ft. He wishes to build a cement walk around the lawn.

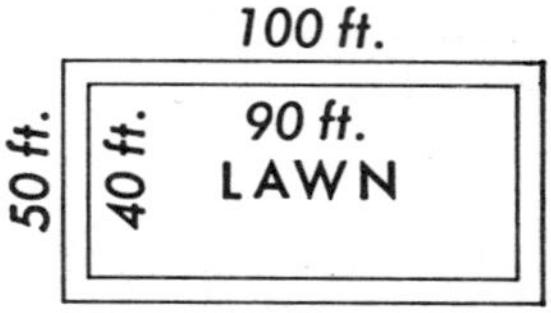

a. How many square feet of ground will the cement walk cover?
b. At $.60 per sq. ft., what will it cost him to build the walk?

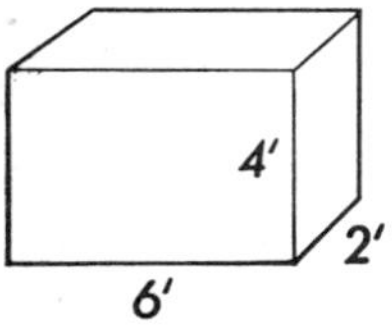

7. The figure at the left represents a closed wooden box.
a. Find the number of square feet of lumber required to make the four sides, the top, and the bottom of the box.
b. If lumber costs $.12 per sq. ft., what is the cost of the lumber needed to make the box?

8. The triangle and rectangle at the right are drawn to a scale of $\frac{1}{8}$ inch equals 1 foot.
   a. Using a ruler, find the base and altitude of each figure.
   b. Find the area of the triangle.
   c. Find the area of the rectangle.

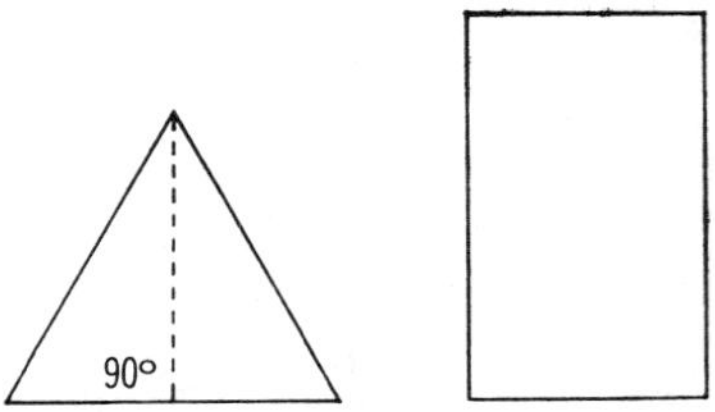

9. The diagram at the right represents a lot upon which Mr. Burns plans to build a gas station.
   a. What is the area of the lot?
   b. What will the lot cost Mr. Burns if he buys it at $1.50 per square foot?

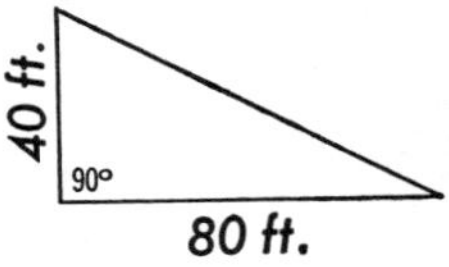

10. MEASURING VOLUMES

UNITS OF CUBIC MEASURE

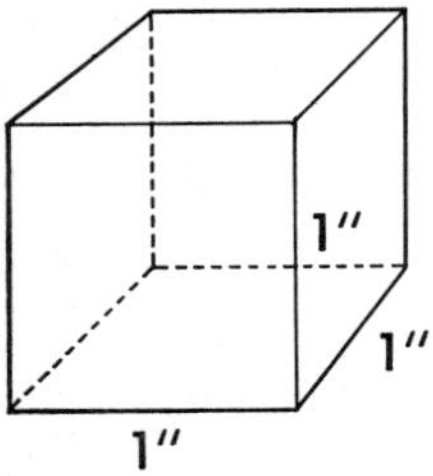

The amount of space that a solid figure contains is called its ***volume.***

Volume is measured by *cubic units.* A *cubic inch* is the space contained in a cube whose length, width, and height are each one inch. Other units used for measuring volume are the *cubic foot,* the *cubic yard,* and the *cubic centimeter.*

THE RECTANGULAR SOLID

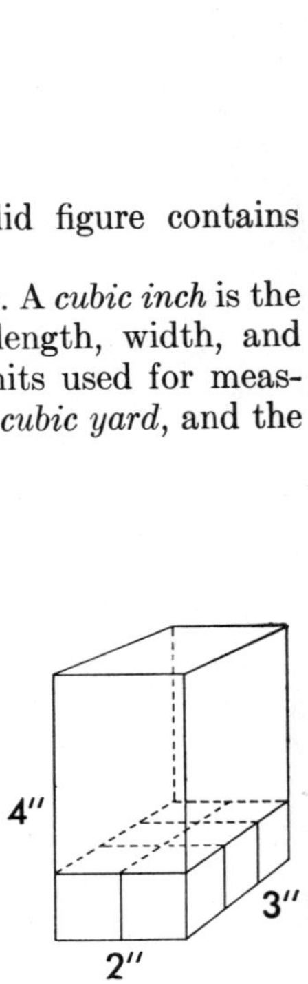

A closed solid figure whose sides are all rectangles is called a ***rectangular solid.*** If we wish to measure the volume of the rectangular solid at the right, we have to find how many units of cubic measure it contains. We notice that there are 6, or 2×3, cubic inches in the first layer, which is 1 inch high. Since the solid is 4 inches high, there will be 4 such layers. Therefore, the volume of the rectangular solid will be 4×6, or 24 cubic inches.

We can now understand the rule for finding the volume of a rectangular solid: "The volume, V, of a rectangular solid is equal to its length, l, multiplied by its width, w, multiplied by its height, h." This rule gives us the following formula:

$$V = lwh$$

MODEL PROBLEM

How many cubic feet are there in a packing case 3 feet long, 4 feet wide, and $3\frac{1}{2}$ feet high?

$V = lwh$ $\qquad l = 3, w = 4, h = 3\frac{1}{2}$

$V = 3 \times 4 \times 3\frac{1}{2}$

$V = 42$ $\qquad$ *Answer:* Volume is 42 cubic feet.

EXERCISES

1. Find the volume of the rectangular solid with the following dimensions:

a. $l = 2'', w = 5'', h = 9''$ *b.* $l = 10'', w = 2'', h = 7''$

c. $l = 12', w = 9', h = 8'$ *d.* $l = 20', w = 6', h = 3'$

e. $l = 8$ in., $w = 6$ in., $h = 15$ in. *f.* $l = 5$ yd., $w = 6$ yd., $h = \frac{1}{3}$ yd.

g. $l = 16$ ft., $w = 4\frac{1}{2}$ ft., $h = 3$ ft. *h.* $l = 12$ in., $w = 8.5$ ft., $h = 5.5$ ft.

2. How many cubic inches are there in a box 14 inches long, 8 inches wide, and 2 inches deep?

3. A box is 1 yard long, $1\frac{1}{2}$ feet wide, and 9 inches deep. What is the volume of the box?

4. How many cubic feet of air are contained in a room 28 feet long, 16 feet wide, and $8\frac{1}{2}$ feet high?

5. A candy box is 2 inches high, 8.4 inches long, and 6.5 inches wide. How many cubic inches does it contain?

6. A concrete floor 3 inches thick is to be laid in a cellar 30 feet by 24 feet. How many cubic feet of concrete are needed?

7. A box contains 144 cubic feet. Find its height if it is 9 feet long and 4 feet wide.

8. One gallon contains 231 cubic inches. How many gallons will a tank hold that is 22 inches long, 14 inches wide, and 12 inches high?

9. The drawing represents an aquarium in the form of a rectangular solid. The figures given are inside dimensions.

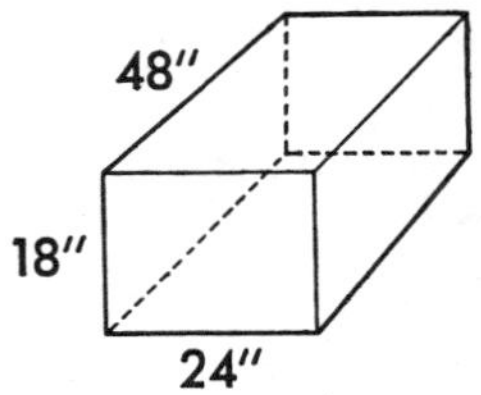

a. What is the volume of the aquarium in cubic inches?

b. Find the greatest number of gallons of water that can be poured into the aquarium without overflowing. Express your answer as a whole number. [231 cubic inches = 1 gallon.]

10. Over a driveway 120 feet long and 9 feet wide, a layer of crushed rock is to be spread to an average depth of 6 inches.

a. Find the number of cubic yards of crushed rock required for the driveway.

b. Find the cost of the crushed rock at $4.50 a cubic yard.

11. The dimensions of a coal bin built in the form of a rectangular solid are 7′ by 10′ by 6′. The bin is level full of coal.

a. How many tons of coal are there in the bin if a ton of coal occupies 35 cu. ft. of space?

b. How much would the coal cost at $18.75 per ton?

12. In a house, the dimensions of the rooms are as follows: kitchen 12 ft. by 10 ft. by 8 ft., living room 18 ft. by 12 ft. by 8 ft., 2 bedrooms each 13 ft. by 10 ft. by 8 ft. The total space in the other rooms is 800 cu. ft.

a. Find the total air space, or volume, of the rooms in the house.

b. What is the weight of the air in the house? [Weight of air = .08 lb. per cu. ft.]

THE CUBE

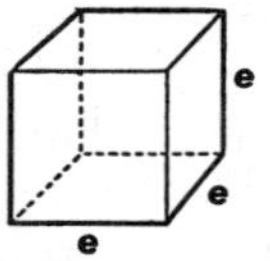

Since a ***cube*** is a rectangular solid all of whose edges are of equal length, to find its volume we use the rule: "The volume of a cube is equal to its length multiplied by its width multiplied by its height."

If the edge of the cube is represented by e and its volume by V, this rule gives us the following formula:

$$V = e \times e \times e \quad or \quad V = e^3$$

MODEL PROBLEM

Find the volume of a cube each of whose edges is 4 inches.

$V = e^3$ $\qquad e = 4$

$V = 4 \times 4 \times 4 = 64$ $\qquad$ *Answer:* Volume is 64 cubic inches.

EXERCISES

1. Find the volume of a cube whose edge is:

a. 2 in.	*b.* 3 ft.	*c.* 5 yd.	*d.* 8 in.	*e.* 10 in.
f. $1\frac{1}{2}$ in.	*g.* 1.9 ft.	*h.* 3.4 yd.	*i.* $\frac{1}{3}$ yd.	*j.* $\frac{3}{4}$ ft.

2. Find the number of cubic inches in 1 cubic foot.

3. Find the number of cubic inches in 1 cubic yard.

4. Find the number of cubic feet in 1 cubic yard.
5. How many cubic feet do 3 cubic yards contain?
6. How many cubic feet does $\frac{2}{3}$ of a cubic yard contain?
7. *a.* Find the volume of a cube whose edge is 3 inches.
   b. Find the volume of a cube whose edge is 6 inches.
   c. How do the two volumes compare?
8. Tell what change takes place in the volume of a cube whose edge is:
   a. doubled *b.* tripled *c.* multiplied by 10 *d.* halved

CONVERTING CUBIC MEASURES

To convert from one unit of cubic measure to another, the table below can be helpful:

Cubic Measure

1728 cubic inches (cu. in.) = 1 cubic foot (cu. ft.)
27 cubic feet = 1 cubic yard (cu. yd.)

EXERCISES

1. 2 cu. ft. = _____ cu. in.
2. $\frac{1}{2}$ cu. ft. = _____ cu. in.
3. $2\frac{1}{3}$ cu. ft. = _____ cu. in.
4. 3 cu. yd. = _____ cu. ft.
5. $\frac{1}{3}$ cu. yd. = _____ cu. ft.
6. $1\frac{2}{3}$ cu. yd. = _____ cu. ft.
7. 1 cu. yd. = _____ cu. in.
8. $\frac{1}{3}$ cu. yd. = _____ cu. in.
9. $1\frac{2}{3}$ cu. yd. = _____ cu. in.
10. 81 cu. ft. = _____ cu. yd.
11. 3456 cu. in. = _____ cu. ft.
12. 117 cu. ft. = _____ cu. yd.

THE CYLINDER

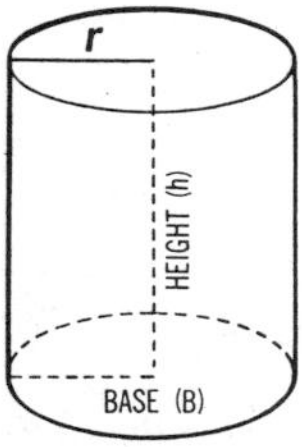

The upper and lower bases of a cylinder are parallel, congruent circles. The perpendicular distance between the two bases is the height of the cylinder.

The rule for finding the volume of a cylinder is: "The volume, V, of a cylinder is equal to the area of the base, B, multiplied by the height, h."

This rule gives us the following formula:

$$V = Bh$$

Since the base is a circle whose area is πr^2, another formula for the volume of a cylinder is:

$$V = \pi r^2 h$$

MODEL PROBLEM

The radius of a cylindrical can is $3\frac{1}{2}$ inches. Its height is 8 inches. Find its capacity in cubic inches. [Use $\pi = \frac{22}{7}$.]

$V = \pi r^2 h$ $\qquad\qquad$ $r = 3\frac{1}{2}$ or $\frac{7}{2}$, $h = 8$

$$V = \frac{22}{\cancel{7}_{1}} \times \frac{\overset{1}{\cancel{7}}}{\cancel{2}_{1}} \times \frac{7}{\cancel{2}_{1}} \times \overset{2}{\cancel{\overset{\cancel{4}}{8}}} = 308$$

Answer: Volume is 308 cubic inches.

EXERCISES

1. Find the volume of each of the following cylinders: [Use $\pi = \frac{22}{7}$.]

a. $r = 21$ in., $h = 10$ in. $\qquad$ *b.* $r = 35$ in., $h = 12$ in.
c. $r = 28$ ft., $h = \frac{3}{4}$ ft. $\qquad$ *d.* $r = 1.4$ ft., $h = 6$ ft.
e. $r = 10$ in., $h = 4.2$ in. $\qquad$ *f.* $r = 18$ in., $h = 5.6$ in.

2. Find the volume of a cylindrical can if the area of the base is 50 sq. in. and the height is 4.5 inches.

3. Find the volume of a cylindrical oil tank if the diameter of the base is 7 feet and the height is $10\frac{1}{2}$ feet. [Use $\pi = \frac{22}{7}$.]

4. A farmer has a cylindrical silo whose radius is 14 ft. and whose height is 40 ft. Find its cubic contents. [Use $\pi = \frac{22}{7}$.]

5. A cylindrical tank used for storing water has a diameter of 12 feet and a height of 14 feet. How many gallons of water will it hold? [1 cubic foot contains 7.5 gallons.]

6. Two cylinders have equal bases. The height of one cylinder is twice the height of the other. How do their volumes compare?

7. Two cylinders have equal heights. The radius of the first cylinder is twice the radius of the other. How do their volumes compare?

VOLUME, CAPACITY, AND WEIGHT IN THE METRIC SYSTEM

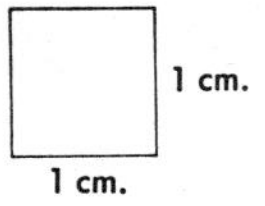

We know that the *centimeter* (cm.) is a unit of length in the metric system. A *square centimeter* (sq. cm. or cm.^{2}), which is a square each of whose sides is 1 cm., is a unit of square measure in the metric system.

A *cubic centimeter* (c.c. or cu. cm.), which is a cube whose length is 1 cm., whose width is 1 cm., and whose height is 1 cm., is a unit of cubic measure in the metric system.

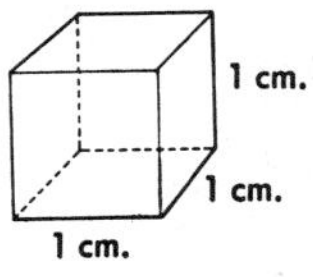

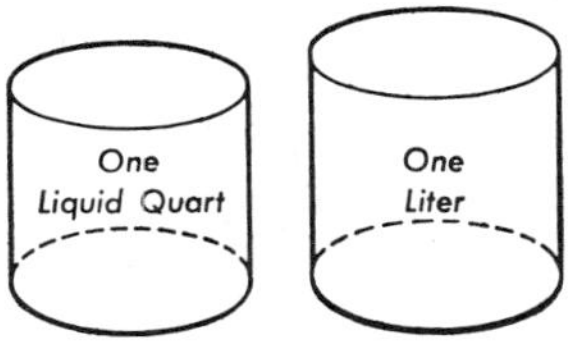

We have learned that the unit used to measure liquid capacity in the metric system is the *liter*. A liter is a solid figure whose volume is 1000 c.c. Such a solid figure, which is usually pictured in the form of a cylinder, contains approximately 1.1 liquid quarts.

We have learned that the unit of weight in the metric system is the *gram*, which is a small weight equal to about .035 ounce. Actually, a gram is the weight of 1 c.c. of water under certain standard conditions. The *kilogram*, which is 1000 grams, is therefore the weight of 1000 c.c. of water, or 1 liter of water. A kilogram turns out to be approximately 2.2 pounds.

MODEL PROBLEM

A fish tank is in the form of a rectangular solid whose length is 50 cm., whose width is 40 cm., and whose depth is 36 cm.

a. Find the volume of the tank.
b. Find the number of liters of water it will hold.

Solution:

a. Volume $= lwh$ $\quad l = 50$ cm., $w = 40$ cm., $h = 36$ cm.
Volume $= 50 \times 40 \times 36 = 72{,}000$
Volume of the tank is 72,000 cubic centimeters. *Ans.*

b. Since 1 liter contains 1000 c.c., we divide 72,000 by 1000 to find the number of liters of water the tank will hold.

$$72{,}000 \div 1000 = 72$$

The tank will hold 72 liters of water. *Ans.*

EXERCISES

1. Find the volume of a rectangular solid with the following dimensions:
a. $l = 3$ cm., $w = 5$ cm., $h = 6$ cm. *b.* $l = 1$ m., $w = 4$ m., $h = 2$ m.
c. $l = 3$ m., $w = 2$ m., $h = 50$ cm.

2. Find the volume of a cube each of whose edges is:
a. 10 cm. *b.* 15 cm. *c.* 20 cm. *d.* 75 cm. *e.* 3 m. *f.* 1.2 m.

3. Find the volume of a cylinder if the area of its base is 77 sq. cm. and its height is 25 cm.

4. Find the volume of a cylinder if the radius of its circular base is 14 cm. and its height is 40 cm.

5. Find the number of cubic centimeters in:

a. 4 liters *b.* 20 liters *c.* 6.5 liters *d.* 9.8 liters *e.* 1.75 liters

6. Find the number of liters in:

a. 8,000 c.c. *b.* 15,000 c.c. *c.* 500 c.c. *d.* 750 c.c. *e.* 2,250 c.c.

7. A box is in the shape of a rectangular solid whose dimensions are $l = 150$ cm., $w = 100$ cm., and $h = 40$ cm. Find the number of liters of air the box will hold.

8. A milk container is in the form of a cylinder the area of whose circular base is 600 sq. cm. and whose height is 30 cm. Find the number of liters of milk the container will hold.

11. SYMMETRY

In nature, in art, and in industry, we find many forms that have a pleasing and attractive appearance because of the *balanced* arrangement of their parts. We say that such forms have ***symmetry.***

LINE SYMMETRY

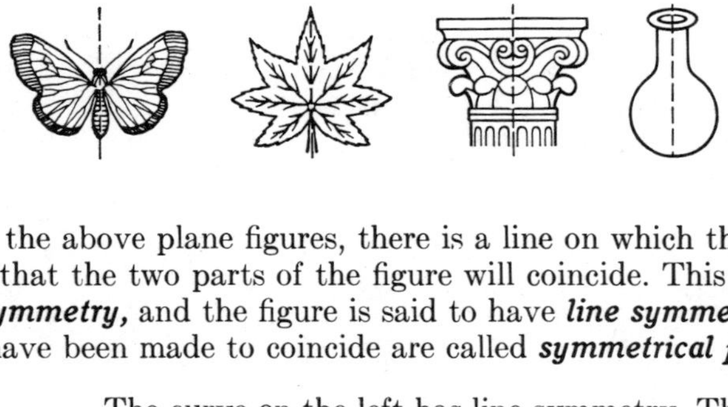

In each of the above plane figures, there is a line on which the figure may be folded so that the two parts of the figure will coincide. This line is called the ***axis of symmetry,*** and the figure is said to have ***line symmetry.*** Any two points that have been made to coincide are called ***symmetrical points.***

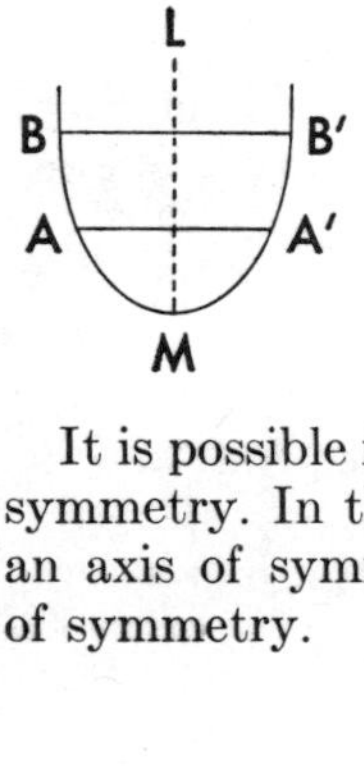

The curve on the left has line symmetry. The dotted line LM is the axis of symmetry. Notice that when the axis of symmetry is drawn, it is perpendicular to and bisects the straight line that joins any two symmetrical points. If A and A', and B and B' are pairs of symmetrical points, then the axis of symmetry, LM, is perpendicular to and bisects line AA' and line BB'.

It is possible for a figure to have more than one axis of symmetry. In the rectangle at the right, the line XY is an axis of symmetry and the line VW is also an axis of symmetry.

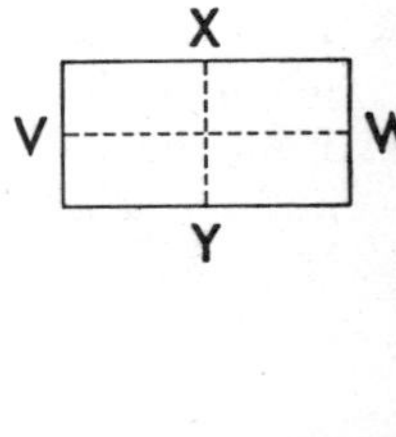

EXERCISES

1. Draw a diagram of a heart that shows its symmetry. Draw the axis of symmetry.
2. Name three objects that have line symmetry.
3. Does an open book have line symmetry? What part of the book represents the axis of symmetry?
4. Draw a diagram of a kite. Draw the axis of symmetry.
5. Draw an isosceles triangle. Draw its axis of symmetry.
6. Does a scalene triangle have symmetry?
7. Draw an equilateral triangle. How many axes of symmetry does it have? Draw them.
8. Draw a square. How many axes of symmetry does it have? Draw them.

POINT SYMMETRY

The earring at the right shows us another type of symmetry. Notice that the design is built around the center of the earring. For each diamond on one side of the center, there is another diamond directly opposite it on the other side of the center. This is an example of ***point symmetry.*** The center on opposite sides of which we find ***corresponding points*** is called the ***center of symmetry.***

The curve on the left has point symmetry. The center of symmetry is point O. Notice that any line segment passing through point O and ending in the curve is divided into two equal parts by point O. If line AB is drawn through the center of symmetry, O, ending in the curve at points A and B, then $OA = OB$.

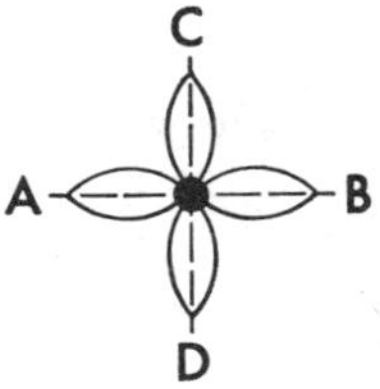

It is possible for a figure to have both line symmetry and point symmetry. The figure on the right, which resembles the petals of a flower, illustrates this. Lines AB and CD are axes of symmetry. Their point of intersection is the center of symmetry.

EXERCISES

1. Draw a circle. Does it have point symmetry? Why?
2. Name three objects that have point symmetry.
3. Draw a rectangle. Does it have point symmetry? If your answer is "yes," find the center of symmetry.
4. Tell whether each of the following has point symmetry, line symmetry, both types of symmetry, or no symmetry:

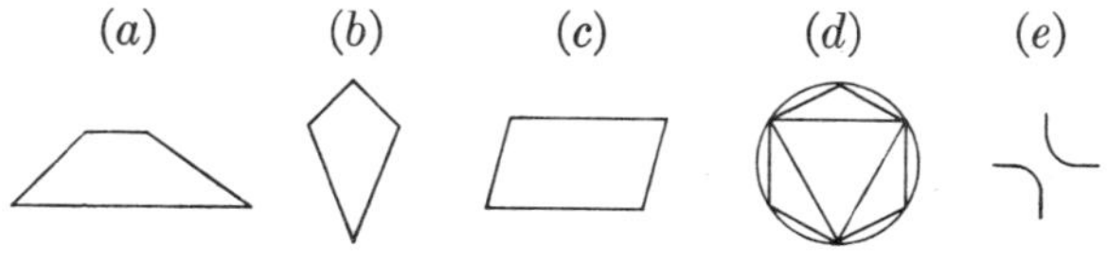

12. CONSTRUCTIONS USING A STRAIGHTEDGE AND COMPASSES

TO BISECT A GIVEN LINE SEGMENT

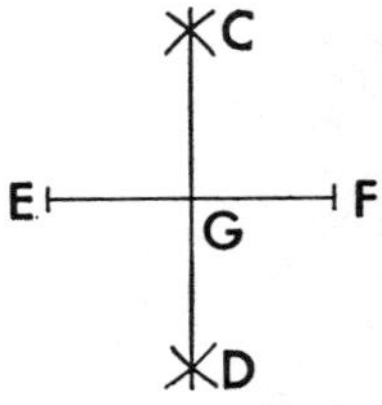

To bisect the line segment *EF*: (1) Open the compasses so that the radius is more than half of *EF*. (2) Using point *E* as a center, draw one arc above *EF* and one arc below *EF*. (3) Using the same radius and point *F* as a center, draw another pair of arcs, one of which is above *EF* and the other below *EF*. (4) Name with the letters *C* and *D* the points where the arcs intersect. (5) Draw line *CD*, which bisects line segment *EF*. (6) Name with the letter *G* the point where *CD* intersects *EF*. Line segment *EG* equals line segment *GF*.

EXERCISES

1. Using a straightedge and compasses, bisect a line segment that has the same length and position as each of the following line segments:

(*a*) (*b*) (*c*) (*d*)

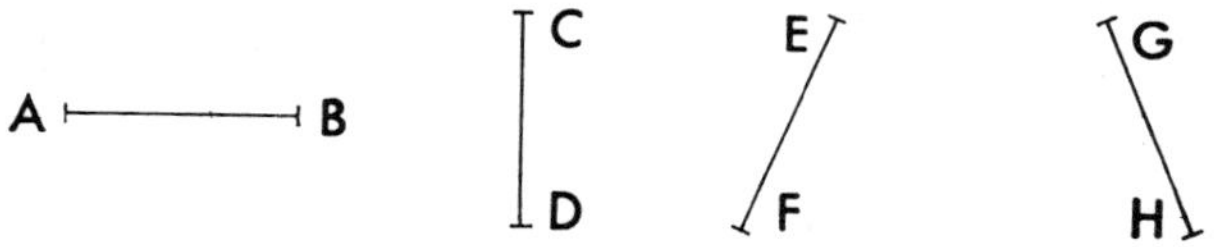

2. Using a straightedge and compasses, bisect line segments whose lengths are:

a. 3 inches *b.* $1\frac{1}{2}$ inches *c.* $2\frac{7}{8}$ inches *d.* $\frac{1}{3}$ foot

3. Draw a line segment 6 inches in length and bisect it. Then bisect each half. Into how many equal parts has the original segment been divided?

4. Line segment *AB* represents the perimeter of a square. Construct a side of the square.

5. Draw a triangle. Bisect each of the sides. Extend the bisectors until they intersect. (*Note:* All the bisectors should intersect at the same point.)

TO BISECT AN ANGLE

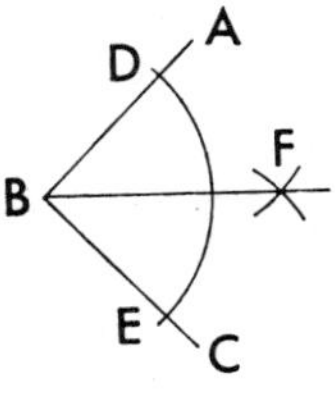

To bisect angle *ABC*: (1) Use any convenient radius and *B* as a center to draw an arc that intersects lines *AB* and *BC* at points *D* and *E*. (2) With point *D* and point *E* as centers, and using equal radii of sufficient length, draw arcs that intersect at point *F*. (3) Draw line *BF*, which bisects angle *ABC*. Angle *ABF* is equal to angle *CBF*.

EXERCISES

1. Draw an angle in the same position and having about the same measure as each of the following angles. Then, using a straightedge and compasses, bisect each of the angles you have drawn.

2. With a protractor, draw the following angles and use compasses and a straightedge to bisect each angle:

   a. 40° *b.* 120° *c.* 90° *d.* 180° *e.* 158°

3. Draw a triangle. Bisect each of the angles. Extend the bisectors until they intersect. (*Note:* All the bisectors should intersect at the same point.)

TO CONSTRUCT A LINE PERPENDICULAR TO A GIVEN LINE THROUGH A POINT ON THE LINE

To construct a line perpendicular to line AB at point P: (1) With point P as a center and with any convenient radius, draw an arc that intersects AB at points C and D. (2) Then, with points C and D as centers and a radius larger than the one used before, draw arcs that intersect at point E. (3) Draw line PE, which is perpendicular to line AB.

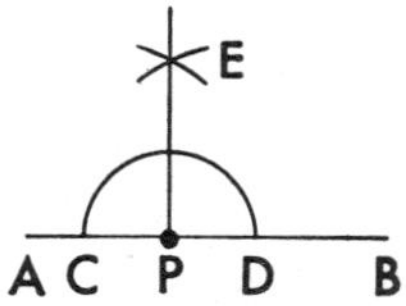

EXERCISES

1. Draw a line that has the same position as each of the following lines, and use a straightedge and compasses to erect a perpendicular at point P:

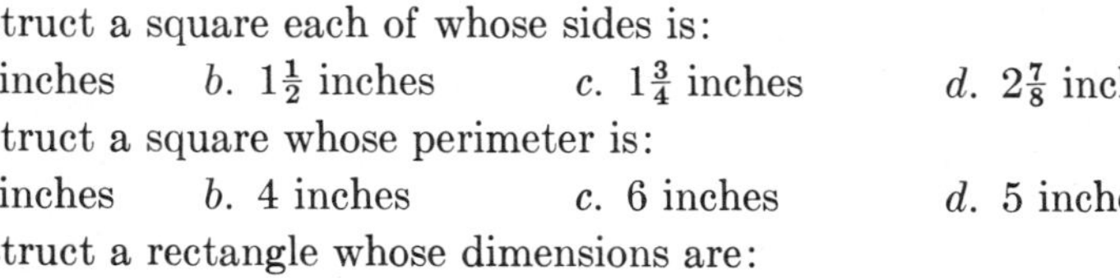

2. Construct a square each of whose sides is:

   a. 2 inches *b.* $1\frac{1}{2}$ inches *c.* $1\frac{3}{4}$ inches *d.* $2\frac{7}{8}$ inches

3. Construct a square whose perimeter is:

   a. 8 inches *b.* 4 inches *c.* 6 inches *d.* 5 inches

4. Construct a rectangle whose dimensions are:

   a. 1″ by 2″ *b.* $1\frac{1}{2}$″ by $\frac{1}{2}$″ *c.* 3″ by $1\frac{3}{4}$″ *d.* $1\frac{1}{4}$″ by $\frac{7}{8}$″

TO CONSTRUCT A LINE PERPENDICULAR TO A GIVEN LINE THROUGH A GIVEN POINT OUTSIDE THE LINE

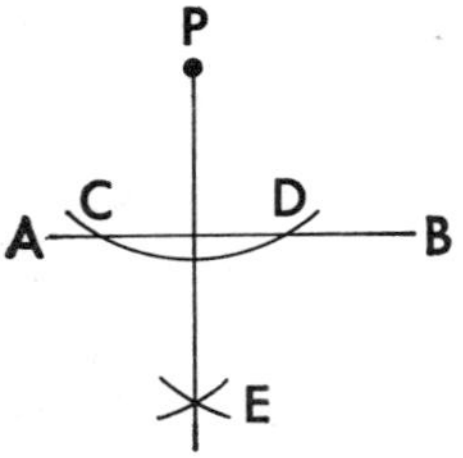

To construct a perpendicular from point P to line AB: (1) With point P as a center and with a sufficiently long radius, draw an arc that intersects line AB at points C and D. (2) Using points C and D as centers and a radius of sufficient length, draw arcs that intersect at point E. (3) Draw line PE, which is perpendicular to line AB.

EXERCISES

1. Draw a line that has the same position as each of the following lines, and use a straightedge and compasses to drop a perpendicular from point P to each line.

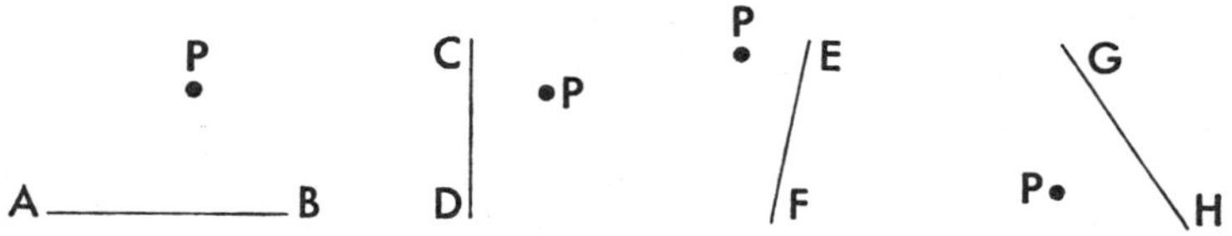

2. Draw a horizontal line AB. Select a point P below line AB. Through point P, construct a line perpendicular to AB.
3. In acute triangle ABC, construct a perpendicular from each vertex to the opposite side. (*Note:* All the perpendiculars should intersect at the same point.)

TO CONSTRUCT AN ANGLE EQUAL TO A GIVEN ANGLE

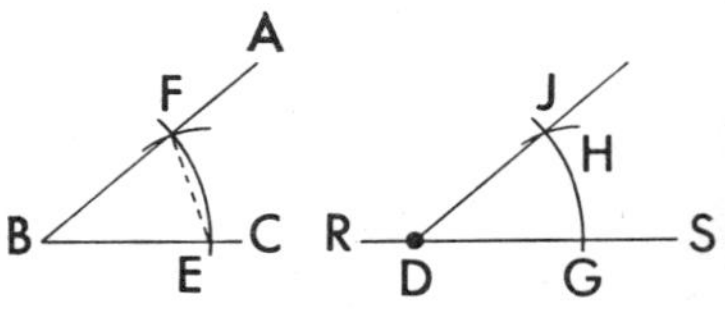

To construct an angle equal to $\angle ABC$ at point D: (1) Through point D, draw line RS. (2) With point B as a center and with any convenient radius, draw an arc that intersects BC at point E and BA at point F. (3) Draw line segment FE. (4) Using point D as a center and the same radius as in step 2, draw arc GJ, which intersects RS at point G. (5) Measure FE with your compasses. With point G as a center and FE as a radius, draw an arc that intersects arc GJ at point H. (6) Draw HD. $\angle HDS = \angle ABC$.

EXERCISES

1. Draw an angle in the same position and having about the same measure as each of the following angles. Then, using a straightedge and compasses, construct an angle equal to each of the angles you have drawn.

2. Draw any two acute angles. Use a straightedge and compasses to construct an angle that is equal to their sum.

3. Draw any triangle.

a. Use a straightedge and compasses to construct an angle that is equal to the sum of the three angles of the triangle.

b. What do you notice about the sum?

TO CONSTRUCT A LINE PARALLEL TO A GIVEN LINE THROUGH A GIVEN POINT

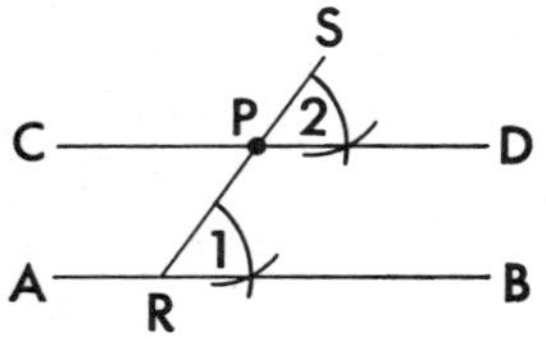

To construct a line parallel to line AB through point P: (1) Through point P, draw a line that intersects line AB at point R. (2) At point P, construct $\angle 2 = \angle 1$, being careful to make $\angle 2$ and $\angle 1$ a pair of corresponding angles. CD passes through point P and is parallel to AB.

EXERCISES

1. Draw a line that has the same position as each of the following lines. Then use a straightedge and compasses to construct through point P a line parallel to each of the given lines.

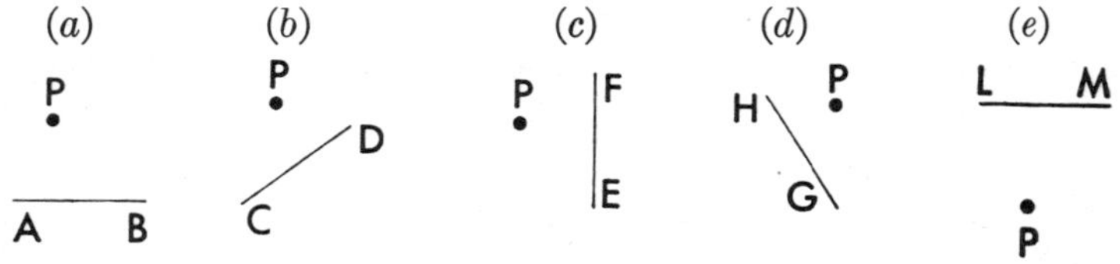

2. *a.* Through point C, construct a line parallel to AB.

b. Through point B, construct a line parallel to AC.

c. Name with the letter D the point where the lines drawn in part a and part b intersect. What type of quadrilateral is $ABDC$? Why?

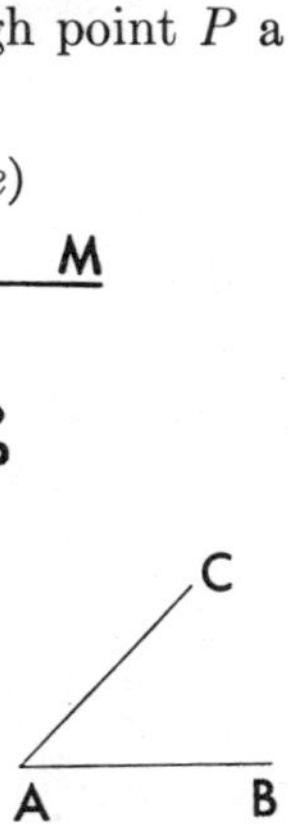

Chapter XII

GRAPHS

I. INTERPRETING AND MAKING PICTOGRAPHS

Pictures are often used in business reports, magazines, newspapers, and posters to present numerical facts clearly, quickly, and colorfully. A graph that uses pictures to show numerical facts is called a ***pictograph.*** Each picture, or ***symbol,*** represents a certain number of the quantity whose story is being told in the graph.

UNITED STATES AIR FORCE PERSONNEL

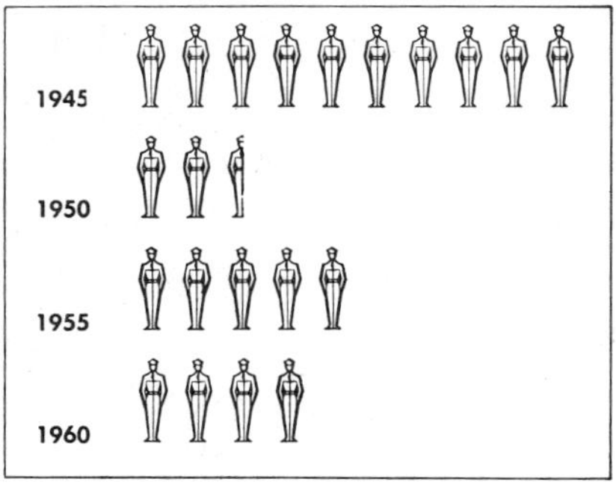

EACH SYMBOL REPRESENTS 200,000 MEN

From the pictograph at the left, we can tell in round numbers how many men were in the Air Force in 1945, 1950, 1955, and 1960. For example, since there are 10 symbols for the year 1945 and each symbol represents 200,000 men, there were $10 \times 200{,}000$ or 2,000,000 men in the Air Force in 1945. Since there are $2\frac{1}{2}$ symbols for the year 1950, there were $2\frac{1}{2} \times 200{,}000$ or 500,000 men in the Air Force in 1950.

From the graph, it is very easy to make comparisons for the years that are pictured. Since there are 10 symbols for 1945 and 5 symbols for 1955, we see that there were twice as many men in the Air Force in 1945 as there were in 1955. Since there are $2\frac{1}{2}$ symbols for 1950 and 5 symbols for 1955, we see that there was an increase of 100% in the strength of the Air Force from 1950 to 1955.

SUGGESTIONS FOR MAKING A PICTOGRAPH

1. Obtain the numerical information required for the graph.
2. If necessary, round off the numerical information.
3. Decide upon a symbol related to the subject of the graph.
4. Indicate on the graph the number that each symbol is to represent.

5. Find out how many symbols are needed to picture each numerical fact.
6. Picture each numerical fact by attractively drawing the required number of symbols.
7. Give the pictograph a title.

EXERCISES

1. Each small ship represents 2,000,000 tons of cargo carried by ship. How many tons are represented by the symbols shown?

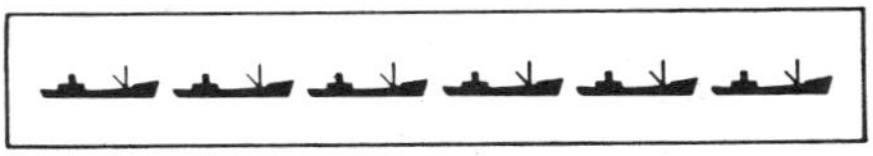

2. Each small horse represents 100,000 horses. How many horses are represented by the symbols shown: 400,500; 405,000; or 450,000?

3. In an army, there are 891,650 men. Round off this number to the nearest hundred thousand. Using a man as a symbol to represent 200,000 men, draw a picture to represent this number.

4. A pictograph may be used to study growth over a period of years. The Atlas Shoe Company showed the growth of its production for the period of 1957–1961 using the pictograph at the right.

PAIRS OF SHOES PRODUCED, 1957-1961

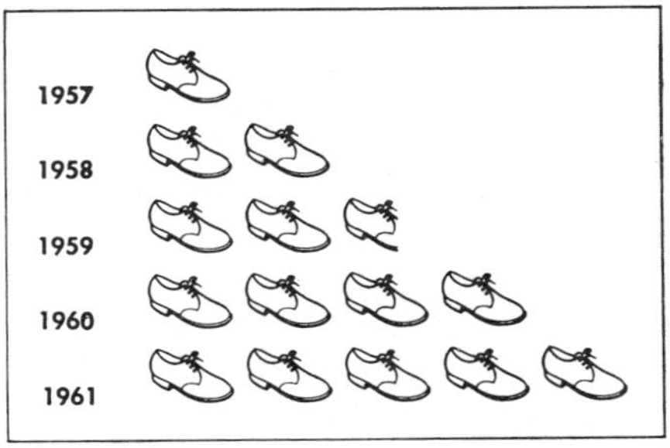

EACH SYMBOL REPRESENTS 200,000 PAIRS OF SHOES

a. How many pairs of shoes were produced during each year?
b. What was the total number of pairs of shoes produced during the five-year period?
c. How many more pairs of shoes were produced in 1960 than in 1959?
d. The number of pairs of shoes produced in 1961 was how many times the number produced in 1959?
e. In 1961, the number of pairs of shoes produced was what per cent of the number of pairs of shoes produced in 1957?

5. The number of eggs produced by a flock of hens in three days was as follows: Monday—9 dozen, Tuesday—12 dozen, Wednesday—10 dozen. Make a pictograph to represent this information.

6. The number of motor vehicles registered in the United States, rounded off to the nearest 5,000,000, was as follows: 1945—30,000,000; 1950—50,000,000; 1955—65,000,000; 1960—75,000,000. Make a pictograph to represent this information.

2. INTERPRETING AND MAKING BAR GRAPHS

To show comparisons of numerical facts about different things, or to show numerical facts about one thing at different times, we can use a ***bar graph.*** In such a graph, the length of a bar is used to represent a numerical fact. The length of the bar depends upon the size of the number that is to be represented and the number that each unit on the ***scale*** represents. The scale, which always appears on the graph, must begin with zero. A graph on which the bars appear in a vertical position is called a ***vertical bar graph.*** A graph on which the bars appear in a horizontal position is called a ***horizontal bar graph.***

Dan had a 10-question arithmetic test on each school day during one week. At the end of the week, he made a graph of the number of problems he had solved correctly. From the graph, he was able to learn the following information:

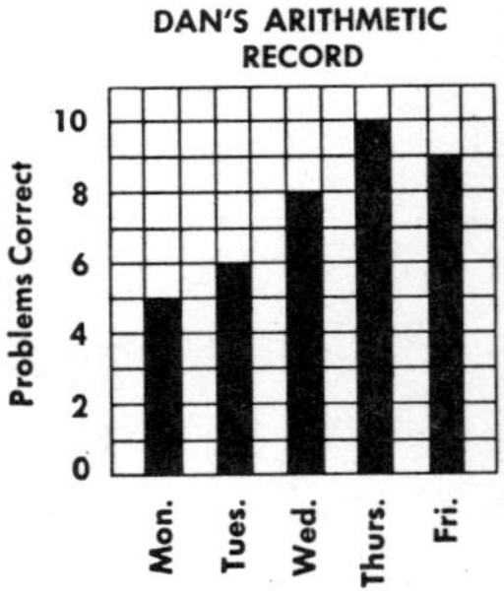

1. On Tuesday, Dan solved 6 questions correctly.
2. Thursday was Dan's best day. Monday was his worst day.
3. During the week, Dan solved 38 questions correctly.
4. On Thursday, Dan correctly solved twice as many problems as he did on Monday.
5. On Tuesday, Dan correctly solved $\frac{6}{8}$, or 75%, of the number of problems he correctly solved on Wednesday.

SUGGESTIONS FOR MAKING A BAR GRAPH

1. Obtain the numerical information required for the graph.
2. If necessary, round off the numerical information.
3. Decide whether to make a vertical or horizontal bar graph.
4. Find a suitable scale to use.
5. Starting at zero, mark off and label this scale on graph paper.
6. Draw the proper bars all having the same width. Make the graph as attractive as possible.
7. Give the graph a title.

EXERCISES

1. What amount of money is represented by the bar in the diagram at the right?

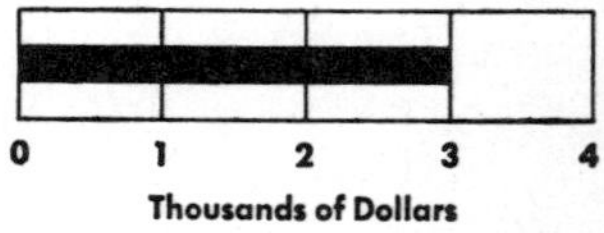

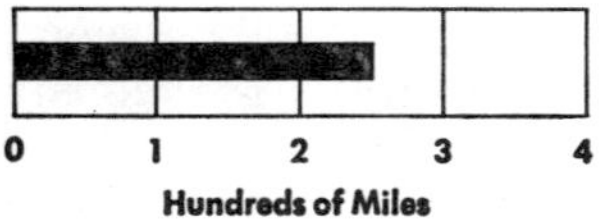

2. What number of miles is represented by the bar in the diagram at the left: 205, 250, or 225?

3. The scale on a bar graph is "$\frac{1}{4}$ inch represents 1000 people." How long should a bar be made to represent 8000 people?
4. The graph below shows the number of acres of different vegetables that a farmer planted during one season.

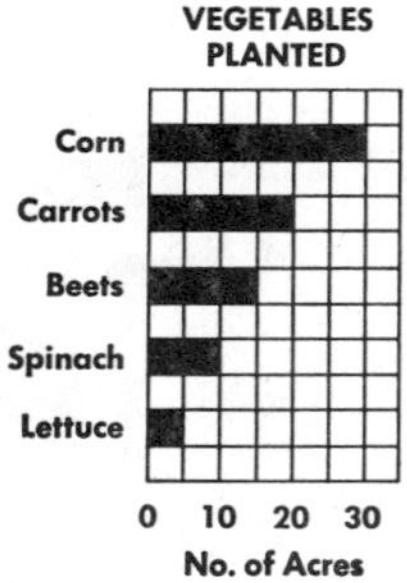

   a. How many acres of carrots were planted?
   b. Which vegetable had the greatest number of acres planted?
   c. Which vegetable had the smallest number of acres planted?
   d. What was the ratio of the number of acres of beets planted to the number of acres of spinach planted?
   e. What was the total number of acres planted?
   f. What per cent of the total number of acres planted was the number of acres of carrots planted?
5. In 1961, a large aircraft manufacturing company used the graph at the right to report to its stockholders how it had used each dollar of sales income during 1960.

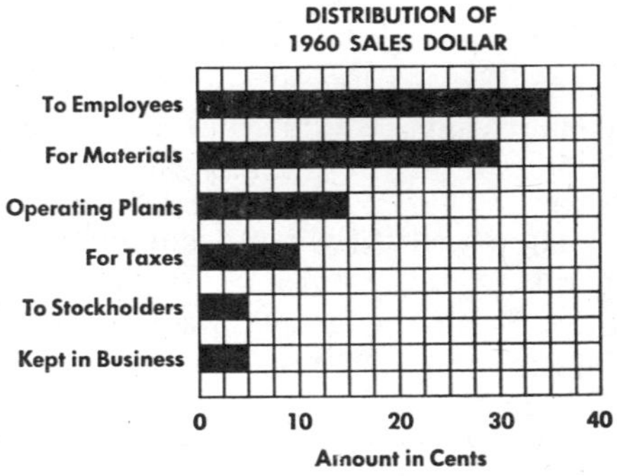

   a. How many cents of each dollar of the company's sales went to its employees?
   b. How many more cents of each sales dollar went to the employees than went for materials?
   c. What was the total number of cents of each sales dollar that went to the employees, for materials, and for operating the plants?
   d. The amount that went for taxes was how many times the amount that went to the stockholders as dividends?
   e. What per cent of each sales dollar was retained in the business?
6. In a spelling test of twenty words given in a class, four students spelled the following numbers of words correctly: Alan, 14; Robert, 20; Thomas, 19; Stanley, 16. Make a vertical bar graph to represent this information.
7. Four schools took part in a track meet. Winfield High scored 42 points; Bridgton, 18; Houston, 27; Padway, 13. Make a horizontal bar graph to represent this information.

8. Harriet received the following marks at midterm: English, 90%; science, 75%; mathematics, 85%; history, 80%. Make a bar graph to compare her marks in these four subjects.

9. Make a bar graph to represent the heights of the following five dams: Boulder, 730 ft.; Kensico, 310 ft.; Shasta, 600 ft.; Grand Coulee, 450 ft.; Gatun, 120 ft.

10. The number of automobile factory sales made in the United States from 1957 to 1960 was as follows:

1957—7,220,520	1959—6,728,629
1958—5,135,106	1960—7,869,271

Round these numbers off to the nearest hundred thousand. Make a bar graph to show the approximate number of automobiles made in the years mentioned. [Use $\frac{1}{16}$ in. to represent 100,000 automobiles.]

3. INTERPRETING AND MAKING LINE GRAPHS

To show how something such as temperature or population changes, we can use a ***broken line graph,*** which is also called a ***line graph.*** In a line graph, we use one or more lines to represent the changes that take place in numerical facts. A rising line in a graph clearly shows that the quantity being studied is increasing. A falling line in a graph shows that the quantity is decreasing. A line graph also pictures the trend that the changing quantity is following.

The graph at the right shows how the outdoor temperature changed from 11 A.M. to 4 P.M. on a certain day. From the graph, we can learn the following information:

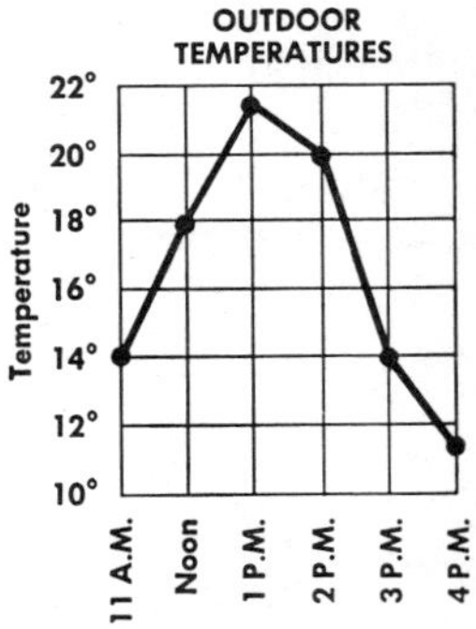

1. We can find the temperature at any hour of the period studied. For example, at noon the temperature was 18°.
2. The lowest temperature occurred at 4 P.M., the highest at 1 P.M.
3. Between the hours of 11 A.M. and 1 P.M., the temperature was rising.
4. Between the hours of 1 P.M. and 4 P.M., the temperature was falling.
5. Between 11 A.M. and noon, the temperature rose 4°.
6. Between the hours of 2 P.M. and 3 P.M., the temperature fell more rapidly than it did between 1 P.M. and 2 P.M.

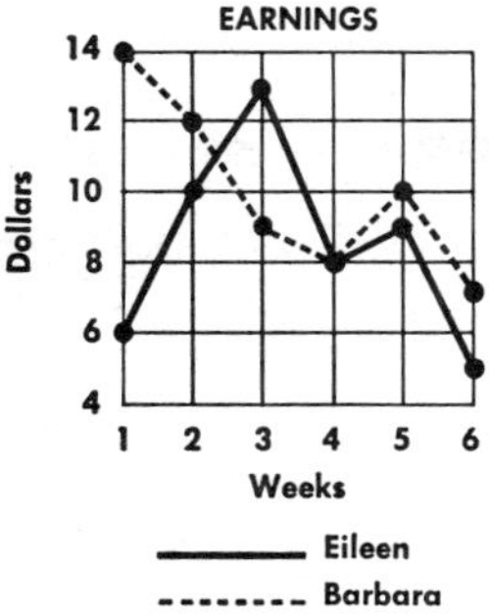

The graph at the left pictures the record of Eileen and Barbara's earnings during a period of six weeks. The solid line on the graph represents Eileen's earnings and the dotted line represents Barbara's earnings. From the graph, we can learn the following information:

1. During weeks 1, 2, 5, and 6, Barbara's earnings were more than Eileen's earnings.
2. During the 4th week, Barbara and Eileen earned the same amount of money, $8.
3. In the first week, the difference between their earnings, $8, was greater than the difference in any other week.
4. Barbara's greatest salary, $14, was $1 more than Eileen's greatest salary, $13.

SUGGESTIONS FOR MAKING A LINE GRAPH

1. Obtain the numerical information required for the graph.
2. If necessary, round off the numerical information.
3. Plan the vertical and horizontal scales so that the graph will fit the paper.
4. Mark off and label the vertical and horizontal scales. The scale need not begin at zero.
5. Place a point on the graph for each set of numbers in the given numerical information.
6. Draw straight lines between consecutive points, making the graph as attractive as possible.
7. Give the graph a title.

EXERCISES

1. The graph on the right shows the cash dividends per share that a large company paid to its stockholders during the years 1952 through 1961.

a. About how much was the dividend per share during 1957?

b. In what year was the lowest dividend paid?

c. Between which two years was there a decrease in the dividend?

d. During which two years was the dividend increase greatest?

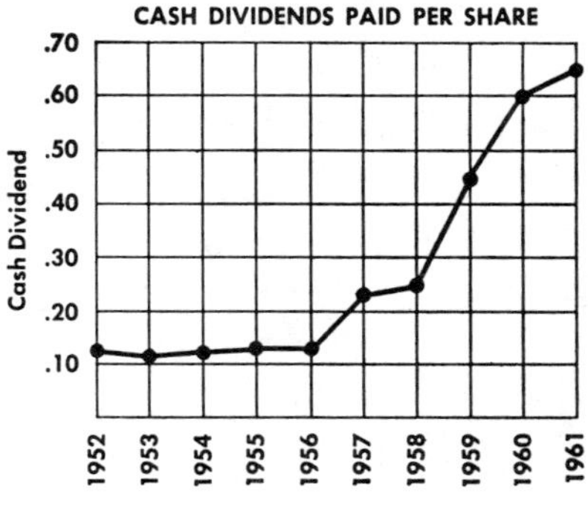

e. The dividend in 1959 was approximately how many times the dividend in 1956?

f. The dividend in 1955 was approximately what per cent of the dividend in 1960?

g. If a man received the dividends on a share of stock for the years 1957 through 1961 inclusive, approximately how much money did he receive?

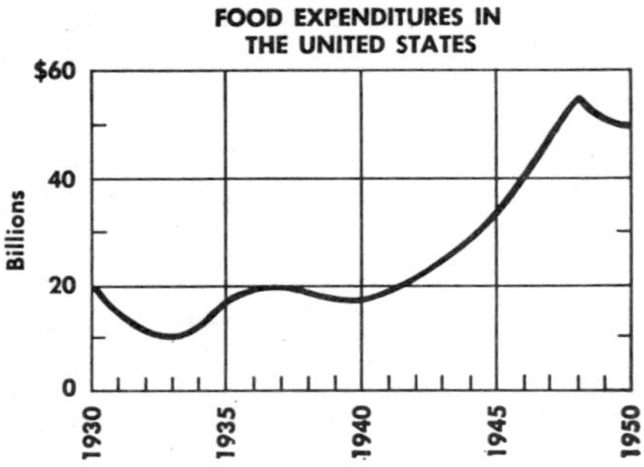

2. The graph on the left shows food expenditures in the United States for each of the years 1930 to 1950 inclusive.

a. During what year were expenditures for food the lowest?

b. During what year were expenditures for food the highest?

c. Approximately how many billions of dollars were spent for food in 1950?

d. In what year did expenditures for food first reach 40 billion dollars?

e. During which one of these five-year periods (1930–35, 1935–40, 1940–45, 1945–50) did expenditures for food remain about the same?

3. To show what had happened to employment costs during the period of 1952–1961, a company prepared the graph on the right.

a. Approximately how many employees did the company have in 1952? In 1961?

b. Approximately how much did the company pay in salaries, wages, and benefits in 1952? In 1961?

c. The number of employees in 1961 was approximately how many times the number of employees in 1952?

d. The employment costs during the year 1961 were approximately how many times the employment costs in 1952?

e. Between which two years did the employment costs increase although the number of employees decreased?

4. Martin's marks in history during the month of November were: Nov. 1, 50%; Nov. 8, 65%; Nov. 15, 75%; Nov. 22, 70%; Nov. 29, 85%. Make a line graph to show this record.

5. During the first six months of a year, the Supreme Sales Corp. reported the following sales: Jan., \$55,000; Feb., \$40,000; March, \$60,000; Apr., \$75,000; May, \$100,000; June, \$85,000. Make a line graph to show these sales.

6. On a certain day in January, the Weather Bureau recorded the following temperatures:

Hour	12 NOON	1 P.M.	2 P.M.	3 P.M.	4 P.M.	5 P.M.	6 P.M.
Temp.	20°	24°	25°	22°	17°	13°	10°

a. Make a line graph of these temperatures.
b. Between what hours was the temperature rising?
c. Between what hours was the temperature falling?

7. The population of the United States from 1880 to 1960, correct to millions, is given below. Make a line graph to show this population growth.

1880, 50 million; 1890, 63 million; 1900, 76 million;
1910, 92 million; 1920, 106 million; 1930, 123 million;
1940, 132 million; 1950, 151 million; 1960, 179 million.

8. In the winter of 1959–1960, the amount of snowfall in inches in New York City was as follows:

Oct.—0.0 Nov.—0.5 Dec.—8.0 Jan.—2.6
Feb.—1.6 Mar.—16.0 Apr.—1.0 May—0.0

a. Make a line graph to show this record.
b. In which month was the snowfall greatest?
c. What was the total snowfall in the winter of 1959–1960?

4. INTERPRETING AND MAKING CIRCLE GRAPHS

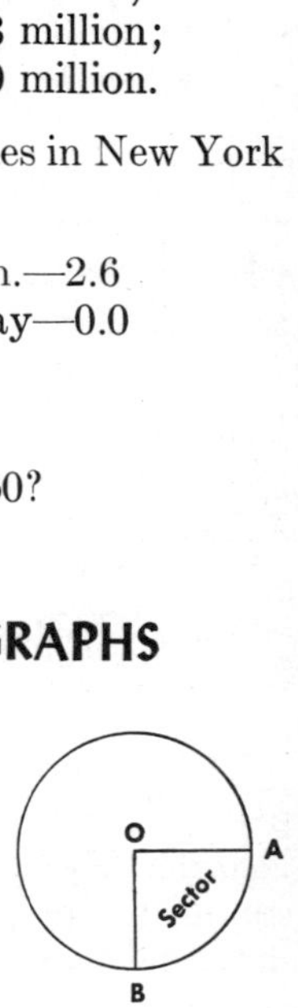

To show how a certain quantity such as a class of pupils or a budget has been divided into several parts, as well as to show the comparison between these parts, we can use a ***circle graph.*** In such a graph, a circle is used to represent the whole quantity. Each of the parts into which the whole quantity has been divided is represented by a part of the circle, called a sector. A ***sector*** of a circle is a part of a circle whose boundaries are two radii and the arc that they cut off on the circumference of the circle. The angle formed by the two radii, $\angle AOB$, is called a ***central angle.***

In a circle graph, the ratio of a sector to the whole circle must be equal to the ratio of the quantity that the sector represents to the whole quantity. The size of the sector is determined by the number of degrees in its central angle. To discover the number of degrees in the central angle for each sector of the graph, we find the fractional part of the whole quantity that each sector is supposed to represent. We then find this fractional part of 360°.

The graph at the left was drawn after determining the ages of the 36 boys in an 8th-grade class. From the graph, we can discover the following information:

1. If we measure the central angle for the 13-years' sector, we find it to be 90°. Therefore, the central angle for the 14-years' sector must be 360° − 90°, or 270°.
2. Since the 13-years' sector is $\frac{90}{360}$ or $\frac{1}{4}$ of the circle, $\frac{1}{4}$ or 25% of the boys in the class are 13 years old. Since the 14-years' sector is $\frac{270}{360}$ or $\frac{3}{4}$ of the circle, $\frac{3}{4}$ or 75% of the boys in the class are 14 years old.
3. Since there are 36 boys in the class, $\frac{1}{4}$ of 36, or 9 boys, are 13 years old. Also, $\frac{3}{4}$ of 36, or 27 boys, are 14 years old.
4. The ratio of the number of boys who are 14 years old to the number of boys who are 13 years old is 270°:90°, or 3:1.

The graph at the right shows the uses to which a merchant put each sales dollar in his business. During the year, the total sales were $120,000. From the graph, we can discover the following information:

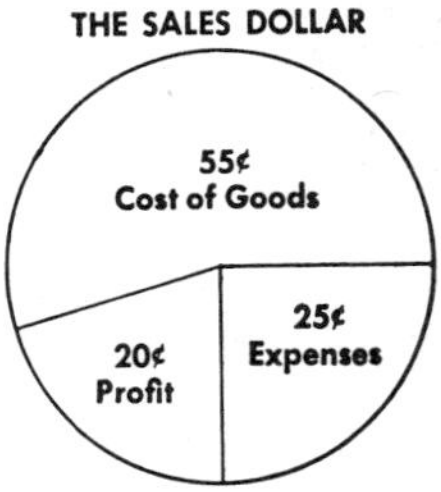

1. The expenses were 25 cents out of $1.00, or 25% of each sales dollar. Also, profit was 20% and cost of merchandise was 55% of each sales dollar.
2. During the year, the expenses were 25% of $120,000, or $30,000. The cost of merchandise was 55% of $120,000, or $66,000. The profit was 20% of $120,000, or $24,000.

The students in a health education class voted as follows in selecting their play-period activity: baseball, 18; basketball, 20; swimming, 12; track, 10. In order to present this information to the class, the members of the activities committee prepared a circle graph on a large poster. First they worked out the table shown on the next page.

Activity	*No. of Pupils*	*Part of Total*	*No. of Degrees in Central Angle*
Baseball	18	$\frac{18}{60}$	$\frac{18}{60} \times 360° = 108°$
Basketball	20	$\frac{20}{60}$	$\frac{20}{60} \times 360° = 120°$
Swimming	12	$\frac{12}{60}$	$\frac{12}{60} \times 360° = 72°$
Track	10	$\frac{10}{60}$	$\frac{10}{60} \times 360° = 60°$
Total	60	$\frac{60}{60}$	360°

PLAY PERIOD ACTIVITIES

Then they drew a large circle in which they made for each activity a sector with the proper number of degrees in the central angle.

SUGGESTIONS FOR MAKING A CIRCLE GRAPH

1. Obtain the numerical information required for the graph.
2. If necessary, round off the numerical information.
3. Make a table that will enable you to find the number of degrees in each sector of the circle.
4. Draw a circle large enough to allow you to fit in all the parts.
5. Using a protractor, measure the central angle for each sector.
6. Label each sector, making the graph as attractive as possible.
7. Give the graph a title.

EXERCISES

1. Tell what fraction and what per cent of the circle each of the following unshaded sectors is:

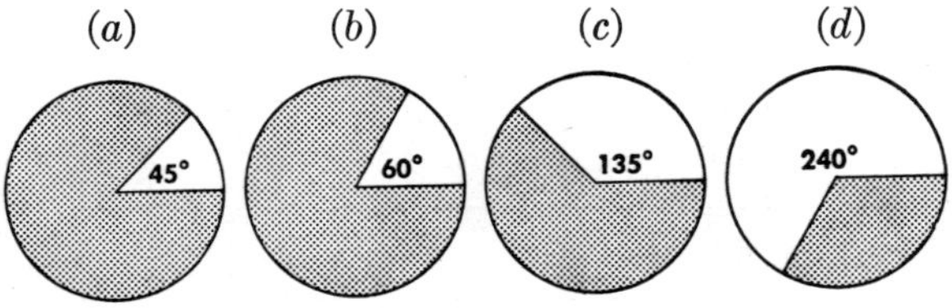

2. Mr. Baxter made a circle graph to show the monthly family budget. He drew a sector of 90° to show the rent. What per cent of the total amount to be spent during the month did he plan to spend for rent?

3. The circle graph below shows how Margaret planned to spend her \$6 weekly allowance.

MARGARET'S BUDGET

a. Use a protractor to measure the central angle of each sector.
b. What fraction of her weekly allowance did she plan to spend on each item in her budget?
c. How much did she plan to spend on each item in her budget?

4. The safety council in a community published the graph at the right to show where accidents to school children occur. Use the graph to answer the following questions:

ACCIDENTS TO SCHOOL CHILDREN

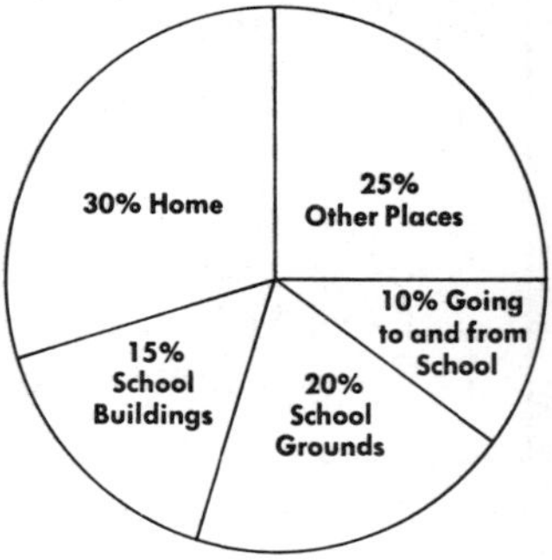

a. Where do the greatest number of accidents to school children occur?
b. What fractional part of the accidents to school children occur on school grounds?
c. How many times as often do accidents occur in the home as in school buildings?
d. How many degrees of the circle should be used to represent accidents in other places?
e. What per cent of accidents occur on school property?

5. A circle is to be divided into 3 sectors that are to be $\frac{1}{4}$, $\frac{1}{3}$, and $\frac{5}{12}$ of the circle. Find the number of degrees in the central angle of each sector.
6. A circle is to be divided into 4 sectors that are to be $33\frac{1}{3}\%$, $16\frac{2}{3}\%$, 20%, and 30% of the circle. How many degrees are there in the central angle of each sector?
7. The books in Mr. Hartley's home library are distributed as follows: fiction, $33\frac{1}{3}\%$; science, 25%; history, $16\frac{2}{3}\%$; plays, 15%; music, 10%. Use a circle whose radius is 2 inches to make a circle graph which shows the distribution of Mr. Hartley's books.
8. Mr. Stolz plans the following budget: rent, 20% of his income; food, 30%; expenses, 40%. He expects to save the rest of his income.
   a. Make a circle graph to show Mr. Stolz's budget.
   b. What per cent of his income does he expect to save?
9. Mr. Robbins found that he had the following expenses in running his house for one year: taxes, \$350; interest, \$100; heating, \$250; insurance, \$100; repairs, \$280; miscellaneous, \$120. Make a circle graph to show the distribution of these expenses.
10. During one year, the causes of fatal automobile accidents were: inattention, 30%; speeding, 28%; traffic violations, 25%; intoxication, 8%; miscellaneous, 9%. Make a circle graph to represent this information.
11. In one year, the world production of grain crops was as follows, in millions of bushels: corn, 5000; wheat, 6000; oats, 5000; barley, 3000; rye, 1000. Make a circle graph to represent this information.

5. INTERPRETING AND MAKING RECTANGLE GRAPHS

To show how a certain quantity has been divided into several parts, we can also use a ***rectangle graph***, sometimes called a ***divided bar graph***. In this type of graph, the whole quantity is represented by a rectangle, or a bar.

The graph below shows the distribution of the ages of the boys in an 8th-year class.

OUR AGES

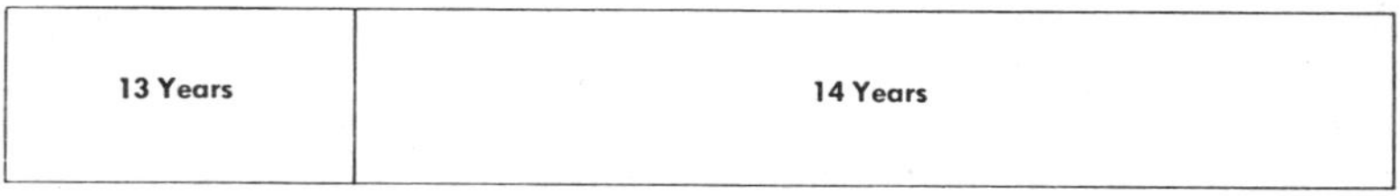

From the above rectangle graph, we can discover the following information:

1. Since the whole rectangle, which represents the number of pupils in the entire class, is 4 inches long and the 13-years' rectangle is 1 inch long, the number of 13-year-old boys in the class is $\frac{1}{4}$, or 25%, of the number of boys in the entire class.
2. Similarly, since the 14-years' rectangle is 3 inches long, the number of 14-year-old boys is $\frac{3}{4}$, or 75%, of the number of boys in the entire class.

Notice that the rectangle graph and the circle graph are used for the same purposes and are interpreted in the same manner. See how the above facts were represented by a circle graph on page 298.

At an airfield, $16\frac{2}{3}\%$ of the planes were fighters, $33\frac{1}{3}\%$ of the planes were bombers, and 50% of the planes were pursuit planes. In order to show how the total number of planes at the airfield was distributed by type, a rectangle graph was made in which the total number of planes at the airfield was represented by a rectangle 3 inches long. A length of 3 inches was used because $16\frac{2}{3}\%$, $33\frac{1}{3}\%$, and 50% of 3 inches would be convenient measurements to make with a ruler.

First the following table was prepared:

Type of Plane	*Per Cent*	*Length of Rectangle*
Fighters	$16\frac{2}{3}$	$16\frac{2}{3}\%$ of 3 in. = $\frac{1}{2}$ in.
Bombers	$33\frac{1}{3}$	$33\frac{1}{3}\%$ of 3 in. = 1 in.
Pursuit Planes	50	50% of 3 in. = $1\frac{1}{2}$ in.
Total	100%	3 in.

Then a 3-in. rectangle was divided to make the following graph:

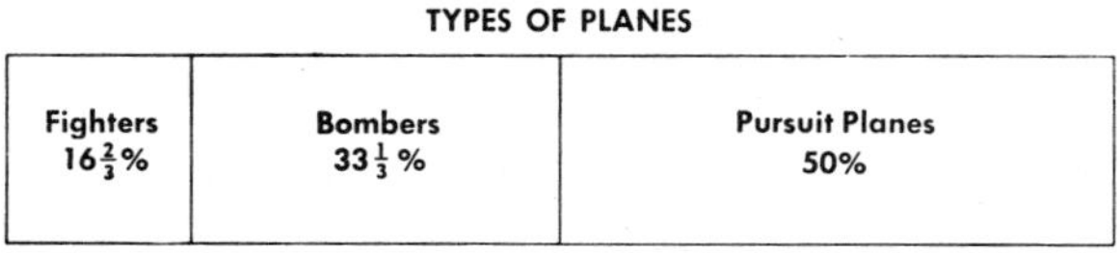

SUGGESTIONS FOR MAKING A RECTANGLE GRAPH

1. Obtain the numerical information required for the graph.
2. Select a convenient length for the rectangle that is to represent the whole quantity.
3. Make a table that will enable you to find the lengths of the rectangles into which the whole rectangle is to be divided.
4. Draw the rectangle that is to represent the whole quantity.
5. Using a ruler, measure off the lengths of the rectangles that are to represent the parts of the whole quantity.
6. Draw and label each rectangle, making the graph as attractive as possible.
7. Give the graph a title.

EXERCISES

1. Mr. Connor invested his money in stocks, bonds, and mortgages. The graph below represents the distribution of his investments. Use the graph to answer the following questions:

MR. CONNOR'S INVESTMENTS

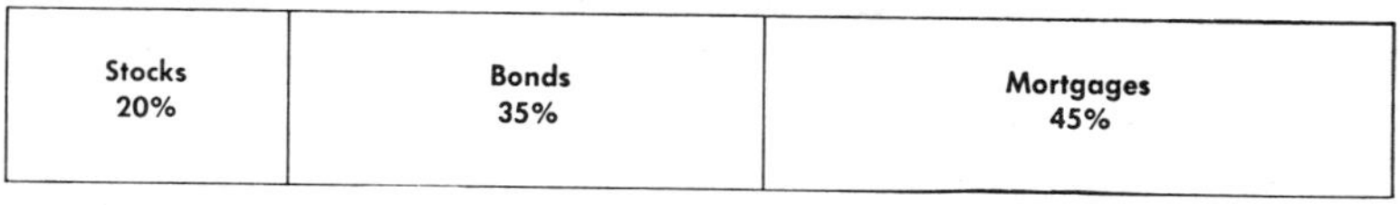

a. What per cent of Mr. Connor's investments are pictured in the graph?

b. What fraction of his investments is in stocks?

c. What fraction is nearest to the part he invested in bonds: $\frac{2}{5}$, $\frac{1}{3}$, or $\frac{3}{8}$?

d. If Mr. Connor invested a total of \$50,000, how much did he invest in stocks? in bonds? in mortgages?

2. In the Southdown School, every pupil is required to join a club. In a class of 36 pupils, 12 joined the Sports Club, 9 joined the Science Club, 9 joined the Music Club, and the rest joined the Art Club. Draw a rectangle graph to represent these facts.

3. One year, the United States Bureau of the Budget estimated that, correct to the nearest 5 cents, the average dollar that the government would receive during the next fiscal year would come from the following sources: individual income taxes—50 cents; customs and other receipts—10 cents; corporation income taxes—25 cents; excise taxes—15 cents. Make a rectangle graph to represent these facts.

4. The Bright Copper Company reported that 30% of its total sales was used to pay for materials, 40% for labor, 20% for expenses, 5% for taxes, and the remainder was profits. Using graph paper, make a rectangle graph to represent these facts.

6. GRAPHING A FORMULA

We have learned that a formula can be used to express a relationship between two quantities. For example, $D = 20t$ represents the relationship between t, the number of hours a car travels, and D, the number of miles the car travels in that time, if the car is traveling at an average rate of 20 miles per hour. We have also learned that if we assume a value for t, we can substitute that value in the formula $D = 20t$ and find the corresponding value of D as shown on the next page.

If $t = 0$, $D = 20 \times 0$, $D = 0$.
If $t = 1$, $D = 20 \times 1$, $D = 20$.
If $t = 2$, $D = 20 \times 2$, $D = 40$.
If $t = 3$, $D = 20 \times 3$, $D = 60$.
If $t = 4$, $D = 20 \times 4$, $D = 80$.

These values can be placed in a table as shown at the right.

t	0	1	2	3	4
D	0	20	40	60	80

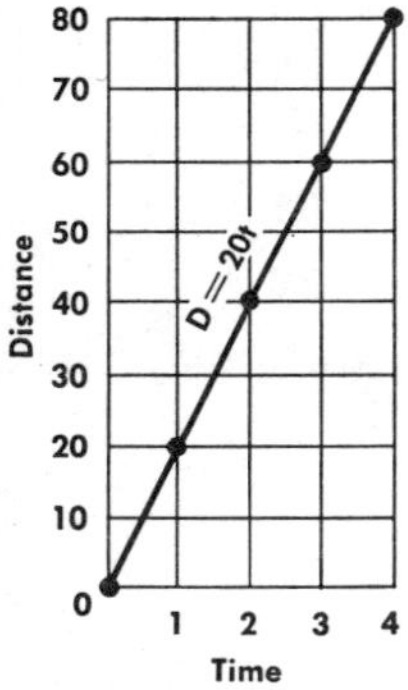

Now we can draw a pair of perpendicular lines called ***axes.*** On the horizontal axis, we can mark off a scale for the number of hours traveled, the *time*. On the vertical axis, we can mark off a scale for the number of miles traveled, the *distance*. We use each pair of numbers in the table to locate a point on the graph. We get the first point on the graph by plotting the set of values $t = 0$, $D = 0$; the second point by using the pair of values $t = 1$, $D = 20$, etc. We draw a straight line that passes through all the points on the graph. The graph that we have drawn pictures clearly the relationship expressed by the formula $D = 20t$.

SUGGESTIONS FOR GRAPHING A FORMULA

1. Make a table by assuming values for one of the literal numbers and substituting these values in the formula to find the corresponding values of the other literal number.
2. Draw and label a horizontal axis and a vertical axis.
3. Select a convenient scale on each axis.
4. Use the pairs of corresponding values in the table to locate the points on the graph.
5. Draw a line through the points on the graph.
6. Write the formula above the line.

EXERCISES

1. The formula for the total cost, C, of n yards of linoleum that cost \$4 per yard is $C = 4n$.

a. Make a table from this formula by giving n values from 0 to 5 inclusive.
b. Make a graph of the formula $C = 4n$.

2. The formula for the perimeter of an equilateral triangle whose side is s is $P = 3s$.

a. Make a table from this formula by giving s values from 0 to 10 inclusive.

b. Make a graph of the formula $P = 3s$.

3. The annual interest, I, that is earned on P dollars invested at 5% is given by the formula $I = .05P$.

a. Make a table from this formula by giving P the values \$0, \$100, \$200, \$300, \$400, and \$500.

b. Make a graph of the formula $I = .05P$.

4. The formula for the distance, D, traveled in t hours by a car whose average rate is 30 miles per hour is $D = 30t$.

a. Make a table from this formula.

b. Make a graph of the formula $D = 30t$.

5. The formula for the number of feet, f, that are contained in y yards is $f = 3y$.

a. Make a table from this formula.

b. Make a graph of the formula $f = 3y$.

Appendix

MODERN MATHEMATICS FOR ENRICHMENT

I. A NUMERAL SYSTEM WHOSE BASE IS FOUR

Suppose that on a distant planet lived people who had only four fingers. Instead of counting by tens as we do, these people might count by fours. They might develop a numeral system whose base is four. Such a system could be called the ***quaternary system,*** from the Latin word "quattuor," which means *four*. Let us see how these people would represent numbers, how they would count, and how they would perform arithmetic operations in their numeration system.

WRITING NUMERALS USING BASE FOUR

When we represented numbers by numerals in base ten, or the decimal system, we worked with groups of ten. In order to represent numbers by numerals in base four, or the quaternary system, we will work with groups of four.

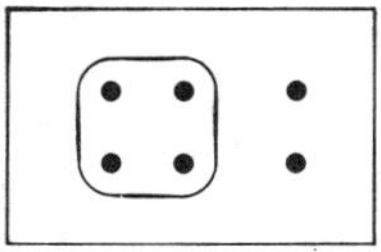

At the left, the dots have been grouped so that there is one group of four dots and two more dots left. The number of dots is represented by the numeral 12_{four}. In this numeral, the 1 shows that there is one group of four, and the 2 shows that there are two ones left. The "lowered" four, also called the "subscript" four, shows that the base four is being used.

The numeral 12_{four} is read as "one, two, base four." We must be careful not to read it as "twelve," because that would mean one group of ten and two ones.

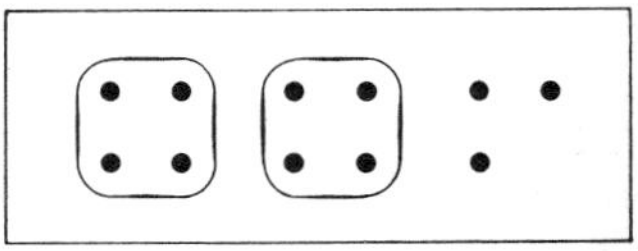

At the right, the dots have been grouped so that there are two groups of four dots and three more dots left. The number of dots is represented by the numeral 23_{four}. In this numeral, the 2 shows that there are two groups of four, and the 3 shows that there are three ones left. The numeral 23_{four} is read "two, three, base four."

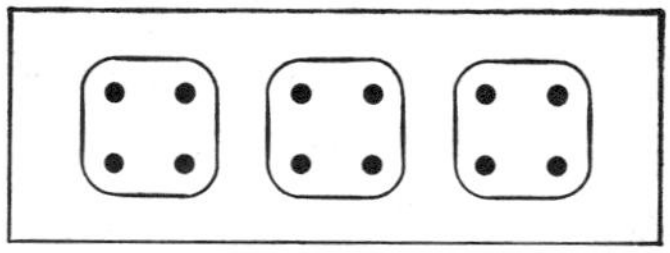

At the right, the dots have been grouped so that there are three groups of four dots, and there are none left. The number of dots is represented by the numeral 30_{four}. In this numeral, the 3 shows that there are three groups of four, and the 0 shows that there are no ones left. The numeral 30_{four} is read, "three, zero, base four."

Notice that when we group in fours, only zero, one, two, or three individual objects can be left. Therefore, we need symbols for only those numbers. We use 0, 1, 2, 3, the same symbols that are used in the decimal system. We will soon see that no other symbols are needed to represent all numbers in the quaternary system.

EXERCISES

In 1–4, copy the set of dots:

a. Draw lines around groups of four.
b. State how many groups of four there are and how many ones are left.
c. Represent the number of dots in the set by a numeral using base four.

1. 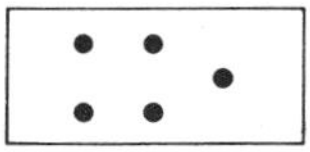

2. 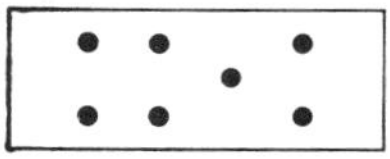

3. 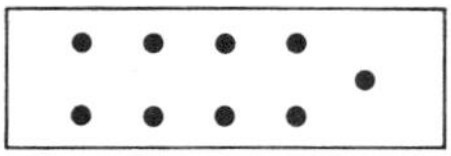

4. 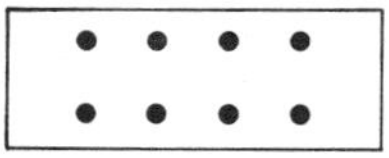

5. At the right, the set of dots has been grouped in tens.

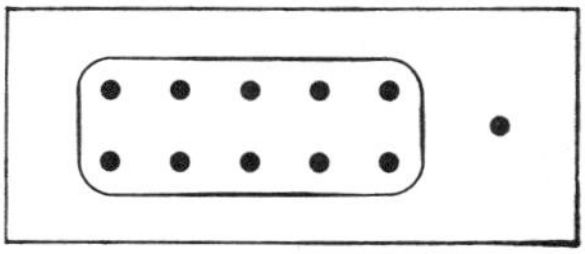

a. State how many groups of ten there are and how many ones are left.
b. Represent the number of dots in the set by a numeral using base ten.

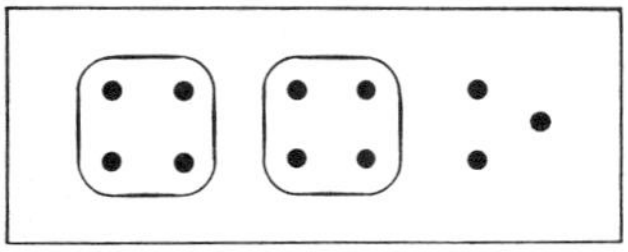

6. At the left, the same set of dots that appears in the previous question has been grouped in fours.

a. State how many groups of four there are and how many ones are left.
b. Represent the number of dots in the set by a numeral using base four.

[*Note:* The answers for exercises 5*b* and 6*b* explain why $11_{\text{ten}} = 23_{\text{four}}$. The base ten numeral and the base four numeral are different symbols for the same number.]

In 7–10, copy the set of dots:

a. Draw lines around groups of ten.
b. State how many groups of ten there are and how many ones are left.
c. Represent the number of dots in the set by a decimal numeral.
d. Draw lines around groups of four.
e. State how many groups of four there are and how many ones are left.
f. Represent the number of dots in the set by a quaternary numeral.
g. Use the results obtained in part *c* and part *f* to write an equation stating that the number of dots represented by the decimal numeral is equal to the number of dots represented by the quaternary numeral.

7. 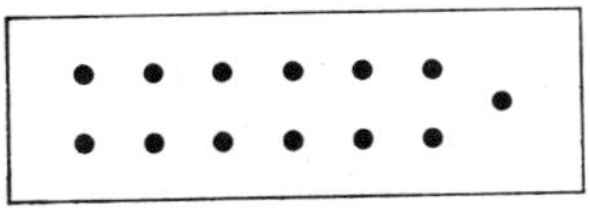

8. 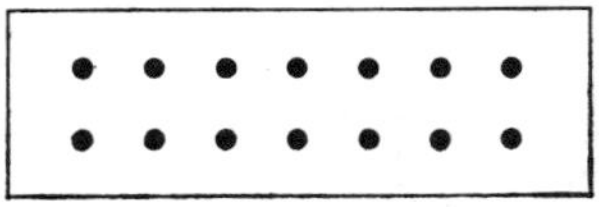

9. 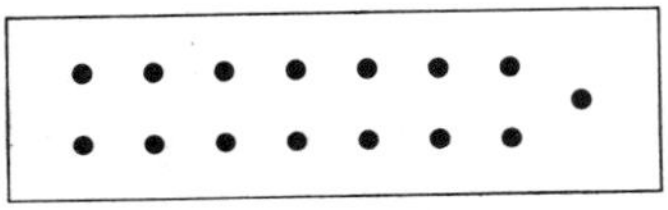

10. 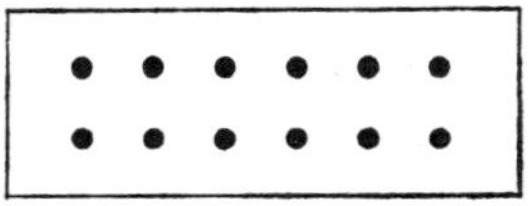

COUNTING IN BASE FOUR

We know that when we count in base ten, we use the number names "one," "two," "three," "four," "five," "six," "seven," "eight," and "nine" and their corresponding number symbols, or numerals, 1, 2, 3, 4, 5, 6, 7, 8, and 9. The number ten, which is the base, is represented by the number name "ten" and by the numeral 10. In all, only the ten number symbols 0, 1, 2, 3, 4, 5, 6, 7, 8, and 9 are needed to represent any number in base ten.

Now let us learn how to count in base four, or the quaternary system. Below are sets of dots arranged so that each set after the first has exactly one more dot than the set before it. At the right of each set are the number name in base four representing the number of dots in the set and the numeral in base four representing that number.

SET OF DOTS	NUMBER NAME	NUMERAL
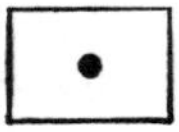	one	1_{four}
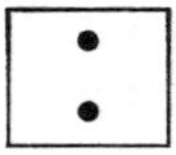	two	2_{four}
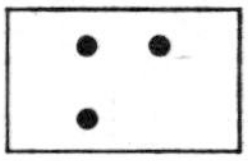	three	3_{four}

SET OF DOTS	NUMBER NAME	NUMERAL
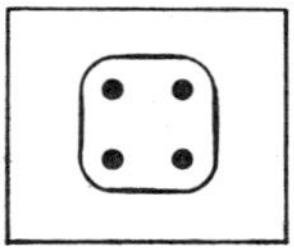	one, zero	10_{four}
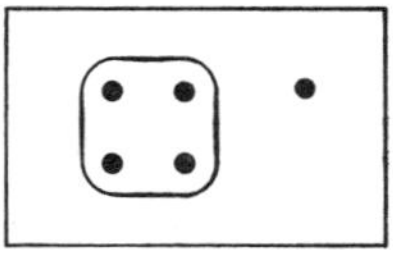	one, one	11_{four}
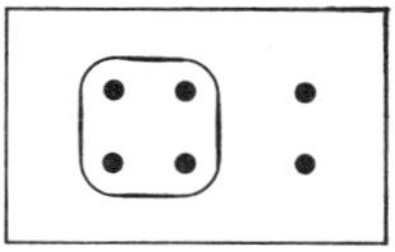	one, two	12_{four}
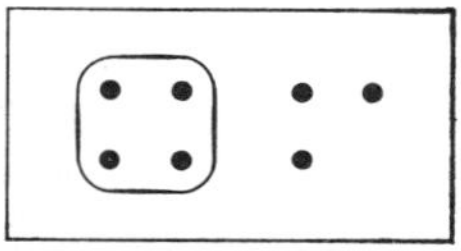	one, three	13_{four}
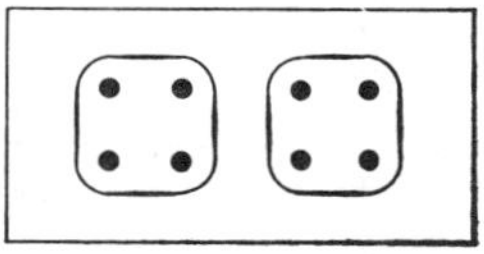	two, zero	20_{four}
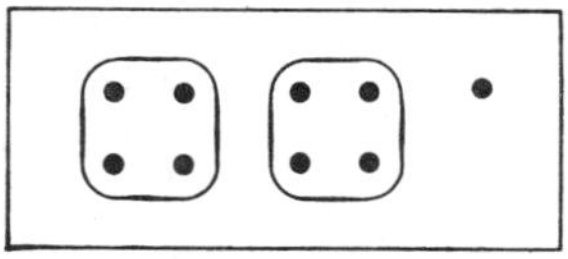	two, one	21_{four}
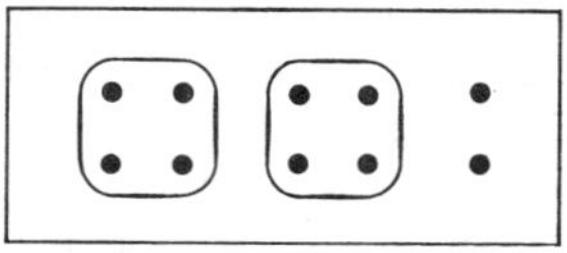	two, two	22_{four}
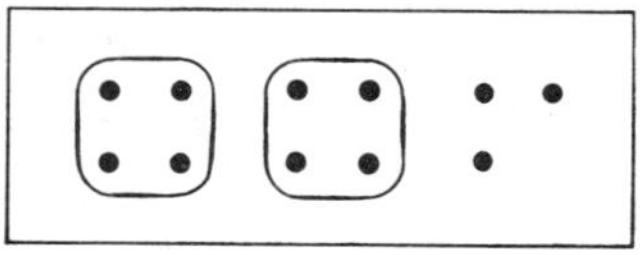	two, three	23_{four}
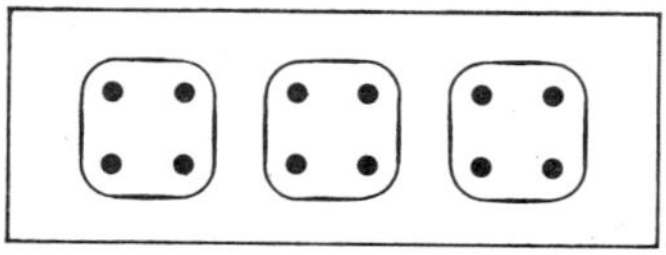	three, zero	30_{four}

SET OF DOTS	NUMBER NAME	NUMERAL
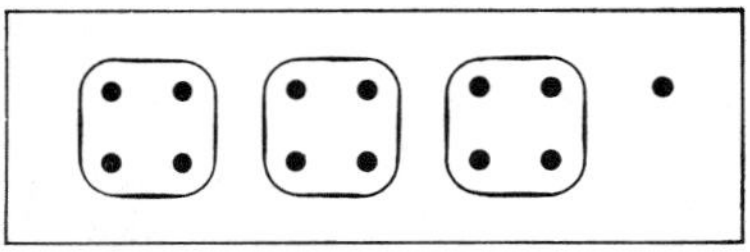	three, one	31_{four}
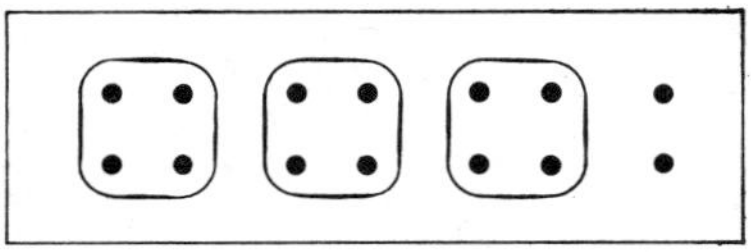	three, two	32_{four}
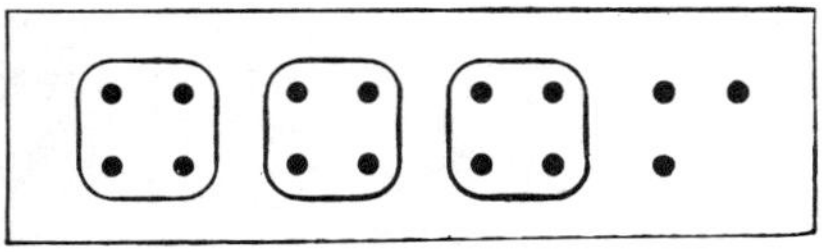	three, three	33_{four}

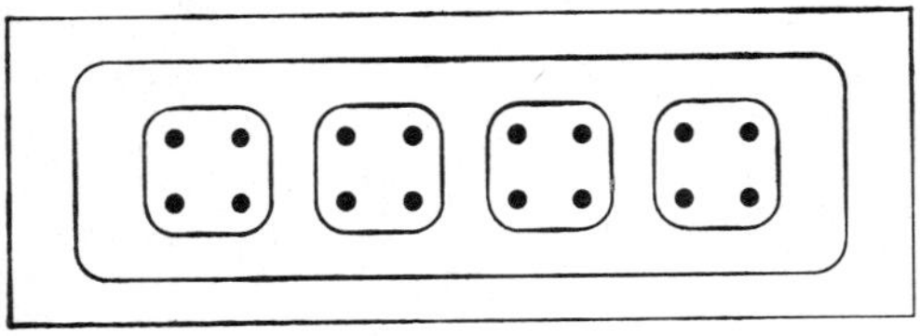

The set of dots at the left has one more dot than the set immediately above it. It has four groups of four, that is, four fours. In order to represent the number of dots in the set using only the symbols 0, 1, 2, 3, we form a new group that has four × four dots—the $(\text{four})^2$ group. The number name in base four representing the number of dots in this set is one, zero, zero. The numeral that represents this number name is 100_{four}. This means that in the set there is one $(\text{four})^2$ group, zero (four) groups, and zero ones.

As we continue in our counting, we reach the number name three, three, three, represented by the numeral 333_{four}. This is the numeral for a set of dots in which there are three $(\text{four})^2$ groups, three (four) groups, and three ones.

The set of dots having one more dot than this last set has four $(\text{four})^2$ groups. In order to represent the number of dots in this set using only the symbols 0, 1, 2, 3, we form a new group that has four × four × four dots—the $(\text{four})^3$ group. The number name in base four representing the number of dots in this set is one, zero, zero, zero. The numeral that represents this number name is 1000_{four}. This means that in the set there is one $(\text{four})^3$ group, zero $(\text{four})^2$ groups, zero (four) groups, and zero ones.

If we continue counting in a similar manner, we find that all counting numbers in base four can be represented by numerals using only the symbols 0, 1, 2, 3.

EXERCISES

In 1–3, group the dots and represent the number of dots as a numeral in base four:

1. 

2. 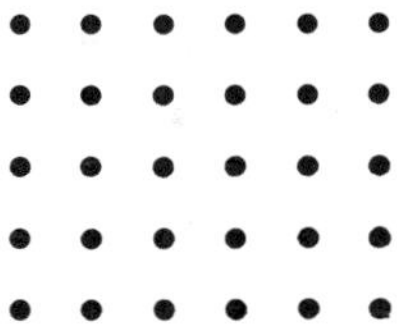

3. 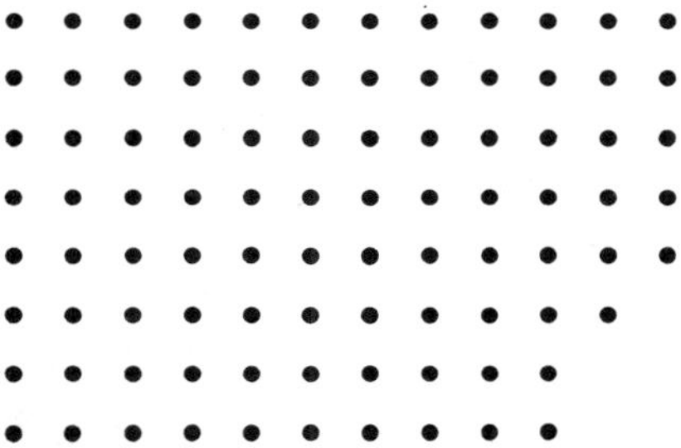

In 4–8, draw the number of dots represented by the given numeral. Group the dots to show the meaning of the numeral.

4. 23_{four} **5.** 30_{four} **6.** 123_{four} **7.** 210_{four} **8.** 1231_{four}

In 9–20:

a. Name the counting number in base four immediately following the number represented by the given numeral.
b. Write the numeral in base four that represents the answer given in part *a*.

9. 3_{four} **10.** 22_{four} **11.** 31_{four} **12.** 33_{four}
13. 130_{four} **14.** 333_{four} **15.** 1012_{four} **16.** 1023_{four}
17. 1233_{four} **18.** 2303_{four} **19.** 2333_{four} **20.** 3333_{four}

21. Which number is greater, 231_{four} or 213_{four}?
22. Which number is greater, 3012_{four} or 333_{four}?
23. Which number is smaller, 2201_{four} or 3000_{four}?
24. Which number is smaller, 1230_{four} or 100000_{four}?

PLACE VALUES IN BASE FOUR

We have learned that in the decimal, or base ten, system of writing numerals, the position of each digit in the numeral determines the number expressed by that digit. The first place at the right is the ones place. Starting at the ones place and moving left, each place is given a value ten (the base) times as large as the value of the place at its right.

In the numeral 222_{ten}, the place values of the digits are:

The first 2 at the right means:	2 ones, or $2 \times$ (one). . . .	2_{ten}
The second 2 from the right means:	2 tens, or $2 \times (\text{ten})^1$. . . .	20_{ten}
The third 2 from the right means: 2 hundreds, or 2 (ten $\times$ ten)'s, or $2 \times (\text{ten})^2$. . . .		200_{ten}
		222_{ten}

In the quaternary, or base four, system of writing numerals, the position of each digit in the numeral also determines the number expressed by that digit. The first place at the right is again the ones place. Also starting at the ones place and moving left, we give each place a value four (the base) times as large as the value of the place at its right.

In the numeral 22222_{four}, the place values of the digits are:

The first 2 at the right means:	2 ones, or $2 \times$ (one). . . .	2_{four}
The second 2 from the right means:	2 fours, or $2 \times (\text{four})^1$. . . .	20_{four}
The third 2 from the right means:	2 (four $\times$ four)'s, or $2 \times (\text{four})^2$. . . .	200_{four}
The fourth 2 from the right means:	2 (four $\times$ four $\times$ four)'s, or $2 \times (\text{four})^3$. . . .	2000_{four}
The fifth 2 from the right means:	2 (four $\times$ four $\times$ four $\times$ four)'s, or $2 \times (\text{four})^4$. . . .	20000_{four}
		22222_{four}

From the above discussion, we see that the value of a digit in a base four numeral depends upon the place the digit holds in the numeral. The chart below summarizes the place values in base four.

PLACE VALUES IN THE QUATERNARY SYSTEM

$(\text{four})^4$	$(\text{four})^3$	$(\text{four})^2$	$(\text{four})^1$	(ones)
2	2	2	2	2

EXERCISES

Give the place value of each digit in each of the following numerals:

1. 23_{four} **2.** 31_{four} **3.** 10_{four} **4.** 321_{four} **5.** 103_{four}

6. 333_{four} **7.** 201_{four} **8.** 2000_{four} **9.** 31202_{four} **10.** 130200_{four}

Give the place value of the 3 in each of the following numerals:

11. 13_{four} **12.** 320_{four} **13.** 1132_{four} **14.** 231020_{four} **15.** 312002_{four}

16. In the numeral 3210_{four}, select the digit that is related to:

a. $(\text{four})^2$ *b.* $(\text{four})^1$ *c.* $(\text{four})^3$

CHANGING A BASE FOUR NUMERAL TO A BASE TEN NUMERAL

POWERS OF FOUR		
$(\text{four})^1 =$	$(4)^1 =$	4
$(\text{four})^2 =$	$(4)^2 = 4 \times 4 =$	16
$(\text{four})^3 =$	$(4)^3 = 4 \times 4 \times 4 =$	64
$(\text{four})^4 =$	$(4)^4 = 4 \times 4 \times 4 \times 4 =$	256

We have learned that when a number is represented by a numeral in base four, the place values of the digits are related to $(\text{four})^1$ or $(\text{four})^2$ or $(\text{four})^3$, etc. That is, the place values are related to powers of four. To change a base four numeral to a base ten numeral, we must know the decimal numerals for the powers of four. Become familiar with the powers of four shown above.

The grouping in the picture below shows that the number of dots in the set may be represented by the numeral 132_{four}. Therefore,

$$132_{four} = (1 \times \text{four} \times \text{four}) + (3 \times \text{four}) + (2 \times \text{one}), \text{ or}$$
$$132_{four} = 1 \times (\text{four})^2 + 3 \times (\text{four})^1 + 2 \times (\text{one})$$

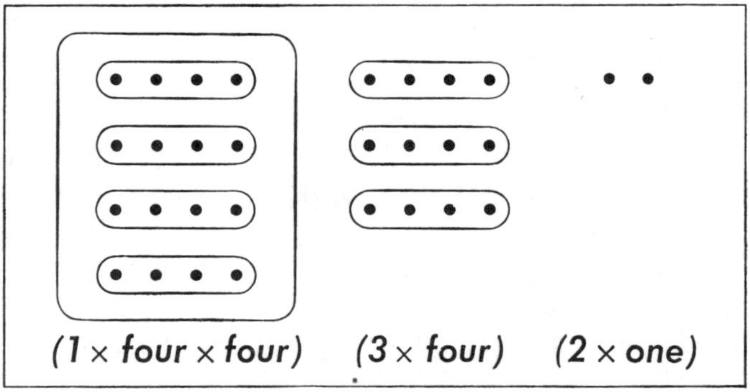

Notice that if we had made use of the place values of the digits of the numeral 132_{four}, we would have obtained the same expanded form as above:

$$132_{four} = 1 \times (\text{four})^2 + 3 \times (\text{four})^1 + 2 \times (\text{one})$$

If we wish to represent the number of dots in the set above by a decimal numeral, we can change the numeral 132_{four} to a base ten numeral in the following manner:

$$\begin{aligned}
132_{four} &= 1 \times (\text{four})^2 + 3 \times (\text{four})^1 + 2 \times (\text{one}) \\
132_{four} &= 1 \times (4)^2 \quad + 3 \times (4)^1 \quad + 2 \times (1) \\
132_{four} &= (1 \times 16) \quad + (3 \times 4) \quad + (2 \times 1) \\
132_{four} &= 16 \quad + 12 \quad + 2 \\
132_{four} &= 30, \text{ or } 132_{four} = 30_{ten}
\end{aligned}$$

It is also possible to represent the number of dots in the previous set by a decimal numeral by grouping the dots in groups of ten. There would be three groups of ten dots and no dots left over. Therefore, the number of dots may also be represented by the decimal numeral 30_{ten}.

Since both 132_{four} and 30_{ten} are numerals that represent the same number, $132_{four} = 30_{ten}$.

> [*Note:* 30_{ten} may be written 30. When no base is written, the base is understood to be ten.]

MODEL PROBLEM

Change 32013_{four} to a base ten, or decimal, numeral.

$$\begin{aligned}
32013_{four} &= 3 \times (\text{four})^4 + 2 \times (\text{four})^3 + 0 \times (\text{four})^2 + 1 \times (\text{four})^1 + 3 \times (\text{one}) \\
32013_{four} &= 3 \times (4)^4 \quad + 2 \times (4)^3 \quad + 0 \times (4)^2 \quad + 1 \times (4)^1 \quad + 3 \times (1) \\
32013_{four} &= (3 \times 256) + (2 \times 64) + (0 \times 16) + (1 \times 4) + (3 \times 1) \\
32013_{four} &= 768 \quad + 128 \quad + 0 \quad + 4 \quad + 3 \\
32013_{four} &= 903, \text{ or } 32013_{four} = 903_{ten} \quad \textit{Ans.}
\end{aligned}$$

KEEP IN MIND

To change a quaternary numeral to a decimal numeral, first express each digit of the quaternary numeral in expanded form as a power of four; then perform the indicated arithmetic.

EXERCISES

Write each of the following numerals in expanded form using powers of four:

1. 33_{four} **2.** 321_{four} **3.** 230_{four} **4.** 303_{four}
5. 1323_{four} **6.** 2100_{four} **7.** 13123_{four} **8.** 132102_{four}

Change each of the following quaternary numerals to a decimal numeral:

9. 23_{four} **10.** 30_{four} **11.** 123_{four} **12.** 203_{four}
13. 333_{four} **14.** 100_{four} **15.** 1213_{four} **16.** 3002_{four}
17. 13213_{four} **18.** 20132_{four} **19.** 121032_{four} **20.** 200123_{four}

CHANGING A BASE TEN NUMERAL TO A BASE FOUR NUMERAL

The grouping by tens in the picture below shows that the number of dots in the set can be represented as (4 × ten) and (2 × one), or by the decimal numeral 42_{ten}, which may be written 42.

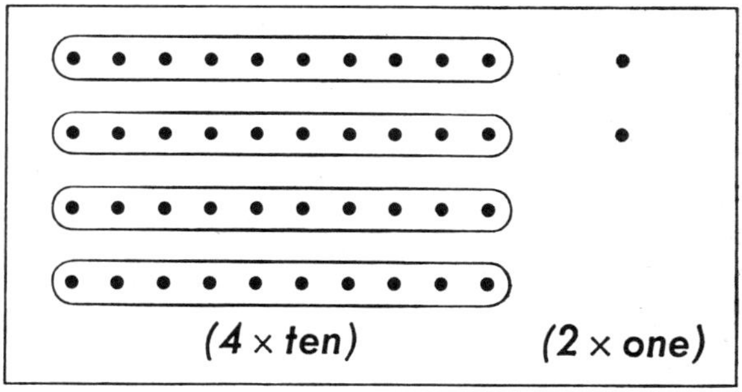

If we group the same set of dots as above in groups of fours, we see that the number of dots in the set can be represented by (2 × four × four), (2 × four), and (2 × one), or by the quaternary numeral 222_{four}.

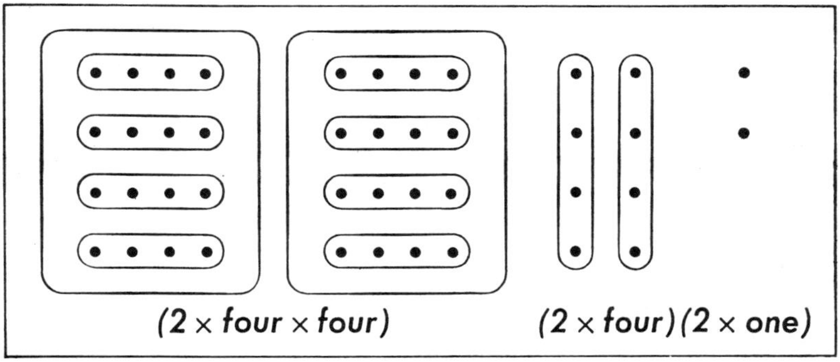

Since 42_{ten} and 222_{four} represent the same number, $42_{ten} = 222_{four}$. Notice that we have changed 42_{ten}, or 42, which is a base 10 or decimal numeral, to 222_{four}, which is a base four or quaternary numeral.

It is also possible to change 42_{ten} to a numeral with base four by using the place value principles in the base four numeration system. To find a numeral in base four that represents the same number as the decimal numeral 42, we must first discover the number of places, or the number of digits, that the base four numeral will have. As we have already seen, the place values in base four are related to powers of four. From the table of powers of four on page 313, we discover that $(\text{four})^3 = (4)^3 = 64$. Since 64 is greater than 42, the base four numeral that we are trying to find cannot have a digit in the $(\text{four})^3$ place. However, $(\text{four})^2 = (4)^2 = 16$ is smaller than 42. Therefore, the numeral we are trying to find will have a digit in the $(\text{four})^2$ place. It will also have a digit in the $(\text{four})^1$ place and in the ones place. Now we know that our base four numeral will have three digits.

Step 1. To find the digit in the $(\text{four})^2$ place, we must find how many $(\text{four})^2$'s, $(4)^2$'s, or 16's are contained in 42. To do this, we divide 42 by 16. The quotient, 2, tells us that 16 is contained in 42 two times. Therefore, the digit in the $(\text{four})^2$ place is 2.

$$\begin{array}{r} 2 \\ 16\overline{)42} \\ \underline{32} \\ 10 \end{array}$$

Step 2. To find the digit in the $(\text{four})^1$ place, we must find how many $(\text{four})^1$'s, $(4)^1$'s, or 4's are contained in 10, the remainder in step 1. To do this, we divide 10 by 4. The quotient, 2, tells us that 4 is contained in 10 two times. Therefore, the digit in the $(\text{four})^1$ place is 2.

$$\begin{array}{r} 2 \\ 4\overline{)10} \\ \underline{8} \\ 2 \end{array}$$

Step 3. The remainder, 2, in step 2 tells us that two ones remain. Therefore, the digit in the ones place is 2.

Step 4. At the right, see how the result of changing the decimal numeral 42 to a base four numeral is written.

$(\text{four})^2$	$(\text{four})^1$	(ones)
2	2	2

Answer: 222_{four}

MODEL PROBLEMS

1. Express the decimal numeral 108 as a numeral in the quaternary system.
Solution:

Step 1. The highest power of four that is less than 108 is $(4)^3$, or 64. Divide 108 by 64. The quotient, 1, is the digit in the $(\text{four})^3$ place.

$$\begin{array}{r} 1 \\ 64\overline{)108} \\ \underline{64} \\ 44 \end{array}$$

Step 2. Divide 44, the remainder in step 1, by $(4)^2$, or 16. The quotient, 2, is the digit in the $(\text{four})^2$ place.

$$\begin{array}{r} 2 \\ 16\overline{)44} \\ \underline{32} \\ 12 \end{array}$$

Step 3. Divide 12, the remainder in step 2, by $(4)^1$, or 4. The quotient, 3, is the digit in the $(\text{four})^1$ place. Since there are no ones left, the digit in the ones place is 0.

$$\begin{array}{r} 3 \\ 4\overline{)12} \\ \underline{12} \\ 0 \end{array}$$

Step 4. Write the digits in the place value table to express the numeral that is the answer.

$(\text{four})^3$	$(\text{four})^2$	$(\text{four})^1$	(ones)
1	2	3	0

Answer: 1230_{four}

2. Change the base ten numeral 710 to a numeral in the base four system.
Solution:

Step 1. The highest power of four that is less than 710 is $(4)^4$, or 256. Divide 710 by 256. The quotient, 2, is the digit in the $(\text{four})^4$ place.

$$\begin{array}{r} 2 \\ 256\overline{)710} \\ \underline{512} \\ 198 \end{array}$$

Step 2. Divide 198, the remainder in step 1, by $(4)^3$, or 64. The quotient, 3, is the digit in the $(\text{four})^3$ place.

$$\begin{array}{r} 3 \\ 64\overline{)198} \\ \underline{192} \\ 6 \end{array}$$

Step 3. Divide 6, the remainder in step 2, by $(4)^2$, or 16. The quotient, 0, is the digit in the $(\text{four})^2$ place.

$$\begin{array}{r} 0 \\ 16\overline{)6} \\ \underline{0} \\ 6 \end{array}$$

Step 4. Divide 6, the remainder in step 3, by $(4)^1$, or 4. The quotient, 1, is the digit in the $(\text{four})^1$ place. Since there are 2 ones left, the digit in the ones place is 2.

$$\begin{array}{r} 1 \\ 4\overline{)6} \\ \underline{4} \\ 2 \end{array}$$

Step 5. Write the digits in the place value table to express the numeral that is the answer.

$(\text{four})^4$	$(\text{four})^3$	$(\text{four})^2$	$(\text{four})^1$	(ones)
2	3	0	1	2

Answer: 23012_{four}

KEEP IN MIND

To change a decimal numeral to a quaternary numeral, we must first find the highest power of four which is less than the decimal numeral so that this power of four can be used as a divisor.

EXERCISES

Find the value of each of the following powers of 4:

1. 4^2 **2.** 4^3 **3.** 4^4 **4.** 4^5 **5.** 4^6 **6.** 4^8

Find the highest power of four that is smaller than the given decimal numeral:

7. 12 **8.** 38 **9.** 72 **10.** 115 **11.** 138
12. 280 **13.** 437 **14.** 692 **15.** 837 **16.** 1200
17. 92 **18.** 3126 **19.** 947 **20.** 100 **21.** 793

Change each of the following decimal numerals to a quaternary numeral:

22. 6 **23.** 9 **24.** 12 **25.** 27 **26.** 50
27. 40 **28.** 89 **29.** 165 **30.** 199 **31.** 414
32. 587 **33.** 1657 **34.** 2189 **35.** 16 **36.** 256

37. Write a base four numeral that represents the number of inches in a foot.

38. Write a quaternary numeral that represents the number of months in 10 years.

39. Write a numeral in the base four system that represents the number of feet in a mile.

In 40–42, tell whether the statement is true or false:

40. There are 30_{four} months in a year.

41. There are 223_{four} days in a week.

42. There are 210_{four} inches in a yard.

ADDITION IN BASE FOUR

When we learned to add in the decimal system, we became familiar with the 100 basic addition facts representing the sums of all possible combinations of any two of the first ten whole numbers, 0, 1, 2, 3, 4, 5, 6, 7, 8, 9.

In order to be able to add in the base four system, we should become familiar with the sums of all possible combinations of any two of the first four whole numbers, 0, 1, 2, 3.

The table at the right shows how the sum of any two of the first four whole numbers, 0, 1, 2, 3, is expressed as a numeral in base four. Each number in the top horizontal row to the right of the "+" sign is added to each number in the vertical column under the "+" sign. The sum of each pair of numbers is represented in the table by a base four numeral. For example, the sum 1 + 0 is 1, as shown by the arrows in the table. Similarly, the sum 1 + 1 is 2 and 1 + 2 is 3.

SUMS IN BASE FOUR

+	0	1	2	3
0	0	1	2	3
1	1	2	3	10
2	2	3	10	11
3	3	10	11	12

If we wish to express 1 + 3 as a base four numeral, we cannot write the sum as 4 because there is no digit 4 in the base four system. The sum must be expressed as the base four numeral 10 (one, zero).

Similarly, if we wish to express 2 + 3 as a base four numeral, we cannot write the sum as 5 because there is no digit 5 in the base four system. The sum must be expressed as the base four numeral 11 (one, one).

Now we are ready to add in base four. See how we find the sum of 21_{four} and 10_{four}.

$$\begin{array}{rl} 21_{four} & = 2 \text{ fours} + 1 \text{ one} \\ +\ 10_{four} & = 1 \text{ four} + 0 \text{ ones} \\ \hline \end{array}$$

$$21_{four} + 10_{four} = 3 \text{ fours} + 1 \text{ one} = 31_{four} \quad \textit{Ans.}$$

Let us find the sum of 112_{four} and 203_{four}.

$$\begin{array}{rl} 112_{four} & = 1\ (four)^2 \quad + 1\ four \ + \ 2\ ones \\ +\ 203_{four} & = 2\ (four)^2\text{'s} + 0\ fours + \ 3\ ones \\ \hline 112_{four} + 203_{four} & = 3\ (four)^2\text{'s} + 1\ four \ + 11\ ones \\ & \\ 112_{four} + 203_{four} & = 3\ (four)^2\text{'s} + 2\ fours + \ 1\ one \\ 112_{four} + 203_{four} & = 321_{four} \quad \textit{Ans.} \end{array}$$

In the base four table of sums, we find:
$1 + 2 = 3$,
$1 + 0 = 1$,
$2 + 3 = 11$ (one, one)
(Since $11 = 1$ four + 1 one)

As shown at the right, we can check the result by changing the base four numerals to decimal numerals and then adding in base ten.

BASE FOUR		BASE TEN
112	=	22
203	=	35
321	=	57

MODEL PROBLEMS

Add:

1. $\begin{array}{r} 22_{four} \\ \underline{10_{four}} \\ 32_{four} \end{array}$

2. $\begin{array}{r} 12_{four} \\ \underline{31_{four}} \\ 103_{four} \end{array}$

3. $\begin{array}{r} 132_{four} \\ \underline{203_{four}} \\ 1001_{four} \end{array}$

4. $\begin{array}{r} 120_{four} \\ 211_{four} \\ \underline{32_{four}} \\ 1023_{four} \end{array}$

EXERCISES

Add each of the following. Check the sum by changing the base four numerals to decimal numerals and then adding in base ten.

1. $\begin{array}{r} 11_{four} \\ \underline{20_{four}} \end{array}$

2. $\begin{array}{r} 130_{four} \\ \underline{202_{four}} \end{array}$

3. $\begin{array}{r} 1002_{four} \\ \underline{2310_{four}} \end{array}$

4. $\begin{array}{r} 13220_{four} \\ \underline{20103_{four}} \end{array}$

5. $\begin{array}{r} 11_{four} \\ \underline{13_{four}} \end{array}$

6. $\begin{array}{r} 202_{four} \\ \underline{123_{four}} \end{array}$

7. $\begin{array}{r} 102_{four} \\ \underline{212_{four}} \end{array}$

8. $\begin{array}{r} 2103_{four} \\ \underline{123_{four}} \end{array}$

9. $\begin{array}{r} 31_{four} \\ \underline{10_{four}} \end{array}$

10. $\begin{array}{r} 122_{four} \\ \underline{130_{four}} \end{array}$

11. $\begin{array}{r} 212_{four} \\ \underline{331_{four}} \end{array}$

12. $\begin{array}{r} 312_{four} \\ \underline{232_{four}} \end{array}$

13. *a.* Add: $2_{four} + 1_{four}$
b. Add: $1_{four} + 2_{four}$
c. Is $2_{four} + 1_{four} = 1_{four} + 2_{four}$ a true statement?
d. We have learned that when two decimal numbers are added, changing the order in which they are added does not change their sum. Does

this *commutative principle for addition* operate when the numbers are represented by numerals in base four?

14. *a.* Add: $(2_{four} + 1_{four}) + 3_{four}$ *b.* Add: $2_{four} + (1_{four} + 3_{four})$

c. Is $(2_{four} + 1_{four}) + 3_{four} = 2_{four} + (1_{four} + 3_{four})$ a true statement?

d. We have learned that when we find the sum of three or more numbers, changing the way in which the numbers are grouped does not change the sum. Does this *associative principle for addition* operate when the numbers are represented by numerals in base four?

In 15–22, add:

15. $2_{four} + 1_{four} + 1_{four}$

16. $2_{four} + 3_{four} + 1_{four}$

17. $12_{four} + 21_{four} + 10_{four}$

18. $32_{four} + 13_{four} + 2_{four}$

19. 21_{four}	**20.** 33_{four}	**21.** 310_{four}	**22.** 212_{four}
23_{four}	22_{four}	23_{four}	31_{four}
$\underline{11_{four}}$	$\underline{1_{four}}$	$\underline{121_{four}}$	$\underline{3_{four}}$

SUBTRACTION IN BASE FOUR

When we studied subtraction in base ten, we learned that if $3 + 6 = 9$, then $9 - 6 = 3$ and $9 - 3 = 6$. We saw that if one addend is subtracted from the sum of two addends, the remainder is the other addend. We said that the operation of subtraction is the *inverse* of the operation of addition. In base four also, the operation of subtraction is the inverse of the operation of addition, as illustrated in the following discussion.

From the base four table of sums, we learn that $2_{four} + 3_{four} = 11_{four}$. If we wish to find $11_{four} - 2_{four}$, we can get the answer by answering the question, "What number added to 2_{four} gives the result 11_{four}?" The base four table of sums can help us answer this question. In the table, the third horizontal row under the "+" sign gives the results of adding different numbers to 2. In this horizontal row, 11 is in the last vertical column, which is headed by 3. Therefore, $11_{four} - 2_{four} = 3_{four}$. By using a similar procedure, we can find that $11_{four} - 3_{four} = 2_{four}$.

Since $2_{four} + 3_{four} = 11_{four}$ and $11_{four} - 2_{four} = 3_{four}$ and $11_{four} - 3_{four} = 2_{four}$, we have an illustration of the fact that in base four the operation of subtraction is the inverse of the operation of addition.

Before trying to solve more difficult problems of subtraction in base four, let us recall what we did when we subtracted in base ten. For example, let us find $52 - 25$.

$$\begin{array}{lll} 52 & = 5 \text{ tens} + 2 \text{ ones} & = 4 \text{ tens} + 12 \text{ ones} \\ \underline{-\ 25} & = 2 \text{ tens} + 5 \text{ ones} & = \underline{2 \text{ tens} + \ \ 5 \text{ ones}} \\ 52 - 25 & & = 2 \text{ tens} + \ \ 7 \text{ ones} = 27 \quad \textit{Ans.} \end{array}$$

Since 5 ones could not be subtracted from 2 ones, one of the 5 tens had to be exchanged for 10 ones, thus changing 5 tens + 2 ones to 4 tens + 12 ones.

We will use a similar procedure to find $31_{four} - 12_{four}$.

$$\begin{array}{lll} 31_{four} & = 3 \text{ fours} + 1 \text{ one} & = 2 \text{ fours} + 11 \text{ ones} \\ -\ 12_{four} & = 1 \text{ four} + 2 \text{ ones} & = 1 \text{ four} + 2 \text{ ones} \\ \hline 31_{four} - 12_{four} & & = 1 \text{ four} + 3 \text{ ones} = 13_{four} \quad \textit{Ans.} \end{array}$$

Since 2 ones could not be subtracted from 1 one, one of the 3 fours had to be exchanged for four ones, written 10 ones in base four. Thus, 3 fours + 1 one became 2 fours + 11 ones. To find 11 ones − 2 ones, the base four table of sums can be used.

To check a subtraction example, we can change the base four numerals to decimal numerals and then subtract in base ten.

MODEL PROBLEMS

Subtract:

1. $\begin{array}{r} 23_{four} \\ \underline{11_{four}} \\ 12_{four} \end{array}$ **2.** $\begin{array}{r} 30_{four} \\ \underline{11_{four}} \\ 13_{four} \end{array}$ **3.** $\begin{array}{r} 313_{four} \\ \underline{120_{four}} \\ 133_{four} \end{array}$ **4.** $\begin{array}{r} 212_{four} \\ \underline{33_{four}} \\ 113_{four} \end{array}$

EXERCISES

1. If $12_{four} + 21_{four} = 33_{four}$, find:
a. $33_{four} - 21_{four}$ *b.* $33_{four} - 12_{four}$

2. If $23_{four} - 10_{four} = 13_{four}$, find:
a. $13_{four} + 10_{four}$ *b.* $23_{four} - 13_{four}$

Use the base four table of sums to find:

3. $3_{four} - 1_{four}$ **4.** $10_{four} - 1_{four}$ **5.** $10_{four} - 3_{four}$
6. $2_{four} - 1_{four}$ **7.** $11_{four} - 3_{four}$ **8.** $12_{four} - 3_{four}$

In 9–13, subtract and check by changing to decimal numerals and subtracting in base ten:

9. $\begin{array}{r} 32_{four} \\ \underline{21_{four}} \end{array}$ **10.** $\begin{array}{r} 30_{four} \\ \underline{12_{four}} \end{array}$ **11.** $\begin{array}{r} 313_{four} \\ \underline{120_{four}} \end{array}$ **12.** $\begin{array}{r} 302_{four} \\ \underline{113_{four}} \end{array}$ **13.** $\begin{array}{r} 210_{four} \\ \underline{23_{four}} \end{array}$

In 14–16, subtract and check:

14. $23_{four} - 12_{four}$ **15.** $302_{four} - 121_{four}$ **16.** $2100_{four} - 322_{four}$

MULTIPLICATION IN BASE FOUR

When we learned to multiply in the decimal system, we became familiar with the basic multiplication facts representing the products of all possible combinations of any two of the first ten whole numbers, 0, 1, 2, 3, 4, 5, 6, 7, 8, 9.

In order to be able to multiply in the base four system, we should be able to obtain the product of all possible combinations of any two of the first four whole numbers, 0, 1, 2, 3.

To find $2_{four} \times 2_{four}$, we may think of this as $2_{four} + 2_{four}$, which gives 10_{four}.

To find $2_{four} \times 3_{four}$, we may think of this as $3_{four} + 3_{four}$, which gives 12_{four}.

To find $3_{four} \times 2_{four}$, we may think of this as $2_{four} + 2_{four} + 2_{four}$, which gives 12_{four}.

Notice that since $2_{four} \times 3_{four} = 12_{four}$ and $3_{four} \times 2_{four} = 12_{four}$, then $2_{four} \times 3_{four} = 3_{four} \times 2_{four}$. We see, therefore, that the *commutative principle for multiplication* operates in base four.

To find $3_{four} \times 3_{four}$, we may think of this as $3_{four} + 3_{four} + 3_{four}$, which gives 21_{four}.

The table at the right shows how the product of any two of the first four whole numbers, 0, 1, 2, 3, is expressed as a numeral in base four. The table is organized in the same manner as the table of sums in base four.

PRODUCTS IN BASE FOUR

×	0	1	2	3
0	0	0	0	0
1	0	1	2	3
2	0	2	10	12
3	0	3	12	21

Let us see whether the distributive principle for multiplication operates in base four.

$$3_{four} \times (2_{four} + 1_{four}) = 3_{four} \times 3_{four} = 21_{four}$$
$$(3_{four} \times 2_{four}) + (3_{four} \times 1_{four}) = 12_{four} + 3_{four} = 21_{four}$$

Since $3_{four} \times (2_{four} + 1_{four}) = (3_{four} \times 2_{four}) + (3_{four} \times 1_{four})$, we see that the *distributive principle for multiplication* operates in base four.

When we multiply a two-digit number in base four by a one-digit number in base four, we use the distributive principle. For example:

$2_{four} \times 31_{four} = 2_{four} \times (30_{four} + 1_{four})$
$2_{four} \times 31_{four} = (2_{four} \times 30_{four}) + (2_{four} \times 1_{four})$
$2_{four} \times 31_{four} = 120_{four} + 2_{four}$
$2_{four} \times 31_{four} = 122_{four}$

To find $2_{four} \times 30_{four}$, first multiply 0_{four} by 2_{four} and then multiply 3_{four} by 2_{four}. Use the base four table of products to get the result, 120_{four}.

The above multiplication can also be arranged vertically as shown at the right.

$$\begin{array}{r} 31_{four} \\ \times\ 2_{four} \\ \hline 122_{four} \end{array}$$

In order to multiply in base four, we proceed as we did when multiplying in base ten, remembering to make the proper exchanges when they become necessary.

Let us multiply 120_{four} by 32_{four}.

$$\begin{array}{rl} 120_{four} & \\ \times\ 32_{four} & \\ \hline 300 & (2_{four} \times 120_{four}) \\ 1020 & (30_{four} \times 120_{four}) \\ \hline 11100_{four} & (32_{four} \times 120_{four}) \end{array}$$

11100_{four} *Ans.*

Now we will test to see whether the associative principle for multiplication operates in base four.

$$(2_{four} \times 3_{four}) \times 2_{four} = 12_{four} \times 2_{four} = 30_{four}$$
$$2_{four} \times (3_{four} \times 2_{four}) = 2_{four} \times 12_{four} = 30_{four}$$

Since $(2_{four} \times 3_{four}) \times 2_{four} = 2_{four} \times (3_{four} \times 2_{four})$, the *associative principle for multiplication* operates in base four.

EXERCISES

Multiply:

1. $\begin{array}{r} 21_{four} \\ \times\ 2_{four} \\ \hline \end{array}$ **2.** $\begin{array}{r} 101_{four} \\ \times\ 3_{four} \\ \hline \end{array}$ **3.** $\begin{array}{r} 210_{four} \\ \times\ 3_{four} \\ \hline \end{array}$ **4.** $\begin{array}{r} 331_{four} \\ \times\ 2_{four} \\ \hline \end{array}$

5. $\begin{array}{r} 12_{four} \\ \times\ 21_{four} \\ \hline \end{array}$ **6.** $\begin{array}{r} 132_{four} \\ \times\ 13_{four} \\ \hline \end{array}$ **7.** $\begin{array}{r} 303_{four} \\ \times\ 20_{four} \\ \hline \end{array}$ **8.** $\begin{array}{r} 233_{four} \\ \times\ 23_{four} \\ \hline \end{array}$

9. Multiply 310_{four} by 13_{four}. Check the answer by first changing the base four numerals to base ten numerals and then multiplying in base ten.

DIVISION IN BASE FOUR

To do division in base four, we follow a procedure similar to the one used in division in base ten. In the following examples, all the numerals are in base four. We must remember to use the tables for the sums and products in base four and to do the subtractions in base four.

Let us divide 10_{four} by 2_{four}.

Step 1. Since we are dividing by 2_{four}, we go to the base four table of products to find in the vertical column headed by 2 the largest product that is equal to or less than 10_{four}, the dividend. $2_{four} \times 2_{four} = 10_{four}$. Therefore, the quotient is 2_{four}.

$$\begin{array}{r} 2_{four} \\ 2_{four}\overline{)10_{four}} \end{array} \quad \textit{Ans.}$$

Step 2. To check, we multiply the quotient, 2_{four}, by the divisor, 2_{four}. The product should be equal to the dividend, 10_{four}.

$$\begin{array}{r} 2_{four} \\ \times\ 2_{four} \\ \hline 10_{four} \end{array}$$

Let us divide 222_{four} by 3_{four}.

Step 1. Since we are dividing by 3_{four}, we go to the base four table of products to find in the vertical column headed by 3 the largest product that is equal to or less than 22_{four}. This product is 21_{four}. Since $3_{four} \times 3_{four} = 21_{four}$, the first numeral in the quotient is 3_{four}.

$$\begin{array}{r} 32_{four} \\ 3_{four}\overline{)222_{four}} \\ \underline{21}\phantom{_{four}}\ \\ 12\phantom{_{four}} \\ \underline{12}\phantom{_{four}} \end{array} \quad \textit{Ans.}$$

Step 2. Multiply $3_{four} \times 3_{four}$ and write the result, 21_{four}, under the first two digits in the dividend, 22_{four}.

Step 3. Subtract: $22_{four} - 21_{four} = 1_{four}$.

Step 4. Bring down the next digit in the dividend, 2_{four}.

Step 5. In the base four table of products, we find in the vertical column headed by 3 the largest product that is equal to or less than 12_{four}. Since $3_{four} \times 2_{four} = 12_{four}$, the next numeral in the quotient is 2_{four}.

Step 6. Continue as in steps 2 and 3 above.

Step 7. To check, we multiply the quotient, 32_{four}, and the divisor, 3_{four}. The product should be equal to the dividend, 222_{four}.

$$\begin{array}{r} 32_{four} \\ \times\ 3_{four} \\ \hline 222_{four} \end{array}$$

Study the following solution and the check for dividing 1302_{four} by 12_{four}.

Solution

$$\begin{array}{r} 103_{four} \\ 12_{four}\overline{)1302_{four}} \\ \underline{12}\phantom{02_{four}} \\ 102\phantom{_{four}} \\ \underline{102}\phantom{_{four}} \end{array} \quad \textit{Ans.}$$

Check

$$\begin{array}{r} 103_{four} \\ \times\ 12_{four} \\ \hline 212\phantom{_{four}} \\ \underline{103}\phantom{_{four}} \\ 1302_{four} \end{array}$$

EXERCISES

Divide and check:

1. 2_{four} by 2_{four}
2. 12_{four} by 2_{four}
3. 12_{four} by 3_{four}
4. 21_{four} by 3_{four}
5. 100_{four} by 2_{four}
6. 13_{four} by 2_{four}
7. 2130_{four} by 3_{four}
8. 1012_{four} by 2_{four}
9. 12023_{four} by 2_{four}
10. 312_{four} by 21_{four}
11. 222_{four} by 12_{four}
12. 3222_{four} by 31_{four}
13. 13002_{four} by 21_{four}
14. 1023_{four} by 11_{four}
15. 20202_{four} by 32_{four}

2. OTHER NUMERAL SYSTEMS

DECIMAL SYSTEM

To develop the decimal numeral system, we work with groups of ten. In this numeral system, ten symbols are used in writing the numerals. These symbols are: 0, 1, 2, 3, 4, 5, 6, 7, 8, 9. Also, each place is given a value ten (the base) times as large as the value of the place at its right.

QUATERNARY SYSTEM

To develop the quaternary system, we work with groups of four. In this system, four symbols are used in writing the numerals. These symbols are: 0, 1, 2, 3. Also, each place is given a value four (the base) times as large as the value of the place at its right.

QUINARY SYSTEM

To develop a numeral system with base five, the quinary system, we work with groups of five. In such a numeral system, five symbols are used in writing the numerals. These symbols are: 0, 1, 2, 3, 4. Each place has a value five (the base) times as large as the value of the place at its right.

SEPTENARY SYSTEM

If we work with groups of seven, we can develop a numeral system with base seven, the septenary system. In this system, seven symbols are used in writing the numerals. These symbols are: 0, 1, 2, 3, 4, 5, 6. Each place has a value seven (the base) times as large as the value of the place at its right.

BINARY SYSTEM

To develop a numeral system with base two, the binary system, we work with groups of two. In the binary system, two symbols are used in writing the numerals. These symbols are: 0, 1. Each place has a value two (the base) times as large as the value of the place at its right. The binary system plays an important role in the operation of modern electronic computers.

GROUPING THE SAME SET IN DIFFERENT WAYS

Decimal system—grouping by tens

Number of dots is (3 × ten) + (7 × one).

Decimal numeral is 37_{ten} or 37.

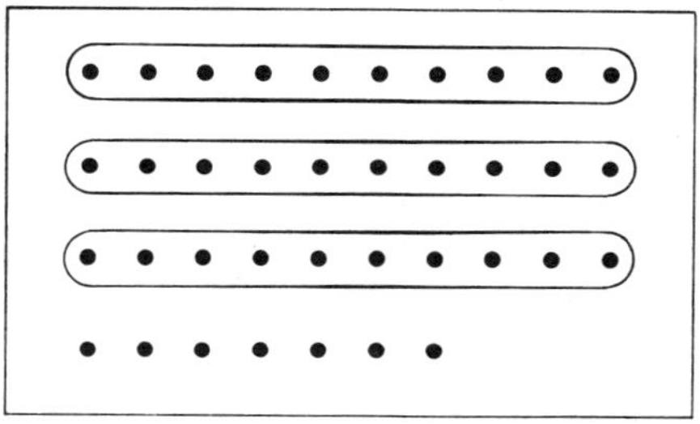

Septenary system—grouping by sevens

Number of dots is (5 × seven) + (2 × one).

Septenary numeral is 52_{seven}.

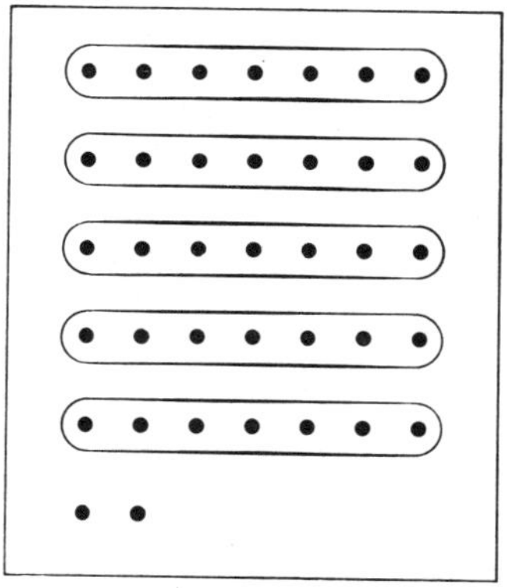

Quinary system—grouping by fives

Number of dots is (1 × five × five) + (2 × five) + (2 × one).

Quinary numeral is 122_{five}.

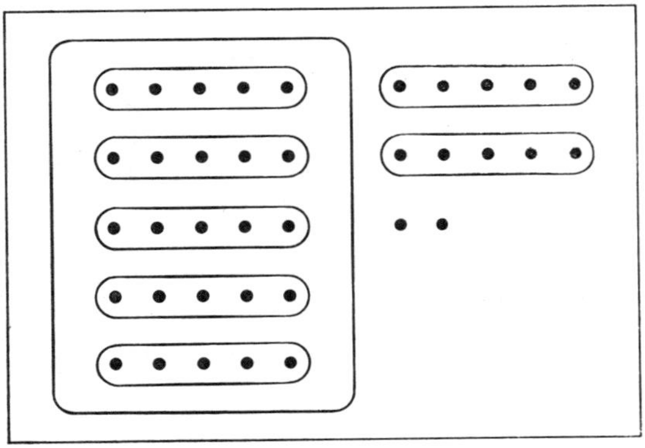

Quaternary system—grouping by fours

Number of dots is (2 × four × four) + (1 × four) + one.

Quaternary numeral is 211_{four}.

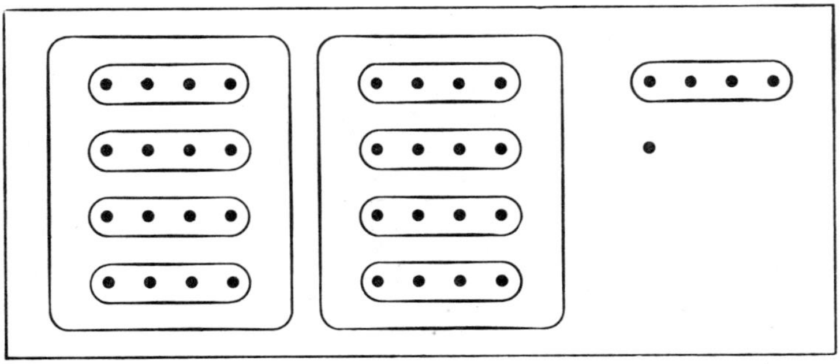

Binary system—grouping by twos

Number of dots is (1 × two × two × two × two × two)
+ (0 × two × two × two × two) + (0 × two × two × two)
+ 1 (two × two) + (0 × two) + (1 × one).
Binary numeral is 100101_{two}.

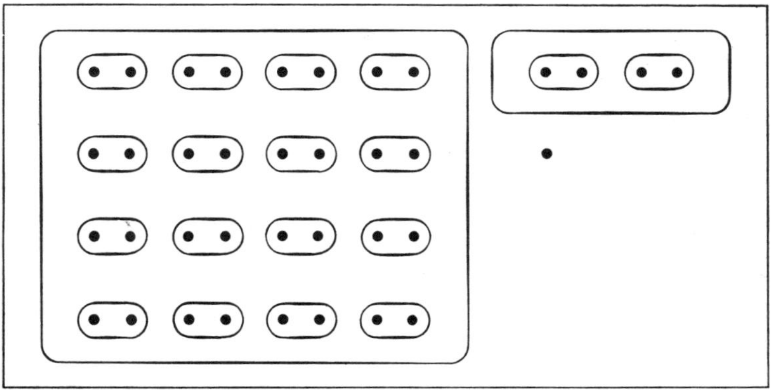

The numerals 37_{ten}, 52_{seven}, 122_{five}, 211_{four}, and 100101_{two} all represent the same number.

EXERCISES

In 1–5, copy the picture at the right:

1. *a.* Group the dots in groups of seven.
b. Represent the number of dots by a numeral in the septenary system.

2. *a.* Group the dots in groups of five.
b. Represent the number of dots by a numeral in the quinary system.

3. *a.* Group the dots in groups of two.
b. Represent the number of dots by a numeral in the binary system.

4. *a.* In groups of how many should the dots be grouped in order to write the numeral in base six that would represent the number of dots?
b. Represent the number of dots by a numeral using base six, the senary system.

5. *a.* In groups of how many should the dots be grouped in order to write the numeral in base eight that would represent the number of dots?
b. Represent the number of dots using base eight, the octonary system.

COUNTING IN OTHER SYSTEMS

We have learned the basic principles for counting in the decimal system and the quaternary system. In order to count in any other numeral system, we follow the same basic principles. See, for example, how these principles are followed in counting in base five, the quinary system.

SET OF DOTS	NUMBER NAME	NUMERAL
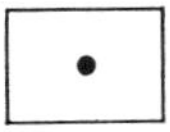	one	1_{five}
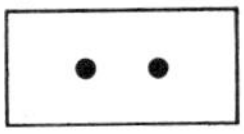	two	2_{five}
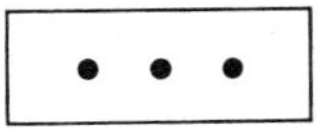	three	3_{five}
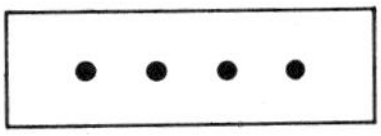	four	4_{five}
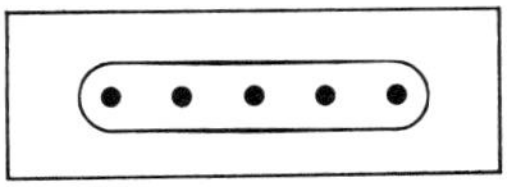	one, zero	10_{five}
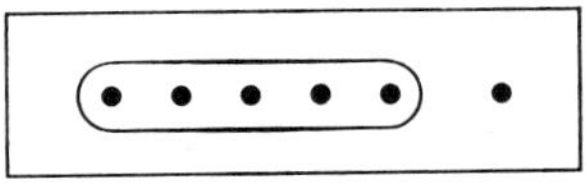	one, one	11_{five}
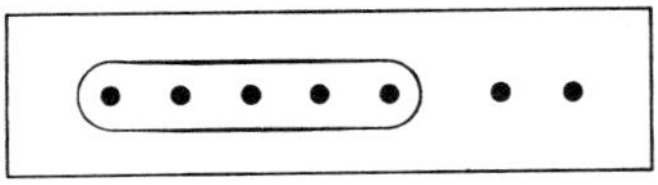	one, two	12_{five}
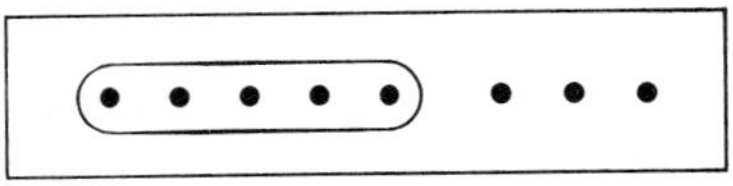	one, three	13_{five}
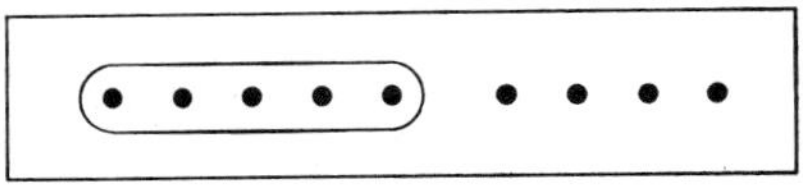	one, four	14_{five}

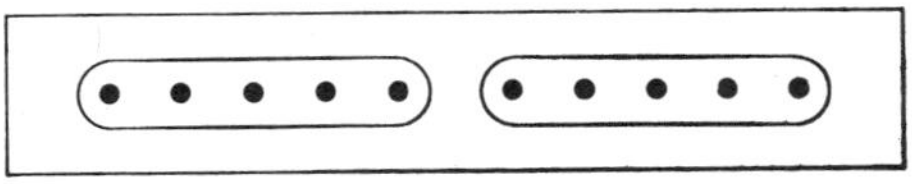

two, zero 20_{five}

EXERCISES

1. Write the numerals representing the numbers one through thirty in the:

a. quinary system
b. septenary system
c. binary system
d. octonary system

In 2–13:

a. Using the same base, name the counting number that immediately follows the number represented by the given numeral.
b. Using the same base, write the numeral representing the answer to *a.*

2. 5_{six} **3.** 12_{seven} **4.** 11_{two} **5.** 7_{eight}
6. 16_{seven} **7.** 17_{eight} **8.** 145_{six} **9.** 188_{nine}
10. 1011_{two} **11.** 444_{five} **12.** 607_{eight} **13.** 888_{nine}

PLACE VALUES IN OTHER SYSTEMS

We know that in the decimal system and in the quaternary system the position of each digit in the numeral determines the number expressed by that digit. The same fact is true in other numeral systems. Below we see the place values in base five.

PLACE VALUES IN THE QUINARY SYSTEM

$(\text{five})^4$	$(\text{five})^3$	$(\text{five})^2$	$(\text{five})^1$	(ones)
3	3	3	3	3

In the numeral 33333_{five}, the place values of the digits are:

The first 3 at the right means: 3 ones, or $3 \times (\text{one})$.... 3_{five}

The second 3 from the right means:
3 fives, or $3 \times (\text{five})^1$.... 30_{five}

The third 3 from the right means:
3 (five × five)'s, or $3 \times (\text{five})^2$.... 300_{five}

The fourth 3 from the right means:
3 (five × five× five)'s, or $3 \times (\text{five})^3$ .... 3000_{five}

The fifth 3 from the right means:
3 (five × five × five × five)'s, or $3 \times (\text{five})^4$.... 30000_{five}

33333_{five}

EXERCISES

1. Make a place value table for five places in each of the following numeral systems:

a. senary system (base six)
b. septenary system (base seven)
c. octonary system (base eight)
d. binary system (base two)
e. nonary system (base nine)
f. ternary system (base three)

2. Give the place value of each digit in each of the following numerals:

a. 17_{eight} *b.* 21_{six} *c.* 10_{two} *d.* 231_{seven} *e.* 203_{nine}
f. 120_{three} *g.* 6014_{seven} *h.* 4126_{eight} *i.* 10342_{five} *j.* 10010_{two}

3. Give the place value of the 1 in each of the following numerals:

a. 10_{two} *b.* 132_{five} *c.* 1430_{six} *d.* 62103_{seven} *e.* 10220_{three}

CHANGING NUMERALS IN OTHER BASES TO BASE TEN NUMERALS

We have learned that we make use of the place values of the digits of the numeral 132_{four} when we write the numeral in expanded form:

$$132_{\text{four}} = 1 \times (\text{four})^2 + 3 \times (\text{four})^1 + 2 \times (\text{one})$$

Similarly, we make use of the place values of the digits of the numeral 132_{five} when we write this numeral in expanded form:

$$132_{\text{five}} = 1 \times (\text{five})^2 + 3 \times (\text{five})^1 + 2 \times (\text{one})$$

If we wish to find the decimal numeral that represents the same numeral as 132_{five}, we proceed as follows:

$$\begin{aligned}
132_{\text{five}} &= 1 \times (\text{five})^2 + 3 \times (\text{five})^1 + 2 \times (\text{one}) \\
132_{\text{five}} &= 1 \times (5)^2 \quad + 3 \times (5)^1 \quad + 2 \times (1) \\
132_{\text{five}} &= (1 \times 25) \quad + (3 \times 5) \quad + (2 \times 1) \\
132_{\text{five}} &= 25 \quad + 15 \quad + 2 \\
132_{\text{five}} &= 42, \text{ or } 132_{\text{five}} = 42_{\text{ten}}
\end{aligned}$$

MODEL PROBLEM

Change 2053_{seven} to a decimal numeral.

Solution:

$$\begin{aligned}
2053_{\text{seven}} &= 2 \times (\text{seven})^3 + 0 \times (\text{seven})^2 + 5 \times (\text{seven})^1 + 3 \times (\text{one}) \\
2053_{\text{seven}} &= 2 \times (7)^3 \quad + 0 \times (7)^2 \quad + 5 \times (7)^1 \quad + 3 \times (1) \\
2053_{\text{seven}} &= (2 \times 343) \quad + (0 \times 49) \quad + (5 \times 7) \quad + (3 \times 1) \\
2053_{\text{seven}} &= 686 \quad + 0 \quad + 35 \quad + 3 \\
2053_{\text{seven}} &= 724, \text{ or } 2053_{\text{seven}} = 724_{\text{ten}} \quad \textit{Ans.}
\end{aligned}$$

EXERCISES

Write each of the following numerals in expanded form using powers of the indicated base:

1. 11_{two}	**2.** 43_{five}	**3.** 210_{three}	**4.** 541_{six}
5. 444_{seven}	**6.** 705_{eight}	**7.** 6085_{nine}	**8.** 10010_{two}

Change each of the following numerals to a decimal numeral:

9. 34_{five}	**10.** 20_{three}	**11.** 321_{six}	**12.** 560_{seven}
13. 666_{eight}	**14.** 100_{nine}	**15.** 1100_{two}	**16.** 7005_{nine}
17. 15200_{six}	**18.** 10010_{two}	**19.** 301230_{four}	**20.** 400123_{five}

CHANGING BASE TEN NUMERALS TO NUMERALS IN OTHER BASES

When we changed 42_{ten} to a numeral with base four, we used the place value principles in the base four system. Similarly, to change 42_{ten} to a numeral with base five, we use the place value principles in the base five system.

Since $(\text{five})^2 = (5)^2 = 25$ is the highest power of five that is smaller than 42, the numeral we are trying to find will have a digit in the $(\text{five})^2$ place. It will also have a digit in the $(\text{five})^1$ place and in the ones place. Thus, we know that our base five numeral will have three digits.

The procedure that we will use to find the digits in the various places of the answer will be the same as the one we used when we worked with base four.

Step 1. To find the digit in the $(\text{five})^2$ place, divide 42 by $(5)^2$, or 25. The quotient, 1, tells us that the digit in the $(\text{five})^2$ place is 1.

$$\begin{array}{r} 1 \\ 25\overline{)42} \\ \underline{25} \\ 17 \end{array}$$

Step 2. To find the digit in the $(\text{five})^1$ place, divide 17, the remainder in step 1, by $(5)^1$, or 5. The quotient, 3, tells us that the digit in the $(\text{five})^1$ place is 3.

$$\begin{array}{r} 3 \\ 5\overline{)17} \\ \underline{15} \\ 2 \end{array}$$

Step 3. The remainder, 2, in step 2 tells us that two ones remain. Therefore, the digit in the ones place is 2.

Step 4. At the right, see how the result of changing the decimal numeral 42 to a base five numeral is written.

$(\text{five})^2$	$(\text{five})^1$	(ones)
1	3	2

Answer: 132_{five}

KEEP IN MIND

> To change a decimal numeral to a numeral in another base, first find the highest power of the base that is less than the decimal numeral so that this power of the base can be used as a divisor.

EXERCISES

Make a table of the first six powers of each of the following:

1. 2 **2.** 3 **3.** 5 **4.** 6 **5.** 7 **6.** 8 **7.** 9

Find the highest power of the number in parentheses that is smaller than the given decimal numeral:

8. 19 (two)	**9.** 23 (three)	**10.** 135 (five)	**11.** 242 (six)
12. 438 (seven)	**13.** 600 (eight)	**14.** 492 (nine)	**15.** 158 (two)
16. 2000 (five)	**17.** 1983 (seven)	**18.** 3502 (nine)	**19.** 3025 (six)

Change each of the following decimal numerals to a numeral in the base shown in parentheses:

20. 3 (two)	**21.** 17 (three)	**22.** 25 (five)	**23.** 45 (six)
24. 89 (seven)	**25.** 72 (eight)	**26.** 146 (nine)	**27.** 256 (two)
28. 179 (three)	**29.** 689 (six)	**30.** 1304 (five)	**31.** 1143 (seven)

32. Represent the number of months in a year by a numeral in base:
a. two *b.* three *c.* five *d.* six *e.* seven *f.* eight *g.* nine

33. Represent the number of years in a century by a numeral in base:
a. nine *b.* eight *c.* seven *d.* six *e.* five *f.* three *g.* two

COMPUTING IN OTHER NUMERAL SYSTEMS

When we learned how to add, subtract, multiply, and divide in base four, we found it helpful to make a table of sums containing all possible sums of any two of the first four whole numbers, 0, 1, 2, 3, and a table of products containing all possible products of any two of the first four whole numbers, 0, 1, 2, 3.

If we wish to compute in base five, we will also find it helpful to make similar tables for all possible sums and products of any two of the first five whole numbers, 0, 1, 2, 3, 4.

SUMS IN BASE FIVE

+	0	1	2	3	4
0	0	1	2	3	4
1	1	2	3	4	10
2	2	3	4	10	11
3	3	4	10	11	12
4	4	10	11	12	13

PRODUCTS IN BASE FIVE

×	0	1	2	3	4
0	0	0	0	0	0
1	0	1	2	3	4
2	0	2	4	11	13
3	0	3	11	14	22
4	0	4	13	22	31

When we compute in base five, we use exactly the same procedures as we did when we computed in base four, remembering to refer to the tables of sums and products in base five.

MODEL PROBLEMS

1. Find the sum of 24_{five} and 13_{five}.
Solution:

$$\begin{array}{rcl} 24_{\text{five}} & = & 2 \text{ fives} + 4 \text{ ones} \\ +\ 13_{\text{five}} & = & 1 \text{ five} + 3 \text{ ones} \\ \hline 24_{\text{five}} + 13_{\text{five}} & = & 3 \text{ fives} + 12 \text{ ones} \\ 24_{\text{five}} + 13_{\text{five}} & = & 4 \text{ fives} + 2 \text{ ones} \\ 24_{\text{five}} + 13_{\text{five}} & = & 42_{\text{five}} \quad \textit{Ans.} \end{array}$$

In the base five table of sums, we find:
$2 + 1 = 3$,
$4 + 3 = 12$ (one, two)
(Since 12 ones = 1 five + 2 ones)

Check by the same method used in base four.

2. From 41_{five} subtract 23_{five}.
Solution:

$$\begin{array}{rcl} 41_{\text{five}} & = 4 \text{ fives} + 1 \text{ one} & = 3 \text{ fives} + 11 \text{ ones} \\ -\ 23_{\text{five}} & = 2 \text{ fives} + 3 \text{ ones} & = 2 \text{ fives} + 3 \text{ ones} \\ \hline 41_{\text{five}} - 23_{\text{five}} & & = 1 \text{ five} + 3 \text{ ones} = 13_{\text{five}} \quad \textit{Ans.} \end{array}$$

Check by the same method used in base four.

3. Multiply 103_{five} by 43_{five}.
Solution:

$$\begin{array}{rl} 103_{\text{five}} & \\ \times\ 43_{\text{five}} & \\ \hline 314 & (3_{\text{five}} \times 103_{\text{five}}) \\ 422 & (40_{\text{five}} \times 103_{\text{five}}) \\ \hline 10034_{\text{five}} & (43_{\text{five}} \times 103_{\text{five}}) \qquad 10034_{\text{five}} \quad \textit{Ans.} \end{array}$$

Check by the same method used in base four.

4. Divide 2043_{five} by 41_{five}.
Solution:

$$\begin{array}{r} 23_{\text{five}} \quad \textit{Ans.} \\ 41_{\text{five}}\overline{)2043_{\text{five}}} \\ \underline{132} \\ 223 \\ \underline{223} \end{array}$$

Check by the same method used in base four.

EXERCISES

GROUP A—QUINARY SYSTEM

1. Add and check:

a. $10_{\text{five}} + 21_{\text{five}}$ *b.* $1240_{\text{five}} + 2103_{\text{five}}$ *c.* $12_{\text{five}} + 23_{\text{five}}$ *d.* $221_{\text{five}} + 142_{\text{five}}$

e. $43_{\text{five}} + 24_{\text{five}}$ *f.* $123_{\text{five}} + 142_{\text{five}}$ *g.* $423_{\text{five}} + 41_{\text{five}}$ *h.* $1342_{\text{five}} + 2413_{\text{five}}$

i. $3_{\text{five}} + 2_{\text{five}} + 1_{\text{five}}$ *j.* $43_{\text{five}} + 24_{\text{five}} + 30_{\text{five}}$ *k.* $231_{\text{five}} + 410_{\text{five}} + 304_{\text{five}}$ *l.* $2103_{\text{five}} + 423_{\text{five}} + 14_{\text{five}}$

2. *a.* Add: $3_{\text{five}} + 4_{\text{five}}$
b. Add: $4_{\text{five}} + 3_{\text{five}}$
c. Is $3_{\text{five}} + 4_{\text{five}} = 4_{\text{five}} + 3_{\text{five}}$ a true statement?
d. Does the *commutative principle for addition* operate when the numbers to be added are represented by numerals in base five?

3. *a.* Find: $(3_{\text{five}} + 12_{\text{five}}) + 2_{\text{five}}$
b. Find: $3_{\text{five}} + (12_{\text{five}} + 2_{\text{five}})$
c. Is $(3_{\text{five}} + 12_{\text{five}}) + 2_{\text{five}} = 3_{\text{five}} + (12_{\text{five}} + 2_{\text{five}})$ a true statement?
d. Does the *associative principle for addition* operate when the numbers to be added are represented by numerals in base five?

4. Since $31_{\text{five}} + 12_{\text{five}} = 43_{\text{five}}$, find:
a. $43_{\text{five}} - 12_{\text{five}}$ *b.* $43_{\text{five}} - 31_{\text{five}}$

5. Since $34_{\text{five}} - 21_{\text{five}} = 13_{\text{five}}$, find:
a. $13_{\text{five}} + 21_{\text{five}}$ *b.* $34_{\text{five}} - 13_{\text{five}}$

6. Use the base five table of sums to find:
a. $4_{\text{five}} - 1_{\text{five}}$ *b.* $10_{\text{five}} - 2_{\text{five}}$ *c.* $12_{\text{five}} - 4_{\text{five}}$

7. Subtract and check:

a.	*b.*	*c.*	*d.*	*e.*
34_{five}	40_{five}	412_{five}	401_{five}	320_{five}
$\underline{21_{five}}$	$\underline{12_{five}}$	$\underline{230_{five}}$	$\underline{214_{five}}$	$\underline{32_{five}}$

8. Multiply and check:

a.	*b.*	*c.*	*d.*	*e.*
21_{five}	310_{five}	421_{five}	234_{five}	431_{five}
$\underline{2_{five}}$	$\underline{4_{five}}$	$\underline{3_{five}}$	$\underline{23_{five}}$	$\underline{34_{five}}$

9. *a.* Find: $3_{five} \times 4_{five}$
b. Find: $4_{five} \times 3_{five}$
c. Is $3_{five} \times 4_{five} = 4_{five} \times 3_{five}$ a true statement?
d. Does the *commutative principle for multiplication* operate when the numbers to be multiplied are represented by numerals in base five?

10. *a.* Find: $(2_{five} \times 4_{five}) \times 1_{five}$
b. Find: $2_{five} \times (4_{five} \times 1_{five})$
c. Is $(2_{five} \times 4_{five}) \times 1_{five} = 2_{five} \times (4_{five} \times 1_{five})$ a true statement?
d. Does the *associative principle for multiplication* operate when the numbers to be multiplied are represented by numerals in the quinary system?

11. *a.* Find: $2_{five} \times (3_{five} + 1_{five})$
b. Find: $2_{five} \times 3_{five} + 2_{five} \times 1_{five}$
c. Is $2_{five} \times (3_{five} + 1_{five}) = 2_{five} \times 3_{five} + 2_{five} \times 1_{five}$ a true statement?
d. Does the *distributive principle for multiplication* operate when the numbers to be multiplied are represented by numerals in the quinary system?

12. Divide and check:

a. 10_{five} by 2_{five} *b.* 22_{five} by 3_{five} *c.* 31_{five} by 4_{five}
d. 341_{five} by 31_{five} *e.* 2121_{five} by 42_{five} *f.* 24120_{five} by 32_{five}

GROUP B—SEPTENARY SYSTEM

13. Prepare a table of sums in base seven. Use this table in exercises 14–25, which deal with numbers represented by numerals in the septenary system.

14. In each of the following, add and check:

a.	*b.*	*c.*	*d.*	*e.*
30_{seven}	24_{seven}	142_{seven}	645_{seven}	5431_{seven}
$\underline{24_{seven}}$	$\underline{13_{seven}}$	$\underline{453_{seven}}$	$\underline{324_{seven}}$	$\underline{6352_{seven}}$

f.	*g.*	*h.*	*i.*	*j.*
3_{seven}	21_{seven}	35_{seven}	124_{seven}	6120_{seven}
2_{seven}	14_{seven}	52_{seven}	350_{seven}	543_{seven}
$\underline{1_{seven}}$	$\underline{22_{seven}}$	$\underline{46_{seven}}$	$\underline{426_{seven}}$	$\underline{25_{seven}}$

15. *a.* Add: $4_{seven} + 5_{seven}$
b. Add: $5_{seven} + 4_{seven}$

c. Is $4_{seven} + 5_{seven} = 5_{seven} + 4_{seven}$ a true statement?
d. Does the *commutative principle for addition* operate when the numbers to be added are represented by numerals in base seven?

16. *a.* Find: $(3_{seven} + 5_{seven}) + 2_{seven}$
b. Find: $3_{seven} + (5_{seven} + 2_{seven})$
c. Is $(3_{seven} + 5_{seven}) + 2_{seven} = 3_{seven} + (5_{seven} + 2_{seven})$ a true statement?
d. Does the *associative principle for addition* operate when the numbers to be added are represented by numerals in the septenary system?

17. Since $4_{seven} + 6_{seven} = 13_{seven}$, find:

a. $13_{seven} - 6_{seven}$ *b.* $13_{seven} - 4_{seven}$

18. Use the base seven table of sums to find:

a. $6_{seven} - 4_{seven}$ *b.* $11_{seven} - 5_{seven}$ *c.* $14_{seven} - 6_{seven}$

19. Subtract and check:

a.	*b.*	*c.*	*d.*	*e.*
53_{seven}	60_{seven}	542_{seven}	512_{seven}	450_{seven}
21_{seven}	13_{seven}	250_{seven}	364_{seven}	63_{seven}

20. Prepare a table of products in base seven. Use this table in exercises 21–25, which deal with numbers represented by numerals in the septenary system.

21. Multiply and check:

a.	*b.*	*c.*	*d.*	*e.*
32_{seven}	410_{seven}	521_{seven}	321_{seven}	465_{seven}
2_{seven}	5_{seven}	4_{seven}	32_{seven}	43_{seven}

22. *a.* Find: $5_{seven} \times 6_{seven}$
b. Find: $6_{seven} \times 5_{seven}$
c. Is $5_{seven} \times 6_{seven} = 6_{seven} \times 5_{seven}$ a true statement?
d. Does the *commutative principle for multiplication* operate when the numbers to be multiplied are represented by numerals in base seven?

23. *a.* Find: $(3_{seven} \times 2_{seven}) \times 4_{seven}$
b. Find: $3_{seven} \times (2_{seven} \times 4_{seven})$
c. Is $(3_{seven} \times 2_{seven}) \times 4_{seven} = 3_{seven} \times (2_{seven} \times 4_{seven})$ a true statement?
d. Does the *associative principle for multiplication* operate when the numbers to be multiplied are represented by numerals in the septenary system?

24. *a.* Find: $2_{seven} \times (3_{seven} + 4_{seven})$
b. Find: $2_{seven} \times 3_{seven} + 2_{seven} \times 4_{seven}$
c. Is $2_{seven} \times (3_{seven} + 4_{seven}) = 2_{seven} \times 3_{seven} + 2_{seven} \times 4_{seven}$ a true statement?
d. Does the *distributive principle for multiplication* operate when the numbers to be multiplied are represented by numerals in the septenary system?

25. Divide and check:

a. 26_{seven} by 4_{seven} *b.* 33_{seven} by 6_{seven} *c.* 42_{seven} by 5_{seven}
d. 606_{seven} by 42_{seven} *e.* 2426_{seven} by 31_{seven} *f.* 35616_{seven} by 63_{seven}

GROUP C—BINARY SYSTEM

26. Prepare a table of sums in base two. Use this table in exercises 27–31, which deal with numbers represented by numerals in the binary system.

27. Add and check:

a.	*b.*	*c.*	*d.*	*e.*
1_{two}	10_{two}	1001_{two}	1101_{two}	11011_{two}
1_{two}	11_{two}	101_{two}	1111_{two}	10111_{two}

f.	*g.*	*h.*	*i.*	*j.*
10_{two}	111_{two}	101_{two}	1101_{two}	10011_{two}
11_{two}	100_{two}	10_{two}	110_{two}	1101_{two}
11_{two}	101_{two}	1_{two}	11_{two}	10_{two}

28. Subtract:

a.	*b.*	*c.*	*d.*	*e.*
11_{two}	10_{two}	101_{two}	1101_{two}	10000_{two}
1_{two}	1_{two}	10_{two}	1011_{two}	1111_{two}

29. Prepare a table of products in base two. Use this table in exercises 30–31.

30. Multiply and check:

a.	*b.*	*c.*	*d.*	*e.*
11_{two}	101_{two}	1000_{two}	1010_{two}	1111_{two}
1_{two}	10_{two}	11_{two}	101_{two}	111_{two}

31. Divide and check:

a. 11_{two} by 1_{two} *b.* 1111_{two} by 11_{two} *c.* 10001_{two} by 11_{two}

32. Modern electronic computers make use of the binary system. In a computer, there are electrical switches. A switch has only two positions—either open, "on," or closed, "off." In the picture at the right, the five circles represent light bulbs on the panel of a computer. When the current is flowing, the light bulb that is "on," shown as ○, represents the digit "1." When the current is not flowing, the light bulb that is "off," shown as ●, represents the digit "0." If the bulbs on the panel are given place values like the digits of the numerals in the binary system, the panel represents the binary numeral 10011_{two}. Find the decimal numeral that the binary numeral 10011_{two} represents.

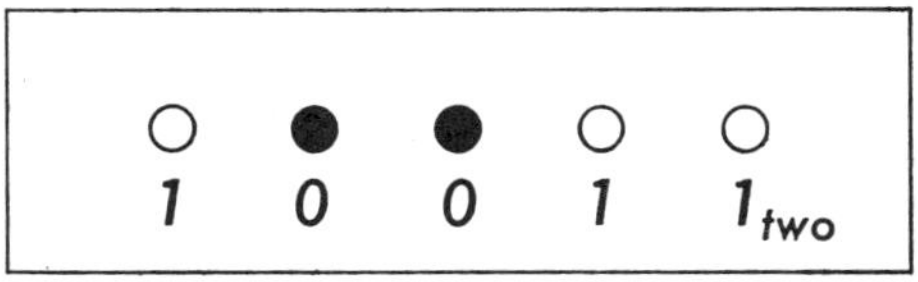

33. Each of the following panels of bulbs represents a number. Represent each number in the binary system and in the decimal system.

a.

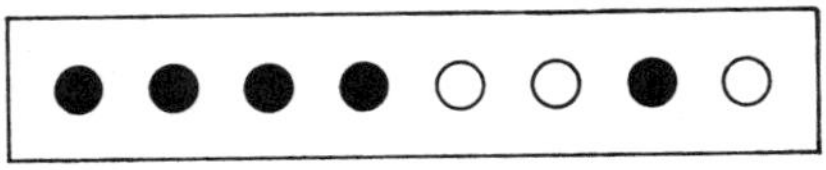

b.

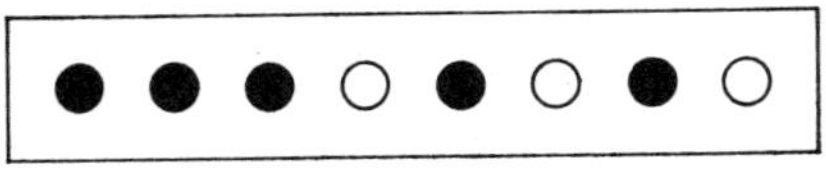

c.

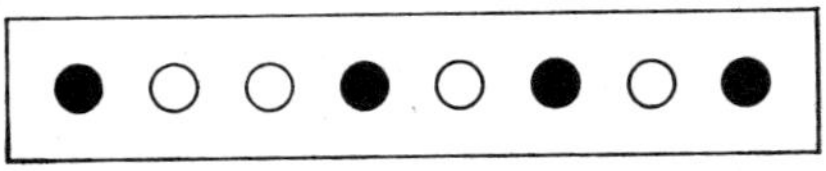

d.

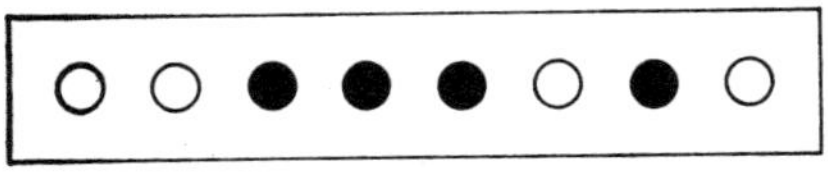

34. *a.* Write the binary numeral for the largest number that can be represented by a panel having eight bulbs.
b. Represent this binary numeral as a decimal numeral.

In 35–39, show how each binary numeral would be represented using a binary computer panel with eight bulbs.

35. 101_{two} **36.** 1100_{two} **37.** 111001_{two} **38.** 1001001_{two} **39.** 11001010_{two}

In 40–42, tell whether the statement is true or false:

40. There are 11000_{two} hours in a day.

41. There are 4000_{five} centimeters in a meter.

42. There are 21253_{seven} feet in a mile.

3. USING POWERS OF TEN TO EXPRESS THE VALUES OF DECIMALS

We have seen that when a whole number is represented by a numeral in the decimal system, the place value of each digit in the numeral is ten times the place value of the digit at its right and one-tenth of the place value of the digit at its left. For example, in the numeral 666, the place value of the middle 6 is 10, the place value of the 6 at its right is 1, and the place value of the 6 at its left is 100. Thus, the place value of the middle 6 is ten times the place value of the 6 at its right and one-tenth of the place value of the 6 at its left.

We have also seen that in a decimal the place value of any digit is ten times the place value of the digit at its right and one-tenth of the place value of the digit at its left. For example, in the decimal .666, the place value of the middle 6 is hundredths ($\frac{1}{100}$), the place value of the 6 at its right is thousandths ($\frac{1}{1000}$), and the place value of the 6 at its left is tenths ($\frac{1}{10}$). Thus, the place value of the middle 6 is ten times the place value of the 6 at its right, since $\frac{1}{100}$ is ten times $\frac{1}{1000}$, and one-tenth of the place value of the 6 at its left, since $\frac{1}{100}$ is one-tenth of $\frac{1}{10}$.

In the decimal .666:

The first 6 at the right of the decimal point means 6 tenths.	$.6 = \frac{6}{10} = 6 \times \frac{1}{10^1}$
The second 6 at the right of the decimal point means 6 hundredths.	$.06 = \frac{6}{100} = 6 \times \frac{1}{10} \times \frac{1}{10} = 6 \times \frac{1}{10^2}$
The third 6 at the right of the decimal point means 6 thousandths.	$.006 = \frac{6}{1000} = 6 \times \frac{1}{10} \times \frac{1}{10} \times \frac{1}{10} = 6 \times \frac{1}{10^3}$

Observe that each of the decimals .6, .06, and .006 has been expressed as the product of a whole number and a fraction whose numerator is 1 and whose denominator is a power of 10. In each case, the exponent of 10 in the denominator is equal to the number of places at the right of the decimal point. The exponent of 10 in the denominator is also equal to the number of zeros in the denominator of the common fraction that expresses the value of the decimal.

MODEL PROBLEMS

1. Express .00003 as the product of a whole number and a fraction whose numerator is 1 and whose denominator is a power of 10.

Solution:

Step 1. .00003 means $\frac{3}{100,000}$.

Step 2. In the decimal .00003, the number of places at the right of the decimal is 5. The number of zeros in the denominator of the common fraction $\frac{3}{100,000}$ is also 5.

Step 3. Therefore, $.00003 = 3 \times \frac{1}{10^5}$. *Ans.*

2. Express .0047 as the product of a whole number and a fraction whose numerator is 1 and whose denominator is a power of 10.

Solution:

Step 1. .0047 means $\frac{47}{10,000}$.

Step 2. The number of places in the decimal .0047 is 4, and the number of zeros in the denominator of the common fraction $\frac{47}{10,000}$ is also 4.

Step 3. Therefore, $.0047 = 47 \times \frac{1}{10^4}$. *Ans.*

3. Express the value of each of the digits in the decimal .286 as the product of a whole number and a fraction whose numerator is 1 and whose denominator is a power of 10.

Solution:

Step 1. The digit 2 means 2 tenths = .2. Therefore, the value of the digit 2 is $2 \times \frac{1}{10^1}$.

Step 2. The digit 8 means 8 hundredths = .08. Therefore, the value of the digit 8 is $8 \times \frac{1}{10^2}$.

Step 3. The digit 6 means 6 thousandths = .006. Therefore, the value of the digit 6 is $6 \times \frac{1}{10^3}$.

EXERCISES

Express each of the following as the product of a whole number and a fraction whose numerator is 1 and whose denominator is a power of 10:

1. $9 \times \frac{1}{10} \times \frac{1}{10}$

2. $7 \times \frac{1}{10} \times \frac{1}{10} \times \frac{1}{10} \times \frac{1}{10}$

3. $2 \times \frac{1}{10} \times \frac{1}{10} \times \frac{1}{10}$

4. $13 \times \frac{1}{10} \times \frac{1}{10} \times \frac{1}{10}$

5. $21 \times \frac{1}{10} \times \frac{1}{10} \times \frac{1}{10} \times \frac{1}{10} \times \frac{1}{10}$

6. $53 \times \frac{1}{10} \times \frac{1}{10} \times \frac{1}{10} \times \frac{1}{10}$

Express each of the following decimals as the product of a whole number and a fraction whose numerator is 1 and whose denominator is a power of 10:

7. .9 **8.** .003 **9.** .0006 **10.** .000004 **11.** .0000005

12. .019 **13.** .0037 **14.** .0041 **15.** .000089 **16.** .0000067

17. .137 **18.** .0516 **19.** .00129 **20.** .000307 **21.** .0000501

Express the value of each digit in each of the following numerals as the product of a whole number and a fraction whose numerator is 1 and whose denominator is a power of 10:

22. .213 **23.** .5712 **24.** .1627 **25.** .51863 **26.** .912376

Express each of the following products as a decimal:

27. $3 \times \frac{1}{10^2}$ **28.** $5 \times \frac{1}{10^5}$ **29.** $9 \times \frac{1}{10}$ **30.** $7 \times \frac{1}{10^4}$ **31.** $2 \times \frac{1}{10^6}$

32. $13 \times \frac{1}{10^3}$ **33.** $73 \times \frac{1}{10^5}$ **34.** $43 \times \frac{1}{10^2}$ **35.** $125 \times \frac{1}{10^3}$ **36.** $203 \times \frac{1}{10^6}$

In 37–46:

a. Express the value of each digit in the given numeral as the product of a whole number and a fraction whose numerator is 1 and whose denominator is a power of 10.
b. Express each of the products found in the answer to part *a* as a decimal.
c. Find the sum of all the decimals found in part *b*.
d. Tell whether the answer in part *c* is the same as the numeral originally given in the exercise.

37. .13 **38.** .257 **39.** .306 **40.** .4592 **41.** .68043
42. .22 **43.** .333 **44.** .5555 **45.** .88888 **46.** .999999

In 47–51:

a. Find the value of each digit in the given numeral as a whole number or as a decimal.
b. Find the sum of all the values found in part *a*.
c. Tell whether the answer in part *b* is the same as the numeral originally given in the exercise.

47. 3.24 **48.** 35.687 **49.** 146.9827 **50.** 555.555 **51.** 7777.7777

Write each of the following decimals as a fraction whose numerator is 1 and whose denominator is a power of 10:

52. .01 **53.** .001 **54.** .1 **55.** .0001
56. .000001 **57.** .00001 **58.** .0000001 **59.** .000000001

4. SCIENTIFIC NOTATION

Scientists frequently deal with very large numbers that have been rounded off. For example, the greatest distance of the planet Neptune from the sun is approximately 3,000,000,000 miles. Scientists also deal with very small numbers that have been rounded off. For example, the mass of a molecule of water is about .00000000000000000000003 gram. In order to be able to write and work with such numbers more easily, scientists have developed the ***scientific notation system.*** In this system, numbers are expressed as a product of a number between 1 and 10, and the proper power of 10. Also, a number is expressed in scientific notation when it is written just as a power of ten. Both 3.5×10^4 and 10^3 are in scientific notation.

EXPRESSING LARGE NUMBERS IN SCIENTIFIC NOTATION

POWERS OF TEN

$10^1 = 10$	$10^5 = 100{,}000$	$10^9 = 1{,}000{,}000{,}000$
$10^2 = 100$	$10^6 = 1{,}000{,}000$	$10^{10} = 10{,}000{,}000{,}000$
$10^3 = 1000$	$10^7 = 10{,}000{,}000$	$10^{11} = 100{,}000{,}000{,}000$
$10^4 = 10000$	$10^8 = 100{,}000{,}000$	$10^{12} = 1{,}000{,}000{,}000{,}000$

Notice that the number of zeros following the 1 in the value of a power of 10 is equal to the exponent of 10.

If we wish to express 2,000,000 in scientific notation, we must realize that 2,000,000 is equal to $2 \times 1{,}000{,}000$. Since 1,000,000 may be expressed as 10^6, the number 2,000,000 may be expressed in scientific notation as 2×10^6.

MODEL PROBLEMS

1. One year, the national debt of the United States was $300,000,000,000. Express this number in scientific notation.

Solution:

Step 1. $300{,}000{,}000{,}000 = 3 \times 100{,}000{,}000{,}000$

Step 2. Since in 100,000,000,000 there are 11 zeros following the one, $100{,}000{,}000{,}000 = 10^{11}$.

Step 3. Therefore, 300,000,000,000 expressed in scientific notation is 3×10^{11}. *Ans.*

2. The distance from the earth to the sun is approximately 93,000,000 miles. Write this number in scientific notation.

Solution:

Step 1. $93{,}000{,}000 = 9.3 \times 10{,}000{,}000$

Step 2. Since in 10,000,000, there are 7 zeros following the one, $10{,}000{,}000 = 10^7$.

Step 3. Therefore, $93{,}000{,}000 = 9.3 \times 10^7$. *Ans.*

3. The population of the United States in 1960 was 1.8×10^8. Find the number of people in the United States at that time.

Solution:

Step 1. $10^8 = 100{,}000{,}000$

Step 2. $1.8 \times 10^8 = 1.8 \times 100{,}000{,}000$

Step 3. $1.8 \times 10^8 = 180{,}000{,}000$ *Ans.*

EXERCISES

Write as a power of 10:

1. 100,000
2. 1,000,000,000
3. 1,000,000,000,000
4. one million
5. one hundred million
6. ten billion

Find the number that is expressed by each of the following numerals:

7. 10^7
8. 10^{10}
9. 10^{13}
10. 10^{15}
11. 3×10^5
12. 4×10^8
13. 6×10^{14}
14. 9×10^9
15. 1.3×10^4
16. 8.3×10^{12}
17. 1.27×10^3
18. 6.14×10^{10}

In each exercise, tell whether the numeral is in scientific notation. Give the reason for your answer.

19. 4×10^7
20. 13×10^9
21. 9.6×10^4
22. 3×2^{10}
23. $.4 \times 10^7$

Express in scientific notation:

24. 400
25. 6000
26. 30,000
27. 400,000
28. 7,000,000
29. 300,000,000
30. 80,000,000
31. 20,000,000,000
32. 5 thousand
33. 20 million
34. 700 million
35. 30 billion

In each of the following, find the number that can replace the question mark and make the resulting statement true:

36. $120 = 1.2 \times 10^?$
37. $760 = 7.6 \times 10^?$
38. $9300 = 9.3 \times 10^?$
39. $52{,}000 = 5.2 \times 10^?$
40. $5280 = 5.28 \times 10^?$
41. $375{,}000 = 3.75 \times 10^?$
42. $1{,}610{,}000 = 1.61 \times 10^?$
43. $872{,}000{,}000 = 8.72 \times 10^?$

Express in scientific notation:

44. 8400
45. 27,000
46. 54,000,000
47. 320,000,000
48. 6750
49. 81,600
50. 453,000
51. 375,000,000

In 52–57, express each number in scientific notation:

52. The planet Uranus is approximately 2,000,000,000 miles from the sun.
53. The velocity of light is 30,000,000,000 centimeters per second.
54. The distance between the earth and its nearest star, other than the sun, is 26,000,000,000,000 miles.
55. A light year, which is the distance light travels in one year, is approximately 5,900,000,000,000 miles.
56. A star that is about 12,000,000,000,000,000,000,000 miles away can be revealed by the Palomar telescope.
57. The approximate distance of the planet Mars from the sun is 145,000,000 miles.

In 58–60, write the decimal numeral for the number expressed:

58. The diameter of the universe is 2×10^9 light years.

59. The distance from the earth to the moon is 2.4×10^5 miles.
60. The sun weighs about 1.8×10^{27} tons.

EXPRESSING SMALL NUMBERS IN CONVENIENT FORM

We have learned that a decimal may be expressed as the product of a whole number and a fraction whose numerator is 1 and whose denominator is a power of 10. We can use this knowledge and the principles of scientific notation to express conveniently numbers that lie between 0 and 1. In this convenient form, numbers are expressed as the product of a number equal to 1 or greater than 1 but less than 10, and a fraction whose numerator is 1 and whose denominator is a power of 10.

The following table expresses some decimals as fractions whose numerators are 1 and whose denominators are powers of 10:

$.1 = \frac{1}{10^1}$	$.00001 = \frac{1}{10^5}$	$.000000001 = \frac{1}{10^9}$
$.01 = \frac{1}{10^2}$	$.000001 = \frac{1}{10^6}$	$.0000000001 = \frac{1}{10^{10}}$
$.001 = \frac{1}{10^3}$	$.0000001 = \frac{1}{10^7}$	$.00000000001 = \frac{1}{10^{11}}$
$.0001 = \frac{1}{10^4}$	$.00000001 = \frac{1}{10^8}$	$.000000000001 = \frac{1}{10^{12}}$

Notice that in each case the exponent of 10 in the denominator is equal to the number of places to the right of the decimal point.

We have learned that .00003 can be expressed as $3 \times \frac{1}{10^5}\cdot$ Notice that .00003 has been expressed as the product of a number greater than 1 and less than 10, and a fraction whose numerator is 1 and whose denominator is a power of 10.

The decimal .000035 can also be expressed in the same convenient form. Since $.000035 = 3.5 \times .00001$, then $.000035 = 3.5 \times \frac{1}{10^5}\cdot$ Notice that in the decimal .000035, the "3," which is the first non-zero digit to the right of the decimal point, is 5 places to the right of the decimal point. Notice, too, that 5 is the exponent of 10 in the convenient form $3.5 \times \frac{1}{10^5}\cdot$

In a similar manner, any small decimal whose value is between 0 and 1 can be expressed in the convenient form described previously.

MODEL PROBLEMS

1. Express .00000005 in the convenient form described previously.
Solution:

Step 1. $.00000005 = 5 \times .00000001$

Step 2. Since in .00000001 there are 8 places to the right of the decimal point, $.00000001 = \frac{1}{10^8}$.

Step 3. Therefore, $.00000005 = 5 \times \frac{1}{10^8}$. *Ans.*

To check, note that "5," the first non-zero digit in the given decimal, is 8 places to the right of the decimal point. In the answer, the exponent of 10 is also 8.

2. Express .0000029 in the convenient form described previously.
Solution:

Step 1. $.0000029 = 2.9 \times .000001$

Step 2. Since in .000001 there are 6 places to the right of the decimal point, $.000001 = \frac{1}{10^6}$.

Step 3. Therefore, $.0000029 = 2.9 \times \frac{1}{10^6}$. *Ans.*

3. The diameter of a red blood corpuscle is $7.5 \times \frac{1}{10^4}$ centimeter. Find the number of centimeters in the diameter.
Solution:

Step 1. $\frac{1}{10^4} = .0001$ Step 2. $7.5 \times \frac{1}{10^4} = 7.5 \times .0001$

Step 3. $7.5 \times \frac{1}{10^4} = .00075$ .00075 centimeter *Ans.*

[*Note:* The result of dividing 7.5 by 10^4 may be obtained quickly by moving the decimal point in 7.5 four places to the left.]

EXERCISES

Write as a fraction with a numerator 1 and a denominator a power of 10:

1. .01 **2.** .00001 **3.** .00000001
4. .0000000001 **5.** one thousandth **6.** one millionth
7. one ten millionth **8.** one billionth

Find the decimal fraction that is expressed by:

9. $\frac{1}{10^6}$ **10.** $\frac{1}{10^8}$ **11.** $\frac{1}{10^{15}}$ **12.** $\frac{1}{10^{18}}$

13. $4 \times \frac{1}{10^3}$ **14.** $7 \times \frac{1}{10^9}$ **15.** $8 \times \frac{1}{10^{10}}$ **16.** $9 \times \frac{1}{10^{13}}$

17. $1.2 \times \frac{1}{10^4}$ **18.** $3.6 \times \frac{1}{10^5}$ **19.** $7.4 \times \frac{1}{10^{11}}$ **20.** $3.14 \times \frac{1}{10^{14}}$

Express in the convenient form described previously:

21. .002 **22.** .0005 **23.** .000003 **24.** .00000009
25. 3 hundredths **26.** 7 thousandths **27.** 6 millionths **28.** 8 billionths

What number can replace the question mark to make the statement true?

29. $.023 = 2.3 \times \frac{1}{10^?}$ **30.** $.000086 = 8.6 \times \frac{1}{10^?}$

31. $.000000019 = 1.9 \times \frac{1}{10^?}$ **32.** $.000000000041 = 4.1 \times \frac{1}{10^?}$

33. $.00156 = 1.56 \times \frac{1}{10^?}$ **34.** $.000000873 = 8.73 \times \frac{1}{10^?}$

Express in the convenient form described previously:

35. .052 **36.** .00061 **37.** .0000039 **38.** .000000014
39. .156 **40.** .00381 **41.** .0000763 **42.** .000000917

In 43–48, express each decimal fraction in the convenient form described:

43. The approximate diameter of the smallest particle visible to the naked eye is .004 inch.

44. A micro-ampere is .000001 of an ampere.

45. A germ whose diameter is .000006 centimeter can be revealed by this microscope.

46. The radius of an electron is about .0000000000005 centimeter.

47. The diameter of some white blood corpuscles is approximately .0008 inch.

48. The wave length of red light is .000065 centimeter.

In 49–51, write the decimal fraction that the numeral expresses:

49. In a motion picture film, the image of each picture remains on the screen approximately $6 \times \frac{1}{10^2}$ second.

50. It takes light about $2 \times \frac{1}{10^8}$ second to cross a room.

5. UNDERSTANDING AND WORKING WITH SETS

THE MEANING OF A SET

In daily living, people often deal with collections, or groups, of objects. A teacher instructs a group of pupils, a conductor of an orchestra works with a group of musicians, a carpenter works with a collection of woodworking tools. In mathematics, the idea of a collection is very important because the mathematician deals with collections of numbers, collections of points, and collections of geometric figures.

A clearly defined collection of objects is called a ***set.*** For example, a girl's father and mother make up a set of parents; a ruler, compass, and protractor make up a set of geometry instruments; the numbers 1, 3, 5, 7, and 9 make up the set of odd counting numbers that are less than 10.

An object that belongs to a set is called a ***member,*** or an ***element,*** of the set. For example, a violin is a member of the set of string instruments; the number 2 is an element of the set of counting numbers.

A SET WITHIN A SET

The set of instruments making up a school orchestra has as its members violins, cellos, trumpets, trombones, clarinets, and drums. Each member of the set of string instruments (the violins and the cellos) is also a member of the set of instruments making up the whole orchestra. The set of string instruments is a ***subset*** of the set of instruments making up the whole orchestra.

In the set of counting numbers 1, 2, 3, 4, 5, each of the even numbers 2 and 4 is a member of the set. Therefore, the set whose members are the even numbers 2 and 4 is a subset of the set of counting numbers 1, 2, 3, 4, 5.

INDICATING MEMBERSHIP IN A SET

If we wish to talk about a set of objects, we must be able to indicate the set in such a way that we can always tell whether a particular object is a member of the set or is not a member of the set. One way of indicating a set is to list the names of its members between braces $\{\ \}$. If we wish to indicate the set whose members are 2, 4, 6, 8, we can write $\{2, 4, 6, 8\}$. This method of representation is called ***tabulating*** the set.

By using a capital letter to represent the set, we can write $A = \{2, 4, 6, 8\}$. This is read, "A is the set whose members are the numbers 2, 4, 6, and 8." The members of a set may be tabulated in any order. For example, $\{2, 4, 6, 8\}$ and $\{8, 4, 6, 2\}$ represent the same set whose members are tabulated differently.

It is not necessary for the members of a set to be related to one another in any way. If we wish certain objects to be the members of a set, we simply tabulate them as the members of the set. For example, in the set {Monday, baseball, foot, United States}, the members of the set are related to each other in no way other than that they are tabulated as the members of a set.

Sometimes it is inconvenient to tabulate the members of a set. For example, if we wished to indicate the set of even numbers between 1 and 99, we would have to tabulate 49 numbers. In such a case, we can ***describe*** the set. We write {even numbers between 1 and 99}, which is read, "the set of even numbers between 1 and 99." A description of a set must be so definite that we can always decide whether or not an object is a member of the set.

MODEL PROBLEM

Indicate the set of letters *a*, *e*, *i*, *o*, *u* by (*a*) tabulating the members and (*b*) describing the set.

Answer:

a. {*a*, *e*, *i*, *o*, *u*}

b. {vowels in the English alphabet}

EXERCISES

In 1–12, give two examples of sets whose members are:

1. men **2.** women **3.** animals **4.** musical instruments
5. lakes **6.** buildings **7.** automobiles **8.** athletes
9. numbers **10.** triangles **11.** quadrilaterals **12.** solid figures

In 13–18, the members of a set are given. Indicate the set by tabulating its members.

13. John, Mary, Tom
14. December, Monday, Christmas
15. 10, 12, 14, 16
16. hammer, chisel, saw
17. rose, carnation, tulip
18. tent, house, cabin

In 19–21, tell whether the statement is true or false:

19. William is a member of the set {Saul, Bruce, William}.
20. 2 is a member of the set {6, 8}.
21. △ is a member of the set {○, △, □, ▱}.

22. Select from the following those that are members of the set of the days of the week:

a. Tuesday *b.* January *c.* Sunday *d.* Friday *e.* June

23. Select from the following those that are members of the set of counting numbers:

a. 5 *b.* $\frac{1}{2}$ *c.* 13 *d.* f *e.* □

24. Select from the following those that are members of the set of United States coins:

a. nickel *b.* franc *c.* pound *d.* dime *e.* penny

In 25–30, tabulate the members of the set that is described:

25. the set of the days of the week beginning with the letter T
26. the set of U.S. presidents between F. D. Roosevelt and L. B. Johnson
27. the set of letters in the word *triangle*
28. the set of different letters in the word *ever*
29. the set of odd numbers between 40 and 50
30. the set of months that have 31 days

31. How many members are there in the set of the states of the United States?

In 32–38, describe the set that is indicated by tabulation:

32. {Saturday, Sunday}
33. {April, June, September, November}
34. {13, 15, 17, 19}
35. {0, 1, 2, 3, 4, 5, 6}
36. {Erie, Superior, Huron, Michigan, Ontario}
37. $\{u, v, x, y, w, z\}$
38. {Alaska, Arizona, Arkansas, Alabama}

In 39–45, tell whether the statement is true or false:

39. {Mary, Jane, Helen} is the same set as {Jane, Helen, Mary}.
40. {baseball, bat, first base} is the same set as {home plate, bat, baseball}.
41. {5, 10, 15, 20} is the same set as {all counting numbers between 1 and 24 that are exactly divisible by 5}.
42. {14, 16, 18} is a subset of {13, 14, 15, 16, 17, 18, 19}.
43. {circle, rectangle, square} is a subset of {all polygons having four sides}.
44. {all fish} is a subset of {salmon, trout, mackerel}.
45. {original 13 states of the U.S.A.} is a subset of {all states of the U.S.A.}.

COMPARING SETS

We can compare sets by comparing the members of one set with the members of the other. If A = {John, Harry, Sam} and B = {Marion, Helen, Rose}, then set A and set B are different because their members are different. The sets $R = \{a, e, i, o, u\}$ and $S = \{i, o, e, a, u\}$ have the same members tabulated in different orders. Because the two sets have the same members, we say that set R and set S are ***equal,*** and we write $R = S$.

From our study of the meaning of the number of a set, we know that two sets can be compared by matching the members of one set with the members of the other set. Although the sets $A = \{1, 3, 5\}$ and $B = \{2, 4, 6\}$ are not equal because their members are different, set A and set B have the same number of members. We can see this when we pair the members of set A and the members of set B as shown below.

$$\begin{array}{ccc} 1 & 3 & 5 \\ \updownarrow & \updownarrow & \updownarrow \\ 2 & 4 & 6 \end{array}$$

Notice that every member of set A has been paired, or matched, with a member of set B and there is no member in either set A or set B that has not been matched with a member of the other set. This pairing, or matching, of the members of set A and set B is an example of ***one-to-one correspondence.***

Whenever there is a one-to-one correspondence between the members of two sets, we say that the two sets are ***equivalent.*** Thus, set A and set B are equivalent. We can write this equivalence as follows: $A \leftrightarrow B$. This is read, "set A is equivalent to set B." Notice that the arrow points in two directions to show that the one-to-one correspondence means that set A is equivalent to set B, and set B is equivalent to set A.

A simple way to determine whether or not two sets are equivalent is to discover, if possible, the number for each set by counting its members. If both sets have the same number, they are equivalent. Set A, above, has 3 members and set B has 3 members. Therefore, set A and set B are equivalent.

EXERCISES

In 1–5, tell whether or not the sets are equal:

1. {winter, summer, spring, autumn} and {spring, summer, autumn, winter}
2. $\{a, e, i, o, u\}$ and $\{1, 2, 3, 4, 5\}$
3. {5, 10, 15, 20} and {counting numbers less than 25 that are exactly divisible by 5}
4. {days in the week that begin with T} and {Thursday, Tuesday}
5. {odd numbers less than 10} and {2, 4, 6, 8}

In 6–10, tell whether or not the sets can be paired so that they will be in one-to-one correspondence:

6. $\{2, 3, 4\}$ and $\{5, 6, 7\}$
7. $\{x, y, z, w\}$ and $\{r, s, t\}$
8. $\{C, A, M, E\}$ and $\{T, A, M, E\}$
9. $\{2, 3, 4, 5\}$ and $\{\frac{1}{2}, \frac{1}{3}, \frac{1}{4}, \frac{1}{5}\}$
10. {odd numbers between 30 and 40} and {odd numbers between 40 and 50}

In 11–14, give the number of members in each set:

11. {△, □, ☆, ○, ▱}
12. {odd numbers less than 100}
13. {states in the United States}
14. {consonants in the English alphabet}

In 15–19, tell whether or not the given sets are equivalent:

15. {Tom, Henry, John, Sid} and {Mary, Ruth, Rose, Irene}
16. $\{a, e, i, o, u\}$ and $\{m, n, p, r, s\}$
17. {△, ○, □} and {0, 1, 2, 3, 4}
18. {2, 4, 6, 8, 10} and {1, 3, 5, 7, 9}
19. {1, 2, 3, 4, 5, 6, 7} and {1, 4, 9, 16, 25, 36}

KINDS OF SETS

Sometimes the number of members of a set is so large that it is troublesome to tabulate all of them. An example is the set of odd numbers less than 1000. In this case, we indicate the set as follows:

$$\{1, 3, 5, 7, 9, 11, \ldots 999\}$$

The three dots indicate that the members of the set continue in a similar manner until the member 999 is reached. If there is an end to the tabulation of the members of a set, the number of members in the set can be counted. We call such a set a ***finite set.***

Sometimes the tabulation of the members of a set is an endless process. For example, it is impossible to tabulate all members of the set of odd numbers. In this case, we indicate the set as follows:

$$\{1, 3, 5, 7, 9, \ldots\}$$

The three dots after the 9 indicate that the tabulation continues endlessly. If there is no end to tabulating the members of a set, there can be no end to counting the members of the set. We call such a set an ***infinite set.***

It is possible for a set to have no members. For example, the set of men more than 100 feet tall has no members. Such a set is called the ***empty set,*** or ***null set.*** It is usually designated by the symbol Ø. The empty set may also be expressed by a pair of empty braces { }. Other examples of the empty set are the set of days of the week that begin with the letter Z and the set of odd numbers exactly divisible by 2.

EXERCISES

In 1–13, is the set a finite set, an infinite set, or an empty set?

1. {2, 4, 6, 8, . . . 20}
2. {2, 4, 6, 8, 10, 12, 14, 16, 18, 20, . . .}
3. {consonants in the English alphabet}
4. {0}
5. {whole numbers smaller than 15 and greater than 14}
6. {squares}
7. {rectangles having 3 sides}
8. {months of the year having 31 days}
9. {months of the year having 35 days}

10. {3-digit numbers between 100 and 1000}
11. {4-digit numbers less than 100}
12. {whole numbers divisible by 10}
13. {states whose population is more than 50 million}

PICTURING A UNIVERSAL SET AND ITS SUBSETS

The Universal Set *U*

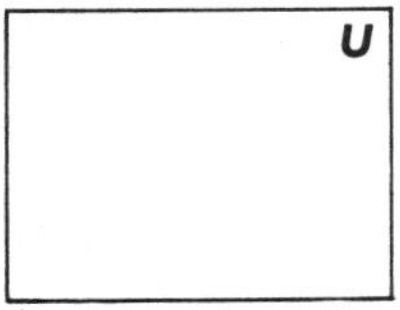

The set of all objects that are being considered in a discussion is called the ***universal set,*** or the ***universe.*** For example, if we are considering the members of the United Nations, we can write the universal set, represented by U, as U = {members of the United Nations}. If we are considering the set of even numbers between 1 and 17, we can write $U = \{2, 4, 6, 8, 10, 12, 14, 16\}$. We can represent the universal set pictorially by a rectangle and its interior.

Let us see how we can picture a subset of a universal set.

If from the universal set $U = \{2, 4, 6, 8, 10, 12, 14, 16\}$ we select the set of numbers, A, each of whose members is exactly divisible by 4, we get $A = \{4, 8, 12, 16\}$. Set A is a subset of set U. We can picture this fact by representing U by a rectangle and its interior, and by representing the subset A by a circle and its interior. Notice that the circle is placed within the rectangle. Such a picture is called a ***Venn diagram.***

Set *A* Is a Subset of Universal Set *U*

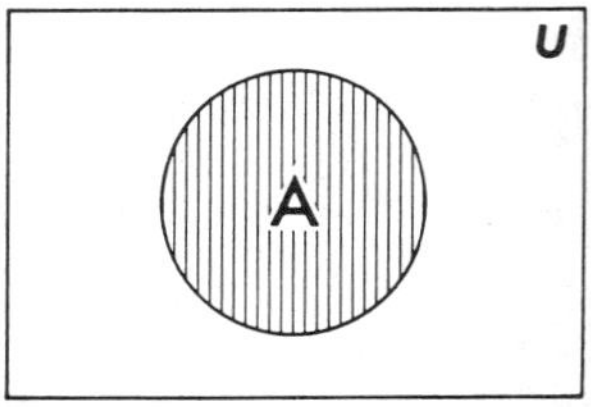

EXERCISES

1. *a.* If the set of objects under consideration is the collection of American coins in daily use, list the members of the universal set U.
b. If set A is a set each of whose members is an American coin worth less than 25 cents, list the members of set A.
c. Make a Venn diagram to show pictorially that set A is a subset of the universal set U.

2. *a.* If the set of objects under consideration is the months of the year, tabulate the members of the universal set U.
b. If set B is a set each of whose members is a month beginning with J, list the members of set B.
c. Make a Venn diagram to show pictorially that set B is a subset of set U.

3. *a.* If the set of objects under consideration is the set of odd numbers less than 20, tabulate the members of the universal set U.
b. If set C is a set each of whose members is an odd number less than 20 that is divisible by 5, list the members of set C.
c. Make a Venn diagram to show pictorially that set C is a subset of set U.

4. If the universal set U is the set of all quadrilaterals and set D is the set of all rectangles, make a Venn diagram to show pictorially that set D is a subset of set U.
5. *a.* If the universal set U is the set of all triangles, describe a set E that is a subset of set U.
   b. Make a Venn diagram to show pictorially that set E is a subset of set U.
6. Two sets under consideration are the set of even numbers less than 5 and the set of natural numbers less than 5. One of these sets is the universal set; the other is a subset of the universal set.
   a. Tabulate the members of the universal set U.
   b. Tabulate the members of the subset A.
   c. Make a Venn diagram to show pictorially that set A is a subset of set U.

INTERSECTION OF SETS

Sometimes we study a situation in which we discuss more than one subset of a universal set. For example, suppose that the universal set is the set of counting numbers, or natural numbers, $U = \{\text{natural numbers}\}$. If A is the set of natural numbers greater than 4 and less than 12, then $A = \{5, 6, 7, 8, 9, 10, 11\}$. If B is the set of natural numbers greater than 8 and less than 15, then $B = \{9, 10, 11, 12, 13, 14\}$. Notice that the numbers 9, 10, 11 are members of both set A and set B. The set whose members are 9, 10, 11 is called the intersection of set A and set B. The ***intersection*** of two sets, A and B, is the set containing those and only those members of set A that are also members of set B.

$A \cap B$ When A and B Intersect

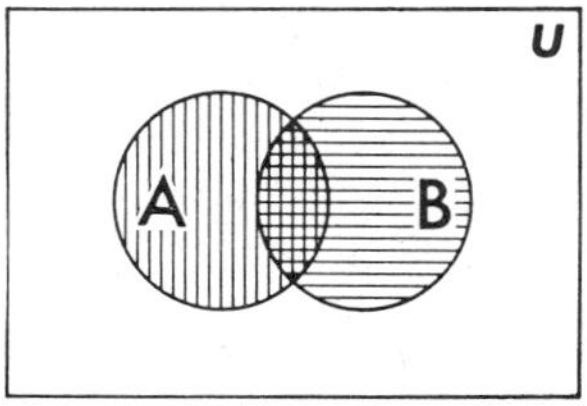

The intersection of set A and set B is represented by the symbol $A \cap B$, read "A cap B." Thus, if $A = \{5, 6, 7, 8, 9, 10, 11\}$ and $B = \{9, 10, 11, 12, 13, 14\}$, then $A \cap B = \{9, 10, 11\}$. The intersection of set A and set B, $A \cap B$, can be shown pictorially by a Venn diagram. The crosshatched part of the circles represents $A \cap B$.

$A \cap B$ When A and B Are Disjoint Sets

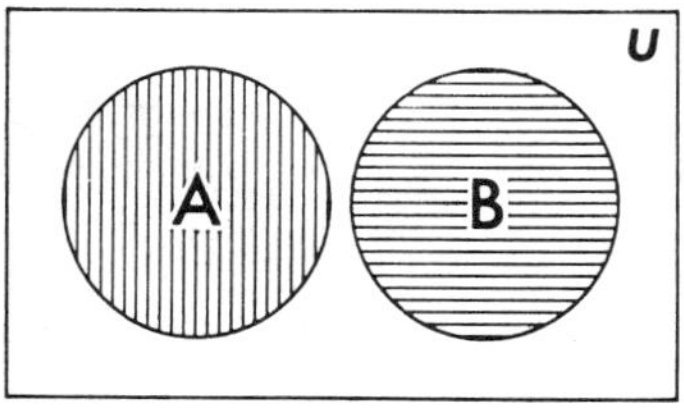

Suppose $U = \{\text{natural numbers}\}$. If A is the set of even natural numbers less than 10, then $A = \{2, 4, 6, 8\}$. If B is the set of odd natural numbers less than 10, then $B = \{1, 3, 5, 7, 9\}$. Notice that set A and set B have no common members. Therefore, the intersection of set A and set B is the empty set, or null set, $\emptyset$ or $\{\ \}$. When two sets such as set A and set B have no common members, they are called ***disjoint sets.*** The intersection of set A and set B, $A \cap B$, when A and B are disjoint sets, can be shown pictorially by a Venn diagram. Since A and B have no common members, no part of the circles is crosshatched.

$A \cap B$ When B Is a Subset of A

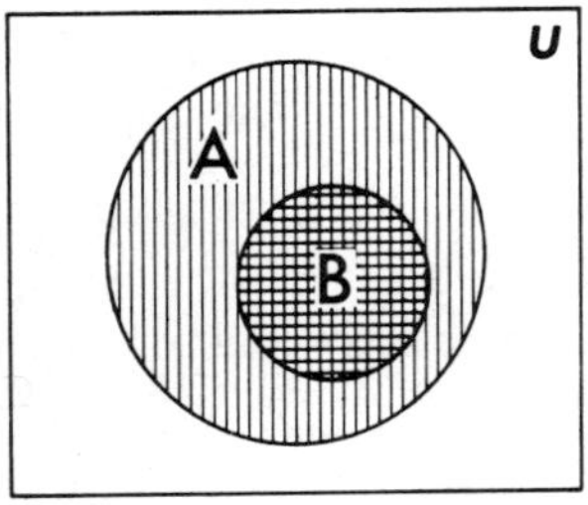

Suppose U = {natural numbers}. If A is the set of natural numbers less than 10, then $A = \{1, 2, 3, 4, 5, 6, 7, 8, 9\}$. If B is the set of odd natural numbers less than 10, then $B = \{1, 3, 5, 7, 9\}$. Notice that set B is a subset of set A. The intersection of set A and set B, $A \cap B$, when set B is a subset of set A, can be shown pictorially by a Venn diagram. The crosshatched part of the circles represents $A \cap B$. When B is a subset of A, then $A \cap B = B$.

EXERCISES

1. U = {natural numbers}, $A = \{10, 11, 12, 13, 14, 15\}$, and $B = \{13, 14, 15, 16, 17\}$.
   a. Tabulate the members of $A \cap B$.
   b. Picture $A \cap B$ with a Venn diagram.

2. U = {natural numbers}, A = {even natural numbers less than 20}, and B = {natural numbers less than 20 that are divisible by 4}.
   a. List the members of set A and set B.
   b. List the members of $A \cap B$.
   c. Picture $A \cap B$ with a Venn diagram.

3. U = {natural numbers less than 20}, A = {natural numbers less than 10}, and B = {natural numbers greater than 10 and less than 20}.
   a. List the members of the universal set U, set A, and set B.
   b. How many members are common to set A and set B?
   c. Describe the set $A \cap B$.
   d. Picture $A \cap B$ with a Venn diagram.

4. U = {people who have hair}, A = {people who have blond hair}, and B = {women who have blond hair}. Picture $A \cap B$ with a Venn diagram.

5. U = {people who have hair}, A = {women who have hair}, and B = {women who have blond hair}. Picture $A \cap B$ with a Venn diagram.

6. U = {people who have hair}, A = {women who have blond hair}, and B = {men who have blond hair}. Picture $A \cap B$ with a Venn diagram.

7. U = {triangles}, A = {isosceles triangles}, and B = {right triangles}. Picture $A \cap B$ with a Venn diagram.

8. U = {triangles}, A = {scalene triangles}, and B = {isosceles triangles}. Picture $A \cap B$ with a Venn diagram.

9. U = {triangles}, A = {isosceles triangles}, and B = {isosceles right triangles}. Picture $A \cap B$ with a Venn diagram.

10. U = {geometric polygons}, A = {quadrilaterals}, and B = {rectangles}. Picture $A \cap B$ with a Venn diagram.

11. U = {geometric polygons}, A = {triangles}, and B = {squares}. Picture $A \cap B$ with a Venn diagram.

12. U = {geometric polygons}, A = {parallelograms}, and B = {rectangles}. Picture $A \cap B$ with a Venn diagram.

13. The universe is the set of men who are members of the Best College athletic teams. A is the set of men on the Best College football team. B is the set of men on the Best College basketball team. C is the set of men on the Best College tennis team. D is the set of men on the Best College baseball team.

a. Make a Venn diagram for $A \cap B$ when some of the Best College men who are on the football team are also part of the basketball team.

b. Make a Venn diagram for $B \cap D$ when none of the Best College men who are on the basketball team are also on the baseball team.

c. Make a Venn diagram for $B \cap C$ when every man on the 10-man tennis team is also a member of the 15-man basketball team.

14. Set U represents the universe and set A and set B are subsets of set U. In a Venn diagram that pictures $A \cap B$, the circle representing set A coincides with the circle representing set B. What relationship must exist between set A and set B?

15. If U = {all pupils in a school}, A = {all pupils in the school who are on the honor roll}, and B = {all pupils in the school who are presidents of their official classes}, tell what each of the following Venn diagrams means:

a.

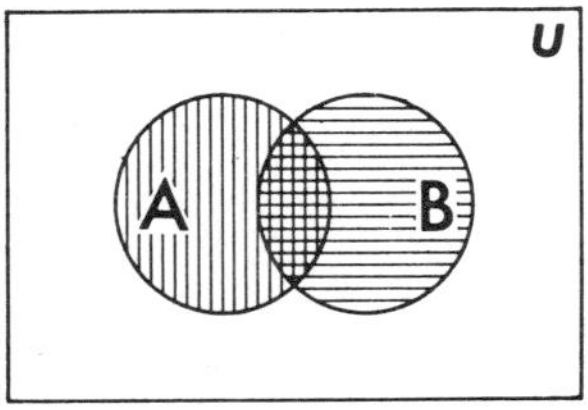

b.

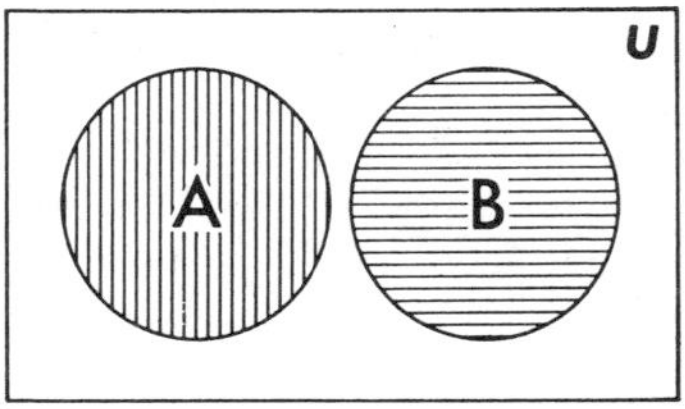

c.

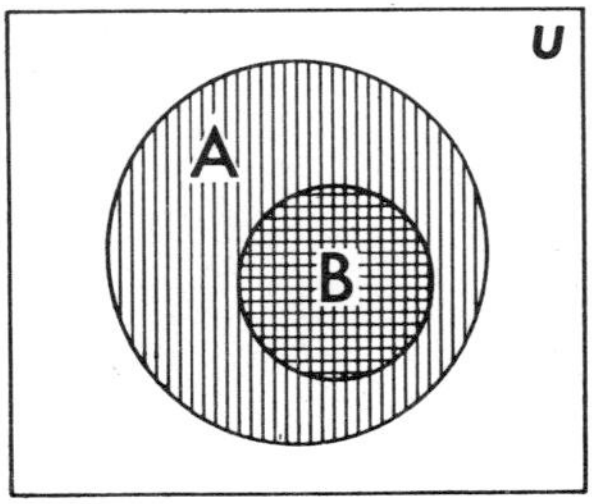

d.

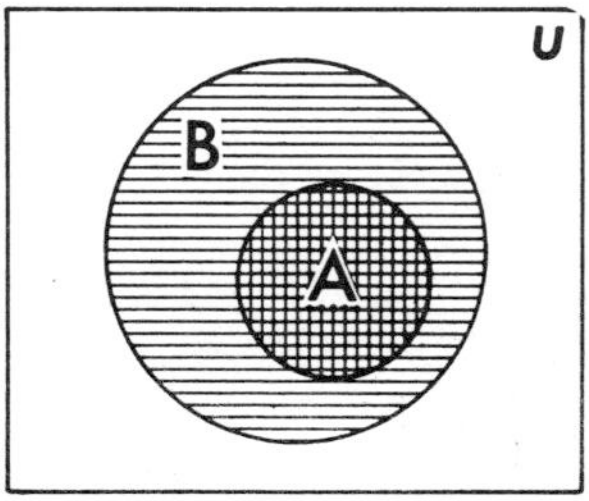

16. If U is the universe, set A is a subset of U, and $\emptyset$ is the null set, represent with a Venn diagram:

a. $A \cap U$ *b.* $A \cap A$ *c.* $A \cap \emptyset$

UNION OF SETS

The ***union*** of two sets, A and B, is the set containing all the members of both set A and set B, any member present in both set A and set B being listed only once. The union of set A and set B is represented by the symbol $A \cup B$, read "A cup B." Thus, if $A = \{5, 6, 7, 8, 9, 10, 11\}$ and $B = \{9, 10, 11, 12, 13, 14\}$, then $A \cup B = \{5, 6, 7, 8, 9, 10, 11, 12, 13, 14\}$. Note that both set A and set B are subsets of the set $A \cup B$. The union of set A and set B, $A \cup B$, can be shown pictorially by a Venn diagram in which U is the set of natural numbers. The shaded region represents $A \cup B$.

$A \cup B$ When A and B Intersect

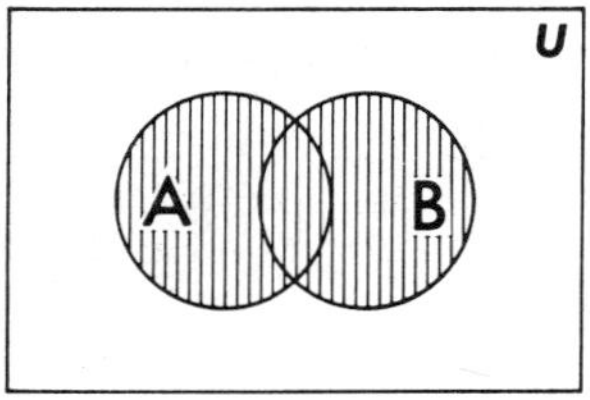

Suppose $U = \{\text{natural numbers}\}$. If A is the set of even natural numbers less than 10, then $A = \{2, 4, 6, 8\}$. If B is the set of odd natural numbers less than 10, then $B = \{1, 3, 5, 7, 9\}$. The union of set A and set B can be written $A \cup B = \{1, 2, 3, 4, 5, 6, 7, 8, 9\}$. The union of set A and set B, $A \cup B$, can be shown pictorially by a Venn diagram. Both shaded regions represent $A \cup B$.

$A \cup B$ When A and B Are Disjoint Sets

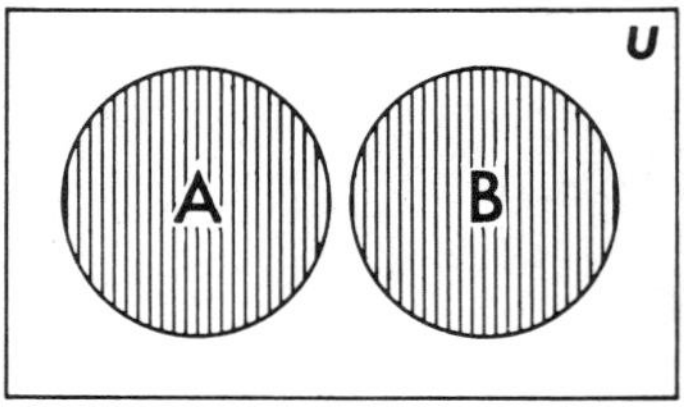

Suppose $U = \{\text{natural numbers}\}$. If A is the set of natural numbers less than 10, then $A = \{1, 2, 3, 4, 5, 6, 7, 8, 9\}$. If B is the set of odd natural numbers less than 10, then $B = \{1, 3, 5, 7, 9\}$. The union of set A and set B can be written $A \cup B = \{1, 2, 3, 4, 5, 6, 7, 8, 9\}$. A Venn diagram can be used to represent $A \cup B$ pictorially. The shaded region represents $A \cup B$. When B is a subset of A, then $A \cup B = A$.

$A \cup B$ When B Is a Subset of A

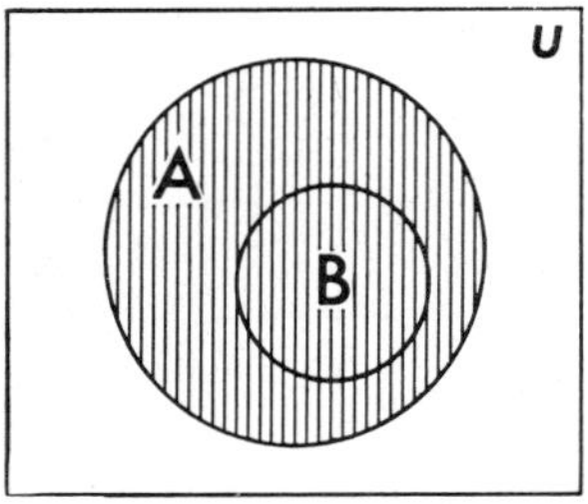

EXERCISES

1. $U = \{\text{natural numbers}\}$, $A = \{10, 11, 12, 13\}$, $B = \{11, 12, 13, 14\}$.

a. Tabulate the members of $A \cup B$.

b. Picture $A \cup B$ with a Venn diagram.

2. U = {letters of the English alphabet}, $A = \{a, e, i, o, u\}$, $B = \{r, s, t\}$.
   a. Tabulate the members of $A \cup B$.
   b. Picture $A \cup B$ with a Venn diagram.
3. U = {natural numbers}, A = {natural numbers less than 30 that are divisible by 5}, B = {natural numbers less than 30 that are divisible by 10}.
   a. List the members of set A, set B, and $A \cup B$.
   b. Picture $A \cup B$ with a Venn diagram.
4. U = {pupils in your school}, A = {boys in your class}, B = {girls in your class}. Picture $A \cup B$ with a Venn diagram.
5. U = {pupils in a class}, A = {girls in the class}, B = {blue-eyed girls in the class}. Picture $A \cup B$ with a Venn diagram.
6. U = {pupils in a class}, A = {blond girls in the class}, B = {blue-eyed, blond girls in the class}. Represent $A \cup B$ with a Venn diagram.
7. If U is the universe, set A is a subset of U, and $\emptyset$ is the null set, represent with a Venn diagram:

   a. $A \cup U$ *b.* $A \cup A$ *c.* $A \cup \emptyset$

6. SETS IN ALGEBRA

As we have seen previously, a sentence that can be judged to be true or false is called a ***statement.*** For example:

"Alaska is the largest state in the U.S.A." is a true statement.
"Maine is south of Georgia" is a false statement.

STATEMENTS OF EQUALITY

A sentence stating that two number expressions are equal, symbolized =, is called an ***equation.*** For example, the sentence $4 + 1 = 3 + 2$ is an equation. An equation may be a statement that is true, or it may be a statement that is false. For example:

$5 + 4 = 9$ is an equation that is a true statement.
$8 - 2 = 4$ is an equation that is a false statement.

EXERCISES

Tell whether each of the following statements is true or false:

1. $8 + 5 = 6 + 7$	**2.** $8 - 3 = 4 + 1$	**3.** $6 \times \frac{1}{2} = 12$
4. $12 \div 3 = 4$	**5.** $7 + 0 = 7$	**6.** $7 \times 0 = 7$
7. $4 \times 1 = 4$	**8.** $9 - 0 = 0$	**9.** $5 + 3 = 3 + 5$
10. $10 - 7 = 7 - 10$	**11.** $8 \div 4 = 4 \div 8$	**12.** $8 \times 4 = 4 \times 8$
13. $\frac{1}{2} + \frac{1}{2} = \frac{1}{4}$	**14.** $\frac{1}{2} + \frac{1}{2} = 1$	**15.** $\frac{1}{2} - \frac{1}{2} = 0$
16. $\frac{1}{4} + \frac{3}{4} = \frac{4}{8}$	**17.** $\frac{2}{4} = \frac{1}{2}$	**18.** $.05 \times 2 = 1$

19. The sum of 8 and 4 equals the product of 2 and 6.

20. The sum of 12 and 2 equals 4 more than 10.

STATEMENTS OF INEQUALITY

The equation $5 + 4 = 7$ is a false statement. We can make a true statement about $5 + 4$ by saying, "$5 + 4$ is not equal to 7." The statement "$5 + 4$ is not equal to 7" may be written $5 + 4 \neq 7$. The symbol $\neq$, which means "is not equal to," is a symbol of ***inequality***. The statement $5 + 4 \neq 7$ is a statement of inequality.

The statement "8 is greater than 5" may be written as $8 > 5$. The symbol $>$ means "is greater than."

The statement "4 is less than 9" may be written as $4 < 9$. The symbol $<$ means "is less than."

Notice that in the inequalities $8 > 5$ and $4 < 9$, the symbol points to the *smaller* number.

EXERCISES

In 1–8, state whether each of the inequalities is true or false:

1. $6 + 9 \neq 12$ **2.** $8 - 2 \neq 6$ **3.** $5 \times 0 \neq 7 \times 0$

4. $8 + 7 \neq 7 + 8$ **5.** $5 \times 1 \neq 5 \div 1$ **6.** $\frac{1}{2} + \frac{1}{2} \neq \frac{1}{2} \times \frac{1}{2}$

7. The product of 5 and 3 is not equal to the sum of 5 and 3.

8. The sum of 10 and 20 is not equal to 40 decreased by 10.

In 9–18, write the inequality using the symbol $>$ or the symbol $<$:

9. 14 is greater than 10. **10.** $8 + 9$ is less than 30.

11. $4 + 2$ is less than $5 + 6$. **12.** 20×2 is greater than $20 + 2$.

13. The sum of 6 and 9 is less than 18.

14. The product of 8 and 9 is greater than 50.

15. The product of 5 and 4 is greater than the sum of 5 and 4.

16. 10 decreased by 4 is less than 12 decreased by 4.

17. $\frac{1}{2}$ of 100 is less than $\frac{3}{4}$ of 100.

18. Twice 40 is greater than 4 times 15.

In 19–24, express each inequality in words:

19. $15 > 12$ **20.** $16 < 20$ **21.** $8 + 3 > 7 + 1$

22. $6 - 2 < 12 - 3$ **23.** $2 \times 0 < 2 + 0$ **24.** $8 + 1 > 8 \times 1$

In 25–32, state whether each of the inequalities is true or false:

25. $5 + 3 > 7$ **26.** $15 - 4 < 13$ **27.** $9 + 1 > 12 - 2$

28. $2 \times 5 < 3 + 6$ **29.** $5 \times 0 > 2 + 1$ **30.** $10 \times 1 > 20 \times .1$

31. The sum of 12 and 4 is less than the product of 12 and 4.

32. The product of 25 and 0 is greater than the product of 10 and 1.

REPRESENTING AND ORDERING NUMBERS ON A NUMBER LINE

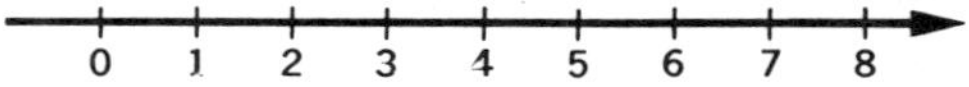

To construct the number line shown above, we start with two points on a line, the one at the left labeled "0" and the one at the right of 0 labeled "1." We use the length of the segment from 0 to 1 as a unit of measure to mark off more points to the right of 1 as shown in the above picture. The numbers 2, 3, 4, etc., are assigned to these points. The arrowhead means that the line extends without end.

A point corresponding to a number is called the ***graph*** of the number. Notice that the graph of a smaller number is to the left of the graph of a larger number. Since the graph of 5 is to the left of the graph of 6, then $5 < 6$. We say that we have ***ordered*** the numbers 5 and 6. Also, since the graph of 4 is to the right of the graph of 1, then $4 > 1$. We have ordered the numbers 1 and 4.

It is also possible to order three or more numbers. Suppose we have the numbers 3, 5, and 7. Then $3 < 5$ and $5 < 7$. Another way to write this is:

$$3 < 5 < 7$$

On the number line, 5 is to the right of 3 and to the left of 7. We say that 5 is "between" 3 and 7.

Notice that if 3 is less than 5 and if 5 is less than 7, then 3 is less than 7.

EXERCISES

1. Point P on the number line at the right is the graph of the number 0. Tell the number for which each of the remaining points on the number line is the graph.

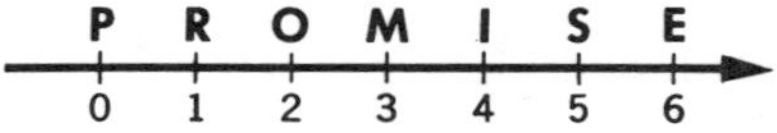

2. How can you tell from the number line that $1 < 5$?

3. How can you tell from the number line that $6 > 3$?

4. Use the number line to tell whether $0 > 4$ or $0 < 4$.

In 5–9, use the symbol $<$ to order the numbers:

5. 6, 10 **6.** 14, 5 **7.** $\frac{1}{2}, \frac{3}{4}$ **8.** 3.5, 2.7 **9.** $1\frac{3}{5}$, 1.8

In 10–14, use the symbol $<$ to order the numbers:

10. 1, 4, 7 **11.** 10, 7, 12 **12.** $\frac{1}{4}, \frac{7}{8}, \frac{2}{3}$ **13.** 2.1, 1.7, 3.2 **14.** $2\frac{1}{4}$, 2.75, $2\frac{1}{3}$

In 15–19, use the symbol $>$ to order the numbers:

15. 12, 8 **16.** $4\frac{1}{2}, 4\frac{7}{8}$ **17.** 6, 12, 15 **18.** 6.4, 5.9, 7.2 **19.** $1\frac{1}{2}$, 2.3, 1.2

In 20–23, state which number is "between" the other two:

20. 14, 18, 10 **21.** $6\frac{1}{3}, 5\frac{1}{2}, 7\frac{1}{4}$ **22.** 3.7, 5.6, 4.3 **23.** $2\frac{7}{8}$, 2.5, $3\frac{1}{4}$

GRAPHING SETS ON A NUMBER LINE

The ***graph of a set of numbers*** is the set of points on the number line corresponding to the numbers of the set.

MODEL PROBLEMS

1. Graph the set {1, 2, 3, 4} on a number line.
Solution:

Step 1. Draw a number line that has natural numbers assigned to one or two points beyond 4.

Step 2. Use a darkened dot to represent the graph of each number in the set.

Step 3. The graph of the set {1, 2, 3, 4} is the set of corresponding points indicated by darkened dots on the number line.

2. Graph {natural numbers greater than 1 and less than 4} on a number line.
Solution:

Step 1. List the set of numbers to be graphed. The set is {2, 3}.

Step 2. Draw a number line that has natural numbers assigned to one or two points beyond 3.

Step 3. Use a darkened dot to represent each number in the set.

Step 4. The graph of the set {2, 3} is the set of corresponding points indicated by darkened dots on the number line.

3. Graph {natural numbers equal to or greater than 1 and less than 5} on a number line.
Solution:

Step 1. List the set of numbers to be graphed. The set is {1, 2, 3, 4}.

Step 2. Draw a number line that has natural numbers assigned to one or two points beyond 4.

Step 3. Use a darkened dot to represent each number in the set.

Step 4. The graph of the set {1, 2, 3, 4} is the set of corresponding points indicated by darkened dots on the number line.

EXERCISES

Graph each of the following sets on a number line:

1. {3, 4, 6} **2.** {0, 1, 3} **3.** {5, 3, 1} **4.** {0}

5. {natural numbers greater than 2 and less than 4}
6. {natural numbers between 4 and 6}
7. {natural numbers greater than 3 and less than 6}
8. {even natural numbers greater than 4 and less than 15}
9. {natural numbers greater than 1 and less than or equal to 4}
10. {natural numbers equal to or greater than 2 and less than 10}
11. {natural numbers equal to or greater than 3 and less than 13}
12. {natural numbers less than 5 or equal to 5}
13. {natural numbers equal to or greater than 1 and equal to or less than 6}
14. {odd natural numbers between 0 and 10}

SOLVING OPEN SENTENCES

Consider the following sentence:

It is the capital of New York State.

We cannot tell whether this statement is true or false until the name of the particular city that is to replace "It" is known. Such a sentence is called an ***open sentence.***

In the open sentence "It is the capital of New York State," if "It" is replaced by Albany, the resulting sentence, "Albany is the capital of New York State," is a true statement. If "It" is replaced by the name of any city other than Albany—for example, Rochester—the resulting sentence, "Rochester is the capital of New York State," is a false statement.

In the sentence "It is the capital of New York State," the symbol "It" does not refer to a definite city, but to a set of cities. "It" is a symbol that holds a place in the sentence for a definite city. Such a symbol is called a ***variable.*** When "It" is replaced by the name of a particular city, the result is a statement that is true or false. The set of cities from which the replacement for the variable is selected is called the ***replacement set*** for the variable, or the ***domain*** of the variable.

Consider the following sentence:

$$x + 5 = 8$$

Since the truth or falsity of this sentence cannot be judged until we know to what definite thing or value x refers, $x + 5 = 8$ is an open sentence. In this sentence, the symbol x is the variable.

Let us replace x in the above open sentence by any member of the set {1, 2, 3, 4, 5}. This set, from which we may select values to replace x, is the replacement set for x, or the domain of x. When x is replaced by each of these values, we get the following series of statements:

$$x + 5 = 8$$

If $x = 1$, then $1 + 5 = 8$ is a false statement.
If $x = 2$, then $2 + 5 = 8$ is a false statement.
If $x = 3$, then $3 + 5 = 8$ is a *true* statement.
If $x = 4$, then $4 + 5 = 8$ is a false statement.
If $x = 5$, then $5 + 5 = 8$ is a false statement.

When the possible replacements for x are members of the set $\{1, 2, 3, 4, 5\}$, the only replacement for x that results in a true statement is 3. We say that the number 3 ***solves*** the open sentence or that the number 3 is the ***solution*** of the open sentence. The solution 3 itself forms a set, $\{3\}$, called the solution set of the open sentence. The ***solution set*** of the open sentence consists of the member or members of the replacement set that change the open sentence to a true statement. Notice that the solution set is a subset of the replacement set.

We have just seen that the number 3 solves the equation $x + 5 = 8$. Therefore, we may say that the solution set of the equation $x + 5 = 8$ is $\{3\}$. We may also say that 3 is the member of the solution set that ***satisfies*** the equation $x + 5 = 8$, and that 3 is a ***root*** of the equation $x + 5 = 8$.

Notice that the equation $x + 5 = 8$ did not become a true statement for all possible replacements from the domain. The equation became a true statement only when the variable x was replaced by 3. An equation that does not become a true statement for all possible replacements of the variable is called a ***conditional equation.***

If we consider the equation $x + 5 = 5 + x$ with the domain $\{1, 2, 3, 4, 5\}$, we notice that when x is replaced by each member of the domain, the open sentence becomes a true statement. An equation that becomes a true statement when each member of the domain replaces the variable is called an ***identity.***

MODEL PROBLEMS

1. Using the replacement set {Japan, France, Greece, Canada, India}, find the replacement or replacements for x that will change the open sentence "x is in North America" into a true statement.

Solution: In the open sentence "x is in North America," replace the variable x by each member of the replacement set.

x is in North America.
Japan is in North America is a false statement.
France is in North America is a false statement.
Greece is in North America is a false statement.
Canada is in North America is a *true* statement.
India is in North America is a false statement.

Answer: Canada

[*Note:* Since Canada is the replacement that changes the open sentence to a true statement, Canada is the solution of the open sentence. Also, the set {Canada} is the solution set of the open sentence.]

2. Using the replacement set {0, 1, 2, 3}, find the solution set for the open sentence $2n > 3$.

Solution: Replace the variable n in the open sentence $2n > 3$ by each member of the replacement set.

$$2n > 3$$

If $n = 0$, then $2 \times 0 > 3$, or $0 > 3$, is a false statement.
If $n = 1$, then $2 \times 1 > 3$, or $2 > 3$, is a false statement.
If $n = 2$, then $2 \times 2 > 3$, or $4 > 3$, is a *true* statement.
If $n = 3$, then $2 \times 3 > 3$, or $6 > 3$, is a *true* statement.

Answer: Since $n = 2$ and $n = 3$ are solutions of $2n > 3$, the solution set is {2, 3}.

3. Using the domain {1, 2, 3}, find the solution set for the open sentence $y + 5 = 9$.

Solution: Replace the variable y in the open sentence $y + 5 = 9$ by each member of the replacement set.

$$y + 5 = 9$$

If $y = 1$, then $1 + 5 = 9$ is a false statement.
If $y = 2$, then $2 + 5 = 9$ is a false statement.
If $y = 3$, then $3 + 5 = 9$ is a false statement.

Answer: Since no member of the replacement set changes the open sentence $y + 5 = 9$ into a true statement, we say that the solution set has no members. In other words, the solution set is the empty set, or null set, symbolized $\emptyset$ or { }.

EXERCISES

In 1–16, tell whether or not the sentences are open sentences:

1. Franklin D. Roosevelt was a president of the United States.

2. She is pretty.

3. Mercury is the smallest planet.

4. X is a member of the United Nations.

5. Columbus discovered Alaska.

6. He hit the most home runs in a season.

7. Y is the world's heavyweight champion.

8. $2 + 3 = 5 + 0$ 9. $6 - 3 = 5 + 1$ 10. $x + 10 = 14$
11. $y - 4 = 12$ 12. $6 + 5 > 10$ 13. $3 + 2 < 10 \times 0$
14. $n > 7$ 15. $r < 5 + 2$ 16. $2t > 5$

In 17–31, tell what is the variable:

17. He is the tallest boy in our class.
18. She has blue eyes.
19. It is the longest river in the United States.
20. He is our principal.
21. X is a senator.
22. The population of New York City is N people.
23. The highest mountain in the world is R.
24. $x + 5 = 9$ 25. $4y = 20$ 26. $r - 6 = 12$ 27. $.5x = 30$
28. $n > 6$ 29. $n + 1 < 3$ 30. $2d > 7$ 31. $14 < h + 9$
32. Using the replacement set {New York, California, Colorado, Florida, Maine}, find the replacement or replacements for Y that will change the open sentence "Y is east of Ohio" to a true statement.
33. Using the domain {Russia, China, United States, France, Argentina}, find the replacement or replacements that will change the open sentence "X has the greatest population in the world" to a true statement.
34. Using the replacement set {baseball, bridge, football, basketball, checkers, chess}, find the replacement or replacements that will change the open sentence "In playing a game of G, a ball is used" to a true statement.

In 35–50, use the replacement set {0, 1, 2, 3, 4, 5} to find all the replacements that will change the open sentence to a true statement. If no replacement will make a true statement, write *None*.

35. $n + 3 = 7$ 36. $5 - n = 2$ 37. $2x = 4$ 38. $5z = 0$
39. $y + 7 = 15$ 40. $2m = 7$ 41. $2x + 1 = 5$ 42. $x - x = 0$
43. $n > 2$ 44. $2n < 6$ 45. $n + 3 > 9$ 46. $2n + 1 < 8$
47. $\frac{n+1}{2} = 2$ 48. $\frac{2n+1}{3} = 4$ 49. $\frac{n}{4} > 1$ 50. $\frac{3x}{2} < x$

In 51–58, using the replacement set {1, 2, 3, 4, 5, 6, 7, 8, 9, 10}, find the solution set:

51. $x + 6 = 9$ 52. $8 - x = 5$ 53. $2x + 1 = 24$ 54. $16 = 18 - x$
55. $y > 9$ 56. $4 < m$ 57. $2m > 17$ 58. $2x - 1 > 50$

In 59–66, using the replacement set {2, $2\frac{1}{2}$, 3, $3\frac{1}{2}$, 4, $4\frac{1}{2}$}, find the solution set:

59. $x + 2 = 4\frac{1}{2}$ 60. $2x = 7$ 61. $5 - r = \frac{1}{2}$ 62. $\frac{x}{2} = 2.25$
63. $y > 4$ 64. $m < 3$ 65. $2x > 8$ 66. $3a < 4.5$

In 67–74, using the domain {1, 2, 3, 4, 5}, find the solution set:

67. $y + 2 = 6$ **68.** $7 - t = 2$ **69.** $5n + 1 = 1$ **70.** $3x - 1 = 7$
71. $q < 2$ **72.** $2s + 1 > 8$ **73.** $4 < n + 1$ **74.** $2t + 3 > 15$

In 75–82, using the domain $\{3, 3\frac{1}{2}, 4, 4\frac{1}{2}, 5, 5\frac{1}{2}\}$, find the solution set:

75. $x + 2 = 6\frac{1}{2}$ **76.** $y - 2 = 3.5$ **77.** $2x - 1 = 5$ **78.** $5\frac{1}{2} = x + \frac{1}{2}$
79. $m > 5$ **80.** $m < 4\frac{1}{2}$ **81.** $2n > 8$ **82.** $2n < 7$

In 83–90, using the domain {2.1, 2.2, 2.3, 2.4, 2.5}, find the solution set:

83. $4n = 10$ **84.** $x + .1 = 2.4$ **85.** $3x - 4 = 2.3$ **86.** $\frac{y}{2} = 3.6$

87. $n > 2.0$ **88.** $y > 2\frac{1}{4}$ **89.** $z < 2\frac{1}{3}$ **90.** $2x + 3 < 6.5$

In 91–98, using the domain {1, 2, 3, 4, 5, 6, 7, 8, 9, 10}, find the root of the equation:

91. $x + 5 = 7$ **92.** $y - 3 = 4$ **93.** $2x + 1 = 9$ **94.** $6 = 3x - 18$

95. $\frac{1}{2}x + 4 = 5$ **96.** $14 - 2x = 2$ **97.** $\frac{x + 8}{4} = 3$ **98.** $3x - 2 = 28$

In 99–104, using the domain {5, 6, 7, 8, 9, 10}, tell whether the equation is a conditional equation or an identity:

99. $x + 3 = 3 + x$ **100.** $x + 3 = 10$ **101.** $y + 3 + 4 = 7 + y$
102. $5a = a \times 5$ **103.** $5a = 40$ **104.** $5 \times 2 \times a = a \times 2 \times 5$

GRAPHING THE SOLUTION SET OF AN OPEN SENTENCE

After we have found the solution set of an open sentence, we can graph the solution set by using the procedure we have already learned for graphing sets on a number line.

MODEL PROBLEMS

1. Using the replacement set {0, 1, 2, 3, 4}, find and graph the solution set for $2x + 1 = 5$.

Solution:

Step 1. Replace the variable x in the open sentence $2x + 1 = 5$ by each member of the replacement set.

$$2x + 1 = 5$$

If $x = 0$, then $2 \times 0 + 1 = 5$ is a false statement.
If $x = 1$, then $2 \times 1 + 1 = 5$ is a false statement.
If $x = 2$, then $2 \times 2 + 1 = 5$ is a *true* statement.
If $x = 3$, then $2 \times 3 + 1 = 5$ is a false statement.
If $x = 4$, then $2 \times 4 + 1 = 5$ is a false statement.

Step 2. Since 2 is the only member of the replacement set that changes the open sentence to a true statement, the solution set is $\{2\}$.

Step 3. Graph the solution set $\{2\}$.

2. Find and graph the solution set of $n + 1 > 2$ when the domain is $\{0, 1, 2, 3, 4\}$.

Solution:

Step 1. Replace the variable n in the open sentence $n + 1 > 2$ by each member of the domain.

$$n + 1 > 2$$

If $n = 0$, then $0 + 1 > 2$ is a false statement.
If $n = 1$, then $1 + 1 > 2$ is a false statement.
If $n = 2$, then $2 + 1 > 2$ is a *true* statement.
If $n = 3$, then $3 + 1 > 2$ is a *true* statement.
If $n = 4$, then $4 + 1 > 2$ is a *true* statement.

Step 2. Since the values 2, 3, and 4 are the members of the domain that change the open sentence to a true statement, the solution set is $\{2, 3, 4\}$.

Step 3. Graph the solution set $\{2, 3, 4\}$.

EXERCISES

In 1–8, find and graph the solution set when the replacement set is $\{0, 1, 2, 3, 4, 5\}$:

1. $2n = 4$ **2.** $n + 3 = 6$ **3.** $2x - 1 = 7$ **4.** $5x = 0$
5. $y > 4$ **6.** $t < 1$ **7.** $t - 3 > 1$ **8.** $3t + 1 > 7$

In 9–16, find and graph the solution set when the domain is $\{0, 1, 2, 3, 4, 5, 6, 7, 8, 9, 10\}$:

9. $4x = 12$ **10.** $y - 6 = 4$ **11.** $2z + 8 = 8$ **12.** $3a - 1 = 26$
13. $t > 5$ **14.** $s + 2 < 11$ **15.** $19 < 3x + 2$ **16.** $14 > 5x + 1$

EQUIVALENT EQUATIONS

Previously (page 201), we learned that equations whose roots are the same are called ***equivalent equations.*** For example, consider the following equations when the domain is the set $\{0, 1, 2, 3, 4, 5, 6, 7, 8, 9, 10\}$:

$$3x + 1 = 19 \qquad 3x = 18 \qquad x = 6$$

All the equations have the same root, 6. Therefore, $3x + 1 = 19$, $3x = 18$, and $x = 6$ are equivalent equations. We can say that the solution set of each of these equations is $\{6\}$. We can also say that equivalent equations are equations having the same solution set of one or more members.

EXERCISES

In 1–7, use the domain {1, 2, 3, 4, 5, 6, 7, 8, 9, 10} to discover whether or not the equations are equivalent equations:

1. $5x = 15 \quad x = 3$
2. $\frac{1}{4}x = 2 \quad x = 8$
3. $x + 5 = 9 \quad x = 4$
4. $x - 6 = 3 \quad x = 7$
5. $2x + 1 = 11 \quad 2x = 10 \quad x = 5$
6. $3x - 5 = 13 \quad 3x = 18 \quad x = 6$
7. $\frac{1}{2}x + 3 = 7 \quad \frac{1}{2}x = 4 \quad x = 2$

In 8–15, using the domain {1, 2, 3, 4, 5, 6, 7, 8, 9, 10}, write two equations equivalent to the given equation:

8. $8x = 64$
9. $\frac{x}{4} = 2$
10. $y + 3 = 8$
11. $y - 6 = 4$
12. $3x + 1 = 13$
13. $4x - 1 = 35$
14. $\frac{y}{2} - 4 = 1$
15. $\frac{y}{3} + 5 = 8$

In 16–24, using all the numbers you know as a domain, find the solution set for each equation and tell whether or not the equations are equivalent:

16. $6x = 42 \quad x = 7$
17. $\frac{x}{4} = 15 \quad x = 60$
18. $2x + 5 = 25 \quad 2x = 20 \quad x = 10$
19. $3x - 1 = 15 \quad 3x = 16 \quad x = 5\frac{1}{3}$
20. $\frac{y}{2} + 7 = 16 \quad \frac{y}{2} = 9 \quad y = 18$
21. $\frac{t}{3} - 4 = 11 \quad \frac{t}{3} = 15 \quad t = 45$
22. $30 = 2x + 4 \quad 26 = 2x \quad x = 13$
23. $17 = 4x - 1 \quad 18 = 4x \quad x = 4.5$
24. $8 + 6x = 32 \quad 6x = 24 \quad x = 4$

In 25–28, using all the numbers you know as a replacement set, write the equation that is not equivalent to the other two equations:

25. $4x = 40 \quad 2x = 38 \quad x = 10$
26. $\frac{x}{2} = 60 \quad \frac{x}{6} = 180 \quad x = 120$
27. $y = 5 \quad 2y + 1 = 11 \quad 2y = 12$
28. $t = 6 \quad 3t = 18 \quad 3t - 1 = 18$

FINDING THE SOLUTION SET OF AN EQUATION

We have seen that the open sentence $x + 5 = 8$ is called an equation because it contains the symbol $=$. When we have discovered the set of numbers each of whose members can replace the variable x in the equation and make the resulting sentence a true statement, we have found the solution set of the equation. We then say that we have ***solved*** the equation. In the case of the equation $x + 5 = 8$, we have seen that when the domain is the set of all the numbers we know, the solution set of the equation $x + 5 = 8$ is {3}. The member of the solution set, 3, ***satisfies*** the equation; that is, when 3 replaces the variable, it changes the open sentence to a true statement. The member of the solution set is called a ***root*** of the equation. Thus, 3 is a root of the equation $x + 5 = 8$.

In chapter X, pages 198–211, we learned how to solve equations, that is, how to find their roots. Actually, we were learning how to find the members of the solution sets of the equations when the domain was the set of all the numbers we knew.

Let us remember that when we are solving an equation, we try to find a simpler equivalent equation whose solution is obvious and reveals the value of the variable that satisfies the equation we are solving.

To find simpler equivalent equations, we used the following principles:

1. If both sides of an equation are divided by the same number (not zero), an equivalent equation results.
2. If both sides of an equation are multiplied by the same number (not zero), an equivalent equation results.
3. If the same number is added to both sides of an equation, an equivalent equation results.
4. If the same number is subtracted from both sides of an equation, an equivalent equation results.

Furthermore, to help us find simpler equivalent equations whose solutions were obvious, we made use of inverse operations. Let us keep in mind that addition and subtraction are inverse operations; also, multiplication and division are inverse operations.

MODEL PROBLEMS

In 1–5, use the set of all the numbers you know as the domain.

1. For the equation $7x = 21$:

a. Find the solution set of the equation.

b. Graph the solution set of the equation.

Solution:

a. $7x = 21$ Since the operation relating 7 and x is multiplication, use the inverse operation, division, and the division principle.

$$\frac{7x}{7} = \frac{21}{7}$$

$$x = 3$$

The solution of $x = 3$ is obviously 3.

To learn whether or not 3 is also the solution of $7x = 21$, replace the variable x by 3.

$$7x = 21$$

$7 \times 3 = 21$ is a true statement.

Therefore, 3 is the solution of $7x = 21$.

The solution set of $7x = 21$ is $\{3\}$. *Ans.*

b. Graph the solution set $\{3\}$.

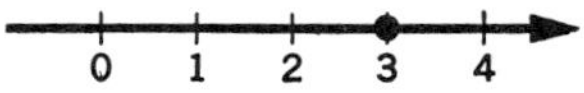

2. For the equation $\frac{x}{2} = 3$:

a. Find the solution set of the equation.
b. Graph the solution set of the equation.

Solution:

a. $\frac{x}{2} = 3$ Since the operation relating x and 2 is division, use the inverse operation, multiplication, and the multiplication principle.

$$2 \times \frac{x}{2} = 2 \times 3$$

$$x = 6$$

The solution of $x = 6$ is obviously 6.

To learn whether or not 6 is also the solution of $\frac{x}{2} = 3$, replace the variable x by 6.

$$\frac{x}{2} = 3$$

$\frac{6}{2} = 3$ is a true statement.

Therefore, 6 is the solution of $\frac{x}{2} = 3$.

The solution set of $\frac{x}{2} = 3$ is $\{6\}$. *Ans.*

b. Graph the solution set $\{6\}$.

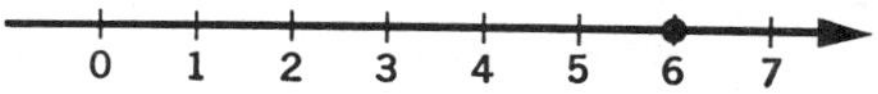

3. For the equation $x - 2 = 3$:

a. Find the solution set of the equation.
b. Graph the solution set of the equation.

Solution:

a. Since the operation relating x and 2 is subtraction, use the inverse operation, addition, and the addition principle.

$$x - 2 = 3$$
$$x - 2 + 2 = 3 + 2$$
$$x = 5$$

The solution of $x = 5$ is obviously 5.

To learn whether or not 5 is also the solution of $x - 2 = 3$, replace the variable x by 5.

$$x - 2 = 3$$

$5 - 2 = 3$ is a true statement.

Therefore, 5 is the solution of $x - 2 = 3$.
The solution set of $x - 2 = 3$ is $\{5\}$. *Ans.*

b. Graph the solution set $\{5\}$.

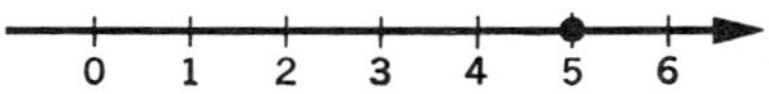

4. For the equation $x + 1\frac{1}{2} = 4$:

a. Find the solution set of the equation.
b. Graph the solution set of the equation.

Solution:

a.

$$x + 1\tfrac{1}{2} = 4$$
$$x + 1\tfrac{1}{2} - 1\tfrac{1}{2} = 4 - 1\tfrac{1}{2}$$
$$x = 2\tfrac{1}{2}$$

Since the operation relating x and $1\frac{1}{2}$ is addition, use the inverse operation, subtraction, and the subtraction principle.

The solution of $x = 2\frac{1}{2}$ is obviously $2\frac{1}{2}$.
To learn whether or not $2\frac{1}{2}$ is also the solution of $x + 1\frac{1}{2} = 4$, replace the variable x by $2\frac{1}{2}$.

$$x + 1\tfrac{1}{2} = 4$$
$$2\tfrac{1}{2} + 1\tfrac{1}{2} = 4 \text{ is a true statement.}$$

Therefore, $2\frac{1}{2}$ is the solution of $x + 1\frac{1}{2} = 4$.
The solution set of $x + 1\frac{1}{2} = 4$ is $\{2\frac{1}{2}\}$. *Ans.*

b. Graph the solution set $\{2\frac{1}{2}\}$.

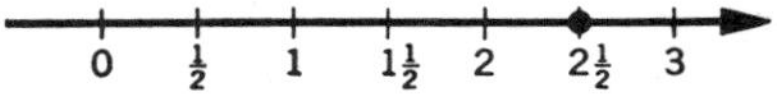

5. For the equation $3x + 1 = 7$:

a. Find the solution set of the equation.
b. Graph the solution set of the equation.

Solution:

a.

$$3x + 1 = 7$$
$$3x + 1 - 1 = 7 - 1$$

Since the operation relating $3x$ and 1 is addition, use the inverse operation, subtraction, and the subtraction principle.

$$3x = 6$$
$$\frac{3x}{3} = \frac{6}{3}$$
$$x = 2$$

Since the operation relating 3 and x is multiplication, use the inverse operation, division, and the division principle.

The solution of $x = 2$ is obviously 2.
To learn whether or not 2 is also the solution of $3x + 1 = 7$, replace the variable x by 2.

$$3x + 1 = 7$$
$$3 \times 2 + 1 = 7 \text{ or } 6 + 1 = 7 \text{ is a true statement.}$$

Therefore, 2 is the solution of $3x + 1 = 7$.
The solution set of $3x + 1 = 7$ is $\{2\}$. *Ans.*

b. Graph the solution set $\{2\}$.

EXERCISES

In 1–56, find and graph the solution set. Use the set of all the numbers you know as the domain.

1. $3x = 15$	**2.** $5y = 15$	**3.** $2a = 5$	**4.** $4b = 9$
5. $4r = 2$	**6.** $.3y = 1.5$	**7.** $1.5x = 6$	**8.** $\frac{2}{3}x = 4$
9. $\frac{x}{2} = 5$	**10.** $\frac{y}{3} = 1$	**11.** $\frac{y}{4} = \frac{1}{2}$	**12.** $\frac{r}{.4} = 10$
13. $x + 2 = 6$	**14.** $y + 8 = 9$	**15.** $a + 7 = 10$	**16.** $b + 2 = 2$
17. $y + 1\frac{1}{4} = 2\frac{3}{4}$	**18.** $x + 3\frac{1}{2} = 7$	**19.** $x + .1 = 1.6$	**20.** $y + 1.5 = 4$
21. $x - 1 = 3$	**22.** $a - 3 = 4$	**23.** $y - 7 = 1$	**24.** $x - 5 = 2$
25. $x - \frac{1}{2} = 3\frac{1}{2}$	**26.** $x - 1\frac{1}{4} = 2\frac{1}{4}$	**27.** $y - .5 = 1.5$	**28.** $z - .7 = 1.8$
29. $3x + 5 = 8$	**30.** $2x + 1 = 13$	**31.** $6x + 2 = 5$	**32.** $4x + 5 = 11$
33. $4x - 1 = 19$	**34.** $3x - 6 = 9$	**35.** $4y - 2 = 4$	**36.** $8m - 1 = 5$
37. $21 = 7r$	**38.** $12 = x + 6$	**39.** $4 = \frac{y}{2}$	**40.** $5 = y - 3$
41. $13 = 3x + 4$	**42.** $19 = 5x - 1$	**43.** $20 = 4x + 6$	**44.** $5 = 8x - 1$
45. $\frac{y}{2} + 1\frac{1}{2} = 5$	**46.** $\frac{y}{2} - 1.7 = 2.8$	**47.** $6 = n + \frac{1}{2}$	**48.** $1.9 = b + 1.6$
49. $3x + 2x = 15$	**50.** $y + 2y = 18$	**51.** $2m + m = 12$	**52.** $20 = 3m + m$
53. $4a - 2a = 8$	**54.** $5x - 3x = 12$	**55.** $6x + x = 14$	**56.** $9y - y = 24$

GRAPHING AND ORDERING SIGNED NUMBERS ON A NUMBER LINE

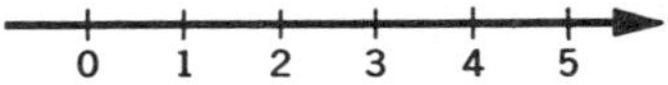

At the right, we see the number line with which we have been working until now. Notice that it extends only to the right.

Below, we see a number line that extends in two directions.

On this number line, we assign positive numbers to the points on the right side of 0, and we assign negative numbers to the points on the left side of 0. Once again, the graph of a smaller number is to the left of the graph of a larger number. Since the graph of -3, read "negative 3," is to the left of the graph of -1, read "negative 1," then $-3 < -1$. We say that we have ordered the numbers -3 and -1. Also, since the graph of $+5$, read "positive 5," is to the right of the graph of -1, then $+5 > -1$. We have ordered the numbers $+5$ and -1.

Since $-3 < -1$ and $-1 < +5$, we can write $-3 < -1 < +5$. That is, -1 is "between" -3 and $+5$.

EXERCISES

1. Point S on the number line at the right is the graph of the number -2. Tell the signed number for which each of the remaining points on the number line is the graph.

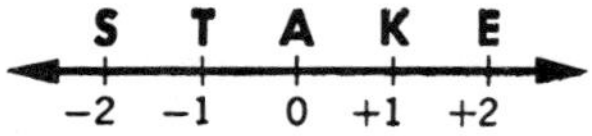

In 2–11, use a darkened dot to represent the signed number on a number line:

2. $+3$ **3.** -2 **4.** $+6$ **5.** -4 **6.** -8
7. $+1.5$ **8.** -2.5 **9.** $+2\frac{1}{2}$ **10.** $-3\frac{1}{2}$ **11.** $-1\frac{1}{2}$

12. How can you tell from a number line that $-2 < +2$?
13. How can you tell from a number line that $-1 > -4$?
14. Use the number line to tell whether $0 > -5$ or $0 < -5$.

In 15–18, use the symbol $<$ to order the numbers:

15. $-3, +7$ **16.** $+5, -4$ **17.** $+1\frac{1}{2}, -1\frac{1}{2}$ **18.** $-3.5, -3\frac{1}{4}$

In 19–26, tell whether or not the sentence is true. If it is not true, make it a true sentence by reversing the inequality symbol.

19. $+5 > +2$ **20.** $-1 < 0$ **21.** $+1 < 0$ **22.** $+8 > -2$
23. $-8 > -2$ **24.** $-4 > +1.5$ **25.** $-1\frac{3}{4} > -1\frac{7}{8}$ **26.** $0 < -\frac{1}{2}$

In 27–30, use the symbol $<$ to order the numbers:

27. $+2, -1, -3$ **28.** $-1, +7, 0$
29. $-2\frac{1}{2}, +6, +2\frac{1}{2}$ **30.** $+8, -2.5, -2\frac{3}{4}$

In 31–34, use the symbol $>$ to order the numbers:

31. $+7, -4$ **32.** $-9, +9$
33. $+2, -2, +4$ **34.** $-1.5, +3\frac{1}{2}, -2\frac{1}{2}$

In 35–38, state which number is "between" the other two:

35. $+6, -1, +3$ **36.** $+8, -8, -2$
37. $+.5, +1.1, -.8$ **38.** $-3\frac{1}{2}, -5\frac{3}{4}, -4\frac{1}{4}$

GRAPHING SETS AND SOLUTION SETS INVOLVING SIGNED NUMBERS ON A NUMBER LINE

The ***graph of a set of signed numbers*** is the set of points on the number line corresponding to the numbers of the set.

MODEL PROBLEMS

1. Graph the set $\{-3, -1, 0, +4\}$ on a number line.

Solution:

Step 1. Draw a number line.

Step 2. Use a darkened dot to represent the graph of each number of the set.

Step 3. The graph of the set $\{-3, -1, 0, +4\}$ is the set of corresponding points indicated by darkened dots on the number line.

2. If the domain is the set $\{-4, -3, -2, -1, 0, +1, +2, +3, +4\}$, graph the solution set of $x > -2$.

Solution:

Step 1. Draw a number line.

Step 2. List the members of the solution set $\{-1, 0, +1, +2, +3, +4\}$. Each of the members of this set, upon replacing the variable x, changes the open sentence $x > -2$ to a true statement.

Step 3. Graph the solution set $\{-1, 0, +1, +2, +3, +4\}$.

3. If the domain is the set of signed numbers, graph the solution set of $x > -2$.

Solution:

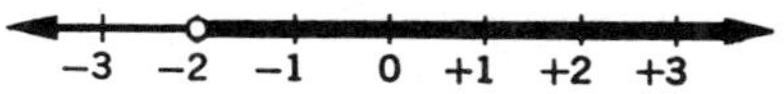

Step 1. Draw a number line.

Step 2. The members of the solution set will be the numbers greater than -2. The number of members in the solution set is infinite.

Step 3. Graph the members of the solution set. The open circle at -2 means that the graph does not contain the point corresponding to -2.

4. If the domain is the set of signed numbers, graph the solution set of $x > -2$ or $x = -2$, meaning x is greater than -2 or equal to -2, which may be written more compactly $x \geq -2$.

Solution:

Step 1. The solution set is the infinite set whose members are greater than -2 or equal to -2.

Step 2. Graph the members of the solution set. The darkened dot at -2 shows that the graph does contain the point corresponding to -2.

5. If the domain is the set of signed numbers, graph the solution set of $-2 < x < +2$.

Solution:

Step 1. The solution set is the infinite set whose members are greater than -2 and less than $+2$.

Step 2. Graph the members of the solution set. The open circles at -2 and at $+2$ mean that the graph does not contain the points corresponding to -2 and $+2$.

EXERCISES

In 1–12, if the domain is the set $\{-4, -3, -2, -1, 0, +1, +2, +3, +4\}$, graph the solution set of each open sentence and state whether the solution set is finite or infinite:

1. $x > 0$ **2.** $x < 0$ **3.** $x \geq 0$
4. $x \leq 0$ **5.** $x > -1$ **6.** $x < +2$
7. $x \geq +1$ **8.** $x \leq -1$ **9.** $-1 < x < +3$
10. $-2 \leq x < +3$ **11.** $-1 < x \leq +2$ **12.** $-2 \leq x \leq +3$

In 13–24, if the domain is the set of signed numbers, graph the solution set of each open sentence and state whether the solution set is finite or infinite:

13. $x > +2$ **14.** $x < -2$ **15.** $x \geq 0$
16. $x \leq 0$ **17.** $x > -3$ **18.** $x < +4$
19. $x \geq +2$ **20.** $x \leq -2$ **21.** $-2 < x < +3$
22. $-1 \leq x < +2$ **23.** $-1 < x \leq +2$ **24.** $-3 \leq x \leq +3$

7. USING THE NUMBER LINE TO ADD SIGNED NUMBERS

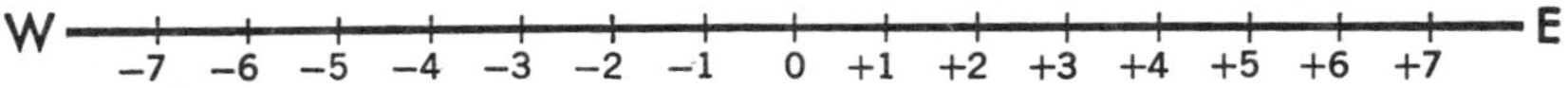

On the east-west line above, 0 represents a starting point. The numbers east of 0 are designated as positive numbers; the numbers west of 0 are designated as negative numbers. Also, if we start at any point on the number line and travel east, we will say that we are traveling in a positive direction; if we start at any point on the number line and travel west, we will say that we are traveling in a negative direction. Thus, if we begin at the starting point, 0, and make a trip 4 miles east, the trip is designated $(+4)$; if we begin at the starting point, 0, and make a trip 4 miles west, the trip is designated (-4).

If a man began a trip at the starting point, 0, and traveled 2 miles east, represented by the directed line segment $(+2)$, and then traveled 4 additional miles east, represented by the directed line segment $(+4)$, which begins where the first directed line segment $(+2)$ ends, he was at $(+6)$ on the number line; that is, he was 6 miles east of the starting point.

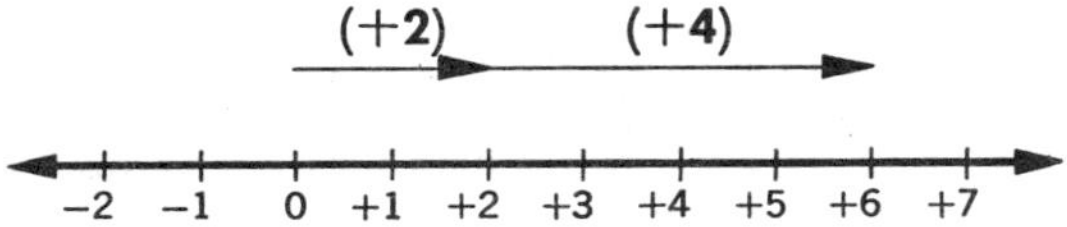

This addition of the signed numbers $(+2)$ and $(+4)$ may be written vertically or horizontally as follows:

$$\begin{array}{r} (+2) \\ +\ (+4) \\ \hline +6 \end{array} \quad \text{or} \quad (+2) + (+4) = +6$$

If a man began a trip at the starting point, 0, and traveled 2 miles west, represented by the directed line segment (−2), and then traveled 4 additional miles west, represented by the directed line segment (−4), which begins where the first directed line segment (−2) ends, he was at (−6) on the number line; that is, he was 6 miles west of the starting point.

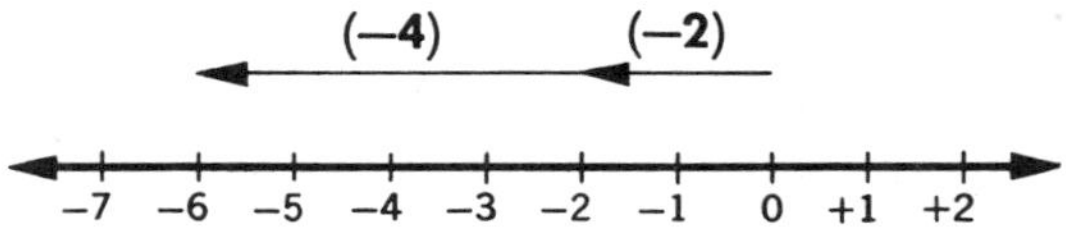

This addition of the signed numbers (−2) and (−4) may be written vertically or horizontally as follows:

$$\begin{array}{r} (-2) \\ +\ (-4) \\ \hline -6 \end{array} \quad \text{or} \quad (-2) + (-4) = -6$$

If a man began a trip at the starting point, 0, and traveled 2 miles east, represented by the directed line segment (+2), and then traveled 4 miles west, represented by the directed line segment (−4), which begins where the first directed line segment (+2) ends, he was at (−2) on the number line; that is, he was 2 miles west of the starting point. This addition of the signed numbers (+2) and (−4) may be written vertically or horizontally as follows:

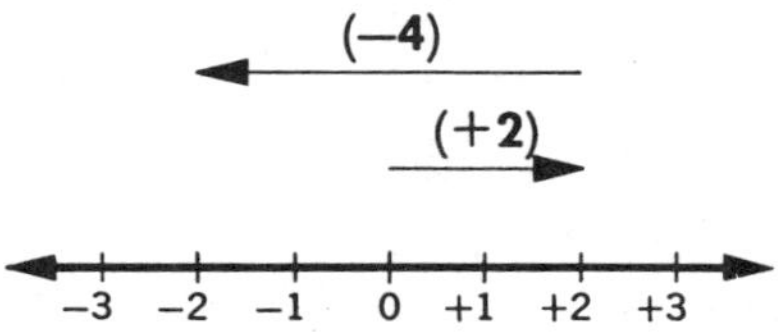

$$\begin{array}{r} (+2) \\ +\ (-4) \\ \hline -2 \end{array} \quad \text{or} \quad (+2) + (-4) = -2$$

If a man began a trip at the starting point, 0, and traveled 2 miles west, represented by the directed line segment (−2), and then traveled 4 miles east, represented by the directed line segment (+4), which begins where the first directed line segment (−2) ends, he was at (+2) on the number line; that is, he

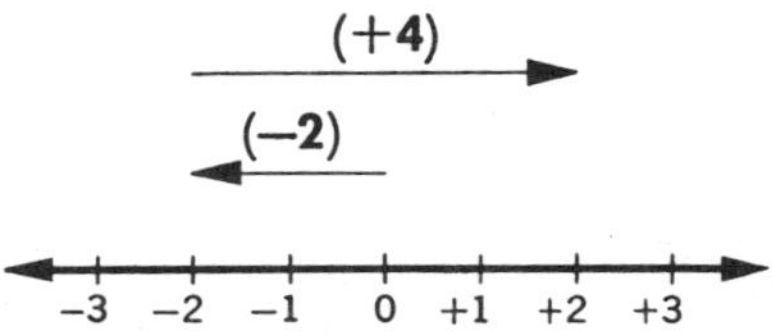

was 2 miles east of the starting point. This addition of the signed numbers (-2) and $(+4)$ may be written vertically or horizontally as follows:

$$\begin{array}{r} (-2) \\ \underline{+\ (+4)} \\ +2 \end{array} \quad \text{or} \quad (-2) + (+4) = +2$$

If a man began a trip at the starting point, 0, and traveled 4 miles east, represented by the directed line segment $(+4)$, and then traveled 4 miles west, represented by the directed line segment (-4), which begins where the first directed line segment $(+4)$ ends, he was back at the starting point, 0. This addition of the signed numbers $(+4)$ and (-4) may be written vertically or horizontally as follows:

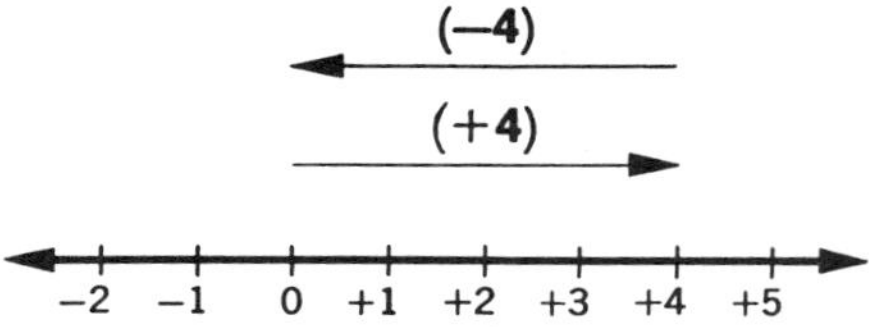

$$\begin{array}{r} (+4) \\ \underline{+\ (-4)} \\ 0 \end{array} \quad \text{or} \quad (+4) + (-4) = 0$$

The rules for adding signed numbers were discussed on pages 224–226.

EXERCISES

In this group of exercises, use a number line whenever it will help you to discover the correct answer.

1. Use the *E–W* number line on page 376 to tell whether a man travels east or west when he travels from:

a. 0 to $(+2)$ *b.* 0 to (-3) *c.* $(+1)$ to $(+5)$ *d.* $(+2)$ to 0
e. (-3) to 0 *f.* (-1) to (-5) *g.* $(+3)$ to (-2) *h.* (-2) to $(+6)$

2. Tell whether a man travels in a positive direction or a negative direction when he travels from:

a. 0 to (-5) *b.* 0 to $(+4)$ *c.* (-4) to 0 *d.* $(+8)$ to 0
e. $(+2)$ to $(+7)$ *f.* $(+2)$ to (-6) *g.* (-1) to $(+3)$ *h.* (-3) to (-7)

In 3–20, add the signed numbers:

3. $(+4) + (+5)$ **4.** $(-3) + (-2)$ **5.** $(-3) + (+6)$
6. $(+4) + (-5)$ **7.** $(+4) + (-4)$ **8.** $(-2) + (+2)$
9. $(-4) + (-7)$ **10.** $(+3) + (-5)$ **11.** $(+6) + (+4)$
12. $(-9) + (+2)$ **13.** $(+8) + (-6)$ **14.** $(-5) + (+7)$
15. $(+2) + (+3) + (+4)$ **16.** $(-5) + (-2) + (-1)$
17. $(+6) + (+8) + (-4)$ **18.** $(-3) + (-7) + (+6)$
19. $(-5) + (+4) + (-2)$ **20.** $(+7) + (+2) + (-9)$

In 21–30, add the numbers that are arranged vertically:

21.	22.	23.	24.	25.
(+8)	(−7)	(+9)	(−12)	(+8)
(+4)	(−8)	(−6)	(+7)	(−8)

26.	27.	28.	29.	30.
(+2)	(−1)	(+5)	(−6)	(+3)
(+6)	(−4)	(+7)	(+4)	(−7)
(+7)	(−8)	(−9)	(−5)	(+4)

31. Select a signed number that will make the statement true:

a. (+7) + (?) = +2 *b.* (+4) + (?) = −3
c. (−4) + (?) = +5 *d.* (−5) + (?) = −8

32. In four successive plays from scrimmage starting at their own ten-yard line, Jefferson High School makes (1) a gain of 7 yards (2) a loss of 4 yards (3) a gain of 3 yards (4) a loss of 5 yards. Using a signed number line, find the position of the ball after the fourth play.

33. A company reports for three months of a year: Jan., $5000 profit; Feb., $2000 loss; Mar., $4000 profit. Using a signed number line (scale unit of 1 = $1000), find the total profit for the first 3 months.

In 34–37, find the sum:

34. (+1) + (−1) **35.** (+5) + (−5) **36.** (−7) + (+7) **37.** (−9) + (+9)

Find a number that must be added to the given number to obtain a sum of 0:

38. (+2) **39.** (−4) **40.** (+9) **41.** (−15) **42.** (−75)

In 43–44, tell whether the statement is true or false:

43. For every positive number there is a negative number such that the sum of the two numbers is 0.

44. For every negative number there is a positive number such that the sum of the two numbers is 0.

45. Two numbers whose sum is 0—for example, (+2) and (−2)—are called ***opposites.*** Find the opposite of:

a. (+1) *b.* (−6) *c.* (+10) *d.* (−15) *e.* (−20) *f.* (+25)

46. The ***absolute value*** of a number is the larger of itself and its opposite. For example, the absolute value of (+3) is (+3), or 3, because (+3), or 3, is larger than its opposite (−3); also, the absolute value of (−4) is (+4), or 4, because (+4), or 4, is larger than its opposite (−4). Find the absolute value of:

a. (+1) *b.* (−2) *c.* (+9) *d.* (−10) *e.* (−8) *f.* (+15)

In 47–51, find the sum:

47. 0 + (+2) **48.** 0 + (−4) **49.** (+10) + 0
50. (−12) + 0 **51.** 0 + (+100)

In 52–53, tell whether the statement is true or false:

52. If a number is added to 0, the sum is the number that was added to 0.

53. If 0 is added to a number, the sum is the number to which 0 was added.

REVIEW EXERCISES

REVIEW EXERCISE 1

Answer all questions. Each question has 2 credits assigned to it; no partial credit will be allowed. Each answer must be reduced to its simplest form.

1. Add 6498; 825; 57; 907; 6870; 8781; 968
2. Multiply \$39.50 by 10
3. Divide 6.9525 by 7.5
4. Subtract 387.5 from 400
5. Add $82\frac{2}{3}$; $15\frac{3}{4}$; $11\frac{1}{6}$
6. Find the product of $16\frac{2}{3}$ and 18
7. If lemons sell at 3 for 19 cents, what is the cost per dozen?
8. In six tests Walter had scores of 80, 90, 100, 70, 90 and 80. What was his average score?
9. The area of a square flower bed is 64 square feet. What is the length of one side?
10. If the temperature dropped from 2 degrees above zero to 10 degrees below zero, how many degrees did it drop?
11. If a boy makes 16 free throws out of 20 tries, what per cent does he make?
12. A certain school has a one-quarter mile track on its playground. How many times around the track would a boy have to run to complete a mile?
13. At 25 cents each, how many school lunches can be bought for \$6?
14. On Thursday, a family had spent \$48, which was 80% of its weekly budget. How much was the budget?
15. A girl was born on June 2, 1937. How old will she be on her next birthday?
16. Billy is making a knot exhibit. If he allows 20 inches of rope for each knot, how many feet of rope will he need to make 18 knots?
17. How far can an airplane travel in 15 minutes if it travels at an average speed of 240 miles an hour?
18. A house was insured for \$4000 at a premium of 60 cents per \$100. What was the premium?
19. Mt. Marcy, the highest peak in New York State, is 5344 feet above sea level. How much higher than a mile above sea level is Mt. Marcy?
20. A piano marked to sell for \$600 is sold for \$400. What is the rate of discount based on the marked price?

21. What is the ratio of a month to a year?
22. Represent the cost of 40 tons of coal at d dollars per ton.
23. Find the value of x in the equation: $5x - 3 = 12$
24. The formula for the area of a sphere is $A = 4\pi r^2$. Find the value of A if $\pi = \frac{22}{7}$ and $r = 7$.
25. Two angles of a triangle measure 75° and 45°. How many degrees are there in the third angle?

REVIEW EXERCISE 2

Answer all questions. Each question has 2 credits assigned to it; no partial credit will be allowed. Each answer must be reduced to its simplest form.

1. Add 4.98; 62.5; 3.865; 587
2. Find the difference between 34,165 and 29,602
3. Multiply 37.5 by .083
4. Add $2\frac{1}{8}$; $5\frac{3}{4}$; $8\frac{1}{2}$
5. At the rate of 48 cents per dozen, how much will 3 eggs cost?
6. Add 13,812; 21,305; 12,429; 16,496; 19,763
7. In a certain school there are 108 boys. How many intramural baseball teams (nine boys on a team) can they form, using all the boys?
8. The total weight of the eleven players on a high school football team is 1716 pounds. What is their average weight?
9. Mr. White's farm is mortgaged for \$2200. At a rate of 5%, what is the annual interest charge for the mortgage?
10. How many times greater is 90 than 15?
11. A boy spent one third of his money for a tennis racquet and one fifth of his money for balls. What fractional part of his money did he spend?
12. Joe used 7 ft. 4 in. of leather lacing to bind his scrapbook and Jim used 9 ft. 2 in. for his. How much more did Jim use than Joe?
13. Which is the greatest: $\frac{3}{4}$; .80; 66%?
14. In a group of 21 girls were 14 Girl Scouts. What per cent of the 21 were Scouts?
15. What is the value of x in the equation $5x + 14 = 24$?
16. If a rectangle 12 feet long has an area of 108 square feet, what is the width of the rectangle?
17. At 10 cents a square foot, what is the cost of cleaning a rug that is 9 feet long and 6 feet wide?
18. If $3x = 18$, what does $2x$ equal?
19. 30 inches of leather will make one belt. How many feet of leather would be needed to make 12 belts?
20. If $x - 2 = 5$, what is the value of x?

21. The ratio of the width of a banner to its length is 11 to 19. If the banner is 44 inches wide, what is its length?
22. The scale of miles on a certain map is 1 in. = 50 miles. How long a line will have to be drawn on the map to show a distance of 1000 miles?
23. Write the formula for the number of minutes (m) in h hours.
24. How many degrees less than a right angle is an angle of 74°?
25. A salesman sold $1850 worth of merchandise on a commission of 35%. How much was his commission?

REVIEW EXERCISE 3

Answer all questions. Each question has 2 credits assigned to it; no partial credit will be allowed. Each answer must be reduced to its simplest form.

1. Add $43.50; $9.95; $7.98; $67.33
2. Subtract .0879 from 3.0008
3. Divide 474.32 by 56
4. Add $3\frac{1}{4}$; $2\frac{3}{5}$; $5\frac{1}{2}$; $3\frac{3}{10}$
5. How many times greater than 6 kilometers is 36 kilometers?
6. Peter had 5 yards of lacing to cut into 20-in. pieces. How many pieces measuring exactly 20 in. each can he cut from the lacing?
7. At the rate of 50 cents per hundred pounds, how much did a license cost for an automobile weighing 3100 pounds?
8. Express $12\frac{1}{2}\%$ as a fraction.
9. If 9 eggs out of 12 hatched, what per cent of the eggs hatched?
10. Which of the following is greater: 35 billion or 35 million?
11. If a discount of 20% is given on an article marked $17.50, what is the selling price?
12. One-half dozen eggs added to four eggs equals how many eggs?
13. If 3 packages of flavor make 1 quart of drink, how many packages will be needed to make one gallon?
14. At 80 cents an hour how much will a boy earn in 5 hours and 15 minutes?
15. How much profit will be made on one gross of pencils if a profit of 1 cent is made on each pencil?
16. From 3 feet 6 inches subtract 1 foot 10 inches.
17. Each edge of a cube is 2 centimeters long. What is the volume of the cube?
18. How much tax must be paid on property assessed for $3800, if the tax rate is $15.65 per thousand?
19. If $a = 4$ and $b = 6$ what does $4a - b$ equal?
20. What per cent of 16 is 8?
21. Find the area of a triangle having a base of 10 feet and an altitude of 8 feet.

22. If coal costs x dollars per ton, how much will $\frac{1}{4}$ ton cost?
23. Find the circumference of a circle whose radius is $10\frac{1}{2}$ inches.
24. How much change should Jane get from a dollar when she buys 6 pounds of potatoes at 2 pounds for 9 cents?
25. Sam picked 36 baskets of apples, which he sold at 85 cents a basket. How much did he receive for the apples?

REVIEW EXERCISE 4

Answer all questions. Each question has 2 credits assigned to it; no partial credit will be allowed. Each answer must be reduced to its simplest form.

1. Find the sum of: .38; 4.01; 37; 24.3
2. Subtract $6\frac{3}{4}$ from $9\frac{1}{2}$
3. Divide 74.4 by .024
4. Find the product of 10,000 and .045
5. Which one of the following numbers has the smallest value: .2; .22; .02; .202; .022?
6. Write in figures: four and one-half million.
7. If a class of sixteen pupils going on a field trip has $8.00 to spend for lunches, how much money will each one be able to spend for lunch?
8. How many slices, each $\frac{1}{4}''$ thick, can be cut from an eighteen-inch piece of meat loaf?
9. After four pieces each 1 ft. 5 in. long were cut from a board, a piece 4 in. long remained. What was the original length of the board?
10. Write $2\frac{1}{2}\%$ as a decimal.
11. The scale of miles on a certain map is 1 inch = 50 miles. What is the distance in miles between two points that are $4\frac{1}{2}$ inches apart on the map?
12. An old-model electric toaster is marked at $22.50. If you receive a 10% discount on the marked price, what will you have to pay for the toaster?
13. An incubator hatched 558 chicks from 600 eggs. What per cent of the eggs hatched?
14. What is the interest on a loan of $450 for 60 days at 6% interest?
15. A man invests $2000 and receives a yearly income of $100. What is the rate of income on his investment?
16. If the last day of school is Friday, June 22, and the first day of school in the fall is Tuesday, September 4, how many days are there in the vacation period?
17. What is the ratio of 6 feet to 3 feet?
18. One day 6% of the pupils enrolled in a school were absent. If 42 pupils were absent, how many pupils were enrolled in the school?
19. The area of the floor of a square room is 144 square feet. What is the length of one side of the floor?

20. How far above the earth, to the *nearest mile,* is a plane flying at an altitude of 15,000 ft.?
21. If the perimeter of an equilateral triangle is 12 inches, what is the length of one side?
22. A man drove 80 miles in 2 hours and then 100 miles in 3 hours. What was his average speed for the 5 hours?
23. How many half-pint bottles can be filled with milk from a 10-gallon can of milk?
24. If y equals 6, what is the value of $3y - 4$?
25. If a board is l inches long and a second board is 6 inches longer, what is the length of the second board in terms of l?

REVIEW EXERCISE 5

Answer all questions. Each question has 2 credits assigned to it; no partial credit will be allowed. Each answer must be reduced to its simplest form.

1. Add 736; 48; 402; 6389
2. Multiply 624 by $83\frac{1}{3}$
3. Add $5\frac{3}{4}$; $2\frac{1}{2}$; $9\frac{3}{8}$
4. Subtract \$53.97 from \$100
5. Divide 245.25 by 2.25
6. Divide 176 by $\frac{22}{7}$
7. Which is larger: 42,000,000 or 42 billion?
8. A friend expected at 5:30 was 1 hour and 40 minutes late. At what time did he arrive?
9. Paul is 5 ft., 4 in. tall and Sam is 62 in. tall. Which boy is taller?
10. If you double the denominator of a fraction, do you increase or decrease the value of the fraction?
11. How much would you pay for 5 oranges at the rate of 60 cents a dozen?
12. On a three-hour bicycle trip a group of boys traveled 9 miles the first hour, 8 miles the second hour and 4 miles the third hour. How many miles per hour did the boys average?
13. If 14% of the stamps in a collection are domestic and the rest are foreign, what per cent are foreign?
14. Jane sold Christmas cards for a total of \$150. Of this she received \$60 as a commission. What was the rate of commission on her sales?
15. A girl had 4 library books that were 3 days overdue. Her fine was 3 cents a day for each book. How much was her total fine?
16. Find the interest on \$200 for 30 days if the rate of interest is 6%.
17. A year's subscription to a weekly magazine costs \$9.75. If a single copy sells for 20 cents, how much will a person save by taking a year's subscription instead of buying a single copy each week?

18. A man owns property assessed at $7000. If the tax rate is $25 per $1000 this year, what is the amount of the tax on his property?
19. In one day an organization raised $60, which was 25% of its Red Cross quota. What was the organization's quota?
20. What measurement is indicated by the arrow on the 6-inch ruler pictured below?

21. Which has the larger area, a square with a side of 10 feet or a circle with a diameter of 10 feet?
22. What does 8^2 equal?
23. How many degrees are there in each angle of an equilateral triangle?
24. The product of two numbers is 120. If one of the numbers is 8, what is the other number?
25. What is the value of x in the equation $3x + 4 = 16$?

REVIEW EXERCISE 6

Answer all questions. Each question has 2 credits assigned to it; no partial credit will be allowed. Each answer must be reduced to its simplest form.

1. Find the sum of: 729, 803, 651
2. Multiply $.98 by $6\frac{1}{2}$
3. Divide 8 by $\frac{2}{3}$
4. Divide 20.25 by .045
5. Add $6\frac{7}{8}$, $3\frac{1}{4}$, $2\frac{1}{2}$
6. Subtract $5\frac{3}{8}$ from 12
7. George and his father caught two fish that weighed 2 lb., 7 oz. and 4 lb., 15 oz. What was the total weight of the two fish?
8. Subtract 2 hours and 50 minutes from 5 hours and 12 minutes.
9. Rewrite 3,756,548 rounded off to the nearest thousand.
10. What per cent of the figure at the right is shaded?

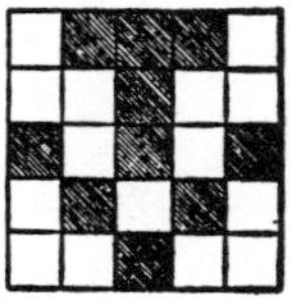

11. During a $33\frac{1}{3}\%$ discount sale, Jack's mother bought a chair that had originally been marked $45. How much did she save?
12. Express in dollars and cents the amount of money you would have if you had 5 quarters, 4 dimes, 6 nickels and 1 penny.
13. What is the interest on $600 for 8 months at 6%?

14. If a machine can print 3000 copies of a certain poster in 1 hour, how many *minutes* will it take, at the same rate of speed, for the machine to print 1000 copies?

15. If the school tax rate in a school district is \$.02516 on each dollar of assessed valuation, what is the school tax on a piece of property having an assessed valuation of \$1000?

16. In the pictogram at the right each complete figure represents 100,000 trees. What is the total number of trees represented by the figures shown?

17. The weights of two anchors are in the ratio 2 to 3. If the smaller anchor weighs 50 pounds, what is the weight of the larger anchor?

18. One morning 2% of the pupils of a school were absent. If 16 pupils were absent, how many pupils were enrolled in the school?

19. What is the area of triangle ABE shown in the diagram if the dimensions of rectangle $ABCD$ are 16 inches by 10 inches?

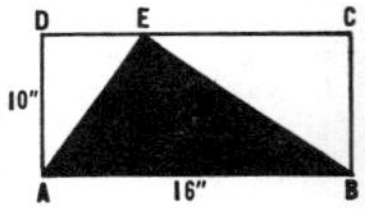

20. Express in terms of s the perimeter of the triangle at the right.

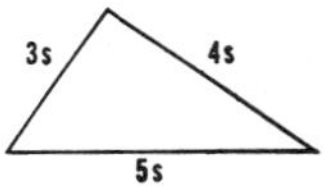

21. By what number do you multiply 9 in order to square it?

22. Name a line in the drawing at the right that represents a radius of the circle.

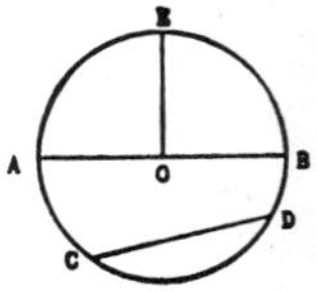

23. Write the following statement in the form of an equation:
If three times a certain number (x) is increased by 2, the result is 8.

24. Find the value of P in the formula $P = 2l + 2w$ when $l = 8$ inches and $w = 6$ inches.

25. Solve the equation: $3x - 6 = 15$

REVIEW EXERCISE 7

Answer all questions. Each question has 2 credits assigned to it; no partial credit will be allowed. Each answer must be reduced to its simplest form.

1. Find the value of $6\frac{1}{2} - 2\frac{1}{3} + 4\frac{1}{12}$

2. Subtract \$4.98 from \$10

3. Multiply 224 by $13\frac{3}{4}$

4. Divide 5.672 by .08
5. On a certain day the sun rises at 7:30 a.m. and sets at 5:15 p.m. How many hours and minutes is it from sunrise to sunset?
6. A ruler is placed against a metal plate as shown in the diagram. What is the length of the plate?

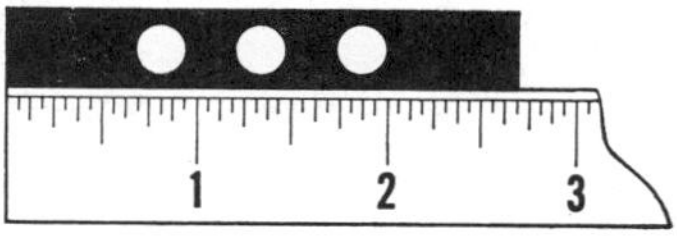

7. According to the census, a certain city had a population of 680,695. Write this number to the *nearest thousand.*

8. How many degrees are there in angle A of triangle ABC shown at the right?

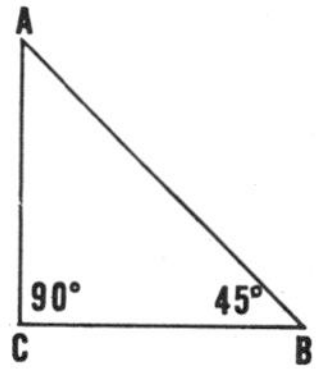

9. If a section 2 feet, 8 inches long is cut from a piece of lumber 10 feet long, what is the length of the part that remains?
10. The total cost of all food supplies needed for a camping trip for a group of Boy Scouts was $186. What was the cost of food for each one of the 20 Scouts who went on the trip?
11. If n represents a certain number, express in terms of n a number that is 3 less than 6 times n.

12. What per cent of the figure at the right is shaded?

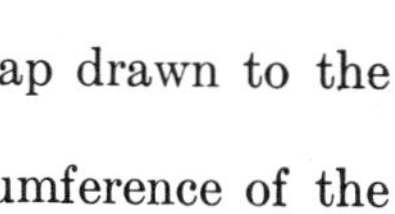

13. A class earned $60 commission from the sale of magazine subscriptions. This amount was equal to 25% commission on all sales. How many dollars worth of magazine subscriptions did they sell?
14. How many miles are represented by $2\frac{1}{4}$ inches on a map drawn to the scale 1 in. = 100 mi.?
15. The diameter of a circle is 28 inches. What is the circumference of the circle? [Use $\pi = 3\frac{1}{7}$.]

16. The large rectangle at the right is made up of equal rectangles with dimensions as indicated. Express algebraically in terms of h and w the perimeter of the large rectangle.

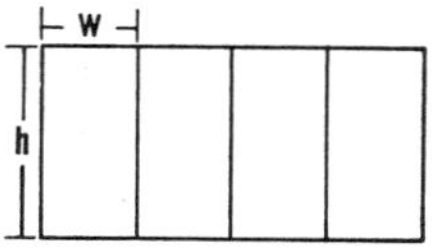

17. A boy is given a $100 bond that pays $2\frac{3}{4}\%$ interest annually. How much interest does he receive at the end of the first year?

18. A woman bought 12 yards of linen toweling. How many towels, each 27 inches long, can be cut from this length of material?

19. If $\frac{n}{3} = 4$, what is the value of n?

20. The hypotenuse of a right triangle is 5 feet in length. One of the other sides of the triangle is 4 feet in length. What is the length of the third side?

21. The formula for finding the surface of a cube is $S = 6e^2$. Find S if e is 3 inches.

22. What amount of money is represented by the bar in the following diagram?

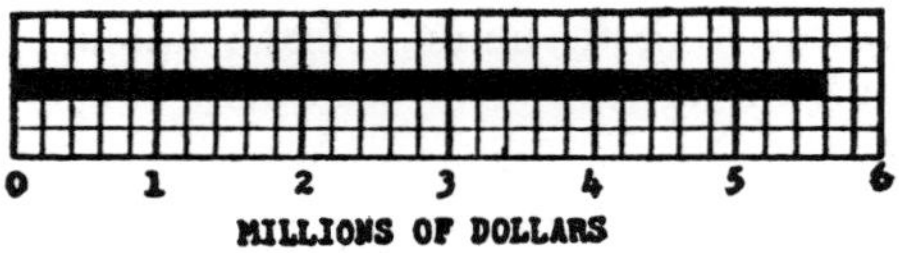

Follow these directions in answering questions 23, 24 *and* 25: Write the letter (*a*, *b*, *c* or *d*) representing the correct answer.

23. The fraction $\frac{2}{3}$ has the same meaning as (*a*) 40% (*b*) 50% (*c*) $66\frac{2}{3}$% (*d*) 150%

24. Lines AD and AB of the rectangle pictured at the right are (*a*) equal (*b*) horizontal (*c*) parallel (*d*) perpendicular

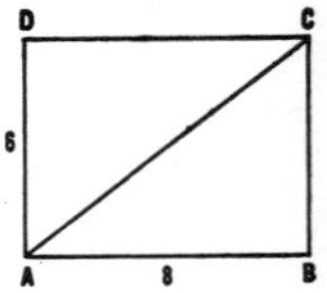

25. For which one of the following geometric solids is it possible to find the volume merely by multiplying its base by its height?

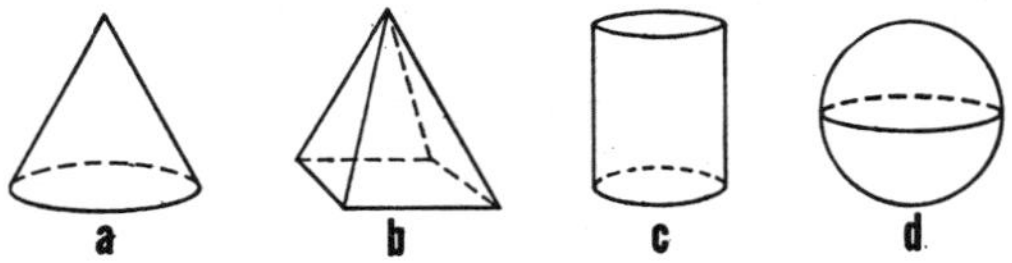

REVIEW EXERCISE 8

Answer all questions. Each question has 2 credits assigned to it; no partial credit will be allowed. Each answer must be reduced to its simplest form.

1. Add \$5.95, \$.18, \$198, \$12.43, \$6.65

2. Subtract 731,969 from 940,614

3. Divide 9.744 by 2.4

4. Multiply $1\frac{1}{2}$ by $1\frac{1}{4}$ by $\frac{2}{3}$

5. Perform the indicated operation: $6\frac{1}{2} + 4\frac{7}{12}$

6. What is the shortest board a man must buy in order to cut three sections from it each 4 feet, 8 inches long?

7. The circle at the right represents a family's total income. The shaded section shows the part of their income allowed for rent. What per cent of their income is allowed for rent?

8. A boy sold \$88.50 worth of stationery. If he received $33\frac{1}{3}\%$ commission, what was the amount of his commission?

9. A man takes out a \$5000 life insurance policy at a yearly rate of \$29.62 per \$1000. What is the yearly premium?

10. On her maiden voyage the *SS United States* made the trip from New York to England in 3 days, 10 hours and 40 minutes, beating the record set by the *SS Queen Mary* in 1938 by 10 hours and 2 minutes. How long did it take the *Queen Mary* to make the trip?

11. In four successive weeks a boy spent 78¢, \$2.04, \$1.53 and 29¢ in buying old stamps for his collection. What was the average cost per week for his hobby?

12. There are 60 pupils in a school who participate in a certain activity. If this is 12% of the student body, how many pupils are there in the school?

13. Two angles of a triangle measure 70° and 40°. How many degrees are there in the third angle?

14. What is the simple interest on \$600 at 4% for 2 years?

15. If three times a certain number increased by 4 is equal to 19, what is the number?

16. The area of a square flower bed is 81 square feet. What is the length of one side of the flower bed?

17. A boy worked on Saturday from 7:30 a.m. until 11:00 a.m. at the rate of 80¢ per hour. How much should he have received for his work?

18. Express algebraically the perimeter of the figure at the right.

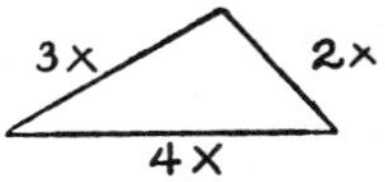

19. What is the value of $6s^2$ when $s = 3$?

20. How many quart bottles can be filled from four 10-gallon cans of milk?

21. The diameter of a bicycle wheel is 28 inches. Express in inches the circumference of the wheel. [Use $\pi = 3\frac{1}{7}$]

22. Find the height of a flagpole which casts a shadow 40 feet long at the same time that a 6-foot pole casts a shadow 10 feet long.

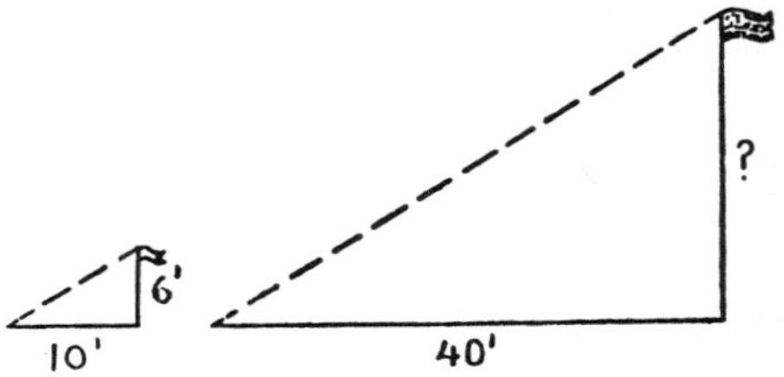

23. A certain brand of vegetables may be purchased in two sizes, *A* or *B*. For size *A* the price is 24¢ for 12 oz. and for size *B* the price is 35¢ for 20 oz. Which is cheaper in cost per oz. for the customer to buy?

24. Which one of the following is the largest real estate rate?
(*a*) \$31.10 per \$1000 (*b*) \$4.23 per \$100 (*c*) 32 mills per \$1

25. Which one of the following has the smallest value?
(*a*) $\frac{5}{8}$ (*b*) .82 (*c*) 75%

REVIEW EXERCISE 9

Answer all questions. Each question has 2 credits assigned to it; no partial credit will be allowed. Each answer must be reduced to its simplest form.

1. Add $18\frac{5}{16}$, $5\frac{3}{8}$, $4\frac{1}{4}$

2. Subtract $6\frac{3}{4}$ from $18\frac{1}{2}$

3. Multiply $31\frac{1}{2}$ by 5

4. Divide 16 by $\frac{3}{5}$

5. Subtract \$4.98 from \$20

6. Round off 97,850 to the *nearest thousand.*

7. Divide 17.28 by 7.2

8. Find $5\frac{1}{2}\%$ of \$2800

9. How many more ounces of cheese are contained in a 10-ounce package than in a half-pound package?

10. A man buys an electric light bulb for 24 cents, which includes a 20% tax. What is the cost of the bulb without tax?

11. If the area of a square is 64 square inches, what is its perimeter?

12. A certain grading job requires 81 cubic feet of topsoil. How many cubic yards of topsoil must be ordered?

13. What is the interest on \$850 at 6% for 3 months?

14. A school team won 16 of the 20 games played. What per cent of the games played did the team win?

15. What is the value of $a^2 + ab$ if $a = 3$ and $b = 5$?

16. If $\frac{7}{8}$ of a class of 32 pupils had perfect attendance, what per cent of the class had perfect attendance?

17. Write the equation showing that if 3 times a certain number is decreased by 2 the result is equal to 34.

18. A house plan is drawn using the scale $\frac{1}{4}$ in. = 1 ft. If on the plan the living room is 7 inches long, what is the actual length of the room?

19. Solve for a: $5a + 8 = 23$

20. The cost of soup to a grocer was 9 cents per can and he sold it for 16 cents per can. If he allowed 25% of the selling price for expenses, what was his profit on each can of soup?

Follow these directions in answering questions 21, 22, 23, 24 *and* 25: Write the letter (a, b or c) representing the correct answer.

21. One quarter of a million is the same as (*a*) 2,500 (*b*) 25,000 (*c*) 250,000

22. If the triangles at the right are similar, the length of side s in the smaller triangle is (*a*) 4″ (*b*) 9″ (*c*) 16″

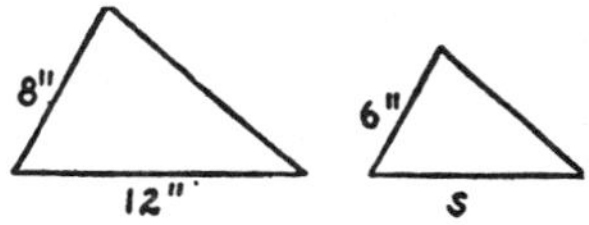

23. The triangle at the right is (*a*) an isosceles triangle (*b*) an equilateral triangle (*c*) a scalene triangle

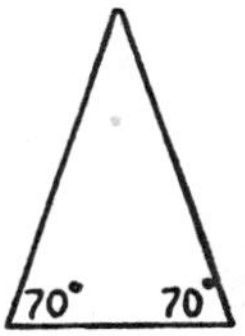

24. Two legs of a right triangle are 3″ and 4″. The length of the hypotenuse of this triangle is (*a*) 1″ (*b*) 5″ (*c*) 7″

25. In the diagram at the right, the line AB was constructed (*a*) bisecting CD (*b*) parallel to CD (*c*) perpendicular to CD at A

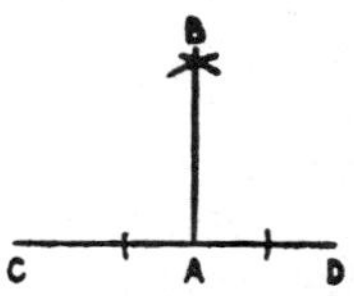

REVIEW EXERCISE 10

Answer all questions. Each question has 2 credits assigned to it; no partial credit will be allowed. Each answer must be reduced to its simplest form.

1. Add $10\frac{1}{6}$, $2\frac{7}{12}$, $7\frac{2}{3}$
2. Subtract $27.95 from $50
3. Divide $3\frac{3}{4}$ by 2
4. Multiply $2.04 by 60.5
5. Divide .768 by .32
6. Multiply 69 by $7\frac{2}{3}$
7. A 29-inch length of ribbon contains how many more inches than a $\frac{3}{4}$-yard length of the same ribbon?
8. A man drove 80 miles in 2 hours and then drove the next 100 miles in 3 hours. How many miles an hour did he average for the 5 hours?
9. What is the length of the piece of cardboard shown in the diagram at the right to the *nearest eighth of an inch?*

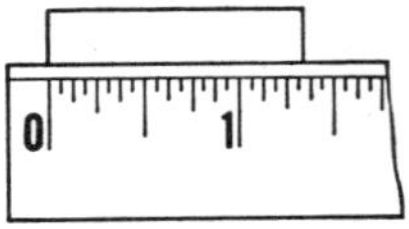

10. What is the simple interest on $460 for 2 years at $2\frac{1}{2}\%$?
11. How much did a boy earn for selling $164.50 worth of greeting cards if he was paid 40% commission on his sales?
12. What is the value of $5a^2$ when $a = 3$?
13. Write the equation for the following statement: If 5 times a certain number is divided by 2, the quotient is 15.
14. A merchant sold hardware for $75, making a profit of $15. What was the per cent of profit on the selling price?
15. How much money must be invested at 5% interest to earn a yearly income of $300?
16. Find the value of x in the following equation: $7x - 4 = 115$
17. At $14.50 per square yard what is the cost of a rug that is 15 feet long and 12 feet wide?
18. Express algebraically the perimeter of the rectangle pictured at the right.

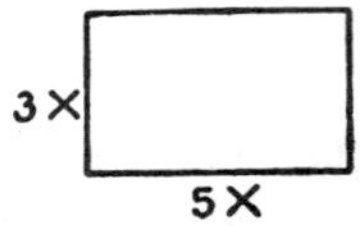

19. A house is insured against fire for $10,000 at 32 cents per $100. What is the premium?

20. If hypotenuse AB in the right triangle pictured is 5 and if side CA is 3, what is the length of side CB?

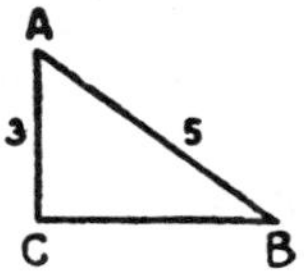

21. A girl is making badges in her homemaking class. She can make one badge from $\frac{1}{8}$ yard of ribbon. How many badges can she make from $1\frac{3}{4}$ yards?

Follow these directions in answering questions 22–25: Write the letter (*a*, *b* or *c*) representing the correct answer.

22. Which one of the following has the smallest value?
(*a*) $\frac{2}{3}$ (*b*) .6 (*c*) $62\frac{1}{2}\%$

23. Which formula should be used to find the circumference of the circle at the right? (*a*) $C = \pi r$ (*b*) $C = \pi r^2$ (*c*) $C = 2\pi r$

24. In a junior high school 6 out of every 10 children enrolled participated in after-school activities. What part of the enrollment participated in after-school activities? (*a*) $\frac{1}{6}$ (*b*) $\frac{3}{5}$ (*c*) $\frac{5}{8}$

25. Which one of the following angles is an obtuse angle?

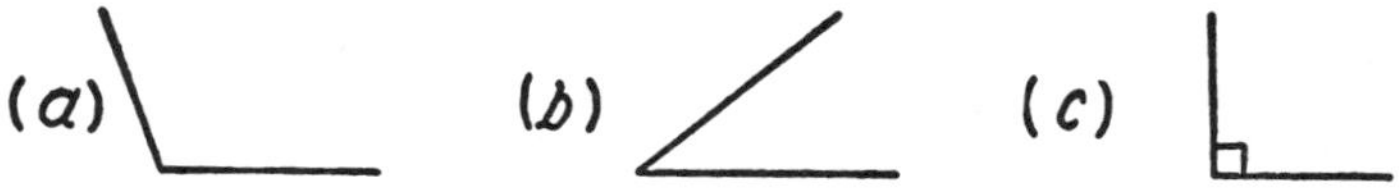

REVIEW EXERCISE 11

Answer all questions. Each question has 2 credits assigned to it; no partial credit will be allowed. Each answer must be reduced to its simplest form.

1. Add $6\frac{1}{2}$, $8\frac{3}{4}$, $5\frac{1}{8}$
2. Subtract $14\frac{2}{3}$ from 56
3. Add \$124.00, \$48.25, \$.98, \$8.09 and \$9.67
4. Find $3\frac{1}{3} \div \frac{2}{3}$
5. Divide 2.064 by .24
6. Multiply \$1.04 by $8\frac{1}{4}$
7. A train scheduled to arrive in Albany at 11:42 a.m. was 45 minutes late. At what time in the afternoon did it reach Albany?
8. How many square yards of linoleum are needed to cover a floor 18 feet long and 15 feet wide?

9. If a scale of $\frac{1}{4}$ inch = 1 foot is used in a house plan, what is the actual number of feet represented by a line $4\frac{1}{2}$ inches long on the plan?

10. Find the number of degrees in angle A of the triangle ABC.

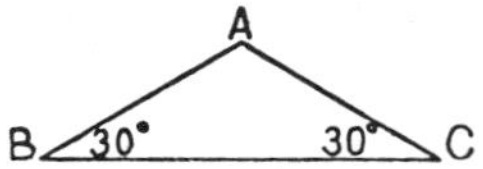

11. The price of eggs increased from 50 cents to 60 cents a dozen. What was the per cent of increase in price?

12. Using the dimensions given on the similar rectangles A and B at the right, find the number of feet in the length of rectangle B.

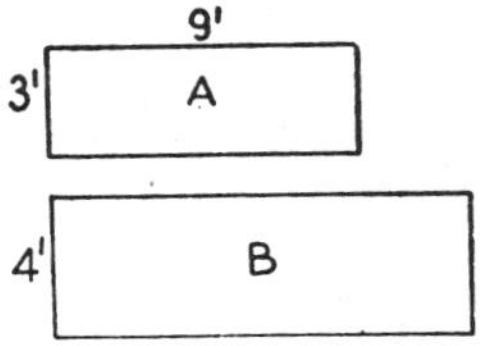

13. By purchasing a coat at a clearance sale, a girl saved \$12, which was a saving of 25% on the original price. What was the original price of the coat?

14. The formula for the area of a triangle is $A = \frac{1}{2}bh$. Find A if $b = 12$ and $h = 10$.

15. Write the equation showing that if twice the number n is increased by 3, the result is 15.

16. What is the simple interest on a loan of \$1600 for 6 months at a yearly rate of 5%?

17. Solve for x: $4x - 8 = 16$

18. A dry cleaner offers a 20% discount on all dry cleaning done during the month of January. If the regular cost of dry cleaning a rug is \$2.25, what is the saving if a rug is cleaned in January at this dry cleaner's?

19. How many packages of candy containing $\frac{3}{4}$ of a pound each can be filled from 15 pounds of candy?

20. Tom spent 2 hours and 30 minutes studying for three classes. What was the average time in *minutes* that he spent in studying for each class?

21. Find the number of feet in the circumference of a circle if the diameter is 84 feet.

22. Find the hypotenuse of a right triangle if the legs are 6 and 8.

23. Each edge of a cube is 4 inches long. Find the number of cubic inches in the volume of the cube.

24. Which of the following has the same value as .5%: $\frac{1}{2}$, .005, $\frac{1}{500}$?

25. Of the following real estate tax rates, which is larger, a or b? (a) \$31.25 per \$1000 (b) \$3.45 per \$100

REVIEW EXERCISE 12

Answer all questions. Each question has 2 credits assigned to it; no partial credit will be allowed. Each answer must be reduced to its simplest form.

1. Add: \$13.95, \$20, \$8.02, \$6.79
2. What is $12\frac{3}{8} - 5\frac{7}{16}$?
3. Divide $4\frac{1}{2}$ by $\frac{3}{4}$
4. Find the value of $2\frac{2}{3} \times 2\frac{1}{4} \times \frac{1}{2}$
5. Divide 1.584 by .12
6. Express 2.5% as a decimal
7. Change $\frac{3}{7}$ to a decimal, correct to the *nearest hundredth*
8. The price of canned corn is \$1.80 per dozen cans or \$.16 each if bought one at a time. How many cents would be saved on each can if a purchase of 12 cans is made by the dozen instead of one at a time?
9. How much does a girl earn for selling \$68 worth of writing paper if she is paid a commission of 40% on her sales?
10. Are the rectangle and the square, in the diagram at the right, equal in area or similar?

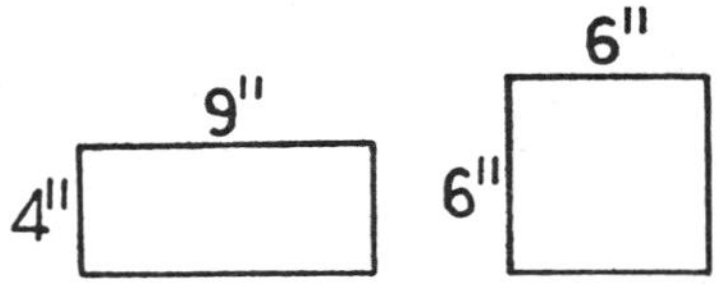

11. A team played 24 games of which it won 18. What per cent of the games played did it win?
12. One man earns \$4050 per *year*. A second man earns \$315 per *month*. How much more does the first man earn in a year than the second man?
13. A girl purchased a blouse for \$2.19. She returned the blouse the next day and selected a better one costing \$2.89. She gave the clerk a dollar bill to pay for the difference in price. How many cents in change should she receive?
14. The daily almanac report for one day during the summer stated that the sun rose at 6:14 a.m. and set at 6:06 p.m. Find the number of hours and minutes in the time between the rising and setting of the sun on that day.
15. In the formula $A = 6s^2$, find the value of A when $s = 3$.
16. What is the perimeter of the figure in the diagram at the right?

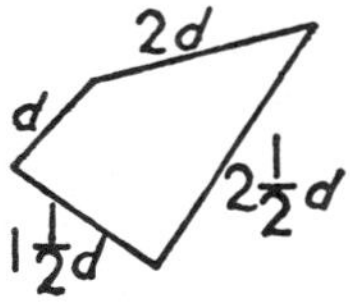

17. In a certain homeroom nine pupils took part in a school program. If this number represented 25% of the total homeroom enrollment, how many pupils were enrolled in the homeroom?

18. In the diagram at the right AC is a diagonal in rectangle $ABCD$. Find the number of square inches in the area of triangle ABC.

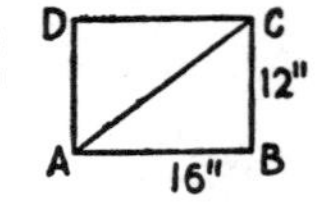

19. Find the simple interest on $4500 for 6 months at a yearly rate of 4%.

20. Find the value of n in the following equation: $3n + 5 = 17$.

21. The yearly income from an investment of $1200 is $60. What yearly rate of interest does this investment pay?

22. On a certain map 1 inch represents 300 miles. How many miles are represented by a line $2\frac{1}{4}$ inches long on this map?

23. If the circumference of the circle whose center is O is found by multiplying $3\frac{1}{7}$ by the length of one of the lines in the figure at the right, which one of the lines, AB, OE or CD, must be used to find the circumference of this circle?

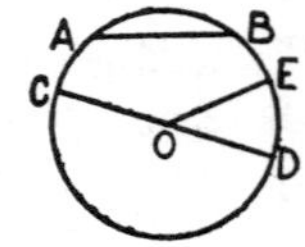

24. What is the length in feet of one side of a square having an area of 100 square feet?

25. During the first 6 weeks of a summer vacation a boy earned $15 per week, and during the next 4 weeks he earned $20 per week. What *average pay per week* did he earn during that 10-week period?

FINAL EXAMINATIONS

FINAL EXAMINATION 1

Part I

Answer all questions in this part. Each question counts 2 credits; no partial credit is allowed. Reduce each answer to its simplest form.

1. Add: \$209, \$8.75, \$.59, \$89, \$76.98
2. Subtract $28\frac{3}{4}$ from 50
3. Multiply 867 by 503
4. Divide 15 by $\frac{5}{6}$
5. Add: $12\frac{5}{8}$, $9\frac{3}{4}$, $8\frac{7}{16}$
6. Divide 124 by .05
7. Change $\frac{4}{15}$ to a decimal correct to the *nearest hundredth.*
8. A steak weighed 2 pounds, 4 ounces. How much did it cost at \$.92 per pound?
9. During his summer vacation a boy earned \$4.50 per day and saved 60% of his earnings. If he worked 45 days, how much did he save during the vacation?
10. From his coin bank a boy took 3 half dollars, 8 quarters, 7 dimes, 6 nickels and 9 pennies to deposit in his school savings account. Express in dollars and cents the total amount of money he deposited.
11. If a roast that requires 1 hour and 40 minutes of roasting time has been in the oven for 55 minutes, how many more minutes of roasting time are required?
12. On the first day of its drive a junior high school raised \$40, which was $33\frac{1}{3}$% of its Junior Red Cross quota. How much was the quota?
13. On a certain map a distance of 10 miles is represented by $\frac{1}{2}$ inch. If two towns are $3\frac{1}{2}$ inches apart on this map, express in miles the actual distance between the two towns.
14. At an annual rate of \$.40 per \$100, what is the fire insurance premium for one year on a house that is insured for \$8000?
15. A meter equals approximately 1.09 yds. How much longer, in *yards,* is a 100-meter dash than a 100-yard dash?
16. A train leaves New York City at 8:10 a.m. and arrives in Buffalo at 4:45 p.m. on the same day. How long, in hours and minutes, does it take the train to make the trip?

17. A jacket that was marked at $12.50 was sold for $10. What was the rate of discount on the marked price?

18. In the formula $S = 6e^2$, find the value of S when $e = 5$.

19. The area of a rectangular room is 1000 square feet. If the width of this room is 25 feet, what is the length of the room in feet?

20. Find the simple interest on $600 for one year at a yearly rate of $3\frac{1}{2}\%$.

21. If n represents a certain number, express in terms of n a number that is 4 more than 5 times n.

22. How many degrees are there in angle A of the triangle at the right?

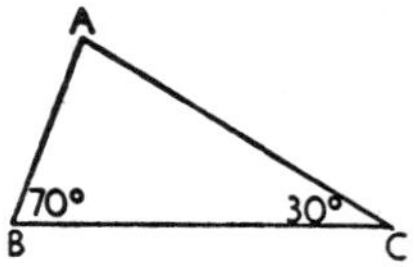

23. Find the value of x in the equation: $2x + 3 = 9$

24. A circular flower garden has a diameter of 21 feet. How many feet of fencing will be required to enclose this garden?

25. If a car averages 18 miles to a gallon of gasoline, how many gallons of gasoline will be used on a trip of 369 miles?

Part II

Answer any five questions from this part. No credit will be allowed unless all necessary operations are given. Reduce each result to its simplest form and mark each answer Ans.

26. In a small community the town tax rate during a recent year was $27.80 per $1000 of the assessed valuation. The school tax rate was $20.30 per $1000 of the assessed valuation. A man's house, valued at $15,000, was assessed for 30% of its value.

a. What is the assessed valuation of the house? [3]

b. What was the total amount the man paid that year in town and school taxes? [7]

27. A girl works in a dress shop where her salary is $25 a week. In addition, she receives a commission of 4% on all sales she makes each week over $250. Her sales for one week amounted to $635 and for the next week, $850. What were her total earnings for this two-week period? [10]

28. A boy had some hens as a project in his 4-H club. The expenses of caring for the hens, including food, averaged $4.20 a week. He collected an average of 24 eggs per day.

a. Find the cost of caring for the hens for 18 weeks. [3]

b. If the boy sold the eggs at an average price of $.50 a dozen, find his total profit during the 18-week period. [7]

29. A farmer harvested potatoes from a rectangular field that was 32 rods long and 45 rods wide. This field produced an average of 360 bushels to the acre, and the potatoes were sold for $1.20 a bushel.

a. How many acres were there in this field? [Use 160 sq. rd. = 1 acre.] [5]

b. How much did the farmer receive for the potatoes harvested from this field? [3]

c. What was the average income *per acre* from this field? [2]

30. Referring to the following figures, answer parts *a* and *b*.

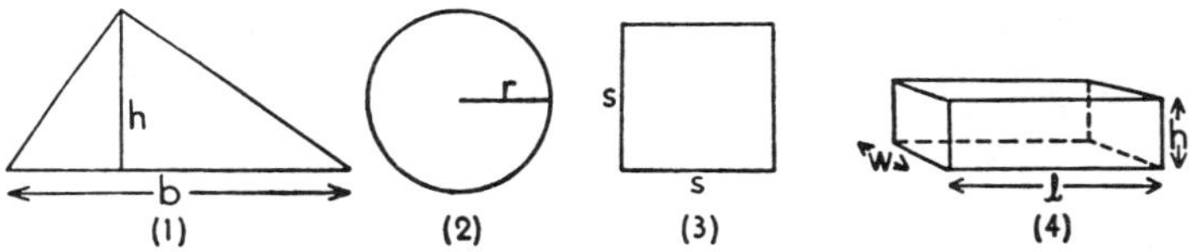

a. The preceding figures are numbered 1-4. Write the numbers 1-4 on your answer paper and after *each* number write the name of the figure. [Choose the name from the list at the right.] [4]

Names of Figures
circle
pentagon
sphere
rectangular prism
triangle
cone
square

b. Write the formula needed to find the

(1) area of figure 2 [2]
(2) perimeter of figure 3 [2]
(3) volume of figure 4 [2]

31. Because of illness a man was unable to work for 35 weeks. During this time he was in a hospital for 6 weeks. The health insurance policy for which he had paid an annual premium of $49.50 guaranteed him $25 a week while he was not able to work, and an additional $12.50 a week while he was in the hospital.

a. How much did he receive from the insurance company for his illness? [6]

b. If this was the first benefit he received from the company during the seven years he had carried the insurance, how much more did he receive from the company that year than he had paid in premiums for the seven-year period? [4]

32. In a homeroom with a total enrollment of 28 pupils, there were 4 more boys than girls. If n represents the number of girls in this homeroom, then

a. express in terms of n the number of boys in the homeroom [1]

b. express in terms of n the total number of pupils in the homeroom [2]

c. write an equation necessary to find the value of n [4]

d. solve this equation for n [2]

e. find the number of boys in this homeroom [1]

33. The following graphs were published by the federal government to show where the tax dollar comes from and where it goes.

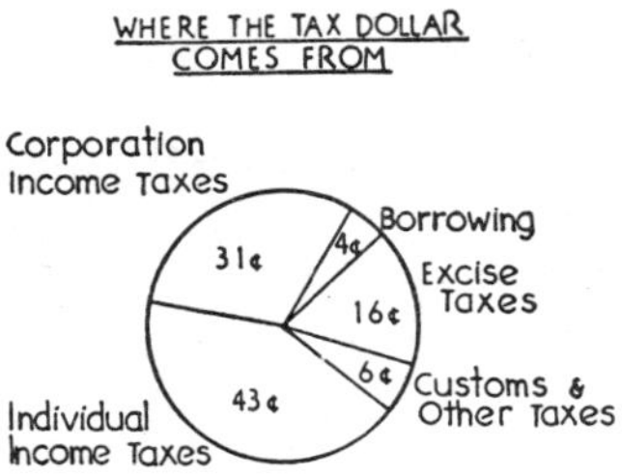

Use the preceding graphs to answer the following questions:

a. What per cent of the federal tax dollar was spent on national security? [1]

b. What per cent *more* money was obtained from individual income taxes than from corporation income taxes? [2]

c. How many dollars, of every million dollars collected in taxes, were obtained from excise taxes? [3]

d. List the *four* sources of income whose total approximately equals the amount spent for national security. [4]

FINAL EXAMINATION 2

Part I

Answer all questions in this part. Questions 1-20 count 2 credits each; no partial credit is allowed. Questions 21-30 count 1 credit each. Reduce each answer to its simplest form.

1. Multiply 687 by 409.

2. Divide 40.5 by .15

3. Subtract $15.36 from $20.

4. Find the value of $7\frac{2}{3} + 5\frac{3}{4} - 4\frac{1}{2}$.

5. Find the value of $1\frac{1}{2} \times 2\frac{1}{4} \times \frac{2}{3}$.

6. What is $44 \div 3\frac{1}{7}$?

7. Change $\frac{5}{16}$ to the *nearest whole per cent.*

8. On a certain road map 1 inch represents 12 miles. How many miles will a length of $2\frac{1}{4}$ inches represent on this map?

9. A newspaper headline read: "Quarter of a million dollars raised for hospital fund." Write this amount of money in numbers.

10. One day the sun rose at 5:43 a.m. and set at 6:19 p.m. How many hours and minutes were there between sunrise and sunset?

11. A girl made a circle graph to show the distribution of report card marks in mathematics in her grade. She drew an angle of 45° to show the per cent of pupils who had a mark of A. What per cent of her grade had a mark of A?

12. A man bought an article and gave the clerk a five-dollar bill. The clerk gave this correct change: 2 pennies, 1 nickel, 1 dime, 1 quarter and 2 one-dollar bills. How much did the article cost?

13. What is the simple interest on a loan of $600 at 5% for 3 months?

14. At an annual rate of $.32 per $100, what is the fire insurance premium for 1 year on a house that is insured for $7,500?

15. Write $3\frac{1}{2}\%$ in decimal form.

16. If ice cream costs $2 per gallon and 1 quart is to serve 8 persons, how much will enough ice cream for 96 persons cost?

17. A junior high school established a school savings plan. During the year the eighth grade saved $950. If this was 50% of the total amount saved by the school, find the total amount saved.

18. How many square yards of linoleum are needed to cover a floor having an area of 270 square feet?

19. On a parking meter was printed: "12 minutes for 1¢. Maximum deposit 10¢." How many hours may a driver legally park at one of the meters for 10 cents?

20. If a box of 24 candy bars is bought for 80¢ and the bars are sold at 5¢ each, what is the per cent of profit on the cost?

Directions (21-30): Indicate the correct completion for *each* of the following by writing the letter *a*, *b* or *c*.

21. If one number is represented by n, a number 10 less than n would be represented by (*a*) $10 - n$ (*b*) $\frac{n}{10}$ (*c*) $n - 10$

22. If $n + 3 = 12$, then the value of n is (*a*) 15 (*b*) 9 (*c*) 4

23. The value of n^2 when $n = 3$ is (*a*) 9 (*b*) 6 (*c*) 4

24. The perimeter of the triangle in the diagram is (*a*) $4a$ (*b*) $5a^2$ (*c*) $5a$

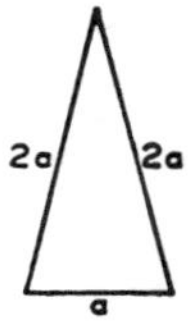

25. The value of $6x$ when $x = \frac{1}{2}$ is (*a*) 3 (*b*) $3x$ (*c*) $6\frac{1}{2}$

26. In triangle ABC at the right, line AD is perpendicular to BC. Angle ADC is (*a*) an obtuse angle (*b*) an acute angle (*c*) a right angle

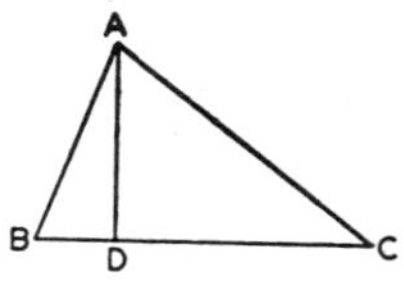

27. The number of equal squares which form the surface of a cube is (*a*) 4 (*b*) 6 (*c*) 8

28. In the diagram, the dotted line is called a
(*a*) bisector (*b*) parallel (*c*) perpendicular

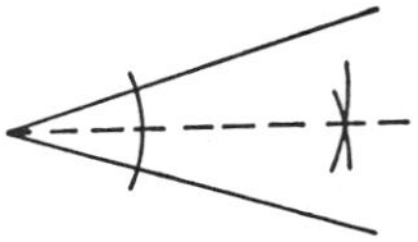

29. The sum of the angles of a triangle is (*a*) 90° (*b*) 180° (*c*) 360°

30. The triangles in the diagram are (*a*) congruent (*b*) equal in area (*c*) similar

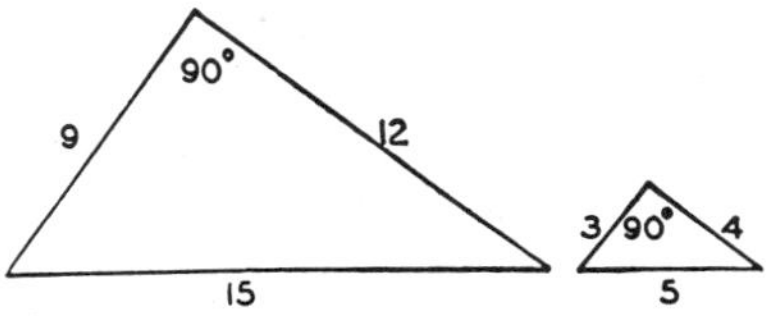

Part II

Answer any five questions from this part. No credit will be allowed unless all necessary operations are given. Reduce each result to its simplest form and mark each answer Ans.

31. The Junior Red Cross sold 1,200 tickets for a fund-raising entertainment. One-third of the tickets were for reserved seats and sold at $.75 each. The remainder of the tickets were for unreserved seats and sold at $.50 each. The expenses were $40.75 for printing and advertising, $90.50 for costumes, $68 for stage equipment, and $26.80 for other expenses. How much profit did the Junior Red Cross make as a result of this entertainment? [10]

32. Typewriters listed at $160 each are sold to a certain club at a $12\frac{1}{2}\%$ discount. In addition, a 2% discount is allowed on the net price if the club pays cash.
a. Find the cash price the club would pay for each typewriter. [7]
b. Find how much the club would save on each typewriter by taking advantage of both discounts. [3]

33. A boy built an aquarium with the dimensions shown in the diagram at the right.
a. Find the volume of the aquarium in cubic inches. [3]
b. Find to the *nearest gallon* the amount of water in the aquarium when it is two-thirds full. [231 cubic inches = 1 gallon.] [7]

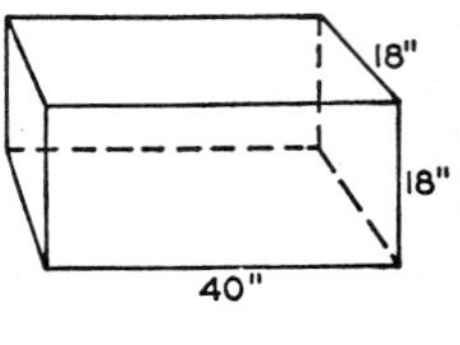

34. In a certain village the school tax was 35 mills per \$1, the village tax was \$1.42 per \$100, and the town tax was \$23.60 per \$1,000 of assessed valuation. Find the total yearly tax paid on property assessed at \$5,000 in that village. [10]

35. In a large industrial plant, a certain employee's rate of pay is \$1.46 per hour for a 40-hour week. For the time worked over 40 hours each week, the hourly rate is $1\frac{1}{2}$ times the regular rate. During one week this employee's time card showed he had worked 8 hours, $9\frac{1}{2}$ hours, $10\frac{1}{2}$ hours, 9 hours, 10 hours, 8 hours. How much did he earn that week? [10]

36. To find the height of a tree on level ground, a boy held a yardstick in a vertical position and found its shadow to be $2\frac{1}{2}$ feet long. At the same time the tree cast a shadow 15 feet long. Draw a diagram and find the height of the tree. [10]

37. *a.* Solve the following problem by the use of a correct algebraic equation: A Girl Scout troop sold cookies to raise money. The first week they sold 20 boxes more than they did the second week. If they sold 180 boxes during the two weeks, how many boxes did they sell each week? [4, 2]

b. In the formula $c^2 = a^2 + b^2$, find the value of c if $a = 9$ and $b = 12$. [4]

38. The following graph once appeared in a publication.

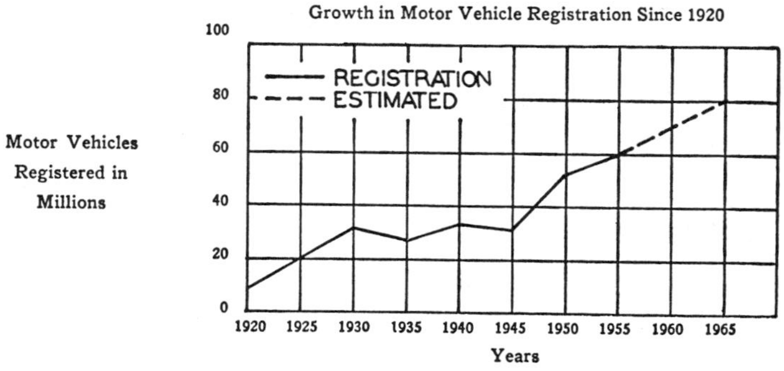

Use the graph to answer the following questions:

a. Approximately how many motor vehicles were registered in 1955? [2]

b. Approximately how many times as many motor vehicles were registered in 1955 as in 1925? [3]

c. According to the estimate made on this graph, how many more motor vehicles will be registered in 1965 than in 1955? [2]

d. What per cent of increase in registration is expected between 1955 and 1965? [3]

FINAL EXAMINATION 3

Part I

Answer all questions in this part. Questions 1-20 count 2 credits each; no partial credit is allowed. Questions 21-30 count 1 credit each. Reduce each answer to its simplest form.

1. Find the sum of: 846, 3215, 42, 937
2. Divide 1.812 by .06
3. Multiply $12\frac{1}{4}$ by $3\frac{1}{7}$.
4. Subtract \$4.63 from \$10.
5. Change $\frac{6}{7}$ to a decimal correct to the *nearest thousandth.*
6. Subtract $5\frac{2}{3}$ from $17\frac{1}{2}$.
7. Write in figures: three million eighty thousand.
8. How many more ounces of cheese are there in a 10-ounce package than there are in a half-pound package?
9. If the total weight of an 11-man football team is 1,815 pounds, what is the average weight in pounds of each man on the team?
10. Which of the following has the largest value: $\frac{5}{6}$, .75, 80%?
11. What is the largest number of containers, each having a capacity of $\frac{1}{2}$ pint, that can be filled from a gallon jug full of milk?
12. A parking lot charges 25 cents for the first half hour of parking and 10 cents for each additional half hour. A man parked his car from 1:45 p.m. to 3:15 p.m. What was his parking charge?
13. Plans are being made to construct a building a mile high. The elevators in this building are expected to go from street level to the top in one minute. Express this speed in feet per second.
14. A chair whose total cost to a dealer was \$32 was sold at a profit of \$18. Find the per cent of profit based on the selling price.
15. A distance of 25 miles is represented on a map by $2\frac{1}{2}$ inches. On this map, how many miles are represented by 1 inch?
16. A woman bought groceries amounting to \$2.73 and gave the clerk 3 one-dollar bills in payment. The clerk could not make the change, so the woman gave him 3 cents more. How much change did she then receive?
17. A \$500 bond pays $3\frac{1}{2}\%$ interest annually. How much interest is paid each year?
18. According to the 1950 census, 14,830,192 people were living in New York State. Round off this number to the *nearest thousand.*
19. A library contains 60 books of biography. If this number is 5% of the total number of books on the shelves, how many books are there in the library?
20. A man has a \$5,000 life-insurance policy on which he pays a premium of \$19.36 per thousand each year. Find his annual premium.

Directions (21-30): Indicate the correct completion for *each* of the following by writing the letter *a, b* or *c*.

21. The angles formed by two perpendicular lines are (*a*) acute angles (*b*) right angles (*c*) obtuse angles

22. In triangle ABC angle A is equal to (*a*) 100° (*b*) 280° (*c*) 60°

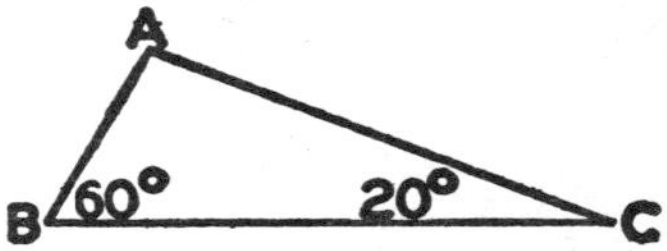

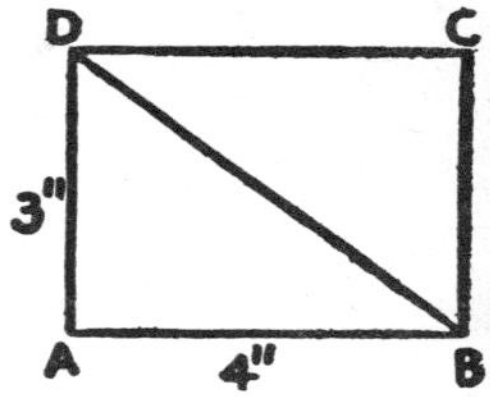

23. The length in inches of diagonal DB of rectangle $ABCD$ is (*a*) 25 (*b*) 7 (*c*) 5

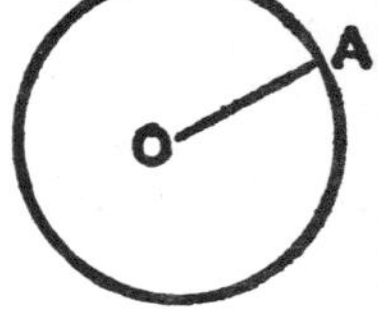

24. In the circle whose center is O, the line OA is a (*a*) diameter (*b*) radius (*c*) chord

25. AB and CO are straight lines. Angle AOC is equal to 140°. The number of degrees in angle COB is (*a*) 40 (*b*) 70 (*c*) 140

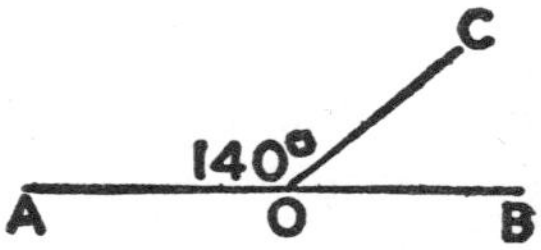

26. The formula for finding the perimeter p of a square whose side is s is (*a*) $p = 4s$ (*b*) $p = s^2$ (*c*) $p = 2s^2$

27. To solve the equation $\frac{a}{3} = 6$, a student should (*a*) subtract 3 from both sides of the equation (*b*) multiply both sides of the equation by 3 (*c*) divide both sides of the equation by 3

28. A certain number is represented by x. A number 5 less than twice this number is (*a*) $2x - 5$ (*b*) $5x - 2$ (*c*) $3x$

29. If $c = 3$, the value of $5c^2$ is (*a*) 225 (*b*) 45 (*c*) 30

30. The expression $4x + 4 - 3x - 2$ is the same as (*a*) $7x + 6$ (*b*) $x + 6$ (*c*) $x + 2$

Part II

Answer any five questions from this part. No credit will be allowed unless all necessary operations are given. Reduce each result to its simplest form and mark each answer Ans.

31. A washing machine priced at \$300 was reduced 25% in price just before the new model came out. In addition, a 10% discount for cash was allowed. What was the actual cash price the buyer paid for this machine? [10]

32. John Doe purchased from the Hobby Supply Shop a supply of balsa wood costing \$2.25. He paid for the purchase by check. The stub and the check shown are from Mr. Doe's personal checkbook. List the letters *a-j* on your answer paper. After *each* letter write the words or numbers exactly as they should appear in the place indicated by the corresponding letter on the stub or check: [10]

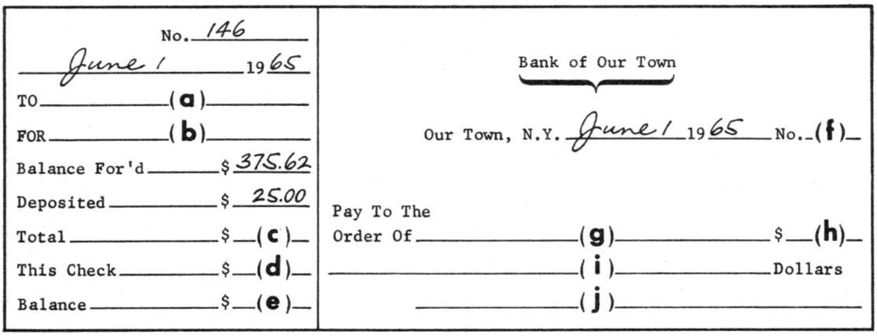

33. A man does piecework in a factory. This means that he is paid a fixed amount for each article he completes. During one week of 5 working days he completed 3,488 motor parts for which he was paid $2\frac{1}{2}$ cents each.

a. Find his gross earnings for the week. [6]

b. Find his average daily rate of pay during that week. [4]

34. The floor plan of the living room and dining area in a new home is shown at the right. The family buying the home wishes to cover the floor completely with carpeting that sells for \$15.75 per square yard.

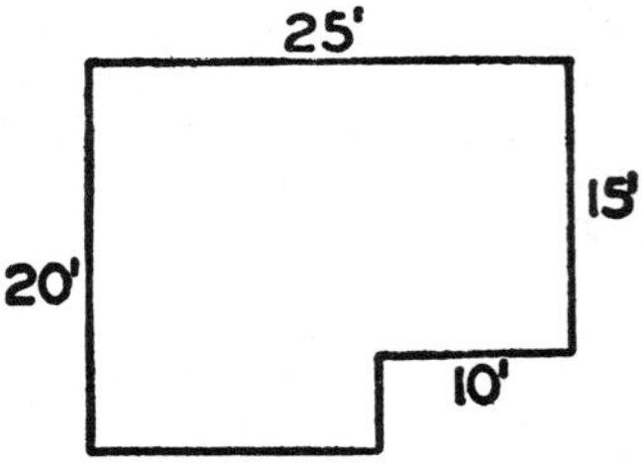

a. How many square yards of carpeting will be needed? [7]

b. What will be the total cost of the carpeting? [3]

35. A man is planning to remake his lawn next fall. The lawn, which is level and in the shape of a rectangle 90 feet long and 30 feet wide, is to be covered with topsoil to a depth of 3 *inches*.

a. How many *cubic feet* of topsoil will the man need? [6]

b. If he carts the topsoil in a small truck which holds a cubic yard (27 cubic feet), how many full loads will he need? [4]

36. In the diagram, O is the center of the circle. Straight lines AD, BE, and CF are drawn through O and divide the circle into 6 equal arcs. AC and FD are drawn.

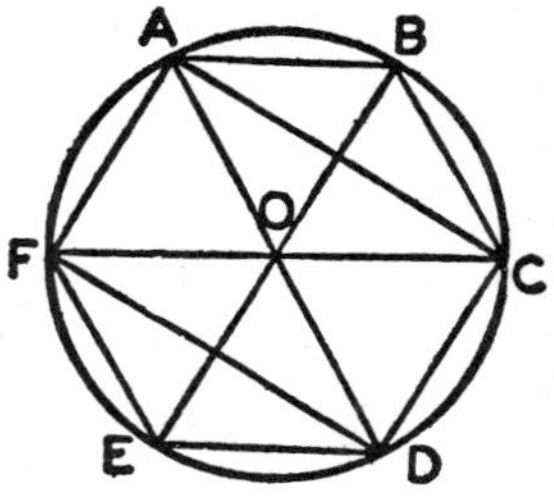

List the numbers 1-5 on your answer paper. After *each* number write *true* if the corresponding statement is true. If the statement is false, write the word that must be substituted for the italicized word to make the statement true. [10]

(1) Figure $ABCDEF$ is a *hexagon.*

(2) Angle BOC is a *right* angle.

(3) AD is a *diameter* of the circle.

(4) Triangle FOD is an *acute* triangle.

(5) Triangle AOC is an *isosceles* triangle.

37. *a.* In an eighth grade election for class president the winner received 65 more votes than the loser. If n represents the number of votes cast for the loser,

(1) express in terms of n the number of votes cast for the winner [1]

(2) express in terms of n the total number of votes cast in the election [2]

b. If 275 votes were cast for both candidates in the election mentioned above,

(1) write a correct equation which can be used to find the value of n [4]

(2) solve this equation for n [2]

(3) find the number of votes cast for the winner [1]

38. In a large city the breakdown of the $28,800,000 raised by means of real estate taxes, for all purposes except schools, is shown in the graph. To raise this sum, the tax rate was set at $21.95 per $1,000 of assessed valuation. Use this information and the graph to answer the following questions.

DISTRIBUTION OF TAX MONEY

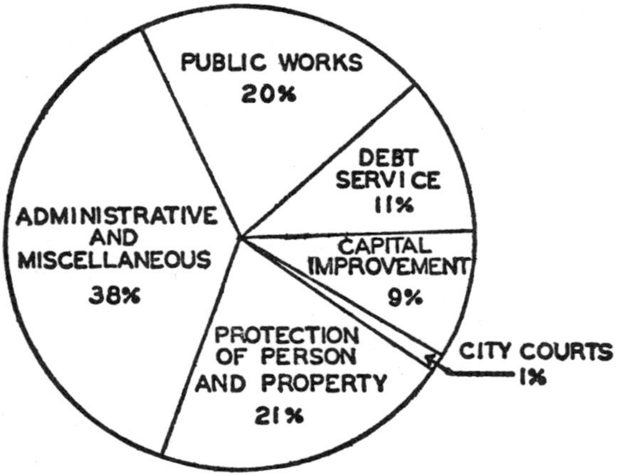

a. Mr. Brown, a taxpayer in this city, owns real estate assessed at $6,000. Find the amount of his real estate tax. [3]

b. On the basis of the tax breakdown shown in the graph, how much does Mr. Brown pay toward the operation of Public Works? [3]

c. Find the amount of money required by this city to maintain "Protection of Person and Property." [4]

FINAL EXAMINATION 4

Part I

Answer all questions in this part. Questions 1-20 count 2 credits each; no partial credit is allowed. Questions 21-30 count 1 credit each. Reduce each answer to its simplest form.

1. Add: $8\frac{1}{3}$, $15\frac{1}{2}$, $6\frac{3}{4}$

2. Subtract $7.56 from $20.

3. Multiply 3.14 by 7.2

4. Divide 9.728 by 1.52
5. Divide $3\frac{9}{10}$ by $3\frac{1}{4}$.
6. Last year a large university in the United States had an enrollment of 40,788 students. Round off this number to the *nearest thousand.*
7. Find the square root of 289.
8. Which fraction has the largest value: $\frac{5}{9}, \frac{5}{8}, \frac{5}{11}$?
9. What is the cost of 9 ounces of cheese at 80 cents per pound?
10. A man borrowed \$3,600 for 1 month at an annual rate of 5%. How much interest did he owe?
11. One piece of wire is 25 feet 8 inches long and another is 18 feet 10 inches long. What is the difference in length?
12. The selling price of an article was \$20. If the cost of the article was \$12 and the overhead expense involved was 15% of the selling price, find the profit.
13. In a recent year the enrollment in a first grade was 150 pupils. The following year the enrollment was 180 pupils. What was the per cent of increase in enrollment?
14. At 15% commission, how much will a salesman's total sales have to be for him to earn \$75?
15. On a road map $\frac{1}{4}$ inch represents 8 miles of actual road distance. How many miles apart are two towns which are represented by points $2\frac{1}{8}$ inches apart on the map?
16. An estimate of the population of the United States was 169,400,000 on December 1, 1957. This is 2,800,000 higher than on December 1, 1956. What was the estimated population on December 1, 1956?
17. In one city the cost of educating each elementary school child is \$432 per year. If there are 180 school days in the year, what is the cost for each child for each day of school?
18. How many square yards are there in the area of a rug which is 15 feet long and 12 feet wide?
19. If the school tax rate in a community is \$33.50 per \$1,000 of assessed valuation, find the amount of tax on property assessed at \$5,000.
20. Write in figures: four hundred twenty and two hundredths.

Directions (21-30): Indicate the correct completion for *each* of the following by writing the letter *a*, *b* or *c*.

21. For which equation is 3 a value of x? (*a*) $2x - 5 = 11$ (*b*) $\frac{x}{4} = 12$ (*c*) $3x + 1 = 10$
22. If $x = 5$ and $y = 4$, the value of $x^2 - 2y$ is (*a*) 17 (*b*) 2 (*c*) 33

23. The perimeter of the rectangle in the diagram is (*a*) $4W + 8$ (*b*) $12W$ (*c*) $2W + 4$

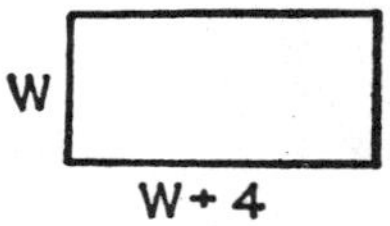

24. The formula for finding the area of a circle is (*a*) $A = 2\pi r$ (*b*) $A = \pi r^2$ (*c*) $A = \pi d$

25. To solve the equation $4w = 12$, a pupil should (*a*) subtract 4 from both sides of the equation (*b*) multiply both sides of the equation by 4 (*c*) divide both sides of the equation by 4

26. The floor plan of a house is represented on a blueprint. This is an example of (*a*) similarity (*b*) congruence (*c*) equality

27. The line AC in the diagram is a (*a*) parallel (*b*) bisector (*c*) diagonal

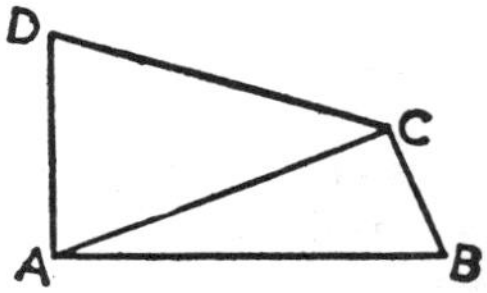

28. A triangle can be constructed having sides whose lengths are (*a*) 6 in., 4 in., 3 in. (*b*) 5 in., 3 in., 8 in. (*c*) 3 in., 2 in., 6 in.

29. In a square whose side is 7 feet, the area in square feet is (*a*) 14 (*b*) 28 (*c*) 49

30. The lengths of the legs of a right triangle are 3 inches and 4 inches. The length of the hypotenuse of this triangle in inches is (*a*) 5 (*b*) 7 (*c*) 25

Part II

Answer any five questions from this part. No credit will be allowed unless all necessary operations are given. Reduce each result to its simplest form and mark each answer Ans.

31. A salesman receives a salary of \$300 a month and a commission of $3\frac{1}{2}\%$ on total sales over \$5,000 made during the month. During one month his total sales were \$14,870. How much in all did he earn that month? [10]

32. A certain store advertised: "All stock reduced 25%. Charge accounts available. No down payment; pay $\frac{1}{2}$ on February 10, $\frac{1}{3}$ on March 10 and the balance on April 10." If a customer used his charge account during this period to buy a coat which originally had been marked \$110,

a. find the reduction in price [2]

b. find the sales price of the coat [2]

c. find the amount due on February 10 [2]

d. find the amount due on March 10 [2]

e. find the balance due on April 10 [2]

33. Following is a table for premiums on ordinary life insurance:

If taken out at age	*Annual premium per $1,000 of insurance*
20	$18.25
25	$20.75
30	$23.75
35	$27.50
40	$32.50

A man took out four of these insurance policies as follows: $1,000 at age 20; $2,000 at age 25; $2,000 at age 30; $5,000 at age 35.

a. Find the total amount of insurance he has now on these four policies. [2]

b. Find the annual premium he pays on *each* of the four policies. [4]

c. Find the *total* premium he pays per year on the four policies. [2]

d. How much more does it cost him per year for the policy taken out at age 30 than for the one taken out at age 25? [2]

34. A room is 30 feet long, 28 feet wide and 12 feet high.

a. Find the number of square feet in its floor area. [4]

b. If its window area is 25% of its floor area, find the number of square feet in its window area. [2]

c. How many cubic feet of space are there in this room? [4]

35. A family of four is planning a trip by car. The main expenses that the members are considering are car expenses, food, and overnight lodging. They first estimated the length of the trip to be 4,000 miles, but later plans increased the total estimated mileage by 25%.

a. How many miles do they expect to cover on the whole trip? [2]

b. If 5 cents a mile will cover car expenses, how much would they spend for car expenses on the trip? [2]

c. Last summer the family opened a vacation club account at the bank and will have paid into it for 20 weeks. How much should have been deposited each week to cover car expenses for the trip? [2]

d. If they plan to average 250 miles per day, how many days will the trip take? [2]

e. If they estimate food and lodging will cost $7 a day for *each* person, how much do they expect to pay for food and lodging for the four people for the entire trip? [2]

36. In a junior high school of 1,200 pupils, there are 50 more girls than boys.

a. If the number of boys is represented by n, represent the number of girls in terms of n. [2]
b. Represent the total number of pupils in terms of n. [2]
c. Write an equation which shows that the total number of pupils in the school is 1,200. [2]
d. Solve the equation to find the number of boys in the school. [2]
e. How many girls are there in the school? [2]

37. In the figure at the right, $ABCD$ is a square. Diagonals AC and BD intersect at O. The following statements about the diagram are either true or false. List the letters a through e on your answer paper. Opposite *each* letter write the word *true* if the corresponding statement is true; write the word *false* if the statement is false. [10]

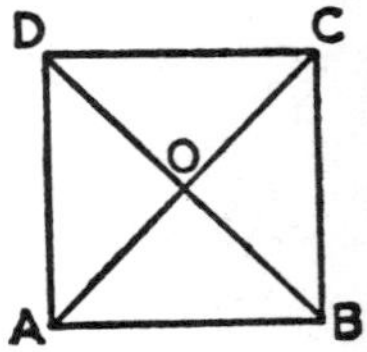

a. $AB = BC$.
b. DC is perpendicular to AB.
c. Angle $CAB = 45°$.
d. Angles DAB, ABC, BCD, and CDA are equal.
e. Angle AOB is acute.

38. List the letters *a-e* on your answer paper. Opposite *each* letter write the *number* 1, 2 or 3, representing the correct answer. [10]

a. A circle graph is to be drawn in which one sector is to be 20% of the circle. The central angle of the sector should contain (1) 36° (2) 72° (3) 90°
b. A circle graph is to be drawn in which the central angle for a sector is 108 degrees. This sector represents what per cent of the circle? (1) 30% (2) 40% (3) 50%
c. Each of the small houses in the pictogram represents 100,000 houses. How many houses are represented by the figures shown? (1) 300,500 (2) 305,000 (3) 350,000

d. What amount of money is represented by the bar in the diagram at the right? (1) $2,050,000 (2) $2,250,000 (3) $2,500,000

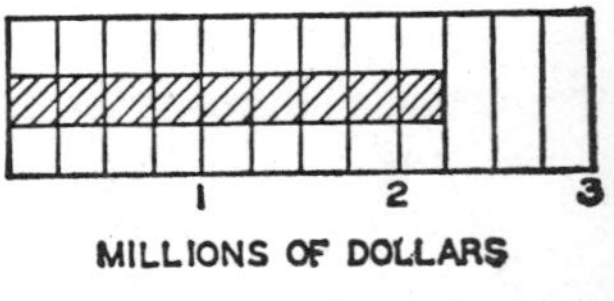

e. The graph in the diagram on the next page shows the population of a suburban town at intervals of 5 years from 1955 through 1965. How much more did the population increase between 1960 and 1965 than it did between 1955 and 1960? (1) 500 (2) 1,000 (3) 1,500

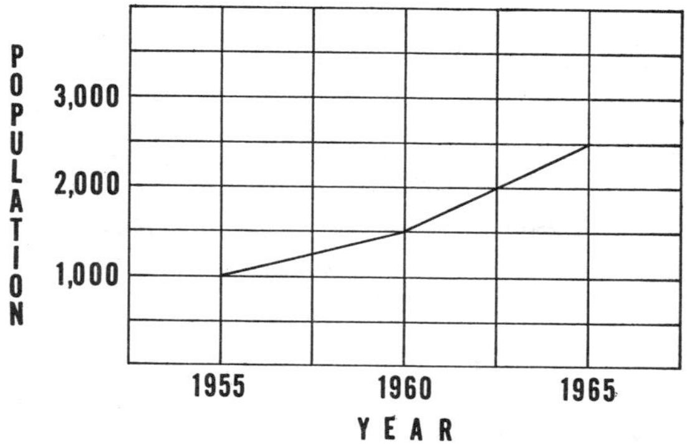

FINAL EXAMINATION 5

Part I

Answer all questions in this part. Questions 1-20 count 2 credits each; no partial credit is allowed. Questions 21-30 count 1 credit each. Reduce each answer to its simplest form.

1. Add: \$245, \$63.25, \$.98, \$7.92, \$10.16
2. Subtract $14\frac{5}{8}$ from 30
3. Divide 12 by $1\frac{1}{2}$
4. Multiply \$1.08 by $6\frac{1}{4}$
5. Divide 73.44 by 2.4
6. Change $\frac{5}{8}$ to a per cent.
7. A certain kind of cloth is a mixture of 50% wool, 37% cotton and the remainder rayon. What per cent of the cloth is rayon?
8. The first man-made satellite is said to have traveled at 18,000 miles per hour. Express this speed in miles per minute.
9. A man drives his car over a toll bridge on his way to and from work. The toll charge is either 15¢ for each one-way crossing, or \$5 for a commuter's book of fifty one-way tickets. For fifty crossings, how much will this man save by buying the book of tickets?
10. In a dictionary, the scale drawing of a bird is $\frac{1}{8}$ actual size. If the drawing is $\frac{3}{4}$ inch high, find in inches the height of the real bird.
11. One year 57,546 schools took part in the federal school lunch program. Round off this number to the *nearest thousand.*

12. An advertisement says that the new box of a detergent contains $33\frac{1}{3}\%$ more than the old 15-ounce box. How many ounces does the new box contain?

13. A game began at 11:45 a.m. and ended at 2:15 p.m. How many hours and minutes did the game last?

14. A boy saved \$5 by buying a jacket at a sale where a 25% discount was given. What was the original price of the jacket?

15. If the area of a square rug is 64 square feet, find in feet the length of each side.

16. A boy sold \$60 worth of magazine subscriptions, for which he was paid \$18. What rate of commission was he paid?

17. Mrs. *A* has two quarts of fruit juice which she is pouring into $\frac{1}{2}$-pint glasses. How many glasses can she fill with the fruit juice?

18. A girl whose salary is \$3,200 a year has $2\frac{1}{4}\%$ of it deducted for Social Security. How much is deducted for this purpose?

19. A team won 25 games and lost 8. What is the ratio of the number of games won to the number of games played?

20. What amount of interest is due at the end of 3 months on a loan of \$2,000 if the yearly interest rate is 5%?

Directions (21-30): Indicate the correct completion for *each* of the following by writing the letter *a*, *b* or *c*.

21. An angle which contains less than 90 degrees is (*a*) an obtuse angle (*b*) a straight angle (*c*) an acute angle

22. In which diagram has a perpendicular been constructed from *A* to *BC*?

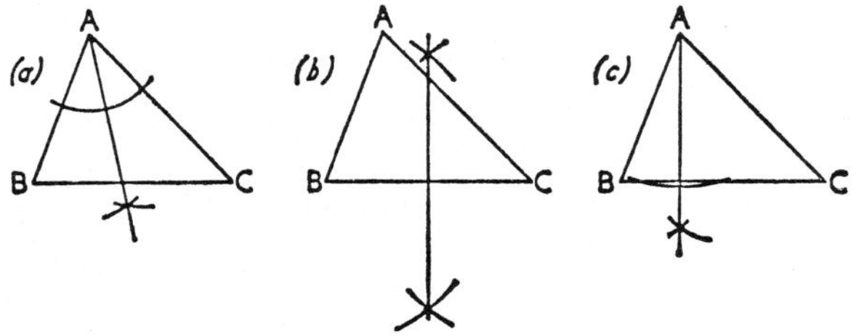

23. A right triangle may be (*a*) equilateral (*b*) isosceles (*c*) equiangular

24. Which geometric figure has more than 5 sides? (*a*) hexagon (*b*) rectangle (*c*) pentagon

25. Printed paper labels are pasted completely around the surface of cylindrical paint cans like the one in the diagram. The shape of such a label before it is pasted on the can is a (*a*) semicircle (*b*) circle (*c*) rectangle

26. The algebraic expression $3n$ means (*a*) $3 + n$ (*b*) $n + n + n$ (*c*) $n \times n \times n$

27. The formula for finding the circumference of a circle is (*a*) $C = \pi r$ (*b*) $C = \pi r^2$ (*c*) $C = 2\pi r$

28. If $n = \frac{1}{2}$, the value of n^2 is (*a*) 1 (*b*) $\frac{1}{4}$ (*c*) 4

29. To solve the equation $n - 3 = 6$, a student should (*a*) divide both sides of the equation by 3 (*b*) add 3 to both sides of the equation (*c*) subtract 3 from both sides of the equation

30. The n students in a class are divided into 5 equal committee groups. The number of students in each group is (*a*) $5n$ (*b*) $n + 5$ (*c*) $\frac{n}{5}$

Part II

Answer any five questions from this part. No credit will be allowed unless all necessary operations are given. Reduce each result to its simplest form and mark each answer Ans.

31. A home freezer can be bought for $450 cash or on the installment plan for $45 down and $28.80 a month for 15 months. How much would a man save if he bought the freezer by paying cash rather than on the installment plan? [10]

32. In a recent year the school tax in a central school district was $20.40 per $1,000 of assessed valuation. In this district, taxes paid within 30 days from the date stamped on the tax bill have 1% of the amount of the bill added as a collection fee. Taxes paid after the 30-day period have a 4% fee added to the 1% collection fee, making a total of 5% to be added. The date stamped on tax bills that year was August 31.

a. A man owned property assessed at $6,500 in this district. If he paid his taxes on September 15, find the total amount of his bill, including the collection fee. [7]

b. On what date that year did taxpayers begin paying the 5% collection fee? [2]

c. What per cent of their tax bills would be saved by all taxpayers who paid within 30 days from August 31 rather than after the 30-day period? [1]

33. To raise money, the members of a club decided to sell novelty desk calendars. They were able to buy the calendars from a local company for $30 per hundred, subject to a 20% discount if 500 or more calendars were ordered. The club ordered 500 calendars.

a. How much did the club pay for the 500 calendars? [3]

b. How many calendars did they have to sell at 50¢ each to meet exactly the cost of their project? [2]

c. If they were able to sell all but 20 of the calendars they bought, how much profit did they make? [3]

d. Using the answer to *c*, find what per cent of the cost of the project was their profit. [2]

34. A certain newspaper ran a contest in which the following prizes were offered:

One prize of $50,000
One prize of $10,000
One prize of $5,000
One prize of $2,000
One prize of $1,000
Ten prizes of $100 each
Ten prizes of $50 each
Ten prizes of $25 each
Fifty prizes of $10 each
Nine hundred fifty prizes of $5 each

a. Find the total amount of prize money offered in this contest. [6]

b. What per cent of the prize money went to the winner of the first prize? [4]

35. Annual Premium and Installment Payments per $1,000 on Ordinary Life Insurance at Ages from 21 through 25

Age	*Annual Premium*	*Semiannual Installment*	*Quarterly Installment*
21	$16.62	$8.48	$4.32
22	17.08	8.71	4.44
23	17.55	8.95	4.56
24	18.04	9.20	4.69
25	18.56	9.47	4.83

Refer to the preceding table to answer the following questions:

a. A man bought a $5,000 ordinary life insurance policy when he was 22 years old.

(1) What annual premium does he pay on this policy? [2]

(2) If he had chosen to pay his premium quarterly, what would each quarterly installment be? [2]

(3) How much does he save each year by making one annual payment rather than by paying his insurance quarterly? [3]

b. If he were to buy another $5,000 ordinary life insurance policy at age 25, how much more would the annual premium on this policy be than that on the policy he took out when he was 22 years old? [3]

36. *a.* In the diagram, angle ABC is a right angle. Angle DBC contains 40°. How many degrees are there in angle ABD? [2]

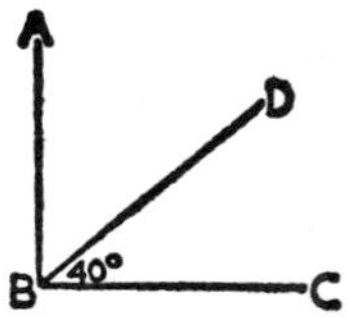

b. O is the center of the circle at the right. If angle AOB in the circle contains 36°, what per cent of the circle is shaded? [2]

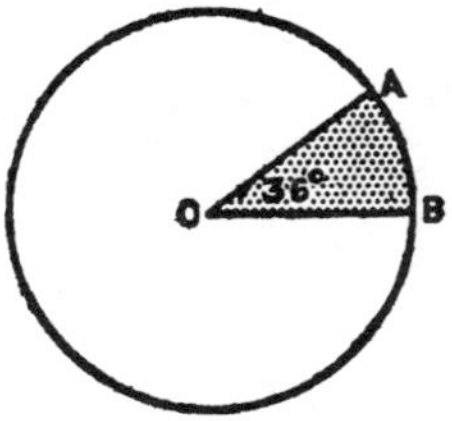

c. The line AB, which is 2 inches long, is the scale length of a distance of 200 feet. Using the same scale, find in inches the length of the line which should be used to represent 350 feet. [2]

d. The three sides of the triangle in the diagram are equal. How many degrees are there in angle B? [2]

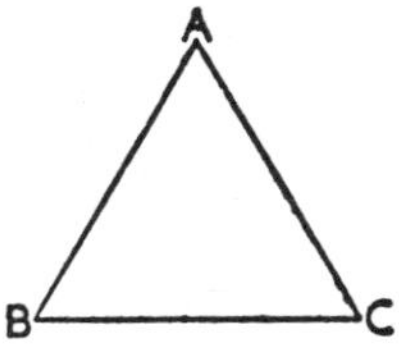

e. Which best completes the statement? The following two rectangles are (1) similar (2) congruent (3) equal in area [2]

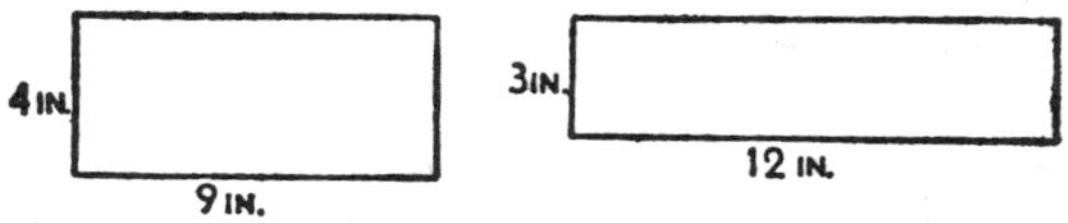

37. The eighth graders sold 75 more tickets for the school play than the seventh graders.

a. If n represents the number of tickets sold by the seventh graders,
(1) express in terms of n the number of tickets sold by the eighth graders [1]
(2) express in terms of n the total number of tickets sold by pupils of both grades [2]

b. (1) If 435 tickets for the play were sold by the seventh and eighth graders together, write a correct equation which can be used to find the value of n. [4]
(2) Solve this equation for n. [2]
(3) Find the number of tickets sold by the eighth graders. [1]

38. The following graph shows the average monthly rainfall for the first six months of a year in a certain city.

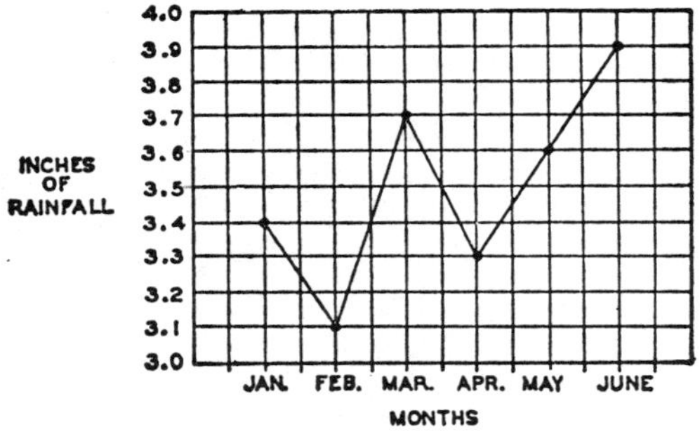

a. In which month was the rainfall greatest? [1]
b. How many inches *less* rain were there in February than in March? [2]
c. List the six months and write the number of inches of rainfall shown for *each* month. [3]
d. Find the total number of inches of rainfall for the six months. [2]
e. Find the number of inches in the average monthly rainfall for the six months. [2]

FINAL EXAMINATION 6

Part I

Answer all questions in this part. Questions 1-20 count 2 credits each; no partial credit is allowed. Questions 21-30 count 1 credit each. Reduce each answer to its simplest form.

1. Subtract \$941.20 from \$1,000
2. Add $8\frac{9}{16}$, $4\frac{1}{2}$, $6\frac{3}{4}$
3. Find the value of $1\frac{1}{2} \times 1\frac{3}{4} \times \frac{2}{3}$
4. Change $\frac{3}{7}$ to a decimal correct to the *nearest hundredth*
5. Multiply \$.86 by $5\frac{1}{2}$
6. Divide 20.25 by .045
7. Which of the following numbers when multiplied by 100 results in 10: .01, .1, 10?
8. On a mathematics test $12\frac{1}{2}\%$ of a class received a mark of D, $37\frac{1}{2}\%$ a mark of C, 25% a mark of B, and the remainder a mark of A. What per cent of the class received a mark of A?
9. On a road map it was explained that 1 inch represented 12 miles. How many miles would a length of $2\frac{1}{4}$ inches represent?

10. A housewife bought a lamp for \$6.75. She returned the lamp the next day and chose a better lamp for \$8.95. She gave the clerk a five-dollar bill to pay for the difference in price. How much change should she receive?

11. At an annual rate of 5%, how much interest is due every 6 months on a loan of \$100?

12. It was reported that 16 pupils in a homeroom of 40 pupils were absent on the day of a blizzard. What per cent of the pupils were absent?

13. A housewife selected two packages of ground beef at a "packaged meat" counter. Together they weighed exactly 2 pounds. If one package she selected weighed 1 pound 4 ounces, what was the weight of the other?

14. By buying a coat during a sale, a girl received a 20% discount. The discount amounted to \$12. What was the original price of the coat?

15. Which has the greatest value: 66%, $\frac{3}{4}$, .80?

16. A boy deposited in his school savings account the money he had saved during the summer. Find the amount of his deposit if he had 10 one-dollar bills, 9 half dollars, 8 quarters, 16 dimes, and 25 nickels.

17. At the annual rate of \$.50 per \$100, how much does it cost for a \$3,000 fire insurance policy for one year?

18. If a car averages 18 miles to a gallon of gasoline, how many gallons of gasoline will be used on a trip of 810 miles?

19. In a recent year 120 pupils were enrolled in the seventh grades of a junior high school. The following year the seventh grade enrollment was 160 pupils. Find the per cent of increase in enrollment.

20. The drawing at the right is part of a bar graph showing the population of a small village. How many people live in the village?

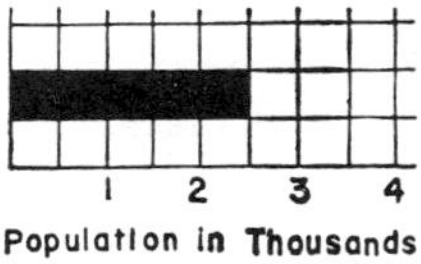

Directions (21-30): Indicate the correct completion for *each* of the following by writing the letter *a*, *b* or *c*:

21. The number of square faces on a cube is (*a*) 12 (*b*) 6 (*c*) 8

22. The figure in which a diagonal has been drawn is

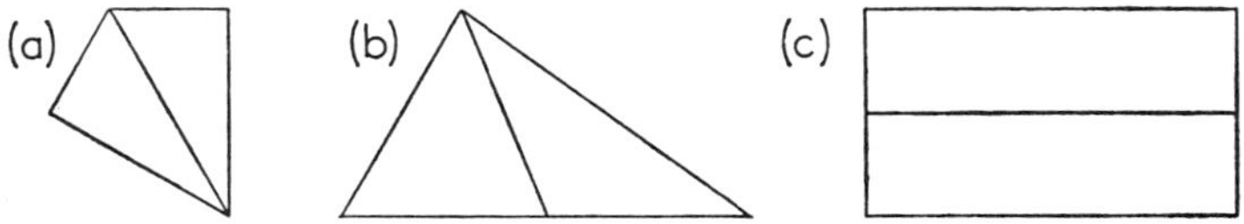

23. The description which best fits congruent figures is that they have (*a*) the same shape but not the same size (*b*) the same size but not the same shape (*c*) the same size and the same shape

24. Two straight lines which meet at right angles are (*a*) parallel (*b*) perpendicular (*c*) vertical

25. The radius of a circle is 10 inches. A point 12 inches from the center of the circle lies (*a*) on the circle (*b*) inside the circle (*c*) outside the circle

26. The equation which expresses the relationship that a certain number n increased by 3 is equal to 9 is (*a*) $n + 3 = 9$ (*b*) $n - 3 = 9$ (*c*) $3n = 9$

27. In the formula $S = 6e^2$, if $e = 3$, then S equals (*a*) 30 (*b*) 36 (*c*) 54

28. If one pencil costs n cents, the cost in cents of 6 pencils is (*a*) $6n$ (*b*) $\frac{n}{6}$ (*c*) $\frac{6}{n}$

29. If $n - 2 = 8$, the value of n is (*a*) 4 (*b*) 6 (*c*) 10

30. The perimeter of the rectangle in the diagram is (*a*) $a + b$ (*b*) $2a + 2b$ (*c*) $2ab$

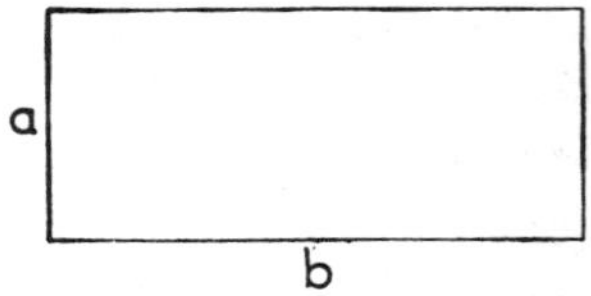

Part II

Answer any five questions from this part. No credit will be allowed unless all necessary operations are given. Reduce each result to its simplest form and mark each answer Ans.

31. A salesman was offered a choice of a monthly salary of $450 or a commission of 8% on his sales. He accepted the commission basis and sold $85,900 worth of goods the first year.

a. How much did the salesman earn during the year? [4]

b. How much would he have earned that year if he had accepted the fixed monthly salary? [3]

c. Did his choice result in a larger earning for the year? How much more or less was it? [1, 2]

32. A man's taxable income is $4,280. The tax instructions tell him to pay 2% on the first $1,000 of his taxable income, 3% on each of the second and third $1,000, and 4% on the remainder. Find the total amount of income tax which he must pay. [10]

33. The inside dimensions of a rectangular storage bin are 10 feet 8 inches long, 6 feet 6 inches wide, and 6 feet deep.

a. When the bin is filled with wheat to a depth of 3 feet, how many cubic feet of wheat does it contain? [5]

b. If one bushel occupies $1\frac{1}{4}$ cubic feet, how many bushels of wheat are there in the bin? [5]

34. Jim buys a bicycle costing $60 on the installment plan. He must pay $20 down and $8 each month thereafter until the bicycle is paid for.

a. What is the balance due after the down payment? [2]
b. What per cent of the cost is required as a down payment? [3]
c. In how many months will the bicycle be paid for? [3]
d. If the cash price of the bicycle was $56.20, how much more will it cost under the installment plan? [2]

35. *a.* Find the value of: $\frac{1}{3} + \frac{1}{4} + \frac{1}{5} + \frac{1}{5}$. [5]
b. Find the product of 48.72 and 12.5, and divide the product by .203. [5]

36. *a.* Solve *each* of the following equations and check *each* result.

(1) $3n + 4 = 19$ [2, 1] (2) $\frac{n}{2} - 5 = 7$ [2, 1]

b. For each of the following problems three equations are given. Indicate the equation which could be used to solve the problem by writing on your answer paper, after the proper number, the letter *a*, *b*, or *c*. [*Do not solve the equation.*] [4]
(1) A girl paid $1 for 25 greeting cards. Find the average cost per card. (*a*) $c + 25 = 100$ (*b*) $\frac{1}{25}c = 100$ (*c*) $25c = 100$
(2) A boy earned $6 more than his brother. Together they earned $14. Find the amount each earned.
(*a*) $n + 6n = 14$ (*b*) $n + n + 6 = 14$ (*c*) $n + 6 = 14$

37. *a.* *AB* and *OC* are straight lines. Find the number of degrees in angle *AOC*. [2]

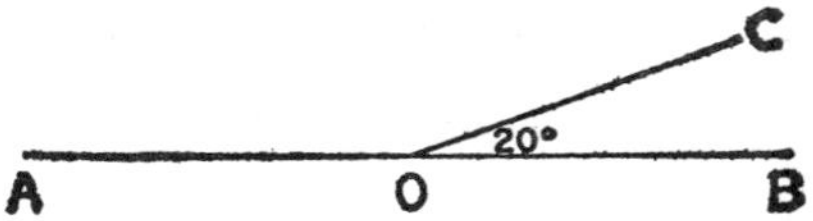

b. How many degrees are there in angle *ABC* in the diagram? [2]

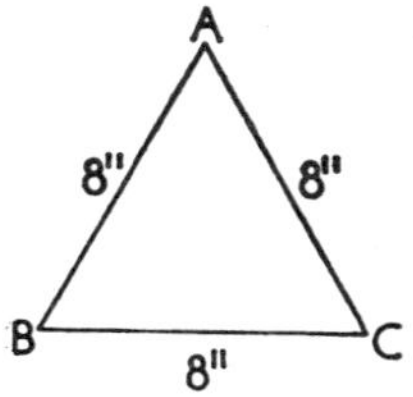

c. Find the length in inches of side *AC* in the triangle in the diagram. [2]

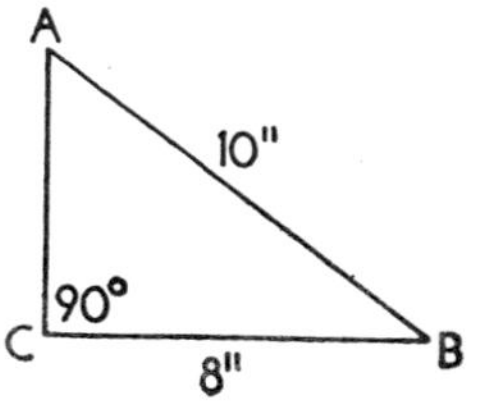

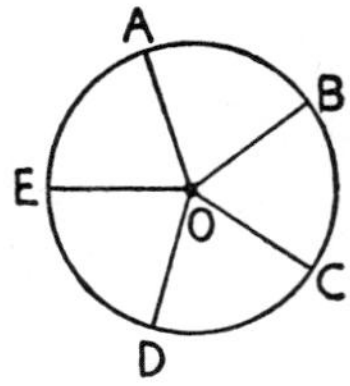

d. In the diagram, arcs AB, BC, CD, DE, and EA are equal. O is the center of the circle. How many degrees are there in angle AOB? [2]

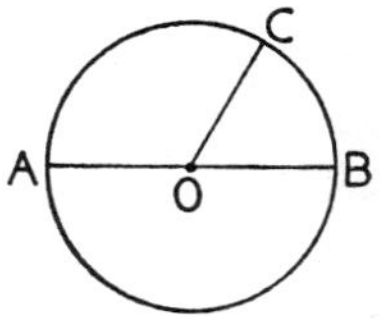

e. O is the center of the circle in the diagram and AB is the diameter. What is the ratio of OC to AB? [2]

38. A family has an income of $500 per month. The accompanying graph shows what part is allotted for each purpose.

a. How much money is set aside each month for rent? [2]

b. How much is allowed for savings and insurance each month? [2]

c. If $25 a month is paid for insurance premiums, how much is left for savings? [2]

d. How many degrees are used to represent operating expenses? [2]

e. The amount spent for health and recreation is how much more than the amount spent for operating expenses? [2]

A Family Budget

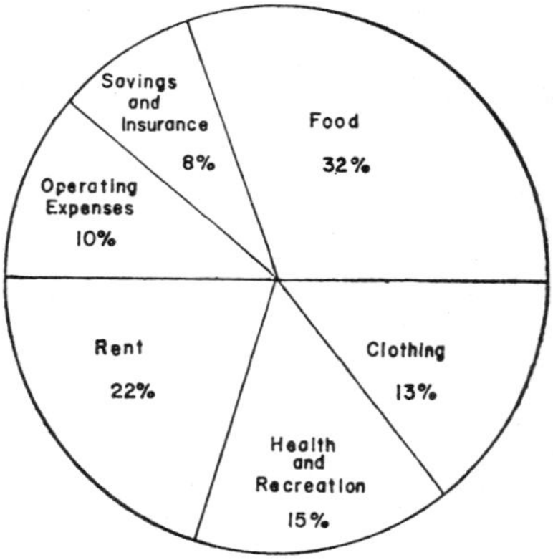

FINAL EXAMINATION 7

Part I

Answer all questions in this part. Questions 1-20 count 2 credits each; no partial credit is allowed. Questions 21-30 count 1 credit each. Reduce each answer to its simplest form.

1. Subtract $94.78 from $360.50

2. Add: $8\frac{3}{8}$, 12, $15\frac{5}{16}$

3. Divide 16 by $2\frac{2}{3}$

4. Multiply 956 by 507

5. Divide 1.672 by .08
6. Which of the following has the same value as $1\frac{1}{2}\%$: 1.5, .15, .015?
7. A group left on a trip at 8:50 a.m. and reached their destination at 3:30 p.m. How long, in hours and minutes, did the trip take?
8. Find the amount of a 2% tax on a purchase amounting to $24.50.
9. On a scale drawing, a line $\frac{1}{4}$ inch long represents a length of 1 foot. On the same drawing, what length represents 4 feet?
10. Last year the postage rate for sending 1 ounce of mail first class was increased from 3 cents to 4 cents. What was the per cent of increase in the postage rate?
11. If 1 pound 4 ounces of meat costs $1.05, what is the cost of the meat per pound?
12. What is the interest for 3 months on a loan of $1,200 at a yearly rate of 5%?
13. How many square yards of linoleum are needed to cover a floor having an area of 270 square feet?
14. A boy took an examination on which there were 20 examples. If he had 18 of them correct, what per cent of the examples did he do correctly?
15. Six girls sold the following number of boxes of cookies: 42, 35, 28, 30, 24, 27. What was the average number of boxes sold by each girl?
16. A porch lounge regularly sells for $29.50. How much money is saved if the lounge is bought at a 20% discount?
17. A man has an $8,000 life-insurance policy on which he pays a premium of $21.75 per $1,000 each year. Find his annual premium.
18. What is the largest number of half-pint bottles which can be filled from a 10-gallon can of milk?
19. What number multiplied by itself equals 144?
20. A boy received 25% commission for selling magazine subscriptions. One week he earned $12. How many dollars worth of magazine subscriptions did he sell that week?

Directions (21-30): Indicate the correct completion for *each* of the following by writing the letter *a*, *b* or *c*.

21. If a diagonal is drawn in a square as shown in the diagram, the two triangles formed are (*a*) equilateral (*b*) right (*c*) acute

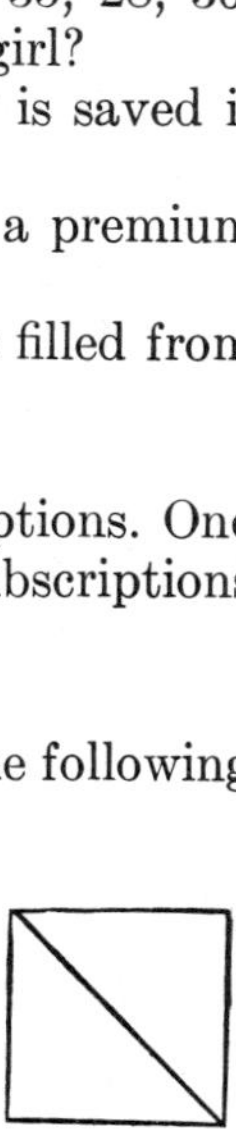

22. If the shaded part of the circle in the diagram is 10% of the circle, angle *AOB* contains (*a*) 10° (*b*) 18° (*c*) 36°

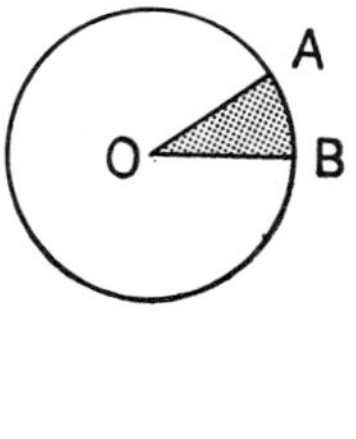

23. In the diagram, the angle represented by x contains (*a*) 40° (*b*) 60° (*c*) 140°

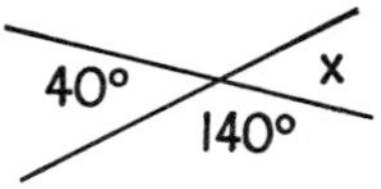

24. The number of equal squares which make up the surface of a cube is (*a*) 4 (*b*) 6 (*c*) 8

25. An octagon is a polygon having exactly (*a*) 5 sides (*b*) 6 sides (*c*) 8 sides

26. If one-third of a number n is equal to 4, the number is (*a*) 3 times 4 (*b*) 3 less than 4 (*c*) 3 greater than 4

27. The expression $5a + 7 - 2a + 9$ is the same as (*a*) $7a + 16$ (*b*) $3a + 16$ (*c*) $19a$

28. The value of $a^2 + 5$ when $a = 3$ is (*a*) 14 (*b*) 11 (*c*) 64

29. The area of the rectangle at the right may be found by using the formula (*a*) $A = 2l + 2w$ (*b*) $A = 2lw$ (*c*) $A = lw$

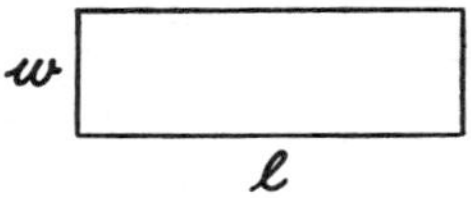

30. Linda has b books and Betty has one-fifth as many. The number of books that Betty has may be represented by (*a*) $b - 5$ (*b*) $\frac{b}{5}$ (*c*) $5 - b$

Part II

Answer any five questions from this part. No credit will be allowed unless all necessary operations are given. Reduce each result to its simplest form and mark each answer Ans.

31. On May 1, a man had a bank balance of $85.35 in his checking account. During May he made deposits of $240, $75.50 and $120. He also wrote checks for $95, $11.75, $180.50, $67.20 and $22.50 which were drawn on his account that month. If the bank deducts from his account a service charge of 10 cents for each check drawn, find the bank balance on June 1. [10]

32. A lady wished to purchase a dinette set which was marked $210. She was told that if she paid cash at the time of purchase, she would receive a 5% discount on the marked price. If she purchased the set on the installment plan, she would be required to make a down payment of 20% of the marked price plus 12 monthly payments of $16.50 each.

a. If the lady paid cash for the set, how much did it cost her? [4]

b. How much would the set have cost her if she had bought it on the installment plan? [4]

c. How much less did she pay by purchasing the set for cash rather than on the installment plan? [2]

33. A building is insured for $12,000 at an annual rate of $.22 per $100.

a. Find the annual premium. [3]

b. It is possible to purchase insurance at a 5-year rate which is 4 times the annual rate. Find the 5-year rate per $100. [2]

c. What would be the total cost of insurance for 5 years at the 5-year rate? [2]

d. How much would be saved in 5 years by taking a 5-year policy instead of five 1-year policies? [3]

34. The diagram at the right represents the section of a man's property which he has graded and seeded. To improve the lawn, he has decided to buy a special type of lawn fertilizer for the entire area.

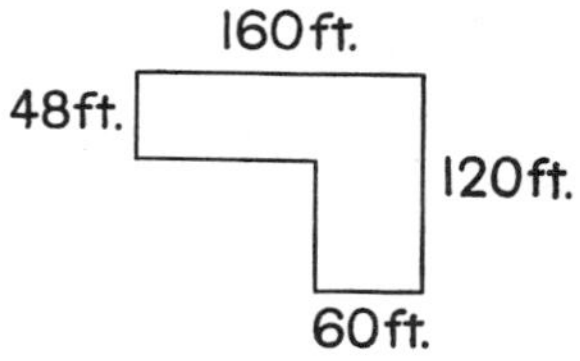

a. Find the number of square feet in the area of the lawn shown in the diagram. [5]

b. The lawn fertilizer he has selected sells for $2.75 a bag. If 1 bag is needed for every 2,000 square feet of lawn, how much will the fertilizer cost for the entire area? [5]

35. *a.* Find the average of 12.46, 24.08, 9.75, 32.24, 10.85 and 35. [5]

b. Find the value of: $10\frac{1}{2} \times 12 \times 8 \times 42\frac{1}{2}$ [5]

36. Indicate the correct completion for *each* of the following by writing the number 1, 2, or 3 opposite the proper letter on your answer paper. [10]

a. A property common to all triangles is that the
(1) three angles are acute
(2) three sides have different lengths
(3) sum of the angles is 180°

b. The two figures below which are equal in area are (1) figures 1 and 2 (2) figures 2 and 3 (3) figures 1 and 3

Fig. 1

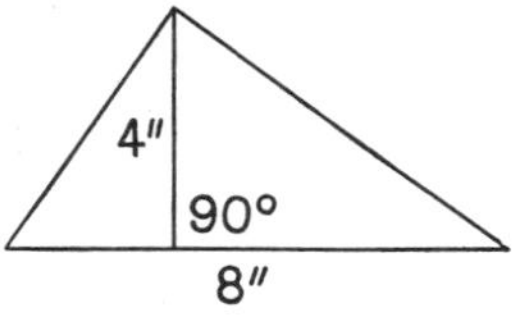

Fig. 2

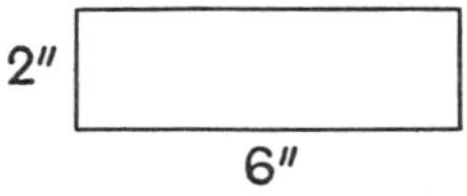

Fig. 3

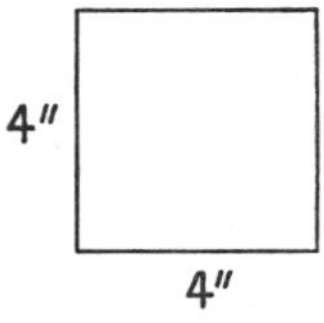

c. If the circumference of a circle is known, the diameter may be found by
(1) multiplying the circumference by π
(2) dividing the circumference by π
(3) adding π to the circumference

d. The best estimate of the number of degrees in angle ABC at the right is
(1) 50° (2) 90° (3) 150°

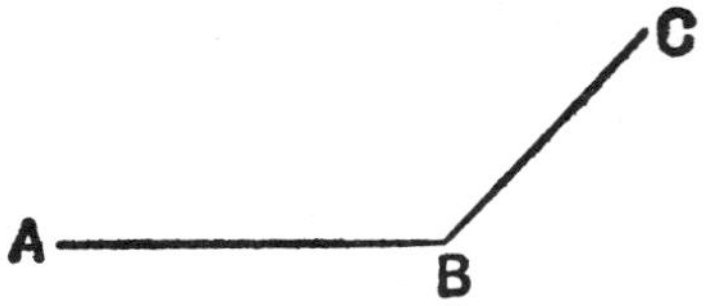

e. The line DE which was constructed in the diagram
(1) bisects side BC
(2) is perpendicular to side AC
(3) bisects angle C

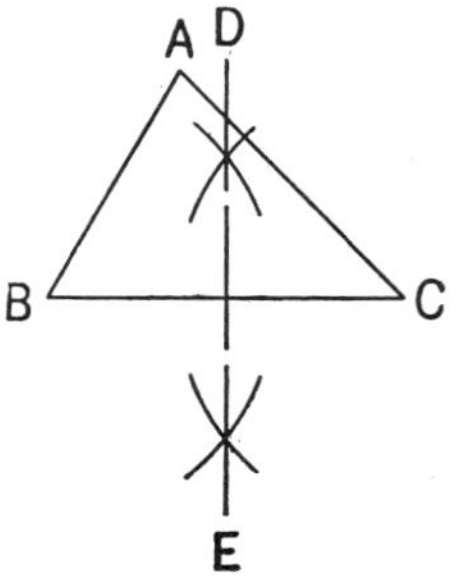

37. *a.* Bill is twice as old as Tom. The sum of their ages is 39. Let x represent Tom's age.
(1) Express Bill's age in terms of x. [1]
(2) Write an equation that can be used to find both boys' ages. [2]
(3) Solve the equation to find the value of x. [2]
(4) How old is Bill? [1]

b. Write an equation for *each* of the following statements, letting n represent the unknown number. [*Solution of the equations is not required.*]
(1) A number increased by 4 equals 12. [1]
(2) Four times a number is 20. [1]
(3) A number divided by 4 is equal to 3. [1]
(4) If 4 is subtracted from a number, the result is 6. [1]

38. The graph on the next page is based on reports obtained from 1,000 owners of a new automobile. Base your answers to the following questions on this information and on the graph:
a. What per cent of the owners reported mileage of 19 miles per gallon? [1]
b. How many times as many owners reported 17 miles per gallon as reported 21 miles per gallon? [2]
c. How many owners reported 20 miles per gallon? [2]
d. What is the sum of all the per cents indicated in the graph? [2]
e. How many more owners reported mileage of 19 miles per gallon than reported mileage of 20 miles per gallon? [3]

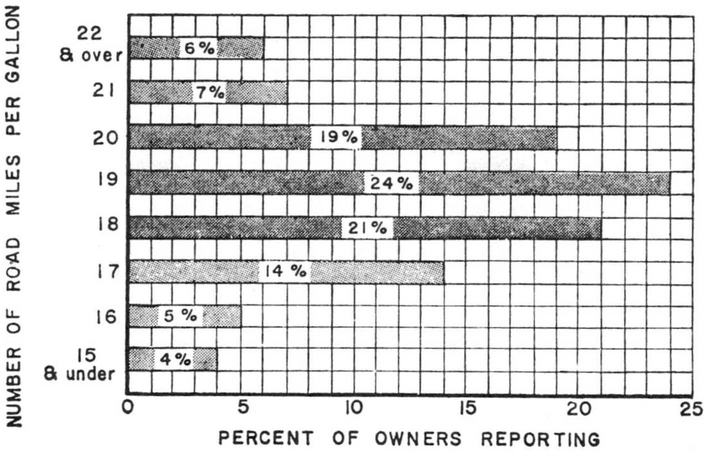
NUMBER OF ROAD MILES PER GALLON
22 & over
6%
21
7%
20
19%
19
24%
18
21%
17
14 %
16
5 %
15 & under
4%
0
5
10
15
20
25
PERCENT OF OWNERS REPORTING

INDEX

CROWN COLLEGE LIBRARY
POWELL, TENNESSEE